T0270921

CAMBRIDGE MONOGRAPHS ON
APPLIED AND COMPUTATIONAL
MATHEMATICS

Series Editors
M. ABLOWITZ, S. DAVIS, J. HINCH,
A. ISERLES, J. OCKENDON, P. OLVER

30   Volterra Integral Equations

The *Cambridge Monographs on Applied and Computational Mathematics* series reflects the crucial role of mathematical and computational techniques in contemporary science. The series publishes expositions on all aspects of applicable and numerical mathematics, with an emphasis on new developments in this fast-moving area of research.

State-of-the-art methods and algorithms as well as modern mathematical descriptions of physical and mechanical ideas are presented in a manner suited to graduate research students and professionals alike. Sound pedagogical presentation is a prerequisite. It is intended that books in the series will serve to inform a new generation of researchers.

A complete list of books in the series can be found at
www.cambridge.org/mathematics.
Recent titles include the following:

15. Collocation methods for Volterra integral and related functional differential equations, *Hermann Brunner*
16. Topology for computing, *Afra J. Zomorodian*
17. Scattered data approximation, *Holger Wendland*
18. Modern computer arithmetic, *Richard Brent & Paul Zimmermann*
19. Matrix preconditioning techniques and applications, *Ke Chen*
20. Greedy approximation, *Vladimir Temlyakov*
21. Spectral methods for time-dependent problems, *Jan Hesthaven, Sigal Gottlieb & David Gottlieb*
22. The mathematical foundations of mixing, *Rob Sturman, Julio M. Ottino & Stephen Wiggins*
23. Curve and surface reconstruction, *Tamal K. Dey*
24. Learning theory, *Felipe Cucker & Ding Xuan Zhou*
25. Algebraic geometry and statistical learning theory, *Sumio Watanabe*
26. A practical guide to the invariant calculus, *Elizabeth Louise Mansfield*
27. Difference equations by differential equation methods, *Peter E. Hydon*
28. Multiscale methods for Fredholm integral equations, *Zhongying Chen, Charles A. Micchelli & Yuesheng Xu*
29. Partial differential equation methods for image inpainting, *Carola-Bibiane Schönlieb*

Volterra Integral Equations

An Introduction to Theory and Applications

HERMANN BRUNNER

Hong Kong Baptist University and
Memorial University of Newfoundland

CAMBRIDGE
UNIVERSITY PRESS

CAMBRIDGE
UNIVERSITY PRESS

University Printing House, Cambridge CB2 8BS, United Kingdom

One Liberty Plaza, 20th Floor, New York, NY 10006, USA

477 Williamstown Road, Port Melbourne, VIC 3207, Australia

4843/24, 2nd Floor, Ansari Road, Daryaganj, Delhi – 110002, India

79 Anson Road, #06–04/06, Singapore 079906

Cambridge University Press is part of the University of Cambridge.

It furthers the University's mission by disseminating knowledge in the pursuit of
education, learning, and research at the highest international levels of excellence.

www.cambridge.org
Information on this title: www.cambridge.org/9781107098725
DOI: 10.1017/9781316162491

First published 2017

A catalogue record for this publication is available from the British Library.

Library of Congress Cataloging in Publication Data
Names: Brunner, H. (Hermann), 1941–
Title: Volterra integral equations : an introduction to theory and applications / Hermann
Brunner, Hong Kong Baptist University.
Description: Cambridge : Cambridge University Press, [2016] | Series: The Cambridge
monographs on applied and computational mathematics
Identifiers: LCCN 2016045368 | ISBN 9781107098725
Subjects: LCSH: Integral equations. | Volterra equations–Numerical solutions.
| Functional analysis.
Classification: LCC QA431 .B7845 2016 | DDC 515/.45–dc23
LC record available at https://lccn.loc.gov/2016045368

ISBN 978-1-107-09872-5 Hardback

Contents

Preface

This monograph presents an introduction to the theory of linear and non-linear Volterra integral equations, ranging from Volterra's fundamental contributions and the resulting classical theory to more recent developments. The latter include Volterra functional integral equations with various kinds of delays, Volterra integral equations with highly oscillatory kernels, and Volterra integral equations with non-compact operators. One of the aims of the book is to introduce the reader to the current state of the art in the theory of Volterra integral equations. In addition, it illustrates – by means of a representative selection of examples – the increasingly important role Volterra integral equations play in the mathematical modelling of phenomena where memory effects play a key role.

The book is intended also as a 'stepping stone' to the literature on the advanced theory of Volterra integral equations, as presented, for example, in the monumental and seminal monograph by Gripenberg (1990). The notes at the end of each chapter and the annotated references point the reader to such papers and books.

We give a brief outline of the contents of the various chapters. As will be seen, Chapters 1, 2, 3 and 6 describe what might be called the classical theory of linear and non-linear Volterra integral equations, while the other chapters are concerned with more recent developments. Those chapters will also reveal that the theory of Volterra integral equations is by no means complete, and that many challenging problems (many of which are stated as Research Problems in the Exercise sections at the ends of the chapters) remain to be addressed.

Chapter 1 contains an introduction to the basic theory of the existence and uniqueness of solutions of linear Volterra integral equations of the first and second kind, including equations with various types of integrable kernel singularities. It also gives a brief introduction to the ill-posed nature of first-kind equations (which will play a key role in Chapter 5 when analysing systems of integral-algebraic equations).

The focus of *Chapter 2* is on the regularity properties of the solutions of the linear Volterra integral equations discussed in Chapter 1. It also contains an introduction to linear functional Volterra integral equations with vanishing or non-vanishing delays: here, the regularity of the solution depends strongly on the type of delay.

Chapter 3 is concerned with the classical theory of non-linear Volterra integral and functional integral equations. In addition, we study non-linear equations whose solutions blow up in finite time (that is, the solution becomes unbounded at some finite time), or which quench in finite time (meaning that the solution remains bounded but its first derivative blows up at some finite time). We also look at some classes of 'non-standard' Volterra integral equations, including auto-convolution equations of the second kind and implicit Volterra integral equations. The chapter concludes with a short discussion of non-linear Volterra functional integral equations: we show that in the analysis of such problems with state-dependent delays, many problems are still open.

In *Chapter 4* we turn to the relatively new subject of Volterra integral equations with highly oscillatory kernels. The focus is on the oscillation properties of the solutions of such equations of the first and second kinds, and on their asymptotic expansions as the oscillation parameter tends to infinity.

Chapter 5 begins with an introduction to the theory of singularly perturbed Volterra integral equations. It then focuses on the theory of systems of so-called integral-algebraic equations: such a system may be viewed as a non-local analogue of a system of differential-algebraic equations.

In *Chapter 6* we present a selection of results on the asymptotic behaviour of resolvent kernels and solutions of linear Volterra integral equations. For equations with convolution kernels, this theory has its origin in the celebrated theorem of Paley and Wiener (1934), while for non-convolution kernels it goes back the late 1960s and then leads into functional analysis and abstract VIEs (studied in Chapter 8).

Chapter 7 contains an introduction to the recently developed theory of so-called cordial Volterra integral equations, especially equations whose underlying integral operator is a non-compact operator (e.g. on the Banach space of continuous functions). This theory also allows for an elegant analysis of the existence of solutions for certain classes of VIEs of the third kind. The chapter concludes with a brief study of cordial Volterra integral equations with highly oscillatory kernels.

Chapter 8 is on Volterra integral operators in the setting of Banach spaces. It focuses on the mapping properties of these operators on the space of continuous functions, on Hölder spaces, and on L^p spaces. It then deals with the quasi-nilpotency and with singular values of Volterra integral operators. In addition, we present an introduction to the asymptotic behaviour of the norms of powers of Volterra integral operators.

Volterra integral equations of various kinds have been playing an increasingly important role in the mathematical modelling of physical, biological and other phenomena that are governed by memory effects. Thus, the aim of *Chapter 9* is to describe, by means of representative examples (and additional references), the wide spectrum of applications of VIEs

Finally, in the *Appendix* we collect definitions and theorems from Functional Analysis, especially from Banach space and operator theory, which are used in earlier chapters of this book, for example, in Chapter 5 (in the analysis of integral-algebraic equations), in Chapter 7 (analysis of cordial Volterra integral equations), and in Chapter 8.

Owing to limitations of space, there are various topics that could not be included in this book. Among these are Volterra–Stieltjes integral equations and stochastic Volterra integral equations (which, due to recent progress in their theory and applications, would merit a separate monograph). However, a short guide to some key references can be found at the end of the Notes to Chapter 9.

At the end of each chapter the reader will find exercises and extensive notes. The *Exercises* range from 'hands-on' problems (intended to illustrate and complement the theory of the respective chapter) to research topics of various degrees of difficulty; these will often include challenging *Research Problems*. The objectives of the *Notes* are twofold: they contain remarks complementing the contents of the given chapter (giving e.g. the sources of original results); and they serve as a guide to papers and books on the more advanced theory of VIEs, and to the literature on aspects of VIEs not treated in the book. At the end of the Notes we also point out a selection of key references on the numerical analyses of the VIEs treated in that particular chapter and identify open problems that may be of interest to the numerical analyst.

Many of the chapters should also be of interest to numerical analysts, since the derivation of a 'good' numerical method (in the sense of reflecting the essential properties of the given problem, especially the regularity of the solution) for solving Volterra integral equations depends crucially on a thorough understanding of the underlying theory. Thus, a secondary aim of this monograph (especially Chapters 2, 3, 4 and 7) is to provide this basis.

The list of *References* is intended to be representative rather than exhaustive. Also, in order to make this extensive list more useful and give it a guiding role (especially to more advanved topics not treated in his book), many of its items have been annotated, so as to enhance the Notes given at the end of each chapter: the brief comments are either cross-references to related work, or they give an idea of the main content of a paper, or point to books and survey articles containing large bibliographies complementing the information in this monograph.

In view of the stated aims of the book, and for the sake of clarity and exposition, we shall, instead of stating a particular theorem and its proof in full generality, often present a representative particular case, so as more clearly to provide insight into the key features (and the basic mathematical techniques underlying the proof of the theorem's general version), and then guide the reader to references where he/she can find the general versions and the detailed proofs. Finding a good balance between describing technical details and conveying the spirit of theorems and their proofs is also to some extent governed by the limitation of space.

This monograph is also intended for senior undergraduate and postgraduate students who wish to acquire a thorough understanding of the basic theory of Volterra integral equations and their role in mathematical modelling. Thus, the book can be used as a textbook for an upper undergraduate or a graduate course on Volterra integral equations. An undergraduate course could, for example, be based on Chapter 1 (except possibly Section 1.6), Chapter 2 (except Section 2.3), Chapter 3 (except Section 3.6), Chapter 4, Chapter 6 (e.g. Sections 6.1, 6.2.1, 6.2.2), and selected sections of Chapter 9. A postgraduate course would likely focus on Chapters 2, 3, 5, 6, 7 and 8.

The motivation for writing this book has its origin in a series of lectures I was invited to present during a two-week postgraduate summer school at Harbin Institute of Technology (HIT) in July 2010, which was attended by some 80 PhD students and by a number of junior and established researchers from across mainland China. The numerous questions raised during the discussions pointed both to the lack of a book providing a comprehensive introduction to the basic theory of Volterra integral equations and to recent advances in this field (including applications), and hence to a lack of senior undergraduate or postgraduate lecture courses (complementing the ubiquitous courses on the theory of ordinary differential equations). I am grateful to Professors Lin Qun (Chinese Academy of Sciences), Liu Mingzhu and Song Minghui (HIT) for giving me the opportunity to present these lectures, which eventually led to this book: I hope that it will contribute to closing the above-mentioned gaps.

The many inspiring discussions with colleagues in Europe and China have allowed me to gain deeper – and often quite unexpected – new insights into various aspects of my research on the theory and numerical analysis of Volterra integral equations. In particular, I wish to express my gratitude to Professor Arieh Iserles (DAMTP, University of Cambridge), especially for inviting me to participate in the six-month programme on *Highly Oscillatory Problems* at the Isaac Newton Institute during the first half of 2007; to Professor Gennadi Vainikko (University of Tartu, Estonia) for discussions that led to the theory of cordial VIEs; and to Professor Roswitha März (Humboldt University, Berlin) for many illuminating conversations on the theory and numerical analysis of differential-algebraic equations. My interest in delay differential equations, and hence in Volterra functional integral equations, began during my visits to the University of Trieste (Italy) and my subsequent collaboration with Professors Alfredo Bellen, Lucio Torelli and, especially, Stefano Maset. I owe them many thanks, as I do my colleagues and friends in China, in particular Professors Lin Qun, Yan Ningning, Xie Hehu and Zhou Aihui (Academy of Mathematics and Systems Science, Chinese Academy of Sciences (CAS), Beijing); Professor Han Houde (Tsinghua University, Beijing); and Professor Huang Qiumei (Beijing University of Technology). It is a pleasure, too, to thank them all for the generous hospitality they extended to me during numerous visits.

I would also like to express my gratitude to my collaborators in China: Professors Zhang Ran (Jilin University, Changchun), Liang Hui (Heilongjiang University, Harbin), Dr Yang Zhan-wen (Academy of Fundamental and Interdisciplinary Science, Harbin Institute of Technology), Professor Xie Hehu (Academy of Mathematics and Systems Sciences, CAS, Beijing), and Professor Huang Qiumei (Beijing University of Technology), as well as Dr Sonia Seyed Allaei (Instituto Superior Tecnico, University of Lisbon). They all visited me for extended periods of time at Hong Kong Baptist University.

Professor Gustav Gripenberg (Aalto University, Finland), Professor Gennadi Vainikko (University of Tartu, Estonia) and Professor Martin Stynes (Beijing Computational Science Research Center) kindly agreed to read the manuscript of this book. I am deeply indebted to them for their valuable comments and suggestions. I am also pleased to thank Professor Liang Hui and Dr Yang Zhanwen for their careful checking of the manuscript.

Most of my research on topics described in this book (especially in Chapters 3, 4, 5 and 7) was carried out while I held a Visiting Professorship (2006–2011) and a Research Professorship (2011–) at Hong Kong Baptist University (HKBU). It is a pleasure to thank the Department of Mathematics and HKBU for their generous financial and technical support. I also wish gratefully to acknowledge the hospitality and the technical support I have received from the Department of Mathematics and Statistics at the University of Strathclyde (Glasgow, Scotland) since I became a Visiting Professor in 2003, and from the Department of Mathematics and Statistics at Dalhousie University (Halifax, Nova Scotia, Canada), where I have been an Adjunct Professor since 2006. The superb collection of mathematical journals at the Killam Library of my home university, Memorial University of Newfoundland, St John's, Canada, has been an invaluable help in carrying out my research work and in writing this book.

A significant part of my research that led to this monograph has been made possible by the Natural Sciences and Engineering Research Council (NSERC) of Canada and the Hong Kong Research Grants Council (RGC) by means of numerous individual research grants. I am particularly grateful to the RGC, whose generous GRF grants made it possible to extend financial support to my research collaborators listed above during their visits to HKBU.

Finally, I would like to thank the Cambridge University Press editorial staff for their kind and invaluable support and advice: David Tranah for precious initial advice (especially on the organisation of the introductory Chapters 1 and 2), and Sam Harrison and Clare Dennison for their encouragement and for guiding the manuscript to the present printed form.

I dedicate this book to Ruth, who has opened her beautiful home *Sunnezyt* and its enchanted garden to me for many years.

1

Linear Volterra Integral Equations

Summary

This chapter presents an introduction to the history and the classical theory of linear Volterra integral equations of the first and second kinds, including equations with weakly singular kernels. The focus of the presentation is on the existence and uniqueness of solutions of such equations. More advanced properties (e.g. the regularity of solutions) of such integral equations will be studied in Chapter 2. There, we shall also present an introduction to the theory of linear Volterra functional integral equations with various types of delay arguments.

1.1 Introduction

Vito Volterra (1860–1940) presented his celebrated theory of integral equations that now bear his name in four papers in 1896. In his *Nota I* he observes the lack of a systematic means of 'inverting definite integrals', except for particular cases, and that neither does there appear to be a systematic way to determine the existence and uniqueness of a solution of this problem. By inversion of a definite integral he means the problem of finding, for given continuous (real-valued) functions $H = H(x, y)$ and $f = f(y)$, a continuous function $\varphi = \varphi(y)$ satisfying the equation

$$f(y) - f(a) = \int_a^y \varphi(x) H(x, y) \, dx$$

on a given interval $[a, a + A]$, assuming that $H(y, y) \neq 0$ on this interval.

In *Nota II* he extends the analysis to unbounded kernels $H(x, y) = G(x, y)/(y - x)^\lambda$ with $0 < \lambda < 1$, still under the assumption that $H(y, y)$ does not vanish on $[a, a + A]$.

This assumption is dropped in *Nota III*: $H(y, y)$ is now allowed to vanish for $y = a$ but is non-zero at any other point in $(a, a + A]$. In *Nota IV* he derives sufficient conditions for the existence of a unique continuous solution when $H(y, y)$ again vanishes only for $y = a$, and f and H are of the form

$$f(y) = y^{n+1} f_1(y), \qquad H(x, y) = \sum_{i=0}^{n} a_i x^i y^{n-i} + H_1(x, y),$$

for some integer $n \geq 0$ and smooth functions f_1 and H_1.

Switching now to a notation that is commonly employed today, let $I :=$ $[0, T]$ be a given bounded and closed interval, and suppose that $f = f(t)$, $g = g(t)$, $a = a(t)$ and $K = K(t, s)$ are given functions with respective domains I and $D := \{(t, s) : 0 \leq s \leq t \leq T\}$. The equation

$$u(t) = f(t) + \int_0^t K(t, s) u(s) \, ds, \quad t \in I \qquad (1.1.1)$$

is called a Volterra integral equation (VIE) of the *second kind* with *kernel K*. A Volterra integral equation of the *first kind* has the form

$$\int_0^t K(t, s) u(s) \, ds = g(t), \quad t \in I. \qquad (1.1.2)$$

If the given function a vanishes at *some* points in I (but does not vanish identically on I), the integral equation

$$a(t) u(t) = f(t) + \int_0^t K(t, s) u(s) \, ds, \quad t \in I \qquad (1.1.3)$$

is called a Volterra integral equation of the *third kind*. Unless stated otherwise, it will be assumed throughout this book that the given functions in (1.1.1), (1.1.2) and (1.1.3) are real-valued.

1.2 Second-Kind VIEs with Smooth Kernels

1.2.1 Existence and Uniqueness of Solutions

For a given interval $I := [0, T]$ let $\mathcal{V} : C(I) \to C(I)$ denote the linear Volterra integral operator defined by

$$(\mathcal{V}\phi)(t) := \int_0^t K(t, s) \phi(s) \, ds, \quad t \in I, \qquad (1.2.1)$$

where the kernel $K = K(t, s)$ is continuous on $D := \{(t, s) : 0 \leq s \leq t \leq T\}$. We are interested in the existence and uniqueness of solutions $u \in$

$C(I)$ of the linear integral equation

$$u(t) = f(t) + (\mathcal{V}u)(t), \quad t \in I, \tag{1.2.2}$$

when f is a given continuous function on I. In later sections we shall admit kernels $K(t, s)$ that are non-smooth or possess integrable singularities. Moreover, in Chapter 8 (Section 8.1) we shall also briefly look at the solvability of the VIE (1.2.2) in other function spaces; for example, in the space $L^2(I)$ of square-integrable functions.

As we have already mentioned in the previous section, the classical theory of linear VIEs is due to the Italian mathematician Vito Volterra. The starting point in his *Nota I* (Volterra, 1896a) is the problem of 'inverting the integral'

$$(\mathcal{V}u)(t) = g(t), \quad t \in I, \quad \text{with } g(0) = 0, \tag{1.2.3}$$

in $C(I)$ when g is a given (differentiable) function. Using the terminology introduced by Lalesco (1908, p. 126), the problem is that of solving a *Volterra integral equation* of the *first kind*. Volterra showed that under certain conditions on its kernel K (see Theorem 1.4.1) the first-kind VIE (1.2.3) is equivalent to a second-kind VIE to which *Picard iteration* (introduced in Picard, 1890) can be applied. This iteration process leads, via the Neumann series associated with the kernel K in (1.2.1), to the *resolvent kernel* and hence to the *resolvent representation* of the solution u.

To be more precise, set $u_0(t) := f(t)$ and define the infinite sequence $\{u_n(t)\}$ by

$$u_n(t) := f(t) + (\mathcal{V}u_{n-1})(t), \quad t \in I, \quad n \geq 1. \tag{1.2.4}$$

Thus,

$$
\begin{aligned}
u_2(t) &= f(t) + \int_0^t K(t, s)u_1(s)\, ds \\
&= f(t) + \int_0^t K(t, s)\left(f(s) + \int_0^s K(s, v)f(v)\, dv \right) ds \\
&= f(t) + \int_0^t K(t, s)f(s)\, ds + \int_0^t \left(\int_v^t K(t, s)K(s, v)\, ds \right) f(v)\, dv \\
&=: f(t) + \int_0^t K_1(t, s)f(s)\, ds + \int_0^t K_2(t, s)f(s)\, ds.
\end{aligned}
$$

A straightforward induction argument then shows that the iterates $u_n(t)$ can be expressed in terms of the *iterated kernels* $K_n = K_n(t, s)$ $(n \geq 1)$, namely,

$$u_n(t) = f(t) + \int_0^t \left(\sum_{\nu=1}^n K_\nu(t, s) \right) f(s)\, ds, \quad n \geq 1, \tag{1.2.5}$$

where $K_1(t, s) := K(t, s)$ and

$$K_n(t, s) := \int_s^t K_1(t, v)K_{n-1}(v, s)\,dv, \quad n \geq 2. \tag{1.2.6}$$

The iterated kernels also satisfy a relationship more general than (1.2.6), as the following result (first established in Volterra, 1896, *Nota I*, p. 316) shows.

Lemma 1.2.1 *Let $K \in C(D)$. Then for any integer r with $1 \leq r < n$ ($n \geq 2$),*

$$K_n(t, s) = \int_s^t K_r(t, v)K_{n-r}(v, s)\,dv, \quad (t, s) \in D. \tag{1.2.7}$$

Proof The above assertion is obviously true for $r = 1$, since $K_1 = K$. Thus, assuming it holds for n, a simple induction argument establishes the result (1.2.7) for $n + 1$. The details are left as an exercise (Exercise 1.7.1). □

Remark 1.2.2 If we associate with a given iterated kernel K_n the linear Volterra operator $\mathcal{V}_n : C(I) \to C(I)$ defined by

$$(\mathcal{V}_n\phi)(t) := \int_0^t K_n(t, s)\phi(s)\,ds, \quad n \geq 1,$$

then the result of Lemma 1.2.1 may be stated in a more general way, by saying that the Volterra integral operators $\mathcal{V}_n$ *commute*:

$$\mathcal{V}_r \circ \mathcal{V}_{n-r} = \mathcal{V}_{n-r} \circ \mathcal{V}_r, \quad 1 \leq r < n \ (n \geq 2). \qquad \diamond$$

Returning to (1.2.6), assuming that $K \in C(D)$ and setting

$$\bar{K} := \max\{|K(t, s)| : (t, s) \in D\},$$

it follows that the uniform bounds

$$|K_n(t, s)| \leq \bar{K}^n \frac{(t - s)^{n-1}}{(n - 1)!} \leq \bar{K}^n \frac{T^{n-1}}{(n - 1)!}, \quad (t, s) \in D,$$

hold for all $n \geq 1$. Thus the *Neumann series* generated by the iterated kernels K_n,

$$\sum_{n=1}^{\infty} K_n(t, s) = \lim_{v \to \infty} \sum_{n=1}^{v} K_n(t, s) =: R(t, s), \quad (t, s) \in D, \tag{1.2.8}$$

converges absolutely and uniformly on D, and this implies that its limit $R(t, s)$, the so-called *resolvent kernel* associated with the given kernel $K(t, s)$, is continuous on D. We therefore obtain

$$u(t) := \lim_{n\to\infty} u_n(t) = f(t) + \lim_{n\to\infty} \sum_{j=1}^{n} (\mathcal{V}_j f)(t)$$

$$= f(t) + \int_0^t R(t,s) f(s)\, ds, \quad t \in I.$$

We will show in Theorem 1.2.3 that this function $u \in C(I)$ is a solution of the VIE (1.2.2).

The uniform convergence of the Neumann series also implies that the resolvent kernel $R(t,s)$ satisfies the *resolvent equation*

$$R(t,s) = K(t,s) + \sum_{n=2}^{\infty} K_n(t,s) = K(t,s) + \sum_{n=2}^{\infty} \int_s^t K(t,v) K_{n-1}(v,s)\, dv,$$

which by (1.2.6) we can write as

$$R(t,s) = K(t,s) + \int_s^t K(t,v) R(v,s)\, dv, \quad (t,s) \in D. \qquad (1.2.9)$$

An analogous relationship between the kernel $K(t,s)$ and its resolvent kernel $R(t,s)$ may be obtained by resorting to the result of Lemma 1.2.1 with $r = n - 1$: it yields the *complementary resolvent equation*

$$R(t,s) = K(t,s) + \int_s^t R(t,v) K(v,s)\, dv, \quad (t,s) \in D. \qquad (1.2.10)$$

We now have all the tools for stating the principal theorem on the existence and uniqueness of continuous solutions to linear VIEs (1.2.2). This result is due to Volterra and can be found in his *Nota I* of 1896.

Theorem 1.2.3 *Let $K \in C(D)$, and assume that R is the resolvent kernel associated with K. Then for any $f \in C(I)$ the second-kind VIE (1.2.2) possesses a unique solution $u \in C(I)$, and this solution can be written in the form*

$$u(t) = f(t) + \int_0^t R(t,s) f(s)\, ds, \quad t \in I. \qquad (1.2.11)$$

Proof Replace t in the VIE (1.2.2) by v, then multiply the equation by $R(t,v)$ and integrate with respect to v over the interval $[0, t]$. Using Dirichlet's formula and the resolvent equation (1.2.10) we obtain

$$\int_0^t R(t,v) u(v)\, dv = \int_0^t R(t,v) f(v)\, dv$$

$$+ \int_0^t R(t,v) \left(\int_0^v K(v,s) u(s)\, ds \right) dv$$

$$= \int_0^t R(t,s)f(s)\,ds$$

$$+ \int_0^t \left(\int_s^t R(t,v)K(v,s)\,dv \right) u(s)\,ds$$

$$= \int_0^t R(t,s)f(s)\,ds + \int_0^t \big(R(t,s) - K(t,s) \big) u(s)\,ds,$$

and this shows that

$$(\mathcal{V}u)(t) = \int_0^t K(t,s)u(s)\,ds = \int_0^t R(t,s)f(s)\,ds, \quad t \in I.$$

The resolvent representation (1.2.11) for the solution of (1.2.2) now follows by substituting the above relation in (1.2.2). Thus, (1.2.11) defines a solution $u \in C(I)$ for the VIE (1.2.2).

In order to show that, under the assumptions of Theorem 1.2.3, this solution is *unique*, assume that $z \in C(I)$ is also a solution. Since

$$z(v) = f(v) + (\mathcal{V}z)(v), \quad v \in I,$$

multiplication of both sides by $R(t,v)$, integration with respect to v over $[0,t]$, and the use of Dirichlet's formula leads to

$$\int_0^t R(t,v)z(v)\,dv = \int_0^t R(t,v)f(v)\,dv$$

$$+ \int_0^t \left(\int_s^t R(t,v)K(v,s)\,dv \right) z(s)\,ds$$

$$= \int_0^t R(t,v)f(v)\,dv + \int_0^t \big(R(t,s) - K(t,s) \big) z(s)\,ds.$$

Here, we have employed the second resolvent equation (1.2.10). Since u and z are continuous solutions of (1.2.2), the above equation reduces to

$$u(t) - f(t) - \int_0^t K(t,s)z(s)\,ds = \big(u(t) - f(t) \big) - \big(z(t) - f(t) \big) = 0, \quad t \in I.$$

The uniqueness of the solution u given by (1.2.11) can also be established directly via *Gronwall's Lemma* (cf. Lemma 1.2.7 in Section 1.2.5). If u and z are two (continuous) solutions of (1.2.2) it follows that

$$u(t) - z(t) = \big(\mathcal{V}(u-z) \big)(t), \quad t \in I.$$

Hence, assuming again that $|K(t,s)| \le \bar{K}$ in D, we obtain

$$|u(t) - z(t)| \le \bar{K} \int_0^t |u(s) - z(s)|\,ds, \quad t \in I.$$

Lemma 1.2.7 and the continuity of u and z thus imply that

$$|u(t) - z(t)| \leq 0 \cdot \exp(\bar{K}t) = 0 \quad \text{for all } t \in I,$$

and this yields $u(t) = z(t)$ for all $t \in I$, as asserted. $\qquad\qquad\qquad\square$

Remark 1.2.4 The Volterra integral equation

$$u(t) = f(t) + \lambda(\mathcal{V}u)(t), \quad t \in I,$$

possesses a unique solution $u \in C(I)$ for *any* (real or complex) parameter λ. This follows directly from the form of the Neumann series corresponding to the kernel $\lambda K(t, s)$, since the nth iterated kernel is now bounded on D by $|\lambda|^n \bar{K}^n T^{n-1}/(n-1)!$. Hence, the inverse of the linear operator $\mathcal{I} - \lambda\mathcal{V}$ exists as a bounded linear operator on $C(I)$ for any kernel $K \in C(D)$ and any (real or complex) number λ. In other words, the *spectrum* $\sigma(\mathcal{V})$ of $\mathcal{V}$ (that is, the set of values λ for which the operator $\mathcal{I} - \lambda\mathcal{V}$ is not invertible in $C(I)$) is given by $\{0\}$. (As we shall see in Section 8.2, an operator possessing this property is called *quasi-nilpotent*.) Thus, for every $\lambda \in \mathbb{R}$ (or $\lambda \in \mathbb{C}$) the VIE $u = f + \lambda\mathcal{V}u$ possesses a unique solution $u \in C(I)$ for any $f \in C(I)$, and this solution is given by

$$u = (\mathcal{I} - \lambda\mathcal{V})^{-1} f = (\mathcal{I} + \lambda\mathcal{R})f,$$

where $\mathcal{R} : C(I) \to C(I)$ is the *resolvent operator* corresponding to $\mathcal{V}$,

$$(\mathcal{R}f)(t) := \int_0^t R(t, s)f(s)\, ds, \quad t \in I.$$

We shall present a more detailed study of the properties of resolvent kernels and resolvent operators in Sections 6.2 and 8.3. $\qquad\qquad\qquad\qquad\diamond$

Remark 1.2.5 The above observations are in general not true for the *Fredholm integral operator*

$$(\mathcal{F}u)(t) := \int_0^T K(t, s)u(s)\, ds, \quad t \in I,$$

with $K \in C(I \times I)$, and the corresponding Fredholm integral equation of the second kind,

$$u(t) = f(t) + \lambda(\mathcal{F}u)(t), \quad t \in I.$$

Here, the operator $\mathcal{I} - \lambda\mathcal{F}$ may not be invertible for all $\lambda \in \mathbb{C}$ (compare e.g. Fredholm, 1903; Cochran, 1972; Gohberg & Goldberg, 1980; or Kress, 1999; see also Appendix A.2). A simple example is the (finite-rank) Fredholm integral operator $\mathcal{F}$ with kernel $K(t, s) = a(t)b(s)$ $(a, b \in C(I))$ and $I = [0, 1]$.

It is readily verified (by setting $c_1 := \int_0^1 b(s)u(s)ds$) that $\mathcal{I} - \lambda \mathcal{F}$ is invertible if, and only if, λ is such that

$$1 - \lambda \int_0^1 a(s)b(s)\,ds \neq 0. \qquad \diamond$$

Theorem 1.2.3 allows us to establish an analogous existence and solution representation result for the linear Volterra *integro-differential equation* (VIDE)

$$u'(t) = a(t)u(t) + f(t) + (\mathcal{V}u)(t), \quad t \in I, \quad \text{with } u(0) = u_0. \qquad (1.2.12)$$

Theorem 1.2.6 *If a, $f \in C(I)$ and $K \in C(D)$, the VIDE (1.2.12) possesses a unique solution $u \in C^1(I)$ for every initial value u_0. This solution can be written in the form*

$$u(t) = S(t,0)u_0 + \int_0^t S(t,s)f(s)\,ds, \quad t \in I,$$

where the (differential) resolvent kernel S (defined in the proof below) lies in $C^1(D)$.

Proof   The initial-value problem (1.2.12) is equivalent to the VIE

$$u(t) = f_0(t) + \int_0^t K_0(t,s)u(s)\,ds, \quad t \in I.$$

Since the functions

$$f_0(t) := u_0 + \int_0^t f(s)\,ds \quad \text{and} \quad K_0(t,s) := a(s) + \int_s^t K(v,s)\,dv$$

satisfy $f_0 \in C^1(I)$ and $\partial K_0/\partial t \in C(D)$, the resolvent kernel $R_0(t,s)$ of $K_0(t,s)$ is also in $C^1(D)$. By Theorem 1.2.3 the solution of this VIE is

$$u(t) = f_0(t) + \int_0^t R_0(t,s)f_0(s)\,ds, \quad t \in I;$$

it can be written as

$$u(t) = \left(1 + \int_0^t R_0(t,s)\,ds\right)u_0 + \int_0^t \left(1 + \int_s^t R_0(t,v)\,dv\right)f(s)\,ds.$$

Setting

$$S(t,s) := 1 + \int_s^t R_0(t,v)\,dv, \quad (t,s) \in D,$$

we obtain the statement of Theorem 1.2.6, since $S \in C^1(D)$. $\qquad \square$

If the given function f in the VIE (1.2.2) is in $C^1(I)$, it is often advantageous (for example, when studying the asymptotic stability of solutions; see Chapter 6) to represent the solution of this VIE in a form resembling the familiar *variation-of-constants formula* for a linear first-order ordinary differential equation. In order to derive this alternative representation, we first observe that the special VIE

$$u(t) = f(t) + \int_0^t a(s)u(s)\,ds, \quad t \in I, \tag{1.2.13}$$

with $f \in C^1(I)$ and $a \in C(I)$, is equivalent to the initial-value problem

$$u'(t) = a(t)u(t) + f'(t), \quad t \in I, \quad u(0) = f(0), \tag{1.2.14}$$

whose solution is given by

$$u(t) = \exp\left(\int_0^t a(v)dv\right) f(0) + \int_0^t \exp\left(\int_s^t a(v)\,dv\right) f'(s)\,ds.$$

Setting

$$Q(t,s) := \exp\left(\int_s^t a(v)dv\right), \quad (t,s) \in D,$$

we obtain the well-known representation of the solution of (1.2.14), namely

$$u(t) = Q(t,0)u(0) + \int_0^t Q(t,s)f'(s)\,ds, \quad t \in I. \tag{1.2.15}$$

On the other hand, we have seen that the resolvent kernel $R(t,s)$ associated with the kernel $K(t,s) = a(s)$ in (1.2.13) satisfies the resolvent equation (1.2.9),

$$R(t,s) = a(s) + \int_s^t a(v)R(v,s)\,dv, \quad (t,s) \in D, \tag{1.2.16}$$

and hence it solves the initial-value problem

$$\frac{\partial R(t,s)}{\partial t} = a(t)R(t,s), \quad R(s,s) = a(s), \quad s \in I. \tag{1.2.17}$$

Its unique solution is

$$R(t,s) = a(s)\exp\left(\int_s^t a(v)\,dv\right), \quad (t,s) \in D.$$

In other words, we have shown that for the *special* Volterra integral equation (1.2.13),

$$\frac{\partial Q(t,s)}{\partial s} = -R(t,s), \quad (t,s) \in D. \tag{1.2.18}$$

We will now prove that the *variation-of-constants formula* (1.2.15) can be extended to the general linear VIE (1.2.2) (compare also Bownds & Cushing, 1973 and Brunner & van der Houwen, 1986, pp. 13–14).

Theorem 1.2.7 *Assume that $f \in C^1(I)$ and $K \in C(D)$. Then the (unique) solution $u \in C(I)$ of the VIE*

$$u(t) = f(t) + \int_0^t K(t, s)u(s)\, ds, \quad t \in I$$

is given by the variation-of-constants formula

$$u(t) = S(t, 0)f(0) + \int_0^t S(t, s)f'(s)\, ds, \quad t \in I, \tag{1.2.19}$$

where $S(t, s)$ is the (unique) continuous solution of

$$S(t, s) = 1 + \int_s^t K(t, v)S(v, s)\, dv, \quad (t, s) \in D. \tag{1.2.20}$$

Moreover, $S(t, s)$ is related to the resolvent kernel $R(t, s)$ of the given kernel $K(t, s)$ by

$$-\frac{\partial S(t, s)}{\partial s} = R(t, s), \quad (t, s) \in D. \tag{1.2.21}$$

Proof It follows from the definition (1.2.20) of $S(t, s)$ and the continuity of $K(t, s)$ that $S(t, t) = 1$ for $t \in I$. Using integration by parts on the right-hand side of (1.2.19) we obtain

$$S(t, 0)f(0) + \left(f(t) - S(t, 0)f(0) - \int_0^t \frac{\partial S(t, s)}{\partial s} f(s)\, ds \right)$$

$$= f(t) - \int_0^t \frac{\partial S(t, s)}{\partial s} f(s)\, ds \ .$$

Thus, we may write (1.2.19) in the form

$$u(t) = f(t) - \int_0^t \frac{\partial S(t, s)}{\partial s} f(s)\, ds, \quad t \in I. \tag{1.2.22}$$

Since u is the unique continuous solution of the VIE (1.2.2), comparison of (1.2.11) and (1.2.22) shows that the relation (1.2.21) between $S(t, s)$ and $R(t, s)$ in (1.2.11) is valid, and hence the solution representation (1.2.19) is true. □

We conclude this section by observing that for certain classes of linear VIEs it is not necessary to resort to Volterra's classical resolvent kernel approach to establish the existence and uniqueness of continuous solutions. An important class consists of Volterra equations that correspond to Volterra operators $\mathcal{V}$ with *finite rank* r; that is, their kernels have the form

$$K(t, s) = \sum_{i=1}^{r} A_i(t) B_i(s), \quad \text{with} \quad A_i, \; B_i \in C(I).$$

If we set

$$z_i(t) := \int_0^t B_i(s) u(s) \, ds \quad (i = 1, \ldots, r),$$

the VIE (1.2.2) corresponding to the above kernel can then be written as an equivalent initial-value problem for a system of r linear ordinary differential equations for the vector $\mathbf{z}(t) := (z_1(t), \ldots, z_r(t))^T$, namely,

$$z_i'(t) = B_i(t) \left(g(t) + \sum_{j=1}^{r} A_j(t) z_j(t) \right) \quad (i = 1, \ldots, r), \quad t \in I,$$

with the initial condition $\mathbf{z}(0) = 0$. Since the functions describing this system are continuous in I, it follows from the theory of linear differential equations that there exists a unique solution $\mathbf{z} \in C^1(I)$ satisfying the given initial condition. Thus, the VIE (1.2.2) with the above finite-rank kernel possesses a unique solution $u \in C(I)$,

$$u(t) = f(t) + \sum_{i=1}^{r} A_i(t) z_i(t) =: f(t) + (\mathbf{A}(t))^T \mathbf{z}(t), \quad t \in I,$$

where $\mathbf{A}(t) := (A_1(t), \ldots, A_r(t))^T$.

1.2.2 Linear VIEs with Convolution Kernels

If the kernel of the VIE (1.2.2) is a *convolution kernel*, that is, if $K(t, s) = k(t - s)$, then the resolvent kernel associated with the VIE (often referred to as the *renewal equation* if $k(z) \geq 0$)

$$u(t) = f(t) + \int_0^t k(t - s) u(s) \, ds, \quad t \in I = [0, T], \tag{1.2.23}$$

inherits this property: we have $R(t, s) = r(t - s)$ for some continuous function r. This is readily seen from the Picard iteration process applied to (1.2.23): according to (1.2.6) the iterated kernels corresponding to $k(t - s)$ are given by

$$k_n(t - s) = \int_0^{t-s} k(t - s - v) k_{n-1}(v) \, dv \quad (n \geq 2), \quad k_1(t - s) := k(t - s),$$

leading to the (absolutely and uniformly convergent) Neumann series

$$r(t - s) := \sum_{n=1}^{\infty} k_n(t - s), \quad 0 \le t - s \le T.$$

It also follows that the resolvent equations (1.2.9) and (1.2.10) assume the forms

$$r(z) = k(z) + \int_0^z k(z - v)r(v)\,dv, \quad z \in I \tag{1.2.24}$$

and

$$r(z) = k(z) + \int_0^z r(z - v)k(v)\,dv, \quad z \in I, \tag{1.2.25}$$

respectively, with $z := t - s$, and therefore for the VIE (1.2.23) Theorem 1.2.3 may be restated in the following form:

Theorem 1.2.8 *Let $k \in C(I)$. Then for any $f \in C(I)$ the linear convolution VIE (1.2.23) possesses a unique solution $u \in C(I)$ which is given by*

$$u(t) = f(t) + \int_0^t r(t - s)f(s)\,ds, \quad t \in I. \tag{1.2.26}$$

Here, the resolvent kernel r is defined by the resolvent equations (1.2.24) or (1.2.25).

The following result (cf. Bellman & Cooke, 1963 or Brunner & van der Houwen, 1986, pp. 16–17) is often useful in applications:

Theorem 1.2.9 *Consider the linear convolution equations*

$$u(t) = f(t) + \int_0^t k(t - s)u(s)\,ds, \quad t \in I \tag{1.2.27}$$

and

$$w(t) = 1 + \int_0^t k(t - s)w(s)\,ds, \quad t \in I. \tag{1.2.28}$$

Assume that $f \in C^1(I)$ and $k \in L^1(I)$. Then the (unique) solutions $u \in C(I)$ and $w \in C(I)$ of (1.2.27) and (1.2.28) are related by

$$u(t) = f(0)w(t) + \int_0^t w(t - s)f'(s)\,ds, \quad t \in I. \tag{1.2.29}$$

If w' is in $L^1(I)$ then we also have

$$u(t) = w(0)f(t) + \int_0^t w'(t - s)f(s)\,ds, \quad t \in I.$$

Proof See Exercise 1.7.5. □

Remark 1.2.10 As we mentioned at the beginning of this section, the linear VIE (1.2.23) with $k(z) \geq 0$ is also known as the *renewal equation*. A detailed discussion of such VIEs can be found in Bellman & Cooke (1963, Chapter 7); see also Section 9.2.1 in the present book. ◇

1.2.3 Adjoint VIEs

For the Volterra integral operator

$$(\mathcal{V}u)(t) = \int_0^t K(t,s)u(s)\,ds, \ \ t \in I := [0,T],$$

we define the *adjoint* operator $\mathcal{V}^*$ by

$$(\mathcal{V}^*u)(t) := \int_t^T K^*(t,s)u(s)\,ds, \ \ t \in I,$$

where $K^*(t,s) := K(s,t)$. For $K^* \in C(D^*)$, with $D^* := \{(t,s) : 0 \leq t \leq s \leq T\}$, the *adjoint VIE*

$$u(t) = f(t) + (\mathcal{V}^*u)(t), \ \ t \in I, \tag{1.2.30}$$

possesses a unique solution $u \in C(I)$ whenever $f \in C(I)$, as shown in the following analogue of Theorem 1.2.3:

Theorem 1.2.11 *Assume that $K^* \in C(D^*)$. Then for all $f \in C(I)$ the adjoint VIE (1.2.30) has a unique solution $u \in C(I)$. This solution is given by*

$$u(t) = f(t) + \int_t^T R^*(t,s)f(s)\,ds, \ \ t \in I. \tag{1.2.31}$$

Here, the resolvent kernel $R^*(t,s)$ of the kernel K^* in (1.2.30) is defined by

$$R^*(t,s) := \sum_{n=1}^{\infty} K_n^*(t,s), \ \ (t,s) \in D^*,$$

with the iterated kernels $K_1^*(t,s) := K^*(t,s)$ and

$$K_n^*(t,s) := \int_t^s K^*(t,v)K_{n-1}^*(v,s)\,dv \ \ (n \geq 2).$$

Proof See Exercise 1.7.6. □

Remark 1.2.12 It is shown in Shaw & Whiteman (1996) that the adjoint Volterra integral operator and the corresponding adjoint second-kind VIE (called the dual VIE in that paper) represent an important tool in the derivation of a posteriori error estimates for discontinuous Galerkin solutions to linear second-kind VIEs.   ◇

1.2.4 Systems of Linear VIEs

Systems (of usually very large dimension) arise naturally in the spatial semi-discretisation of partial Volterra integral equations. A typical example is given by

$$-\nabla^2 w(t, x) = f(t, x) - \int_0^t k(t - s)\nabla^2 w(s, x)\, ds,$$
$$t \in I = [0, T], \quad x \in \Omega,$$
$$w(t, x) = 0, \quad t \in I, \ x \in \partial\Omega,$$

which occurs as a mathematical model in linear quasi-static visco-elasticity problems (see e.g. Shaw, Warby & Whiteman, 1997 and Shaw & Whiteman, 2001, and their lists of references). Here, $\Omega \subset \mathbb{R}^d$ is open and bounded, with (piecewise) smooth boundary $\partial\Omega$. Spatial approximation of the differential operator based on finite element (or, for simple geometries, finite difference) techniques leads to a system of second-kind VIEs of the form

$$\mathbf{u}(t) = \mathbf{f}(t) + \int_0^t \mathbf{K}(t, s)\mathbf{u}(s)ds, \quad t \in I, \qquad (1.2.32)$$

where $\mathbf{u}(t) = \left(u_1(t), \ldots, u_M(t)\right)^T \in \mathbb{R}^M$. The function $\mathbf{f}(t) = \left(f_1(t), \ldots, f_M(t)\right)^T \in \mathbb{R}^M$ is continuous on I, and

$$\mathbf{K}(t, s) = \left[\begin{array}{c} K_{ij}(t, s) \\ (i, j = 1, \ldots, M) \end{array} \right] \in \mathbb{R}^{M \times M}$$

is an M-by-M matrix whose elements K_{ij} are either in $C(I)$ or contain weak singularities. If the system (1.2.32) results from the spatial semi-discretisation of a partial VIE like the one mentioned above, then M will generally be very large.

The theory on the existence and uniqueness of a continuous solutions $\mathbf{u}$ follows in a straightforward way from the theory developed in Section 1.2.1 (it can already be found in Volterra, 1896c and Volterra, 1913, p. 189). In particular, the resolvent kernel $\mathbf{R} = \mathbf{R}(t, s) \in \mathbb{R}^{M \times M}$ satisfies the resolvent equations

$$\mathbf{R}(t,s) = \mathbf{K}(t,s) + \int_s^t \mathbf{K}(t,v)\mathbf{R}(v,s)ds, \quad (t,s) \in D$$

and

$$\mathbf{R}(t,s) = \mathbf{K}(t,s) + \int_s^t \mathbf{R}(t,v)\mathbf{K}(v,s)ds, \quad (t,s) \in D,$$

in analogy to (1.2.9) and (1.2.10).

The following theorem is thus readily proved: the iterated matrix kernels $\mathbf{K}_n(t,s)$ are as in (1.2.6), and the uniform upper bounds for their norms $\|\mathbf{K}_n(t,s)\|_\infty$ ($n \geq 1$) on D are obtained in complete analogy to the scalar case, leading again to absolute and uniform convergence of the Neumann series of the given matrix kernel $\mathbf{K}(t,s)$,

$$\mathbf{R}(t,s) := \sum_{n=1}^\infty \mathbf{K}_n(t,s), \quad (t,s) \in D;$$

cf. (1.2.8). This implies that the matrix resolvent kernel $\mathbf{R}$ inherits the regularity of $\mathbf{K}$.

Theorem 1.2.13 *Assume that the elements K_{ij} of the matrix kernel $\mathbf{K}$ are in $C(D)$, and let $\mathbf{R} \in \mathbb{R}^{M \times M}$ ($M \geq 2$) denote the corresponding matrix resolvent kernel. Then for every $\mathbf{f} \in C(I)$ the system of second-kind Volterra integral equations (1.2.32) possesses a unique solution $\mathbf{u} \in C(I)$, and this solution is given by*

$$\mathbf{u}(t) = \mathbf{f}(t) + \int_0^t \mathbf{R}(t,s)\mathbf{f}(s)ds, \quad t \in I. \tag{1.2.33}$$

If $\mathbf{f} \in C^m(I)$ and $\mathbf{K} \in C^m(D)$ ($m \geq 1$), the solution $\mathbf{u}$ of the linear VIE system (1.2.32) lies in $C^m(I)$.

Example 1.2.14 The VIE with *two variable limits*,

$$u(t) = f(t) + \int_{-t}^t K(t,s)u(s)\,ds, \quad t \in \tilde{I} := [-T, T], \tag{1.2.34}$$

can be reduced to a system of two classical second-kind VIEs (Volterra, 1913), pp. 92–95; see also Ghermanesco, 1961, pp. 37–39). Note that the kernel $K(t,s)$ is now defined on the butterfly-like domain $\tilde{D} := \{(t,s) : -t \leq s \leq t, t \in I\}$. Following Volterra, we first write the above VIE as

$$u(t) = f(t) + \int_{-t}^0 K(t,s)u(s)\,ds + \int_0^t K(t,s)u(s)\,ds.$$

An obvious change of variable then leads to the VIE

$$u(t) = f(t) + \int_0^t K(t, -s)u(-s)\, ds + \int_0^t K(t, s)u(s)\, ds. \qquad (1.2.35)$$

Replacing t by $-t$ and then proceeding as above, we find the VIE

$$u(-t) = f(-t) - \int_0^t K(-t, -s)u(-s)\, ds - \int_0^t K(-t, s)u(s)\, ds. \quad (1.2.36)$$

If we define

$$u_1(t) := u(t), \; u_2(t) := u(-t), \quad f_1(t) := f(t), \quad f_2(t) := f(-t),$$

$$K_{11}(t, s) := K(t, s), \quad K_{12}(t, s) := K(t, -s),$$

$$K_{21}(t, s) := -K(-t, s), \quad K_{22}(t, s) := -K(-t, -s),$$

and $\mathbf{u}(t) := (u_1(t), u_2(t))^T, \; \mathbf{f}(t) := (f_1(t), f_2(t))^T,$

$$\mathbf{K}(t, s) := \begin{bmatrix} K_{11}(t, s) & K_{12}(t, s) \\ K_{21}(t, s & K_{22}(t, s) \end{bmatrix},$$

the VIEs (1.2.35), (1.2.36) can be written as a system (1.2.32) of VIEs with $M = 2$. It thus follows from Theorem 1.2.13 that this system possesses a unique continuous solution $\mathbf{u}(t)$, provided that f and K are continuous on their respective domains. We leave it to the reader to show that the first component $u_1(t)$ of $\mathbf{u}(t)$ solves the original VIE (1.2.34).

We have thus established the first part of the following theorem on the existence of a continuous solution to (1.2.34) on $\tilde{I}$ (see also Volterra, 1897, pp. 163–165; Ghermanescu, 1959, 1961). The proof of the second part is left as an exercise.

Theorem 1.2.15 *Assume that $f \in C(\tilde{I})$ and $K \in C(\tilde{D})$. Then the VIE (1.2.34) possesses a unique solution $u \in C(\tilde{I})$, and there exists a continuous $\tilde{R}(t, s)$ so that this solution can be written in the form*

$$u(t) = f(t) + \int_{-t}^t \tilde{R}(t, s)f(s)\, ds, \; t \in \tilde{I}.$$

Remark 1.2.16 If the kernel $\mathbf{K}(t, s)$ in (1.2.32) is of convolution type,

$$\mathbf{k}(t - s) = \begin{bmatrix} k_{ij}(t - s) \\ (i, j = 1, \dots, M) \end{bmatrix} \in \mathbb{R}^{M \times M},$$

the corresponding VIE with $k_{ij}(z) \geq 0$,

$$\mathbf{u}(t) = \mathbf{f}(t) + \int_0^t \mathbf{k}(t - s)\mathbf{u}(s)\, ds, \; t \in I,$$

is the vector form of the renewal equation (1.2.23). ◇

1.2.5 Comparison Theorems

We begin with a generalisation of the classical result by Gronwall (1919). Its proof can be found, for example, in Quarteroni & Valli (1997, pp. 13–14).

Lemma 1.2.17 *Let* $I := [0, T]$ *and assume that* z, $f \in C(I)$, $k \in C(I)$, *with* $k(s) \geq 0$. *If* z *satisfies the inequality*

$$z(t) \leq f(t) + \int_0^t k(s)z(s)\,ds, \quad t \in I,$$

then

$$z(t) \leq f(t) + \int_0^t k(s)f(s)\exp\left(\int_s^t k(v)dv\right)ds \quad \text{for all } t \in I. \quad (1.2.37)$$

If f *is non-decreasing on* I, *the inequality (1.2.37) reduces to*

$$z(t) \leq \exp\left(\int_0^t k(s)\,ds\right)f(t) \quad \text{for all } t \in I.$$

Remark 1.2.18 Gronwall's original result of 1919 is obtained by setting $k(s) = k_0 > 0$ and $f(t) = at$ with $a \geq 0$. We note also that the continuous function $k = k(s)$ can be replaced by an *unbounded*, but integrable, function, for example, by $k(t - s) = (t - s)^{-\alpha}$ with $0 < \alpha < 1$. We will return to this generalisation in Section 1.3.4 (Lemma 1.3.13). ◇

We now present some representative comparison theorems for solutions of Volterra inequalities (compare also Beesack, 1969, 1985a and Miller, 1971a).

Theorem 1.2.19 *Assume that* $f \in C(I)$ *and* $K \in C(D)$, *with* $f(t) \geq 0$ *and* $K(t, s) \geq 0$ *on* I *and* D, *respectively. Let* $R = R(t, s)$ *be the resolvent kernel corresponding to the kernel* $K = K(t, s)$. *If* $z \in C(I)$ *satisfies the Volterra inequality*

$$z(t) \leq f(t) + \int_0^t K(t, s)z(s)\,ds, \quad t \in I,$$

then

$$0 \leq z(t) \leq f(t) + \int_0^t R(t, s)f(s)\,ds, \quad t \in I,$$

and we have $R(t, s) \geq K(t, s) \geq 0$ *for all* $(t, s) \in D$.

Proof We observe that the non-negativity of K is inherited by its iterated kernels K_n; cf. (1.2.6). Hence, by the uniform convergence of the Neumann series (1.2.8), the resolvent kernel R possesses the same property. This also implies that

$$R(t, s) = \sum_{n=1}^{\infty} K_n(t, s) \geq K_1(t, s) = K(t, s) \geq 0, \quad (t, s) \in D.$$

$\square$

A different kind of comparison result is presented in Theorem 1.2.20. Its proof, as well as variants of this result (including the extension to VIEs in $L^2(I)$), can be found in Beesack (1969, 1975); see also Pachpatte (1998).

Theorem 1.2.20 *Assume that the functions $f_i \in C(I)$ ($i = 1, 2$) satisfy $|f_1(t)| \leq f_2(t)$ on I, and let the analogous inequality $|K_1(t, s)| \leq K_2(t, s)$ $(t, s) \in D$ hold for the kernels K_i. Then the (unique) continuous solutions u_1 and u_2 of the integral equations*

$$u_i(t) = f_i(t) + \int_0^t K_i(t, s)u_i(s)ds, \ \ t \in I \quad (i = 1, 2)$$

are related by

$$|u_1(t)| \leq u_2(t) + |f_1(t)| - f_2(t), \ \ t \in I.$$

If f_1 and K_1 are non-negative on I and D, respectively, the absolute value signs in the above inequality can be omitted.

Additional, and more general, comparison theorems are contained in Section 9.8 of the monograph by Gripenberg, Londen & Staffans (1990). The books by Miller (1971a) and Cochran (1975) contain non-linear analogues of Theorem 1.2.20.

1.3 Second-Kind VIEs with Weakly Singular Kernels

In this section we study the solutions to second-kind VIEs whose kernels contain a weakly singular (i.e. integrable) factor of algebraic or logarithmic type. These VIEs are of the form

$$u(t) = f(t) + (\mathcal{V}_\alpha u)(t), \ \ t \in I = [0, T]; \quad (1.3.1)$$

they correspond to the Volterra integral operator $\mathcal{V}_\alpha : C(I) \to C(I)$ defined by

$$(\mathcal{V}_\alpha u)(t) := \int_0^t k_\alpha(t - s)K(t, s)u(s)\,ds. \quad (1.3.2)$$

Here, the singular part $k_\alpha(t - s)$ is given by

$$k_\alpha(t - s) := \begin{cases} (t - s)^{\alpha-1} & \text{if } 0 < \alpha < 1 \\ \log(t - s) & \text{if } \alpha = 0, \end{cases} \tag{1.3.3}$$

and the kernel K is continuous on D and satisfies $K(t, t) \neq 0$ on I (in order to exclude degeneracy). In the following we shall often use the notation

$$K_\alpha(t, s) := k_\alpha(t - s)K(t, s),$$

and we will refer to VIEs of the form (1.3.1) as *weakly singular VIEs*. Moreover, since (first-kind) VIEs corresponding to the operator $\mathcal{V}_\alpha$ with $0 < \alpha < 1$ and $K(t, s) \equiv 1$ were first studied by Niels Henrik Abel (Abel, 1823, 1826), we will occasionally call the VIE with $0 < \alpha < 1$ a second-kind *Abel–Volterra integral equation*.

Remark 1.3.1 Second-kind VIEs with more general kernel singularities (for example, products of diagonal and boundary singularities, or products of integrable algebraic and logarithmic singularities), will be briefly studied in Sections 2.1.3 and 2.1.4. ◇

We shall see that for continuous f and K the existence and uniqueness result of Theorem 1.2.3 remains valid for (1.3.1) whenever $\alpha \in [0, 1)$ (cf. Theorem 1.3.5, below). However, for smooth K and f (e.g. $K \in C^m(D)$ and $f \in C^m(I)$ with $m \geq 1$) the regularity result in Theorem 1.2.7 is no longer true: for non-trivial smooth K and f the solution of (1.3.1) with $0 < \alpha < 1$ satisfies only $u \in C(I) \setminus C^1(I)$, regardless of the value of m. This will be made precise in Chapter 2 (cf. Theorems 2.1.2 and 2.1.6).

1.3.1 The Mittag-Leffler Function and Weakly Singular VIEs

In order to obtain some initial insight into the existence and uniqueness of a continuous solution of (1.3.1) we first consider a special, but representative, VIE whose kernel contains an integrable algebraic singularity.

The function

$$E_\beta(z) := \sum_{n=0}^{\infty} \frac{z^n}{\Gamma(1 + n\beta)} \quad (z \in \mathbb{C}, \ \beta > 0), \tag{1.3.4}$$

where Γ denotes the gamma function, is known as the *Mittag-Leffler function*; it was introduced early in the twentieth century by the Swedish mathematician Gösta Mittag-Leffler, whose name it now bears (see e.g. Mittag-Leffler, 1903; compare also Wiman, 1905). It is an entire function of order $1/\beta$ for any $\beta > 0$.

Since we have $E_1(z) = e^z$ (an entire function of order 1), the Mittag-Leffler function may be regarded as a generalisation of the exponential function. Other important special cases are

$$E_2(z^2) = \cosh(z), \quad E_2(-z^2) = \cos(z)$$

and

$$E_{1/2}(\pm z^{1/2}) = e^z[1 + \mathrm{erf}(\pm z^{1/2})] = e^z \mathrm{erfc}(\mp z^{1/2}).$$

Here, $z^{1/2}$ is the principal value of the square root of z corresponding to cutting the complex plane along the negative real axis, and

$$\mathrm{erf}(z) := 2(\pi)^{-1/2} \int_0^z e^{-v^2} \, dv, \qquad \mathrm{erfc}(z) := 1 - \mathrm{erf}(z),$$

denote respectively the error function and the complementary error function.

If $0 < \beta \leq 1$, $E_\beta(-z)$ is *completely monotone* on $[0, \infty)$:

$$\frac{(-1)^j d^j E_\beta(-z)}{dz^j} \geq 0, \ z \geq 0 \ (j \in \mathbb{N})$$

(Pollard, 1948). Additional properties and applications can be found, for example, in Erdélyi (1955), the survey paper by Mainardi & Gorenflo (2000) and the books by Mainardi (2010) and Diethelm (2010).

The linear initial-value problem

$$u'(t) = \lambda u(t), \ t \geq 0, \ \text{with} \ u(0) = u_0 \tag{1.3.5}$$

is equivalent to the second-kind VIE

$$u(t) = u_0 + \lambda \int_0^t u(s) \, ds, \quad t \geq 0. \tag{1.3.6}$$

Since its solution $u(t) = e^{\lambda t} u_0 = E_1(\lambda t) u_0$ is an *entire function* of order 1, it is perhaps not surprising that the solution of the special weakly singular VIE

$$u(t) = u_0 + \lambda \int_0^t \frac{(t-s)^{\alpha-1}}{\Gamma(\alpha)} u(s) \, ds, \quad t \geq 0 \quad (0 < \alpha < 1) \tag{1.3.7}$$

can be expressed in terms of the Mittag-Leffler function. The following theorem is due to Hille & Tamarkin (1930; see also Tychonoff, 1934 and Friedman, 1963).

Theorem 1.3.2 *The resolvent kernel corresponding to the kernel $\lambda(t - s)^{\alpha-1}/\Gamma(\alpha)$ in (1.3.7) has the convolution form*

$$r_\alpha(t - s) = \lambda \frac{d}{dt} E_\alpha\big(\lambda(t-s)^\alpha\big), \ 0 \leq s < t \leq T. \tag{1.3.8}$$

For any interval $I = [0, T]$ *the unique solution* $u \in C(I)$ *of the VIE (1.3.7) with* $0 < \alpha < 1$ *is given by*

$$u(t) = \left(1 + \int_0^t r_\alpha(t - s)\, ds\right) u_0 = E_\alpha(\lambda t^\alpha) u_0, \quad t \in I; \qquad (1.3.9)$$

it is completely monotone on $[0, \infty)$.

We do not prove Theorem 1.3.2 here since the result will be obtained as a special case of Theorem 1.3.5 (cf. Corollary 1.3.8).

Remark 1.3.3 For $\alpha = 1$ we recover the (smooth) solution of (1.3.5), $u(t) = E_1(\lambda t) = e^{\lambda t} u_0$. If $\alpha \in (0, 1)$ the solution of (1.3.6) is no longer smooth on I: according to (1.3.9) and (1.3.4), near $t = 0^+$ its first derivative behaves like

$$u'(t) = \left(\frac{\lambda \alpha t^{\alpha-1}}{\Gamma(1+\alpha)} + \frac{2\lambda^2 \alpha t^{2\alpha-1}}{\Gamma(1+2\alpha)} + \cdots\right) u_0,$$

and hence it becomes unbounded as $t \to 0^+$. As we shall see in Section 2.1.2, the above expression for u near $t = 0$ also reflects the general situation: the solutions of linear (and non-linear) second-kind VIEs with algebraic kernel singularity $k_\alpha(t - s)$ $(0 < \alpha < 1)$, but otherwise smooth data, are smooth on $(0, T]$ but have an unbounded first derivative at $t = 0$. The same is true if the kernel singularity is $k_0(t - s) = \log(t - s)$, i.e. if $\alpha = 0$ in (1.3.3); cf. Brunner, Pedas & Vainikko (1999), and Theorem 2.1.6 in the following chapter. ◇

1.3.2 Existence and Uniqueness of Solutions

We first consider the VIE corresponding to $\alpha \in (0, 1)$ in (1.3.3),

$$u(t) = f(t) + (\mathcal{V}_\alpha u)(t), \quad t \in I = [0, T], \qquad (1.3.10)$$

where $\mathcal{V}_\alpha : C(I) \to C(I)$ is given by

$$(\mathcal{V}_\alpha u)(t) := \int_0^t (t - s)^{\alpha-1} K(t, s) u(s)\, ds \quad (0 < \alpha < 1).$$

Lemma 1.3.4 contains a useful tool for the analysis of solution properties of (1.3.10), including the derivation of resolvent kernels associated with weakly singular kernels. Its proof employs the change of the variable of integration $s = v + (t - v)z$.

Lemma 1.3.4 *Assume that* $f \in C(I)$*. Then for any* $\alpha > 0$, $\beta > 0$,

$$\int_0^t (t-s)^{\alpha-1} \left(\int_0^s (s-v)^{\beta-1} f(v)\, dv \right) ds$$

$$= \int_0^t \left(\int_v^t (t-s)^{\alpha-1}(s-v)^{\beta-1}\, ds \right) f(v)\, dv$$

$$= \int_0^t (t-s)^{\alpha+\beta-1} \left(\int_0^1 (1-z)^{\alpha-1} z^{\beta-1}\, dz \right) f(s)\, ds$$

$$= B(\alpha, \beta) \int_0^t (t-s)^{\alpha+\beta-1} f(s)\, ds,$$

where $B(\alpha, \beta) := \Gamma(\alpha)\Gamma(\beta)/\Gamma(\alpha+\beta)$ *denotes the Euler beta function.*

The following existence and uniqueness theorem is the analogue of Theorem 1.2.3.

Theorem 1.3.5 *Assume that* $K \in C(D)$*. Then for any* $f \in C(I)$ *the linear, weakly singular Volterra integral equation (1.3.10) with* $0 < \alpha < 1$ *possesses a unique solution* $u \in C(I)$*. This solution is given by*

$$u(t) = f(t) + \int_0^t R_\alpha(t,s) f(s)\, ds, \quad t \in I. \tag{1.3.11}$$

The resolvent kernel $R_\alpha(t,s)$ *corresponding to* $K_\alpha(t,s) := (t-s)^{\alpha-1} K(t,s)$ *inherits the weak singularity* $(t-s)^{\alpha-1}$*: it has the form*

$$R_\alpha(t,s) = (t-s)^{\alpha-1} Q_\alpha(t,s), \quad 0 \le s < t \le T. \tag{1.3.12}$$

Here, $Q_\alpha(t,s)$ *is defined by the Neumann series*

$$Q_\alpha(t,s) := \sum_{n=1}^{\infty} Q_{\alpha,n}(t,s), \quad (t,s) \in D, \tag{1.3.13}$$

where $Q_{\alpha,1}(t,s) := K(t,s)$ *and*

$$Q_{\alpha,n}(t,s) := (t-s)^{(n-1)\alpha} \Phi_{\alpha,n}(t,s) \quad (n \ge 2),$$

with

$$\Phi_{\alpha,n}(t,s) := \int_0^1 (1-z)^{\alpha-1} z^{(n-1)\alpha-1}$$
$$\times Q_{\alpha,n-1}(t, s+(t-s)z) K(s+(t-s)z, s)\, dz.$$

The functions $Q_{\alpha,n}$ *are continuous on* D*, and the functions* $\Phi_{\alpha,n}$ *have the same regularity as the given kernel* K*.*

Proof In analogy to (1.2.4), the Picard iteration process for (1.3.10) defines the infinite sequence $\{u_n(t)\}$ by choosing $u_0(t) := f(t)$ and then setting

$$u_n(t) := f(t) + (\mathcal{V}_\alpha u_{n-1})(t), \quad t \in I, \quad n \geq 1.$$

This yields

$$u_n(t) = f(t) + \sum_{j=1}^{n} (\mathcal{V}_\alpha^j f)(t), \quad t \in I,$$

and it is easy to see that the j-fold composition $\mathcal{V}_\alpha^j$ of $\mathcal{V}_\alpha$ with itself possesses the iterated kernel

$$K_{\alpha,j}(t, s) := \int_s^t K_\alpha(t, v) K_{\alpha,j-1}(v, s)\, dv, \quad (t, s) \in D \quad (j \geq 2),$$

with $K_{\alpha,1}(t, s) := K_\alpha(t, s)$. Using the variable transformation $v = s + (t-s)z$ and an induction argument, it is easy to establish the following result on the form of these iterated kernels.   □

Lemma 1.3.6 *Let* $0 < \alpha < 1$ *and* $K \in C(D)$, *with* $\bar{K} := \max\{|K(t, s)| : (t, s) \in D\}$. *Then the iterated kernels* $\{K_{\alpha,n}(t, s)\}$ *corresponding to the kernel* $K_{\alpha,1}(t, s) := K_\alpha(t, s)$ *in (1.3.10) can be written as*

$$K_{\alpha,n}(t, s) = (t - s)^{\alpha-1} Q_{\alpha,n}(t, s),$$

with $Q_{\alpha,1}(t, s) := K(t, s)$ *and, for* $n \geq 2$,

$$Q_{\alpha,n}(t, s) := (t - s)^{(n-1)\alpha} \int_0^1 (1 - z)^{\alpha-1} z^{(n-1)\alpha-1} K(t, s + (t - s)z)$$

$$\times Q_{\alpha,n-1}(s + (t - s)z, s)\, dz.$$

The functions $Q_{\alpha,n}$ *are uniformly bounded on* D:

$$|Q_{\alpha,n}(t, s)| \leq \bar{K}^n T^{(n-1)\alpha} \frac{(\Gamma(\alpha))^n}{\Gamma(n\alpha)}, \quad (t, s) \in D \quad (n \geq 1). \qquad (1.3.14)$$

The resulting uniform (and absolute) convergence of the series

$$\sum_{n=1}^{\infty} Q_{\alpha,n}(t, s) =: Q_\alpha(t, s), \quad (t, s) \in D$$

implies that $Q_\alpha \in C(D)$ for all $\alpha \in (0, 1)$. The representation (1.3.11) then follows, thus extending the result (1.2.11) of Theorem 1.2.3 to VIEs with kernels containing a weak algebraic singularity.

In order to show that this solution u is the only solution in $C(I)$, assume that (1.3.10) possesses another solution $z \in C(I)$. This hypothesis implies that

$$u(t) - z(t) = (\mathcal{V}_\alpha(u - z))(t) = \int_0^t K_\alpha(t, s)[u(s) - z(s)]\,ds, \quad t \in I.$$

Hence,

$$|u(t) - z(t)| \le \bar{K} \int_0^t (t - s)^{\alpha-1}|u(s) - z(s)|\,ds, \quad t \in I.$$

Since $0 < \alpha < 1$, the generalised Gronwall inequality dealt with in Lemma 1.3.13 (see Section 1.3.4 below) yields

$$|u(t) - z(t)| \le E_\alpha(\bar{K}t^{1-\alpha}) \cdot 0 = 0 \quad \text{for all } t \in I,$$

since $f(t) \equiv 0$. The uniqueness of u thus follows from the continuity of $|u - z|$. $\qquad\square$

Remark 1.3.7

(i) Lemma 1.3.6 shows that, since the iterated kernel $K_{\alpha,n}(t, s)$ contains the factor $(t - s)^{n\alpha-1}$, the iterated kernels $K_{\alpha,n}(t, s)$ become *bounded* (and hence continuous) on D when $n \ge \lceil 1/\alpha \rceil$.

(ii) We note that the above uniform estimate (1.3.14) for the iterated kernels corresponding to the weakly singular kernel $K_\alpha(t, s)$ can already be found in Tychonoff (1938).

(iii) We shall discuss the form of the solution when the weak singularity $(t - s)^{\alpha-1}$ in (1.3.10) is replaced by the bounded (but non-smooth) term $(t - s)^{\nu+\alpha-1}$ with $\nu \in \mathbb{N}$ ($\nu \ge 2$) in Section 2.1.3. $\qquad\diamond$

The result stated in the following corollary is a generalisation of the one in Theorem 1.3.2.

Corollary 1.3.8 *Let $f \in C(I)$ and $0 < \alpha < 1$. Then the (unique) solution $u \in C(I)$ of the weakly singular VIE*

$$u(t) = f(t) + \lambda \int_0^t \frac{(t - s)^{\alpha-1}}{\Gamma(\alpha)} u(s)\,ds$$

can be written as

$$u(t) = \frac{d}{dt}\left(\int_0^t E_\alpha(\lambda(t - s)^\alpha) f(s)\,ds \right)$$

$$= f(t) + \int_0^t \left(\frac{d}{dt} E_\alpha(\lambda(t - s)^\alpha) \right) f(s)\,ds, \quad t \in I.$$

Therefore the resolvent kernel associated with its kernel $k_\alpha(t, s) = \lambda\dfrac{(t - s)^{\alpha-1}}{\Gamma(\alpha)}$ *is*

$$r_\alpha(t - s) = \lambda \frac{d}{dt} E_\alpha\big(\lambda(t - s)^\alpha\big) = \lambda(t - s)^{\alpha-1} \sum_{n=1}^{\infty} \frac{\lambda^{n-1}}{\Gamma(n\alpha)}(t - s)^{(n-1)\alpha}$$

$$(0 \leq s < t \leq T).$$

For constant $f(t) = u_0$ we can explicitly compute the above integrals (note that the infinite series converges absolutely and uniformly on D), and this allows us to obtain the result stated in Theorem 1.3.2 for the solution of the special weakly singular VIE (1.3.7).

Remark 1.3.9 The existence and uniqueness of solutions of VIEs (1.3.10) whose kernels contain algebraic or logarithmic singularities of the form (1.3.3) can be established within a more general framework, using two key properties of the corresponding Volterra integral operators $\mathcal{V}_\alpha$ (see Section 8.1 and Appendix A.2):

(i) The Volterra integral operator $\mathcal{V}_\alpha : C(I) \to C(I)$ is a *compact* operator for all $\alpha \in [0, 1]$; and
(ii) $\mathcal{V}_\alpha$ is *quasi-nilpotent* for all $\alpha \in [0, 1]$; that is, its spectrum $\sigma(\mathcal{V}_\alpha)$ is given by $\{0\}$ (see also Halmos, 1982, pp. 98–99 and 298–300). ◇

It follows from the *Fredholm Alternative* (cf. Halmos, 1982, pp. 95–96) that a compact operator whose point spectrum is empty is *quasi-nilpotent*. Hence, the operator $\mathcal{I} - \mathcal{V}_\alpha$ underlying the VIE $u = f + \mathcal{V}_\alpha u$ is always invertible and so (in analogy to the discussion in Remark 1.2.4) we have

$$u = (\mathcal{I} - \mathcal{V}_\alpha)^{-1} f = (\mathcal{I} + \mathcal{R}_\alpha)f \quad (0 \leq \alpha < 1).$$

Here, $\mathcal{R}_\alpha$ denotes the resolvent operator associated with the resolvent kernel $R_\alpha(t, s)$ of $\mathcal{V}_\alpha$ which, for $0 < \alpha < 1$, is given by (1.3.12) and (1.3.13); for the logarithmic kernel singularity ($\alpha = 0$) no analogous expression for the resolvent kernel $R_0(t, s)$ appears to be known.

This implies that the existence and uniqueness of a solution $u \in C(I)$ of the VIE (1.3.1) is also guaranteed if the kernel singularity is of *logarithmic* type, $k_0(t - s) = \log(t - s)$ (corresponding to $\alpha = 0$ in (1.3.3)).

Theorem 1.3.10 *Let $\alpha = 0$ and $K \in C(D)$ in (1.3.3). Then for any $f \in C(I)$ the VIE*

$$u(t) = f(t) + \int_0^t \log(t - s) K(t, s) u(s) \, ds, \quad t \in I \qquad (1.3.15)$$

possesses a unique solution $u \in C(I)$. This solution can be written as

$$u(t) = f(t) + \int_0^t R_0(t, s) f(s) \, ds, \quad t \in I,$$

where $R_0(t, s)$ denotes the resolvent kernel associated with the weakly singular kernel $\log(t - s) K(t, s)$.

Proof The analogue of Picard iteration (1.2.4) for the VIE (1.3.15) is

$$u_n(t) := f(t) + \int_0^t \log(t - s) K(t, s) u_{n-1}(s) \, ds, \quad t \in I \ (n \geq 1),$$

with $u_0(t) := f(t)$. By the assumptions on f and K the Picard iterates satisfy $u_n \in C(I)$ for all $n \geq 0$, and they can be shown to converge uniformly on I to a (continuous) solution of (1.3.15). We leave the proof of this, and the one of the uniqueness of u, as an exercise (cf. Exercise 1.7.8).   □

1.3.3 Other Types of Singular Kernels

Certain mathematical modelling processes lead to weakly singular VIEs of the form

$$u(t) = f(t) + \int_0^t (t^2 - s^2)^{-1/2} K(t, s) u(s) \, ds, \quad t \in I, \qquad (1.3.16)$$

with $K \in C(D)$. As we shall see in Section 7.2.1, the existence of a unique solution $u \in C(I)$ of (1.3.16) depends on the value of $K(t, s)$ at $(t, s) = (0, 0)$. If $K(0, 0) = 0$ then (1.3.16) possesses a unique continuous solution for every $f \in C(I)$ (since the underlying Volterra integral operator is compact and quasi-nilpotent; that is, its spectrum is $\{0\}$). However, for any K with $K(0, 0) \neq 0$ the corresponding Volterra operator is no longer compact, and it has an uncountable number of eigenvalues (cf. Section 7.1.2). In this case (1.3.16) possesses a unique continuous solution for every $f \in C(I)$ only if 1 is not an eigenvalue of the Volterra integral operator.

In applications (see, for example, Gorenflo & Vessella, 1991 and Section 9.1 below) one encounters a particular case of (1.3.16) (with $K(0, 0) = 0$), namely

$$u(t) = f(t) + \int_0^t (t^2 - s^2)^{-1/2} s u(s) \, ds, \quad t \in I). \qquad (1.3.17)$$

The following theorem is concerned with the existence of a unique solution $u \in C(I)$ of (1.3.17); see also Becker (2012) for a slightly different solution representation.

Theorem 1.3.11 *The VIE (1.3.17) possesses a unique solution* $u \in C(I)$ *for all* λ *and all* $f \in C(I)$*, and this solution has the representation*

$$u(t) = f(t) + \int_0^t R_{1/2}(t, s) f(s) \, ds, \quad t \in I.$$

The resolvent kernel $R_{1/2}(t, s)$ *inherits the singularity of the kernel: it has the form*

$$R_{1/2}(t, s) = \Gamma(1/2) s (t^2 - s^2)^{-1/2} \sum_{j=1}^{\infty} \frac{(\Gamma(1/2))^{j-1}}{2^{j-1} \Gamma(j/2)} (t^2 - s^2)^{(j-1)/2}, \quad t \in I,$$

where $\Gamma(1/2) = \pi^{1/2}$.

Proof Using the definition of the iterated kernels corresponding to the kernel $s(t^2 - s^2)^{-1/2}$ one readily obtains the Neumann series and hence the solution representation. We leave the details to the reader (see Exercise 1.7.11, also for a generalisation to kernels of the form $s(t^\gamma - s^\gamma)^{-1/\gamma}$, $\gamma > 1$). $\square$

Remark 1.3.12 The singular term of the integrand in (1.3.16) can be written in the form $t^{-1}(1 - z^2)^{-1/2}$. The corresponding Volterra integral operator

$$(\mathcal{V}_\varphi u)(t) := \int_0^t t^{-1} \varphi(s/t) K(t, s) u(s) \, ds,$$

where $\varphi(z) := (1 - (s/t)^\gamma)^{-1/\gamma}$ with $\gamma = 2$, is an example of a so-called *cordial* Volterra integral operator which we will study in Chapter 7. There we shall show that this cordial Volterra integral operator $\mathcal{V}_\varphi$ is *not compact* whenever $K(0, 0) \neq 0$ (for example, when the kernel $K(t, s) = \lambda \neq 0$ is constant). In this case the VIE has a unique solution $u \in C(I)$ only if 1 is not an eigenvalue of the integral operator.   ◇

1.3.4 Comparison Theorems

We conclude this look at the theory of weakly singular VIEs of the second kind by describing a number of generalisations of the comparison theorems of Section 1.2.5. These results (as well as more general variants) are due to McKee (1982), Beesack (1985) and Dixon, Mckee & Jeltsch (1986).

We start with the generalisation of Gronwall's classical result (Lemma 1.2.17 with constant $k(t) = M$) to integral inequalities with weakly singular kernels. The next lemma and its discrete counterpart play an important role in the convergence analysis of numerical methods for weakly singular VIEs of the second kind (see, for example, McKee, 1982; Dixon & McKee, 1984; or Brunner, 2004, Chapter 6).

Lemma 1.3.13 *Assume that*

(i) $f \in C(I)$, $f(t) \geq 0$, *and f is non-decreasing on I;*
(ii) *the continuous, non-negative function z satisfies the inequality*

$$z(t) \leq f(t) + M \int_0^t \frac{(t-s)^{\alpha-1}}{\Gamma(\alpha)} z(s)\, ds, \quad t \in I \ (0 < \alpha < 1), \quad (1.3.18)$$

for some $M > 0$.

Then

$$z(t) \leq E_\alpha(M t^\alpha) f(t), \quad t \in I, \qquad (1.3.19)$$

where E_α denotes the Mittag-Leffler function introduced in Section 1.3.1.

Proof Since f is continuous and non-negative on I, the estimate (1.3.19) can be readily deduced either from (1.3.9) in Theorem 1.3.1 or from (1.3.21) in Theorem 1.3.9 below. We leave the details to the reader. $\square$

The *comparison theorems* of Section 1.2.6 can be extended to integral inequalities with weakly singular kernels. We state the analogue of Theorem 1.2.19 for

$$z(t) \leq f(t) + \int_0^t K_\alpha(t,s) z(s)\, ds, \quad t \in I \ (0 < \alpha < 1) \qquad (1.3.20)$$

with weakly singular kernel

$$K_\alpha(t,s) := (t-s)^{\alpha-1} K(t,s) \ (0 < \alpha < 1).$$

Theorem 1.3.14 *Assume that $f \in C(I)$, with $f(t) \geq 0$ on I, and $K \in C(D)$, with $K(t,s) \geq 0$ on D. Let $R_\alpha(t,s) = (t-s)^{\alpha-1} Q_\alpha(t,s)$ denote the resolvent kernel associated with the given kernel $K_\alpha(t,s) = (t-s)^{\alpha-1} K(t,s)$. If $z \in C(I)$ satisfies the inequality (1.3.20), then*

$$z(t) \leq f(t) + \int_0^t R_\alpha(t,s) f(s)\, ds, \quad t \in I. \qquad (1.3.21)$$

Proof Recall from Theorem 1.3.5 (cf. (1.3.12) and (1.3.13)) that

$$Q_\alpha(t, s) = \sum_{n=1}^{\infty} Q_{\alpha,n}(t, s), \quad (t, s) \in D.$$

Since, by their recursive definition, all terms $Q_{\alpha,n}(t, s)$ are non-negative on D, the assertion (1.3.21) follows immediately. □

Remark 1.3.15 Is there an analogue of the result in Lemma 1.3.8 for inequalities of the form

$$z(t) \leq f(t) + M \int_0^t (t^2 - s^2)^{-1/2} z(s) \, ds, \quad t \in I ?$$

As we have already mentioned, the above inequality can be written in 'cordial form',

$$z(t) \leq f(t) + \int_0^t t^{-1} \varphi(s/t) z(s) \, ds,$$

with core $\varphi(z) = (1 - z^2)^{-1/2}$. We shall study the answer to this question in Section 7.2.1. ◇

1.4 VIEs of the First Kind: Smooth Kernels

1.4.1 Existence and Uniqueness of Solutions

As we have already mentioned in Section 1.2.1, Volterra was led to his theory of VIEs of the second kind when he studied the solvability (*inversione*, as he called it) of the first-kind integral equation

$$\int_0^t H(t, s) u(s) \, ds = g(t), \quad t \in I = [0, T], \tag{1.4.1}$$

under appropriate assumptions on g and the kernel H. Here is Volterra's classical result (Volterra, 1896a, *Nota I*).

Theorem 1.4.1 *Assume that the kernel H of (1.4.1) lies in $C(D)$ and satisfies $\partial H/\partial t \in C(D)$, with $H(t, t) \neq 0$ for all $t \in I$. Then for any $g \in C^1(I)$ with $g(0) = 0$, the integral equation (1.4.1) possesses a unique solution $u \in C(I)$. Its value at $t = 0$ is given by*

$$u(0) = \frac{g'(0)}{H(0, 0)}.$$

Proof Clearly, the condition $g(0) = 0$ is necessary for u to be continuous at $t = 0$. The assumptions for H and g permit the differentiation of both sides of (1.4.1), yielding

$$H(t,t)u(t) + \int_0^t \frac{\partial H(t,s)}{\partial t}u(s)ds = g'(t), \quad t \in I.$$

Since $H(t,t)$ does not vanish on I, (1.4.1) is thus equivalent to the linear second-kind VIE

$$u(t) = f(t) + \int_0^t K(t,s)y(s)ds, \quad t \in I,$$

where the functions $f \in C(I)$ and $K \in C(D)$ describing this equation are

$$f(t) := g'(t)/H(t,t) \quad \text{and} \quad K(t,s) := -\big(\partial H(t,s)/\partial t\big)/H(t,t).$$

The proof is now completed by appealing to Theorem 1.2.3.

The uniqueness of the solution can also be established in the following way. If $u, z \in C(I)$ are solutions of (1.4.1), then

$$\int_0^t H(t,s)\big(u(s) - z(s)\big)ds = 0 \quad \text{for all } t \in I.$$

Since $H(t,t) \neq 0$ for $t \in I$, the continuity of H implies that there exists a $t_1 \in I$ with $t_1 > 0$ so that

$$H(t,s) \neq 0 \quad \text{for} \quad (t,s) \in D_1 := \{(t,s) : 0 \le s \le t \le t_1\}.$$

Hence, $u(t) - z(t) \equiv 0$ on the interval $[0, t_1]$. This process can obviously be continued to the entire interval I. □

Remark 1.4.2 The existence and the uniqueness of solutions of first-kind Volterra integral equations with *convolution kernels*,

$$\int_0^t h(t-s)u(s)ds = g(t), \quad t \in I, \tag{1.4.2}$$

with $h \in C^1(I)$ and $h(0) \neq 0$, follows from Theorem 1.4.1. ◇

Remark 1.4.3 In Section 3.5 (Theorem 3.2.4 with $\alpha = 0$) we shall see an extension of this classical result to *non-linear* first-kind VIEs of *Hammerstein type*,

$$\int_0^t H(t,s)G\big(s,u(s)\big)ds = g(t), \quad t \in I :$$

under appropriate regularity assumptions this problem is equivalent to an *implicit* VIE of the second-kind,

$$G\big(t, u(t)\big) = f(t) + \int_0^t K(t, s)G(s, u(s))\, ds$$

(see also Section 3.4.3), where f and K have the same meaning as in the proof of Theorem 1.4.1. ◇

The assumption in Theorem 1.4.1 that $H(t, t)$ be non-zero for $t \in I$ is not necessary for (1.4.1) to possess a unique continuous solution on I, as the following example shows.

Example 1.4.4 The kernel

$$H(t, s) = \frac{(t - s)^{r-1}}{(r - 1)!}, \quad \text{with } r \in \mathbb{N},\ r \geq 2,$$

vanishes identically along the line $s = t$ $(t \in I)$. However, the VIE (1.4.1) with this kernel and with

$$g \in C^r(I), \quad g^{(j)}(0) = 0 \ (j = 0, \ldots, r - 1),$$

possesses the unique continuous solution $u(t) = g^{(r)}(t)$, $t \in I$. This can be readily verified either by direct substitution or by differentiating both sides of the integral equation r times.

The next example reveals that this is no longer true if $H(t, t)$ in (1.4.2) possesses isolated zeros in I.

Example 1.4.5 Let the kernel $H(t, s)$ in (1.4.1) be given by $H(t, s) = 2t - 3s$. Here we have $H(0, 0) = 0$, and $H(t, t) \neq 0$ whenever $t > 0$. It is easily verified that, for $g(t) = t^2$, $u(t) = 2 + \gamma t$ is a continuous solution of (1.4.1) for any $\gamma \in \mathbb{R}$. The first-kind VIE described by the above data is equivalent to the VIE

$$t u(t) = -2t + \int_0^t 2u(s)\, ds, \quad t \in I.$$

As we shall see in Sections 1.6 and 7.2.3, this is a VIE of the *third kind* since the coefficient of $u(t)$ on the left-hand side vanishes at $t = 0$. If we rewrite it in the form

$$u(t) = -2 + \int_0^t t^{-1} \cdot 2u(s)\, ds, \quad t \in (0, T],$$

it can also be viewed as a *cordial* VIE of the second kind. Such VIEs will be the subject of Section 7.2.

In his *Nota III* and *Nota IV* of 1896, Volterra studied the solvability in $C(I)$ of the first-kind VIE (1.4.1) when $H(0,0) = 0$. His analysis was extended by Holmgren (1900) and, especially, Lalesco (1908, pp. 136–143). We give a brief outline of Lalesco's analysis (which Volterra used in his book of 1913; cf. pp. 63–70). Suppose that the kernel $H(t,s)$ in (1.4.1) is a polynomial in t of degree $n \geq 1$,

$$H(t,s) = a_0(s) + \sum_{j=1}^{n} a_j(s) \frac{(t-s)^j}{j!}, \quad (t,s) \in D,$$

where the functions a_j are arbitrarily smooth on I. If $g \in C^{n+1}(I)$ (with $g^{(i)}(0) = 0$, $n = 0, 1, \ldots, n$), $(n+1)$-fold differentiation of (1.4.1) with respect to t leads to the linear differential equation

$$[a_0(t)u(t)]^{(n)} - [a_1(t)u(t)]^{(n-1)} \pm \cdots + (-1)^n[a_n(t)u(t)] = g^{(n+1)}(t). \quad (1.4.3)$$

Setting $t = 0$ and employing the notation $a_j^{(k)} := a_j^{(k)}(0)$ in the equations corresponding to the intermediate differentiation steps, we obtain a linear system for the n initial values, namely

$$a_0^{(0)}u(0) = g'(0)$$
$$a_0^{(0)}u'(0) + [a_0^{(1)} - a_1^{(0)}]u(0) = g''(0)$$
$$\vdots$$
$$a_0^{(0)}u^{(n-1)}(0) + [(n-1)a_0^{(1)} - a_1^{(0)}]u^{(n-1)}(0) + \cdots + [a_0^{(n-1)}$$
$$+ \cdots + a_{n-1}^{(0)}]u(0) = g^{(n)}(0).$$

If $H(0,0) = a_0(0) \neq 0$, the above system yields a unique set of initial values $u^{(i)}(0)$ ($i = 0, \ldots, n-1$) for the differential equation (1.4.3). However, if $H(0,0) = a_0(0) = 0$, this is no longer true. If we write (1.4.3) in the form

$$u^{(n)}(t) + b_1(t)u^{(n-1)}(t) + \cdots + b_n(t)u(t) = \frac{g^{(n+1)}(t)}{a_0(t)},$$

we see that the coefficients $b_j(t)$ have a pole at $t = 0$. It turns out, as Lalesco (1908, p. 140) observed (see also Volterra, 1913, pp. 63–65) that this differential equation is of *Fuchsian* type (named after the German mathematician Lazarus Fuchs who studied such equations in the mid-1860s). As an illustration, consider the case $n = 2$: here, the coefficients $b_i(t)$ are

$$b_1(t) = \frac{-a_1(t) + 2a_0'(t)}{a_0(t)}, \quad b_2(t) = \frac{a_2(t) - a_1'(t) + a_0''(t)}{a_0(t)}.$$

We will not pursue the analysis of the above Fuchsian differential equation arising from the first-kind VIE (1.4.1) with $H(0, 0) = 0$ any further. The reader interested in the details may wish to consult Davis (1927), which contains a detailed review of the analysis in Volterra's *Nota III* and *Nota IV*, and of the solvability of the above Fuchsian differential equation. But we shall return to first-kind VIEs (1.4.1) with $H(0, 0) = 0$ in Section 1.6 below and show that a natural framework for the analysis of the existence of continuous solutions of (1.4.1) (and of third-kind VIEs) is the theory of *cordial* VIEs.

1.4.2 First-Kind VIEs are Ill-Posed Problems

As we have seen in Theorem 1.4.1 and its proof, the solution of a first-kind VIE

$$(\mathcal{V}u)(t) := \int_0^t H(t, s)u(s)\, ds = g(t), \quad t \in I, \tag{1.4.4}$$

does not depend continuously on the given data H and g: if the condition $g(0) = 0$ is not satisfied, that is, if we change $g(t)$ to $g_\varepsilon(t) := g(t) + \varepsilon g_1(t)$ (where $g(0) = 0$) with some small $\varepsilon \neq 0$ and $g_1 \in C^1(I)$, $g_1(0) \neq 0$, the resulting VIE no longer possesses a continuous solution on the closed interval $I = [0, T]$. In other words, a VIE of the first kind is an example of an *ill-posed problem*.

According to Hadamard (cf. Gorenflo & Vessella, 1991, pp. 5–7, or Kirsch, 1996, pp. 9–13) a problem is *well posed* if it has the following three properties:

(E) (Existence) The problem possesses *at least one* solution.
(U) (Uniqueness) The problem has *at most one* solution.
(S) (Stability) The solution *depends continuously on the data*.

The problem is said to be *ill posed* if at least one of (E), (U), (S) does not hold.

For the reader familiar with the basic theory of linear operators in normed spaces (cf. Appendix A.2 for a review of the relevant definitions and theorems) we make the above definition more precise.

Definition 1.4.6 Assume that X and Y are (real or complex) normed spaces, and let A be a linear operator from X to Y. The equation $Au = f$ is called *well posed* (or: properly posed) if (i) A is bijective and (ii) $A^{-1} : Y \to X$ is continuous. Otherwise, the equation is called *ill posed* (or: improperly posed). ◇

Remark 1.4.7 We shall see in the Appendix (Theorem A.2.11) that if Λ : $X \to Y$ is a linear and compact operator (for example, a linear Volterra integral operator with continuous or weakly singular kernel), then $Au = f$ is ill posed if $\dim X = \infty$.   ◇

Consider now the first-kind VIE (1.4.4). For smooth K and g (with $g(0) = 0$) this equation is equivalent to the second-kind VIE

$$H(t, t)u(t) + \int_0^t \frac{\partial H(t, s)}{\partial t} u(s)\, ds = g'(t), \quad t \in I. \qquad (1.4.5)$$

It follows from Theorem 1.2.3 that if $H(t, t) \neq 0$ on $[0, T]$, this second-kind VIE is a well-posed problem. For kernels H with $H(t, t) \equiv 0$ on $[0, T]$, the above VIE reduces to a first-kind VIE with kernel $\partial H(t, s)/\partial t$. We may again carry out a differentiation step and study the resulting analogue of (1.4.4). However, if the kernel $H(t, s)$ is such that $H(t, t)$ vanishes only for *some* (but not all) points in I, then (1.4.5) is a VIE of the third kind.

A measure for the ill-posedness of the VIE (1.4.4) is given by the degree of ν-smoothing of the underlying Volterra integral operator $\mathcal{V}$ (Lamm, 2000).

Definition 1.4.8 The Volterra operator with smooth (scalar) kernel $H(t, s)$,

$$(\mathcal{V}u)(t) = \int_0^t H(t, s)u(s)\, ds, \quad t \in I,$$

is said to be ν-smoothing if there is an integer $\nu \geq 1$ such that the kernel $H(t, s)$ has the following properties:

(a) $\left. \dfrac{\partial^j H(t, s)}{\partial t^j} \right|_{s=t} = 0, \quad t \in I, \quad j = 0, 1, \ldots, \nu - 2;$

(b) $\left. \dfrac{\partial^{\nu-1} H(t, s)}{\partial t^{\nu-1}} \right|_{s=t} = k_\nu \neq 0, \quad t \in I;$

(c) $\dfrac{\partial^\nu H}{\partial t^\nu} \in C(D) \ (D := \{(t, s) : 0 \leq s \leq t \leq T\}).$

If $\left. \dfrac{\partial^j H(t, s)}{\partial t^j} \right|_{s=t} = 0$ for $t \in I$ and *all* $j \in \mathbb{N}$, then $\mathcal{V}$ is an *infinitely-smoothing* Volterra integral operator. The first-kind Volterra integral equation $\mathcal{V}u = g$ is called a *ν-smoothing problem* if $\mathcal{V}$ is a ν-smoothing operator and $g \in C^\nu(I)$.   ◇

We note that if $K(t, s)$ is a convolution kernel, $K(t, s) = k(t - s)$, the above conditions reduce to

$$k^{(j)}(0) = 0 \ (j = 0, \ldots, \nu - 2); \quad k^{(\nu-1)}(0) \neq 0; \quad k^{(\nu)} \in C(I).$$

In Section 8.4 we shall see that the degree of ill-posedness of a first-kind VIE can also be characterised by the rate of decay of the *singular values* of the underlying Volterra integral operator $\mathcal{V}$.

Example 1.4.9 The Volterra integral operator $\mathcal{V}$ of Example 1.4.4,

$$(\mathcal{V}u)(t) = \int_0^t \frac{(t-s)^{r-1}}{(r-1)!} u(s)\,ds \quad (r \in \mathbb{N}, \ r \geq 1),$$

is a ν-smoothing operator with $\nu = r$. For $r = 1$ solution of the corresponding VIE $(\mathcal{V}u)(t) = g(t)$ (with $g \in C^1(I)$ and $g(0) = 0$) is $u(t) = g'(t)$. It is well known that differentiation is an ill-posed process, and that it becomes more ill posed as r increases.

Example 1.4.10 The classical example of an infinitely-smoothing Volterra integral operator is the one associated with the (spatially one-dimensional) sideways heat equation,

$$(\mathcal{V}u)(t) = \int_0^t k(t-s)u(s)\,ds, \quad \text{with} \quad k(z) = \frac{1}{2\sqrt{\pi}z^{3/2}}e^{-1/4z}.$$

This problem is well known to be severely ill posed (see Lamm, 2000, p. 55 and its references on pp. 76–77; see also Example 9.1.6 in Chapter 9).

Remark 1.4.11 As we shall see in Section 5.2, *systems* of VIEs of the first kind,

$$\int_0^t \mathbf{H}(t,s)\mathbf{u}(s)\,ds = \mathbf{g}(t), \quad t \in I, \tag{1.4.6}$$

where the matrix kernel $\mathbf{H}$ is in $\mathbb{R}^{M \times M}$ ($M \geq 2$), occur in the analysis of *integral-algebraic equations*. Under regularity conditions on $\mathbf{H}$ and $\mathbf{g}$ analogous to the ones for scalar first-kind VIEs (cf. Theorem 1.4.1) – in particular, if $\mathbf{H}(t,t)$ is non-singular for all $t \in I$, and $\mathbf{H}$ and $(\partial/\partial t)\mathbf{H}$ are continuous on D – the system (1.4.6) possesses a unique continuous solution on I. We will extend the definition of ν-smoothing to Volterra integral operators describing the above system (1.4.6) in Section 5.2.2; it will play a key role in the definition of the so-called *tractability index* of a system of integral-algebraic equations. ◇

1.5 VIEs of the First Kind with Weakly Singular Kernels

1.5.1 Abel's Integral Equation

Using a notation consistent with the one to be employed later in this section, we write Abel's integral equation in the form

$$\int_0^t \frac{(t-s)^{\alpha-1}}{\Gamma(\alpha)} u(s)\, ds = g(t), \quad t \in I \ (0 < \alpha < 1), \tag{1.5.1}$$

where the given function g is continuously differentiable on I and satisfies $g(0) = 0$. Abel showed that its solution can be found by means of an *inversion formula*. For future reference (see Section 8.3.2) we will refer to the formula (1.5.2) below as the *resolvent representation* (or *variation-of-constants formula*) of the solution of (1.5.1).

Theorem 1.5.1 *Assume that* $g \in C^1(I)$. *Then for any* $\alpha \in (0, 1)$ *the VIE (1.5.1) possess a unique continuous solution u on* $(0, T]$. *This solution has the form*

$$u(t) = \frac{d}{dt}\left(\int_0^t \frac{(t-s)^{-\alpha}}{\Gamma(1-\alpha)} g(s)\, ds\right) \tag{1.5.2}$$

$$= \frac{g(0)}{\Gamma(1-\alpha)} t^{-\alpha} + \int_0^t \frac{(t-s)^{-\alpha}}{\Gamma(1-\alpha)} g'(s)\, ds, \quad 0 < t \le T. \tag{1.5.3}$$

If g is such that $g(0) = 0$, *the solution of (1.5.1) is continuous on the closed interval* $I = [0, T]$ *and is given by*

$$u(t) = \int_0^t \frac{(t-s)^{-\alpha}}{\Gamma(1-\alpha)} g'(s)\, ds, \quad t \in I. \tag{1.5.4}$$

Proof   We shall establish the above result as a special case of Volterra's 1896 theorem (compare the discussion preceding Theorem 1.5.7).   □

Remark 1.5.2 The representation (1.5.2) of the solution of Abel's integral equation (1.5.1) gives rise to the following *question*: does there exist a *resolvent kernel* $R(t, s)$ corresponding to the kernel $H(t, s)$ of a *first-kind* VIE (particularly if, as in (1.4.2), the kernel is smooth and of convolution type) so that the solution can be written as

$$u(t) = \frac{d}{dt}\left(\int_0^t R(t, s)g(s)\, ds\right), \quad t \in I\ ?$$

We shall deal with this problem in Section 8.3.2 (but see also Theorem 7.2.4 in Section 7.2.2, where the analogue of the representation (1.5.2) is derived for a particular class of cordial VIEs of the first kind).   ◇

In Theorem 1.5.1 we assumed that the given function g in (1.5.1) lies in the space $C^1(I)$. What can be said about the existence/uniqueness of a solution for (1.5.1) if we only have $g \in C(I)$? In other words, what is the 'minimal' function space in which g must lie so that the solution of (1.5.1) is an element of $L^1(0, T)$ or $C(I)$? It is clear that we need more than just $g \in C(I)$.

The answer to this question was given by Tonelli (1928, p. 189); see also Tamarkin (1930, Theorem 4. We present it in the following theorem (cf. Gorenflo & Vessella, 1991, pp. 11–12 and 17–22).

Theorem 1.5.3 *There exists a solution $u \in L^1(0, T)$ of (1.5.1) if, and only if, g is such that $g \in L^1(0, T)$ and the function G_α defined by*

$$G_\alpha(t) := \int_0^t \frac{(t-s)^{-\alpha}}{\Gamma(1-\alpha)} g(s)\,ds$$

is absolutely continuous on $[0, T]$ and satisfies $G_\alpha(0) = 0$.

Proof Assume first that $u \in L^1(I)$ is a solution of the VIE (1.5.1). This implies that

$$\int_0^T |g(s)|\,ds \le \frac{1}{\Gamma(\alpha)} \int_0^T \left(\int_s^T (t-s)^{\alpha-1}\,dt \right) |u(s)|\,ds$$

$$= \frac{1}{\alpha\Gamma(\alpha)} \int_0^T (T-s)^\alpha |u(s)|\,ds$$

$$\le \frac{T^\alpha}{\Gamma(1+\alpha)} \int_0^T |u(s)|\,ds < \infty$$

for $T < \infty$. Hence g must be in $L^1(0, T)$. Moreover, it follows from the definition of $G_\alpha(t)$ and (1.5.1) that

$$G_\alpha(t) = \frac{1}{\Gamma(1-\alpha)} \int_0^t (t-s)^{-\alpha} \left(\int_0^s \frac{(t-v)^{\alpha-1}}{\Gamma(\alpha)} u(v)\,dv \right) ds$$

$$= \frac{1}{\Gamma(1-\alpha)\Gamma(\alpha)} \int_0^t \left(\int_v^t (t-s)^{-\alpha}(s-v)^{\alpha-1}\,ds \right) u(v)\,dv$$

$$= \int_0^t u(v)\,dv,$$

since the inner integral has the value $B(1-\alpha, \alpha) = \Gamma(1-\alpha)\Gamma(\alpha)$ (Lemma 1.3.4). By the assumption on G_α, $\int_0^t u(s)\,ds$ is absolutely continuous, and hence the assertion follows. The proof of the sufficiency part is left as an exercise (Exercise 1.7.18). $\square$

Theorem 1.5.3 is concerned with L^1-solutions for (1.5.1). This raises the question of what (additional) conditions (weaker than the ones in Theorem 1.5.1) on the function g lead to a (unique) solution $u \in C(I)$ of (1.5.1). The following theorem (a particular case of a result in Atkinson, 1974) contains an answer to this question.

Theorem 1.5.4 *Assume that g in (1.5.1) possesses the form*

$$g(t) = t^\beta \tilde{g}(t), \quad with \quad \tilde{g} \in C^1(I) \quad and \quad \beta + 1 - \alpha \geq 1.$$

Then the VIE (1.5.1) has a unique solution $u \in C(I)$, and this solution can be written as

$$u(t) = t^{\beta-\alpha} \tilde{u}(t), \ t \in I,$$

where $\tilde{u}(t) = b + t v(t)$ for some $v \in C(I)$. The constant b is zero if, and only if, $\tilde{g}(0) = 0$.

Proof See Exercise 1.7.19. □

Remark 1.5.5 In Theorem 1.5.11 and Corollary 1.5.12 we shall encounter the extension of Theorem 1.5.4 to first-kind VIEs with integrable kernel singularities that are more general than the one in (1.5.1). ◇

The expression (1.5.3) for the solution of (1.5.1) reveals that for smooth g, e.g. $g \in C^{d+1}(I)$ with $d \geq 1$, the assumption $g(0) = 0$ is not sufficient for the corresponding solution u to lie in $C^d(I)$. We illustrate this in Theorem 1.5.6 for a class of VIEs with bounded kernels. The result (which is readily verified) also confirms an earlier remark that the condition $K(t,t) \neq 0$ on I is not necessary for the existence of a unique solution $u \in C(I)$.

Theorem 1.5.6 *Let $r \in \mathbb{N}$, $r \geq 1$, and assume that the function g in the VIE*

$$\int_0^t \frac{(t-s)^{r+\alpha-1}}{\Gamma(r+\alpha)} u(s)\,ds = g(t), \ t \in I \ (0 < \alpha < 1), \tag{1.5.5}$$

is such that the conditions $g \in C^{r+1}(I)$ and $g^{(j)}(0) = 0 \ (j = 0, \ldots, r)$ are satisfied. Then (1.5.5) possesses a unique solution $u \in C(I)$, and this solution is given by

$$u(t) = \int_0^t \frac{(t-s)^{-\alpha}}{\Gamma(1-\alpha)} g^{(r+1)}(s)\,ds, \ t \in I. \tag{1.5.6}$$

An analogous theorem on the regularity of solutions of *general* first-kind VIEs with weakly singular kernels is presented in Section 2.2.2 (Theorem 2.2.3).

1.5.2 General First-Kind VIEs: Volterra's Nota II of 1896

Nota II, the second of Volterra's 1896 papers, extends the result of his *Nota I* on the existence and uniqueness of a continuous solution for a first-kind VIE to equations of the form

$$\int_0^t (t-s)^{\alpha-1} H(t,s) u(s)\, ds = g(t), \quad t \in I \ (0 < \alpha < 1), \tag{1.5.7}$$

with smooth kernel H. Volterra first shows that this equation is equivalent to a first-kind VIE with *bounded* (and smooth) kernel $H_\alpha(t,s)$ but with non-smooth right-hand side. To see this, rewrite equation (1.5.7) as

$$\int_0^v (v-s)^{\alpha-1} H(v,s) u(s)\, ds = g(v), \quad v \in I,$$

then multiply both sides by $(t-v)^{-\alpha}$ and integrate with respect to v from 0 to t. This yields the equation

$$\int_0^t (t-v)^{-\alpha} \left(\int_0^v (v-s)^{\alpha-1} H(v,s) u(s)\, ds \right) dv = \int_0^t (t-v)^{-\alpha} g(v)\, dv,$$

or, after interchanging the order of integration,

$$\int_0^t \left(\int_s^t (t-v)^{-\alpha} (v-s)^{\alpha-1} K(v,s)\, dv \right) u(s)\, ds = \int_0^t (t-s)^{-\alpha} g(s)\, ds.$$

By using the variable transformation $v = s + (t-s)v$, the inner integral on the left-hand side can be written as

$$\int_s^t (t-v)^{-\alpha} (v-s)^{\alpha-1} H(v,s)\, dv$$

$$= \int_0^1 (1-w)^{-\alpha} w^{\alpha-1} H(s+(t-s)w, s)\, dw =: H_\alpha(t,s).$$

Hence the given VIE (1.5.7) is equivalent to the first-kind VIE

$$\int_0^t H_\alpha(t,s) u(s)\, ds = \int_0^t (t-s)^{-\alpha} g(s)\, ds =: G_\alpha(t), \quad t \in I. \tag{1.5.8}$$

Its kernel $H_\alpha(t,s)$ is bounded and inherits the smoothness of the original kernel $H(t,s)$ in (1.5.7), and it satisfies

$$H_\alpha(t,t) = H(t,t) \int_0^1 (1-w)^{-\alpha} w^{\alpha-1}\, dw = \Gamma(1-\alpha)\Gamma(\alpha) H(t,t) \neq 0, \quad t \in I,$$

where $\Gamma(\cdot)$ denotes the Gamma function (recall Lemma 1.3.4). Note that

$$\Gamma(1 - \alpha)\Gamma(\alpha) = \pi / \sin(\alpha\pi).$$

We observe that for $H(t, s) = 1/\Gamma(\alpha)$, equation (1.5.7) becomes the first-kind VIE (1.5.1) we studied in Section 1.5.1: by (1.5.8) its solution is defined implicitly by

$$\Gamma(1 - \alpha) \int_0^t u(s) \, ds = \int_0^t (t - s)^{-\alpha} g(s) \, ds, \quad t \in I,$$

which for $0 < t \leq T$ yields the variation-of-constants formula (1.5.2).

We have thus established the result stated in the following theorem: it is Volterra's second fundamental result of 1896 (*Nota II*).

Theorem 1.5.7 *Assume that*

(i) $g \in C^1(I)$, *with* $g(0) = 0$;
(ii) $H \in C(D)$, $\partial H / \partial t \in C(D)$, *with* $H(t, t) \neq 0$ *for* $t \in I$.

Then for any $\alpha \in (0, 1)$ *the following statements are true:*

(a) The first-kind VIE (1.5.7) with weakly singular kernel $(t - s)^{\alpha - 1} H(t, s)$ *is equivalent to the first-kind VIE (1.5.8) whose kernel*

$$H_\alpha(t, s) := \int_0^1 (1 - z)^{-\alpha} z^{\alpha - 1} H(s + (t - s)z, s) \, dz$$

is continuous on D *and satisfies* $\partial H_\alpha / \partial t \in C(D)$, *and whose (non-smooth) right-hand side is given by*

$$G_\alpha(t) := \int_0^t (t - s)^{-\alpha} g(s) \, ds.$$

(b) The VIE (1.5.7) possesses a unique solution $u \in C(D)$ *for every* $\alpha \in (0, 1)$. *At* $t = 0$ *the solution has the value* $u(0) = G'_\alpha(0)/H_\alpha(0, 0) = 0$.

Proof We have already established assertion (a). As can be seen from the discussion preceding Theorem 1.5.7, the kernel $H_\alpha(t, s)$ of (1.5.8) inherits the properties of $H(t, s)$ for any $\alpha \in (0, 1)$. Moreover, using integration by parts we may write the right-hand side of (1.5.8) as

$$G_\alpha(t) = \frac{g(0)}{1 - \alpha} t^{1-\alpha} + \frac{1}{1 - \alpha} \int_0^t (t - s)^{1-\alpha} g'(s) \, ds, \quad t \in I.$$

Since $g(0) = 0$ and $g' \in C(I)$ it follows that $G_\alpha \in C^1(I)$, and hence we have

$$G'_\alpha(t) = \int_0^t (t - s)^{-\alpha} g'(s) \, ds, \quad t \in I.$$

This reveals that $G_\alpha(0) = 0$ and $G'_\alpha \in C(I)$; the latter implies that

$$u(0) = G'_\alpha(0)/H_\alpha(0, 0) = 0.$$

In other words, H_α and G_α in (1.5.8) satisfy the assumptions in the existence and uniqueness theorem for first-kind VIEs with bounded kernel (Theorem 1.4.1). These observations lead to assertion (b). □

We shall show in Theorem 2.2.3 that while the solution u of (1.5.7) is continuous on the closed interval $[0, T]$, its derivative near $t = 0^+$ will behave like $t^{-\alpha}$ for any non-trivial C^d-data with $d \geq 2$.

As we have seen in the preceding section, the assumption that $g \in C^1(I)$ is not necessary for the existence of a solution $u \in C(I)$ for (1.5.7): the existence of a continuous u is governed by the regularity of G_α introduced in (1.5.8). In other words: if $g \in C(I) \setminus C^1(I)$ then the VIE (1.5.1) may still possesses a unique solution $C(I)$. The following theorem, a generalisation of Theorem 1.5.4, makes this precise (cf. Gorenflo & Vessella, 1991, pp. 90–91).

Theorem 1.5.8 *Let g in (1.5.7) satisfy the assumptions stated in Theorem 1.5.3, with $\beta + 1 - \alpha \geq 1$. If K is such that $K \in C^1(D)$ and $K(t, t) \neq 0$ for $t \in I$, then the proposition of Theorem 1.5.4 remains true for the solution of (1.5.7).*

Remark 1.5.9 In Section 8.1.2 we shall present an analogous result on the existence and uniqueness of solutions of (1.5.7) in $L^p(I)$ with $1 \leq p < \infty$ (using the fact that $C(I)$ is a closed subspace of $L^\infty(I)$). ◇

In his 1916 paper (Ch. 4) Volterra analysed the solution of first-kind VIEs whose kernels contain weakly singular factors of both algebraic and *logarithmic* types. He showed in particular that the derivation of the solution of the VIE

$$\int_0^t \left(\log(t - s) + \gamma \right) u(s) ds = g(t), \quad t \in I,$$

with $g(0) = 0$, and with

$$\gamma := -\Gamma'(1)/\Gamma(1) = -\int_0^\infty e^{-s} \log(s) ds \doteq 0.5772157,$$

denoting the *Euler constant*, is considerably more complex than in the case $0 < \alpha < 1$. The starting point of the analysis is the fact that this VIE is equivalent to a first-kind equation of the form

$$\int_0^t \lim_{\alpha \to 1} \frac{d}{d\alpha} \left(\frac{1}{\Gamma(\alpha)} (t - s)^{\alpha - 1} \right) u(s) ds = g(t).$$

Details and results on related VIEs can also be found in Krasnov et al. (1977, pp. 141–143), Srivastava & Buschman (1977, p. 87), and in volume 3 of Fenyö & Stolle (1984).

Remark 1.5.10 The ν-smoothing property of the Volterra integral operator

$$(\mathcal{V}_\alpha u)(t) = \int_0^t (t-s)^{\alpha-1} H(t,s)u(s)\,ds, \quad t \in I \ (0 < \alpha < 1)$$

is determined by the ν-smoothing of the Volterra integral operator whose kernel is $H(t,s)$. This is a consequence of the fact that the VIE $(\mathcal{V}_\alpha u)(t) = g(t)$ can be written in the equivalent form

$$\int_0^t H_\alpha(t,s)u(s)\,ds = \int_0^t (t-s)^{-\alpha} g(s)\,ds, \quad t \in I,$$

where the kernel

$$H_\alpha(t,s) := \int_0^1 (1-z)^{-\alpha}(1-z)^{\alpha-1} H(v+(t-v)z, v)\,dz$$

is smooth (recall Theorem 1.5.7). Therefore, we will say that the weakly singular Volterra integral operator $\mathcal{V}_\alpha$ $(0 < \alpha < 1)$ is ν-smoothing if the smooth part $H(t,s)$ of its kernel satisfies the conditions (a), (b) and (c) in Definition 1.4.8.   ◇

1.5.3 Other Types of Kernel Singularities

We briefly touch upon the first-kind Abel-type VIE corresponding to the more general Volterra–Abel integral operator

$$(\mathcal{A}_\alpha \phi)(t) := \int_0^t \big(h(t) - h(s)\big)^{\alpha-1} H(t,s)\phi(s)\,ds, \quad 0 < \alpha < 1, \quad (1.5.9)$$

with $H \in C(D)$, $H(t,t) \neq 0$ $(t \in I)$, and $h \in C^1(I)$, $h'(t) > 0$ $(t > 0)$. In many applications (see e.g. Anderssen, 1977 or Gorenflo & Vessella, 1991) we have $h(t) = t^\gamma$, with $\gamma > 1$. In this case we will denote the corresponding integral operator (1.5.10) by $\mathcal{A}_{\gamma,\alpha}$:

$$(\mathcal{A}_{\gamma,\alpha} \phi)(t) := \int_0^t (t^\gamma - s^\gamma)^{\alpha-1} H(t,s)\phi(s)\,ds, \quad 0 < \alpha < 1. \quad (1.5.10)$$

The reader may also wish to consult Sneddon (1972) and Anderssen (1977) where, in addition to (1.4.11) with $h(t) = t^\gamma$ $(\gamma > 1)$, Volterra–Abel integral equations corresponding to the *adjoint operator*

$$(\mathcal{A}_{\gamma,\alpha}^* \phi)(t) := \int_t^T (s^\gamma - t^\gamma)^{\alpha-1} H(s,t)\phi(s)\,ds \quad (\gamma > 1, \ 0 < \alpha < 1)$$

$$(1.5.11)$$

are studied.

The following theorem (see, for example, Sneddon, 1972, pp. 209–210, or Gorenflo & Vessella, 1991, p. 16) presents the inversion formula for the solution of

$$(\mathcal{A}_\alpha u)(t) = g(t), \quad t \in I,$$

when $H(t,s) = 1/\Gamma(\alpha)$. A similar inversion formula exists for first-kind VIEs corresponding to the adjoint operator $\mathcal{A}_\alpha^*$.

Theorem 1.5.11 *Assume that $h \in C^1(I)$ is non-negative and strictly increasing on I, and let $g \in C^1(I)$. Then the solution of*

$$\frac{1}{\Gamma(\alpha)} \int_0^t \left(h(t) - h(s)\right)^{\alpha-1} u(s)\,ds = g(t) \quad (0 < \alpha < 1) \qquad (1.5.12)$$

is given by

$$u(t) = \frac{1}{\Gamma(1-\alpha)} \frac{d}{dt} \left(\int_0^t \left(h(t) - h(s)\right)^{-\alpha} h'(s) g(s)\,ds \right), \quad 0 < t \le T.$$

$$(1.5.13)$$

On this interval the solution can also be expressed in the form

$$u(t) = \frac{h'(t)g(0)}{\Gamma(1-\alpha)} \left(h(t) - h(0)\right)^{-\alpha} + \frac{1}{\Gamma(1-\alpha)} \int_0^t \left(h(t) - h(s)\right)^{-\alpha} h'(t) g'(s)\,ds.$$

$$(1.5.14)$$

Proof If we employ the new variables

$$\tau := h(t), \ \sigma := h(s), \ \beta := h(0), \ \gamma := h(T),$$
$$\phi(\sigma) := u(s)/h'(s), \ G(\tau) := g(t),$$

the equation (1.5.12) becomes a standard first-kind VIE with weakly singular kernel,

$$\int_\beta^\tau (\tau - \sigma)^{\alpha-1} \phi(\sigma)\,d\sigma = G(\tau), \quad 0 < \tau \le \gamma.$$

According to Theorem 1.5.1 it possesses the solution

$$\phi(\tau) = \frac{1}{\Gamma(1-\alpha)} \frac{d}{d\tau} \int_\beta^\tau (\tau - \sigma)^{-\alpha} G(\sigma)\,d\sigma, \quad \beta < \tau \le \gamma.$$

Hence the desired expression (1.5.13) for the solution of (1.5.12) is obtained
by reverting to the original variables. In order to establish (1.5.14) we use
integration by part in (1.5.13) and then differentiate the resulting equation. □

The following corollary describes the solution representation for a particular, practically important case of $h(t)$ in (1.5.13), namely $h(t) = t^\gamma$ with
$\gamma > 1$.

Corollary 1.5.12 *Consider the integral equation (1.5.12) corresponding to*
$h(t) = t^\gamma$ *with* $\gamma > 1$ *and* $0 < \alpha < 1$. *If* $g \in C^1(I)$, *the unique solution*
$u \in C(0, T]$ *of*

$$\int_0^t \frac{(t^\gamma - s^\gamma)^{\alpha-1}}{\Gamma(\alpha)} u(s)\, ds = g(t), \quad t \in I, \tag{1.5.15}$$

is given by

$$u(t) = \frac{d}{dt}\left(\int_0^t \frac{(t^\gamma - s^\gamma)^{-\alpha}}{\Gamma(1-\alpha)} \gamma s^{\gamma-1} g(s)\, ds \right)$$

$$= \frac{\gamma t^{\gamma(1-\alpha)-1}}{\Gamma(1-\alpha)} g(0) + \gamma t^{\gamma-1}\int_0^t \frac{(t^\gamma - s^\gamma)^{-\alpha}}{\Gamma(1-\alpha)} g'(s)\, ds, \quad t \in (0, T].$$

If $\gamma(1-\alpha) = 1$ *the above expressions reduce to*

$$u(t) = \frac{d}{dt}\left(\int_0^t \frac{(t^\gamma - s^\gamma)^{-(1-1/\gamma)}}{\Gamma(1/\gamma)} \gamma s^{\gamma-1} g(s)\, ds \right)$$

$$= \frac{\gamma}{\Gamma(1/\gamma)} g(0) + \gamma t^{\gamma-1}\int_0^t \frac{(t^\gamma - s^\gamma)^{-(1-1/\gamma)}}{\Gamma(1/\gamma)} g'(s)\, ds, \quad t \in I.$$

Remark 1.5.13 Corollary 1.5.12 yields information on the regularity of the
solution of (1.5.15) at $t = 0$: if $\alpha \in (0, 1)$ and $\gamma > 1$ are such that $\gamma(1-\alpha) \geq$
1, the solution is *bounded* at $t = 0$ even if $g(0) \neq 0$. This is in contrast to
the result for the classical Abel integral equation (1.5.1), where $0 < \alpha <$
1 and $\gamma = 1$ (see Theorem 1.5.1): here, $g(0) \neq 0$ leads to a solution that
is unbounded at $t = 0$. We will state and prove a regularity result (due to
Atkinson, 1974) for a more general class of weakly singular first-kind VIEs
with in Section 2.2.3 (see Theorem 2.2.6).   ◇

We conclude with a result revealing the fundamental difference between the
integral operators $\mathcal{V}_\alpha$ and $\mathcal{A}_{\gamma,\alpha}$ $(0 < \alpha < 1, \gamma > 1)$. We choose, as a simple
example, the integral operator (1.5.10) with $\gamma = 2$, $\alpha = 1/2$ and $H(t, s) \equiv 1$
(see also Atkinson, 1997a, p. 20).

Theorem 1.5.14 *For each real number $r \geq 0$ the function $\phi_r(t) := t^r$ is an eigenfunction of $\mathcal{A}_{2,1/2}$; it corresponds to the eigenvalue*

$$\lambda_r := \int_0^1 (1 - s^2)^{-1/2} s^r \, ds \in [0, \pi/2].$$

Moreover, $(\mathcal{A}_{2,1/2}\phi_r)(0) = 0$, and we have

$$\|\mathcal{A}_{2,1/2}\|_\infty = \max \{\|\mathcal{A}_{2,1/2} x\|_\infty : \|x\|_\infty \leq 1\} = \pi/2$$

(see also Appendix A.2, Definition A.2.2(iii)).

Proof Direct computation shows that, for any $r \geq 0$,

$$(\mathcal{A}_{2,1/2}\phi_r)(t) = \int_0^t (t^2 - s^2)^{-1/2} \phi_r(s) ds = \lambda_r \phi_r(t), \quad t \in I,$$

with $(\mathcal{A}_{2,1/2}\phi_r)(0) := 0$ and

$$\lambda_r := \int_0^1 (1 - s^2)^{-1/2} s^r ds \in [0, \pi/2].$$

Thus, λ_r is an *eigenvalue* of the integral operator $\mathcal{A}_{2,1/2}$ $(0 < \alpha < 1)$, with corresponding eigenfunction $\phi_r(t) = t^r$ $(r \geq 0)$. $\qquad\square$

Remark 1.5.15 We observe that this Volterra integral operator possesses a *continuous spectrum*, namely $\sigma(\mathcal{A}_{2,1/2}) = [0, \pi/2]$. This is implies that, in contrast to the (regular or weakly singular) Volterra integral operators we encountered in the previous sections, $\mathcal{A}_{2,1/2}$ is a *non-compact* integral operator from $C(I) \rightarrow C(I)$ (cf. Atkinson, 1976). We shall discuss VIEs with such non-compact operators in detail in Chapter 7. However, we mention here that the papers by Vainikko (2012, 2014) present the state of the art in the theory of a class of singular first-kind VIEs corresponding to non-compact integral operators. $\qquad\diamond$

1.6 VIEs of the Third Kind (I)

We have seen in Sections 1.4.1 and 1.5.2 that, under appropriate smoothness assumptions on the given functions g and H, the first-kind VIE

$$\int_0^t (t - s)^{\alpha-1} H(t, s) u(s) \, ds = g(t), \quad t \in I = [0, T] \ (0 < \alpha \leq 1), \quad (1.6.1)$$

is equivalent to the integral equation

$$H(t,t)u(t) = g'(t) - \int_0^t H_t(t,s)u(s)\,ds, \quad t \in I$$

(if $\alpha = 1$), or to

$$H_\alpha(t,t)u(t) = G'_\alpha(t) - \int_0^t \frac{\partial H_\alpha(t,s)}{\partial t} u(s)\,ds, \quad t \in I,$$

with

$$H_\alpha(t,t) = \Gamma(1-\alpha)\Gamma(\alpha)H(t,t) \quad \text{and}$$

$$G_\alpha(t) = \int_0^t (t-s)^{-\alpha} g(s)\,ds \quad (0 < \alpha < 1).$$

Here, we have set $H_t(t,s) := \partial H(t,s)/\partial t$. If $H(t,t)$ does not vanish on I this equation is a VIE of the second kind that possesses a unique continuous solutions on I. When $H(t,t)$ vanishes identically on I, the above VIE equation is still of the first-kind VIEs, and we can study its differentiated form. However, if $H(t,t)$ vanishes on some proper subset of I, for example, on a set of finitely many points or on a compact subinterval, the existence of a unique solution $u \in C(I)$ is no longer guaranteed. Here, the basic special case is $H(0,0) = 0$, with $H(t,t) \neq 0$ for $t > 0$. The analysis of the solvability of equations with this property (a special case of VIEs of the *third kind*; see also the Notes at the end of this chapter) forms the content of Volterra's *Nota III* of 1896.

Definition 1.6.1 The integral equation

$$a(t)u(t) = f(t) + \int_0^t (t-s)^{\alpha-1} K(t,s)u(s)\,ds, \quad t \in I \ (0 < \alpha \leq 1) \ (1.6.2)$$

is said to be a VIE of the *third kind* if the function a vanishes on a non-empty *proper subset* of I (for example, a finite number of points or a compact (proper) subinterval). ◇

Example 1.6.2 A basic example of a linear third-kind VIE is

$$(t-c)u(t) = f(t) + \int_0^t K(t-s)u(s)\,ds, \quad t \in I, \qquad (1.6.3)$$

with $c \in [0, T]$. VIEs of this type were studied by Fényes (1967, 1977).

Example 1.6.3 Grandits (2008) studied third-kind VIEs (1.6.2) with $\alpha = 1$ and kernels

$$K(t,s) = K_1(t,s) + (1-s/t)^{1/2} K_2(t,s) \quad (0 \leq s < t \leq T),$$

where K_1 and K_2 are smooth on D. For $a(t) = t$ this VIE is equivalent to

$$u(t) = t^{-1}f(t) + \int_0^t t^{-1}\big(K_1(t,s) + (1-s/t)^{1/2}K_2(t,s)\big)u(s)\,ds, \quad t \in (0,T].$$

As we shall see in Section 7.1 (see also Exercise 7.5.6), the two Volterra integral operators corresponding to the kernels $t^{-1}K_1(t,s)$ and $t^{-1}(1-s/t)^{1/2}K_2(t,s)$ are *cordial* integral operators with *cores* $\varphi(s/t) \equiv 1$ and $\varphi_2(s/t) = (1-s/t)^{1/2}$. If $K_i(0,0) = 0$ $(i = 1,2)$ they are compact operators on $C(I)$, and hence the analysis of the existence of a unique solution $u \in C(I)$ is straightforward. However, if we have $K_i(0,0) \neq 0$ $(i = 1,2)$ the analysis becomes more complex since these Volterra integral operators are no longer compact.

Before pursuing Volterra's theory (and in anticipation of its connection with the theory of *cordial* VIEs to be discussed in Section 7.2.3) we will look at the third-kind VIE

$$tu(t) = f(t) + \int_0^t K(t,s)u(s)\,ds, \quad t \in I, \tag{1.6.4}$$

assuming that f and K are smooth functions on their respective domains. If we write this VIE in the form

$$u(t) = t^{-1}f(t) + \int_0^t t^{-1}K(t,s)u(s)\,ds, \quad t \in (0,T], \tag{1.6.5}$$

we see that the condition $f(0) = 0$ is necessary for the existence of a solution u that is continuous on the closed interval $[0,T]$. The existence of a *unique* solution $u \in C(I)$ will depend on the value $K(0,0)$, as the following theorem shows.

Theorem 1.6.4 *Assume that $f \in C^d(I)$ and $K \in C^d(D)$ for some $d \geq 1$. If in addition these functions are such that $f(0) = 0$ and $K(0,0) = 0$, the third-kind VIE (1.6.4) possesses a unique solution $u \in C(I)$.*

Proof The assumptions on the smoothness of f and K and their values at $t = 0$ and $(t,s) = (0,0)$ imply that the functions $t^{-1}f(t)$ and $t^{-1}K(t,s)$ have continuous derivatives with respect to t of order $d-1$ on I and D, respectively. The assertion then follows from Theorems 1.2.2 and 1.3.3. □

The situation changes dramatically if $K(0,0) \neq 0$, since (as we shall see in Section 7.1) the Volterra integral operator describing (1.6.5) is now no longer a compact operator on $C(I)$, and its spectrum will be seen to be uncountable. Thus, the unique solvability of (1.6.4) in $C(I)$ will depend on whether or not the coefficient $\mu = 1$ of $u(t)$ is in the spectrum of this non-compact operator.

This discussion will of course be highly relevant for the analysis of the first-kind VIE (1.6.1), and its equivalent form

$$H(t, t)u(t) = g'(t) - \int_0^t H_t(t, s)u(s)\,ds, \quad t \in I :$$ (1.6.6)

if $H(0, 0) = 0$ then the latter is a VIE of the third kind. We will return to this problem in Section 7.2.3.

1.7 Exercises and Research Problems

Exercise 1.7.1 Prove Lemma 1.2.1.

Exercise 1.7.2 Let $\mathcal{V}_i : C(I) \to C(I)$ $(i = 1, 2)$ be linear Volterra integral operators given by

$$(\mathcal{V}_i\phi)(t) := \int_0^t K_i(t, s)\phi(s)\,ds, \quad t \in I = [0, T].$$

For which classes of (continuous or weakly singular) kernels $K_i(t, s)$ do these operators commute; that is, when does $\mathcal{V}_1 \circ \mathcal{V}_2 = \mathcal{V}_2 \circ \mathcal{V}_1$ hold?

Exercise 1.7.3 Carry out the proof of Theorem 1.2.15 on the existence (and representation) of a unique solution $u \in C(\tilde{I})$ of the VIE

$$u(t) = f(t) + \int_{-t}^t K(t, s)u(s)\,ds, \quad t \in \tilde{I} := [-T, T].$$

Exercise 1.7.4 Does the result of Theorem 1.2.3 remain valid if the kernel in (1.2.2) is the weakly singular kernel $(t - s)^{\alpha-1}K(t, s)$, with $0 < \alpha < 1$ and $K \in C(D)$?

Exercise 1.7.5 Prove Theorem 1.2.9.

Exercise 1.7.6 Derive the solution representation in Theorem 1.2.11 for the adjoint Volterra integral equation (1.2.30).

Exercise 1.7.7 Prove Theorem 1.3.2 directly, without resorting to the result in Theorem 1.3.5.

Exercise 1.7.8 Complete the proof of Theorem 1.3.10:

(i) Show that the sequence of Picard iterates for the VIE (1.3.15) converges uniformly on I to a solution u.

(ii) Establish the analogue of Lemma 1.3.13 where the inequality (1.3.18) is replaced by

$$z(t) \le f(t) + M \int_0^t \log(t - s)z(s)\,ds, \quad t \in I; \tag{1.7.1}$$

that is, find an upper bound for $z(t)$ analogous to the one in (1.3.19) and use the result to prove the uniqueness of u in (i).

Exercise 1.7.9 Find an explicit expression (analogous to the one in Corollary 1.3.8) for the solution of the VIE

$$u(t) = f(t) + \frac{\lambda}{\Gamma(\nu + \alpha)} \int_0^t (t - s)^{\nu + \alpha - 1} u(s)\,ds, \quad t \in I,$$

when $f \in C(I)$, $\nu \in \mathbb{N}$ ($\nu \ge 2$), $0 < \alpha < 1$.

Exercise 1.7.10 Find an analytic expression for the solution of the weakly singular VIE

$$u(t) = f(t) + \int_0^t \int_0^\tau (t - \tau)^{\alpha - 1}(\tau - v)^{\beta - 1} u(v)\,dv \quad (\alpha > 0,\ \beta > 0),$$

assuming that $f \in C(I)$. Show that for $f(t) \equiv 1$ the result in McKee & Cuminato (2015, Theorem 1) is a trivial consequence of Theorem 1.3.2 in Chapter 1.

Exercise 1.7.11 Carry out the proof of Theorem 1.3.11, and then extend it to obtain the resolvent kernel and corresponding solution representation for the more general Abel-type VIE

$$u(t) = f(t) + \lambda \int_0^t s(t^\gamma - s^\gamma)^{-1/\gamma} u(s)\,ds \quad (\gamma > 1).$$

Exercise 1.7.12 Give a direct proof for the uniqueness of the solution $u \in C(I)$ of the first-kind VIE (1.4.1), under the assumptions stated in Theorem 1.4.1, without resorting to the equivalent second-kind VIE (1.4.2).

Exercise 1.7.13 (Fenyö & Stolle, 1984, vol. 3, pp. 317–318)
Assume that the given functions in the first-kind VIE

$$\int_0^t k(t/s)u(s)\,ds = g(t), \quad t \in I, \tag{1.7.2}$$

are such that $g \in C^1(I)$, $g(0) = 0$, k is absolutely continuous on I, with $k(1) \neq 0$. Does (1.7.2) have a unique solution in $C(I)$?

Exercise 1.7.14 Consider the first-kind VIE (1.4.1) with $g(t) = t^2 g_1(t)$, and suppose that $g_1(t)$ and $K(t, s)$ have continuous derivatives of arbitrary order. Discuss the existence and uniqeness of a unique solution of (1.4.1) under the assumption that $H(t, t)$ vanishes at $t = 0$. (Compare also Volterra, 1896, Nota II, pp. 234–236.)

Exercise 1.7.15 Under which assumptions on K and g does the first-kind VIE

$$\int_{-t}^{t} K(t, s)u(s)\, ds = g(t), \quad t \in \tilde{I} := [-T, T]$$

possess a unique solution $u \in C(\tilde{I})$? (See also Volterra's book of 1913, pp. 97–100.)

Exercise 1.7.16 (*Research Problem*)
Consider the system of first-kind VIEs

$$\int_0^t \mathbf{H}_\alpha(t, s)\mathbf{u}(s)\, ds = \mathbf{g}(t), \quad t \in I,$$

whose weakly singular matrix kernel has the form

$$\mathbf{H}_\alpha(t, s) := \left[\begin{array}{c} (t - s)^{\alpha_{ij} - 1} H_{ij}(t, s) \\ (i, j = 1, \ldots, M) \end{array} \right],$$

where $M \geq 2$ and $0 < \alpha_{ij} \leq 1$, with $\alpha_{ij} < 1$ for at least one index pair (i, j). Extend the result of Theorem 1.5.7 on the existence and uniqueness of a solution $\mathbf{u} \in C(I)$ to the above VIE and analyse the regularity of the solution $\mathbf{u}$. Discuss in particular the case where $H_{ij}(t, s) \equiv 1(i, j = 1, \ldots, M)$.

Exercise 1.7.17 Analyse the solvability of the *adjoint* Abel integral equation

$$\int_t^T \frac{(s - t)^{\alpha - 1}}{\Gamma(\alpha)} u(s)\, ds = g(t), \quad t \in [0, T] \ (0 < \alpha < 1),$$

and find the corresponding inversion formula.

Exercise 1.7.18 Prove the sufficiency part of Theorem 1.5.3.

Exercise 1.7.19 Prove Theorem 1.5.4.

Exercise 1.7.20 Provide a detailed proof for the generalised inversion formula of Theorem 1.5.11.

Exercise 1.7.21 Establish the analogue of Theorem 1.5.11 for the Abel-type first-kind VIE

$$\frac{1}{\Gamma(\alpha)} \int_t^T \left(h(s) - h(t) \right)^{1-\alpha} u(s)\, ds = g(t), \quad t \in I.$$

Exercise 1.7.22 Analyse the existence and uniqueness of a solution $u \in C[0, 1]$ of the third-kind VIE

$$a(t)u(t) = f(t) + \int_0^t K(t, s)u(s)\, ds, \quad t \in [0, 1],$$

with

$$a(t) = \begin{cases} (1 - 2t)^2 & \text{if } 0 \le t < 1/2 \\ 0 & \text{if } 1/2 \le 3/4 \\ 4t - 3 & \text{if } 3/4 \le t \le 1. \end{cases}$$

Assume that f and K are smooth functions.

Exercise 1.7.23 (*Research Problem*)
Assume that f and K are arbitrarily smooth functions. Discuss the solvability (in $C(I)$) of the third-kind VIE

$$|t - c|^\beta u(t) = f(t) + \int_0^t K(t, s)u(s)\, ds, \quad t \in I,$$

where $\beta > 0$ and (i) $c = T$; (ii) $c \in (0, T)$.

Exercise 1.7.24 Let $\beta > 1$, $f \in C(I)$, with $f(0) = 0$, and $K \in C(D)$. Is the third-kind VIE

$$t^\beta u(t) = f(t) + \int_0^t K(t, s)u(s)\, ds, \quad t \in I,$$

an ill-posed problem (in $C(I)$)?

1.8 Notes

Section 1.1: *Introduction*
The name 'integral equation' was introduced by Du Bois-Reymond (1888). In this paper on elliptic partial differential equations he observes: 'Ich schrieb diese Gleichungen nicht hin, als ob sie etwa das Problem lösten oder doch der Lösung näher führten,

sie sollen nur ein Beispiel unter zahllosen sein, dafür, dass man bei Randwertprob-
lemen der linearen partiellen Differentialgleichungen beständig vor diesselbe Gattung
von Aufgaben gestellt wird, welche jedoch, wie es scheint, für die heutige Analysis
im Allgemeinen unüberwindliche Schwierigkeiten darbieten. Ich meine die zweckmäs-
sig *Integralgleichungen* zu nennenden Aufgaben, welche darin bestehen, dass die zu
bestimmende Function, ausser ihrem sonstigen Vorkommen, in ihnen unter bestimmten
Integralen enthalten ist, . . .'. [An English translation of most of this can be found in
Davis (1926, p. 10): 'I write down these equations not as if they solve the problem or
even carry it nearer to a solution; they serve only as examples of the fact that in the
boundary value problem of linear partial differential equations one is continually faced
by this type of problem which still, for the analysis of today, presents in general insurmountable difficulties. I propose to give to these very useful problems the name *integral
equations*.']

These problems that present 'insurmountable difficulties' are integral equations with
fixed limits of integration; known now as *Fredholm integral equations* of the second
kind, they were first studied by Ivar Fredholm in his fundamental paper of 1903. [An
English translation of this paper can be found in Stewart (2011).]

Certain classes of Volterra-type integral equations had already been considered by
Liouville in the 1830s (see e.g. Dieudonné, 1981 and Monna, 1973). The name 'Volterra
integral equation' appears to have been coined by Lalesco (1908, p. 126), following a
suggestion by his teacher, É. Picard. An English translation of parts of Volterra's *Nota
I* and *Nota II* can be found in Birkhoff (1973).

The paper by Lauricella (1908) gives a survey of the early developments of the theory
of integral equations with variable upper limits of integration. The 1909 book by Bôcher
(2nd ed., 1913) was the first monograph dedicated to the theory of integral equations.
It contains a beautiful presentation of Volterra's fundamental results of 1896, as well
as of Fredholm's theory of 1903 on second-kind integral equations with fixed limits of
integration. (A detailed review of the book can be found in the January 1910 issue of
the *Notices of the American Mathematical Society*, pp. 207–213.) Its publication was
followed by the report by Bateman (1910) on the state of the art in their theory, and by
the books of Lalesco (1912, containing a chronological bibliography), Volterra (1913),
(1927, cf. Volterra, 1959). Subsequent reviews of the theory of VIEs can be found in
Hellinger & Toeplitz (1927), Vivanti (1929: this book contains a comprehensive list of
references, including PhD theses), and Davis (1926, 1927, 1930). See also the books
by Kowalewski (1930), Schmeidler (1950) and Tricomi (1957, in which the solvability
of second-kind VIEs in L^2 is discussed), and Cochran (1972). The classic introduction
to the modern theory of (linear and non-linear) VIEs is Miller's monograph of 1971.
The books by Hönig (1975) and Mingarelli (1983) extend the Volterra theory to integral
equations of Volterra–Stieltjes type (not discussed in the present book).

Advanced analyses of linear Volterra integral equations may be found in Gohberg &
Kreĭn (1970) and in the seminal monograph by Gripenberg, Londen & Staffans (1990);
see also Corduneanu (1991) and Prüss (1993). The monograph by Agarwal, O'Regan

& Wong (2013, on constant-sign solutions of systems of integral equations) has several chapters dealing with systems of VIEs; it also contains an extensive list of references.

The papers by Klebanov & Sleeman (1996) and Väth (1998, 1998b), and the book by Väth (1999) beautifully complement the above expositions of Volterra theory: they respectively introduce an axiomatic theory of VIEs and study abstract VIEs by means of topological and algebraic methods. Volterra's work (and life) are described by J. Pérès in volume I of Volterra's *Opere Matematiche* (1954), in the recent books by Goodstein (2007) and Guerraggio & Paoloni (2012), and in the articles by Guerraggio (2006) and Coen (2008: I am grateful to my colleague Renzo Piccinini at Dalhousie University for bringing Coen's paper to my attention).

Section 1.2: *Second-Kind VIEs with Smooth Kernels*
Picard introduced the iteration technique that now bears his name in his *Mémoire* of 1890. Some of the results by Le Roux (1895) predate the ones by Volterra (1896a); however, his convergence analysis for the Neumann series was based on a geometric series and hence is valid only under the condition that $\bar{K}T < 1$.

The name of the resolvent kernel (also occasionally called the 'reciprocal kernel', owing to the similarity of the resolvent representation (1.2.11) with the given VIE (1.2.1)) seems to have been introduced by Volterra in his book of 1913 (cf. p. 47). Theorem 1.2.5 (VIEs with convolution kernels) occurs in Volterra (1913, p. 52); see also Friedman (1963) and the book by Bellman & Cooke (1963: its Chapter 7 contains an extensive discussion of the renewal equations, complemented by a large set of exercises).

Discussions of various properties of resolvent kernels of general linear VIEs can be found in Miller (1971, pp. 191–193), Bownds & Cushing (1973), Ling (1982), Gripenberg, Londen & Staffans (1990), and Burton & Dwiggins (2011). The book by Krasnov, Kissélev & Makarenko (1977, pp. 21–24) contains numerous examples of kernels and corresponding resolvent kernels. Resolvent kernels corresponding to rank-1 kernels $K(t, s) = A(s)B(s)$ (see the end of Section 1.2.1) are discussed in Becker (2013).

The papers by Beesack (1969, 1985) and the 2006 monograph by Pachpatte are good sources on integral inequalities and comparison theorems. Analogous results for second-kind VIEs with weakly singular kernels were established by Tychonoff (1938); see also Brakhage, Nickel & Rieder (1965), McKee (1982), Reynolds (1984), Dixon & McKee (1984), Beesack (1985), Norbury & Stuart (1987), Gorenflo & Pfeiffer (1991), Pedas & Vainikko (2006) and Becker (2011).
Section 1.3: *Second-Kind VIEs with Weakly Singular Kernels*
The 1910 doctoral thesis by G.C. Evans (supervised by Maxime Bôcher at Harvard University) represents the first study of second-kind VIEs with weakly singular kernels; its results were published in 1910 and 1911 (cf. Evans, 1910). A travelling fellowship from Harvard University enabled him to spend two years (1910–1912) in Europe, studying with Volterra in Rome for most of this time. In these papers Evans analysed the solvability of VIEs with rather general kernel singularities, for example,

$$u(t) = f(t) + \int_0^t (t - s)^{-\alpha} t^{-\beta} s^{-\gamma} K(t, s) u(s) \, ds,$$

with $\alpha + \beta + \gamma < 1$. VIEs with algebraic kernel singularities were subsequently studied by Davis (1926; see also the bibliographical notes), Hille & Tamarkin (1930), and Tychonoff (1934). Extensions and additional (historical) references may be found in Reynolds (1984). A formal definition of the meaning of a weakly singular kernel can be found in Miller (1971, p. 252).

The paper by Brakhage, Nickel & Rieder (1965) deals with weakly singular VIEs (1.3.7) where $\alpha = p/q$ is *rational*. It is shown that the corresponding continuous solution can be expressed in terms of a *finite* sum. The paper also gives an explicit expression for the resolvent kernel in the case $\alpha = 1/2$. (The authors do not seem to be aware of the Mittag-Leffler function and the 1930 paper by Hille & Tamarkin.) A comprehensive survey of the theory and applications of Mittag-Leffler functions is Haubold, Mathai & Saxena (2011).

The papers of Becker (2011, 2012) present explicit expressions for resolvent kernels of VIEs with various types of weakly singular kernels; many of these can already been found in earlier papers (see above) and in e.g. Brunner (2004, Ch. 6). Two-dimensional second-kind VIEs with two weakly singular kernels, and their explicit solutions in terms of Mittag-Leffler functions, are studied in McKee & Cuminato (2015).

Properties of fractional integrals were studied in detail in the fundamental paper by Hardy & Littlewood (1928). The monumental monograph by Samko, Kilbas & Marichev (1993) has become the standard work in this field. Illuminating introductions to fractional calculus (and fractional differential equations) can be found in Gorenflo & Mainardi (1997), Diethelm (2010) and Mainardi (2010; but compare also Davis, 1927, pp. 31–43, for an early discussion of the properties of fractional integration). In addition, the reader may wish to consult Diethelm & Ford (2012) on various properties of solutions of fractional VIEs.

Section 1.4: *VIEs of the First Kind: Smooth Kernels*
In Section 2 of his 1940 paper, Smithies studied first-kind VIEs of the form

$$\int_{-\infty}^t k(t - s) u(s) \, ds, \quad t \geq 0,$$

with kernels $k(z) = k_1(z) + k_2(z)$. Here, k_1 is assumed to be of bounded variation on $(0, \infty)$ while $k_2 \in L^1(0, \infty)$. The book by Srivastava & Buschman (1977) deals with linear first-kind Volterra integral equations possessing special kernels of convolution type; it contains an extensive list of equations whose solutions can be found explicitly. A summary of basic properties of Volterra convolutions may be found in de Hoog & Anderssen (2010, pp. 431–435). Fenyö & Stolle present a wide-ranging overview of first-kind VIEs in Volume 3 (cf. Section 13.2). Seminal treatments of the theory of first-kind VIEs (including non-classical ones) can be found in Apartsyn (2003); see also his papers of 2004 and 2008 on multilinear VIEs of the first kind, e.g.

$$\sum_{j=1}^{r} \int_0^t \cdots \int_0^t K_j(t, s_1, \ldots, s_r) \prod_{i=1}^{j} u(s_i) \, ds_i = g(t), \quad t \in I.$$

The 1999 book by Bukhgeim is an authoritative source of information on first-kind VIEs (including non-linear ones and equations in an abstract setting).

Section 1.5: *VIEs of the First Kind: Weakly Singular Kernels*
First-kind VIEs with kernel singularities $(t - s)^{\alpha - 1}$ $(0 < \alpha < 1)$ are often called VIEs of *Abel type* since such equations were first studied by Niels Henrik Abel in 1823 and 1826. In his book of 1913, Volterra presents a comprehensive review of Abel's work on singular first-kind integral equations and of the contents of his Nota II of 1896. See also Goursat (1903), Tonelli (1928a), Rothe (1931), Sneddon (1972), Smarzewski & Malinowski (1978) and Lamb (1985) for a discussion of various aspects of Abel's integral equation and related versions. An extensive list of inversion formulas for first-kind VIEs with various weak kernel singularities can be found in the books by Srivastava & Buschman (1977) and Gorenflo & Vessella (1991, Appendix 1.B, where the formula (1.B.3ii) appears to be incorrect). A very general class of Abel-type integral equations is analysed in Gorenflo, Kilbas & Yakubovich (1994), namely equations with kernels of the form $K\left(t^\alpha s^\beta (t - s)^\gamma\right)(t - s)^{\mu - 1}$, where α, β and μ are complex parameters, with $\operatorname{Re} \mu > 0$.

Section 1.6: *VIEs of the Third Kind*
The name 'integral equation of the third kind' appears to have been introduced independently by Picard (1911: cf. p. 459: 'On peut appeler *équation intégrale de troisième espèce* . . .') and by Hilbert (cf. Ch. 15, p. 195 in Hilbert, 1910: 'so möge allgemein . . . eine *Integralgleichung dritter Art* heissen'). Both Picard and Hilbert studied third-kind integral equations of *Fredholm* type,

$$a(x)u(x) = f(x) + \int_0^X K(x, t)u(t) \, dt, \quad x \in [0, X].$$

While Picard assumed that a is holomorphic on $[0, X]$ and has a finite number of simple zeros in $[0, X]$, Hilbert considers the case where the function a is piecewise constant, with values alternating finitely many times between $+1$ and -1. More general analyses of solutions to third-kind Fredholm integral equations can be found in Bart & Warnock (1973), Schock (1985) and Pereverzev, Schock & Solodky (1999). In addition, see Gabbasov & Zamaliev (2010) and Gabbasov (2011).

In his *Nota III* of 1896, Volterra studied the solvability of third-kind *Volterra* integral equations that arise when the (smooth) kernel $H(t, s)$ of a first-kind VIE has the property that $H(0, 0) = 0$, with $H(t, t) \neq 0$ on $(0, T]$. In *Nota IV* he presented an extension of this analysis to more general kernels. Volterra's results were subsequently refined by Holmgren (1900) and, especially, by Lalesco (1908) and by Evans (who studied third-kind VIEs possessing various types of kernel singularities; see Evans, 1909, pp. 132–134; Evans, 1910, pp. 410–413; Evans, 1911, pp. 430–431). In his book of 1913 (pp. 63–70), Volterra gave a detailed review of Lalesco's analysis. Many of these results, as well as a discussion of the connection between first-kind VIEs with kernels $H(t, s)$ for which $H(t, t)$ vanishes not only for $t = 0$ but also at a finite number of points in $(0, T)$, and differential equations of *Fuchsian* type, may be found in the 1927 survey paper (pp. 10–23) by Davis; see also Tricomi (1957, pp. 18–22).

The first detailed analysis of the solvability of non-linear Volterra integral equations of the third kind is due to Sato (1953b): he considered equations with $a(t) = t$ (cf. Remark 7.3.4 in Section 7.3 below). The more recent papers by Fényes (1967, 1977) deal with VIEs of the form

$$(t + c)u(t) = f(t) + \int_0^t k(t - s)u(s)\, ds,$$

with $c \geq 0$ (1967) or $c < 0$ (1977). In addition, see Reynolds (1984, pp. 236–238), Magnickiĭ (1979), Imanaliev & Asanov (2007 on regularisation of systems). An extension of Evans's work can be found in Grandits (2008). Here, the regularity of solutions to third-kind VIEs with kernels of the form $K(t, s) = 1 + \Gamma(t, s)$, with $\Gamma(t, s) := \Gamma_1(t, s) + (1 - s/t)^{1/2}\Gamma_2(t, s)$ and smooth Γ_i, is studied. Compare also Seyed Allaei, Yang & Brunner (2015: this paper has an extensive list of references). Since there is a close connection between certain classes of third-kind VIEs and *cordial VIEs* we shall present the analysis of such VIEs in Section 7.2.3.

Numerical Analysis of Linear VIEs

Among the books discussing the numerical solution of VIEs are the ones by Baker (1977), Fenyö & Stolle (1984, volume 4), Linz (1985), Brunner & van der Houwen (1986) and Brunner (2004); see also the seminal monograph by Prössdorf & Silbermann (1991) on the numerical analysis of general singular integral equations. The review papers by Anderssen (1977), Anderssen & de Hoog (1990 on Abel-type VIEs), Brunner (1986) and Baker (2000) convey a good picture of the development of numerical methods for VIEs.

The use of adjoint VIEs in the a posteriori error estimation of discontinuous Galerkin methods for second-kind VIEs is described in Shaw & Whiteman (1996), while discontinuous Galerkin methods for a rather general class of third-kind VIEs were analysed in Pereverzev & Prössdorf (1997) and Pereverzev, Schock & Solodky (1999). Collocation methods for a particular class of third-kind VIEs are the subject of the paper by Seyed Allaei, Yang & Brunner (2016).

2

Regularity of Solutions

Summary

Building on the results of Chapter 1, the focus in the first part of the present chapter is on the analysis of the regularity properties of solutions of various types of Volterra integral equations, especially equations with weakly singular kernels. In the second part we extend these results to solutions of Volterra functional integral equations with vanishing and non-vanishing delays.

2.1 VIEs of the Second Kind

2.1.1 VIEs with Smooth Kernels

We have seen in Section 1.2.1 (Theorem 1.2.3) that the solution of the linear second-kind VIE

$$u(t) = f(t) + \int_0^t K(t, s)u(s)\, ds, \quad t \in I = [0, T], \qquad (2.1.1)$$

can be expressed by means of the resolvent kernel of the given kernel $K(t, s)$,

$$u(t) = f(t) + \int_0^t R(t, s)f(s)\, ds, \quad t \in I, \qquad (2.1.2)$$

where the Neumann series $R(t, s) := \sum_{n=1}^{\infty} K_j(t, s)$ depends on the iterated kernels

$$K_n(t, s) := \int_s^t K(t, v)K_{n-1}(v, s)\, dv, \quad (t, s) \in D \ (n \geq 2),$$

with $K_1(t, s) := K(t, s)$. If K is smooth, say $K \in C^m(D)$ for some $m \geq 1$, the corresponding iterated kernels inherit this regularity. The uniform convergence of the Neumann series then implies that the same is true for the resolvent kernel $R(t, s)$: we have $R \in C^m(D)$. Thus, the following theorem is an immediate consequence of the resolvent representation (2.1.2) for the solution u of (2.1.1).

Theorem 2.1.1 *Assume that* $K \in C^m(D)$ *for some* $m \geq 1$. *Then its resolvent kernel R has the same degree of regularity, namely* $R \in C^m(D)$. *Hence for any* $f \in C^m(I)$ *the solution of the Volterra integral equation (2.1.1) is in* $C^m(I)$.

2.1.2 VIEs with Weakly Singular Kernels

The regularity result in Theorem 2.1.1 in general does not hold for the weakly singular analogue of the VIE (2.1.1),

$$u(t) = f(t) + (\mathcal{V}_\alpha u)(t), \quad t \in I \ (0 \leq \alpha < 1), \tag{2.1.3}$$

where the underlying Volterra integral operator $\mathcal{V}_\alpha : C(I) \to C(I)$ given by

$$(\mathcal{V}_\alpha u)(t) := \int_0^t K_\alpha(t, s) u(s) \, ds. \tag{2.1.4}$$

Its kernel $K_\alpha(t, s) := k_\alpha(t - s) K(t, s)$ contains the weak (that is, integrable) *diagonal* singularity

$$k_\alpha(t - s) := \begin{cases} (t - s)^{\alpha-1} & \text{if } 0 < \alpha < 1 \\ \log(t - s) & \text{if } \alpha = 0. \end{cases} \tag{2.1.5}$$

We will assume that $K \in C(D)$ satisfies $K(t, t) \neq 0$ for $t \in I$, in order to avoid degeneracy.

In Theorem 1.3.5 we stated a result on the representation of the solution of the weakly singular VIE (2.1.3) with $\alpha \in (0, 1)$. That result revealed that the regularity result of Theorem 2.1.1 ('smooth f and K lead to an equally smooth solution') is no longer valid if the kernel of the VIE contains such a weak singularity. The following theorem (cf. Lubich, 1983a, Brunner, 1983, Cao, Herdman & Xu, 2003, Brunner, 2004 and Section 6.1.2) presents a precise description of the regularity of the solution of (2.1.3): it shows that u is *Hölder-continuous* (cf. Appendix A.1.2) on I but is in general not in $C^1(I)$.

Theorem 2.1.2 *Assume that* $f \in C^m(I)$ *and* $K \in C^m(D)$, *with* $K(t, t) \neq 0$ *on* I. *Then:*

(a) *For* $\alpha \in (0, 1)$ *and any* $m \geq 1$ *the unique (continuous) solution of the weakly singular VIE*

$$u(t) = f(t) + \int_0^t (t - s)^{\alpha-1} K(t, s) u(s) \, ds, \quad t \in I, \tag{2.1.6}$$

lies in the Hölder space $C^\alpha(I)$ but not in $C^1(I)$. More precisely, its regularity is given by

$$u \in C^m(0, T] \cap C(I), \quad \text{with} \quad u'(t) \sim C_\alpha t^{\alpha-1} \text{ as } t \to 0^+, \quad (2.1.7)$$

where C_α is a non-zero constant.

(b) This solution can be written in the form

$$u(t) = \sum_{(j,n)_\alpha} \gamma_{j,n}(\alpha) t^{j+n\alpha} + U_{m,\alpha}(t), \quad t \in I, \quad (2.1.8)$$

for some $U_{m,\alpha} \in C^m(I)$. Here, the sum is taken over the index set

$$(j, n)_\alpha := \{(j, n) : j, n \in \mathbb{N}, \ j + n\alpha < m\}.$$

Proof We have seen in Theorem 1.3.5 that the solution of (2.1.6) can be written in the form

$$u(t) = f(t) + \int_0^t R_\alpha(t, s) f(s) \, ds, \ t \in I, \quad (2.1.9)$$

where $R_\alpha(t, s) = (t - s)^{\alpha-1} Q_\alpha(t, s)$. The infinite series defining $Q_\alpha(t, s)$,

$$Q_\alpha(t, s) = \sum_{n=1}^{\infty} (t - s)^{(n-1)\alpha} \Phi_{\alpha,n}(t, s),$$

with $\Phi_{\alpha,n} \in C^m(D)$, converges uniformly on D. Hence, for any non-trivial $f \in C^m(I)$ the assertion regarding the regularity of u follows immediately from the presence of the algebraic singularity $(t - s)^{\alpha-1}$ in $R_\alpha(t, s)$.

In order to establish (b), let $G_{\alpha,n}(t, s) := \Phi_{\alpha,n}(t, s) f(s)$ and consider

$$\int_0^t R_\alpha(t, s) f(s) ds = \sum_{n=1}^{\infty} \int_0^t (t - s)^{n\alpha-1} G_{\alpha,n}(t, s) ds. \quad (2.1.10)$$

It follows from the assumed regularity of f and K that $G_{\alpha,n}(\cdot, \cdot) \in C^m(D)$ for all $n \geq 1$. By Taylor's formula, employing multi-index notation with $d := (d_1, d_2)$ ($d_i \in \mathbb{N}$) and

$$|d| := d_1 + d_2, \quad d! := d_1! d_2!, \quad \mathbf{t}^d := t^{d_1} s^{d_2}, \quad D^d := \frac{\partial^{|d|}}{\partial^{d_1} \partial^{d_2}},$$

we may write

$$G_{\alpha,n}(t, s) = \sum_{|d| < m} \frac{1}{d!} D^d G_{\alpha,n}(0, 0) \mathbf{t}^d + \sum_{|d| = m} \frac{1}{d!} G_{\alpha,n}(\zeta_1, \zeta_2) \mathbf{t}^d$$

for some $(\zeta_1, \zeta_2) \in D$. Note that

$$\int_0^t (t-s)^{n\alpha-1} s^j \, ds = t^{j+n\alpha} \int_0^1 (1-v)^{n\alpha-1} v^j \, dv = B(n\alpha, j+1) \cdot t^{j+n\alpha},$$

with $B(\cdot, \cdot)$ denoting the Euler beta function (see Lemma 1.3.4). Upon substituting the above Taylor expansion and the one for f,

$$f(s) = \sum_{j=0}^{m-1} \frac{f^{(j)}(0)}{j!} s^{j-1} + \frac{1}{(m-1)!} \int_0^s (s-v)^{m-1} f^{(m)}(v) \, dv, \quad t \in I,$$

in (2.1.10), it is readily verified that the solution (2.1.9) can be written in the form

$$u(t) = \sum_{(j,n)_\alpha} \gamma_{n,j}(\alpha) t^{j+n\alpha} + U_{\alpha,m}(t), \quad t \in I,$$

where the coefficients $\gamma_{n,j}(\alpha)$ depend on the Taylor coefficients of $G_{\alpha,n}$ and f, and where the term $U_{\alpha,m}(t)$ contains all terms $t^{j+n\alpha}$ with $j + n\alpha \geq m$ and all the Taylor remainder terms (recall that the infinite series for $Q_\alpha(t, s)$ converges uniformly on D). This completes the proof of Theorem 2.1.2.   □

Remark 2.1.3

(i) An alternative proof of (2.1.8) can be found in Cao, Herdman & Xu (2003).

(ii) If the given functions f and K are real and *analytic* in their domains, then it can be shown (see Lubich, 1983a) that there is a function $U = U(z_1, z_2)$, real and *analytic* at $(0, 0)$, such that the solution of the VIE (2.1.6) with $0 < \alpha < 1$ can be written as $u(t) = U(t, t^\alpha)$. Related regularity results can also be found in the paper by Miller & Feldstein (1971).

(iii) If the kernel $K(t, s)$ in (2.1.6) is such that $K(t, t) = 0$ on I (for example, if $K(t, s) = (t-s)^r$ for some $r > 0$, or $K(t, s) = \sin(t-s)$), the solution u of (2.2.1) may be more regular (see also Exercise 2.4.2).   ◇

The effect of a non-smooth non-homogeneous term f on the regularity of the solution of the weakly singular VIE

$$u(t) = f_1(t) + t^\beta f_2(t) + \int_0^t (t-s)^{\alpha-1} K(t, s) u(s) \, ds, \quad t \in I, \quad (2.1.11)$$

with $0 < \alpha < 1$ and $\beta > 0$, $\beta \notin \mathbb{N}$, is described in the following theorem.

Theorem 2.1.4 *Let* $f(t) = f_1(t) + t^\beta f_2(t)$, *with* $f_i \in C(I)$ $(i = 1, 2)$ *and* $\beta > 0$ $(\beta \notin \mathbb{N})$, *and assume that* $K \in C(D)$. *Then the (unique) solution* $u \in C(I)$ *of the VIE (2.1.11) with* $0 < \alpha < 1$ *can be written as*

$$u(t) = f_1(t) + \int_0^t R_\alpha(t, s) f_1(s) \, ds$$

$$+ t^\beta f_2(t) + \int_0^t R_\alpha(t, s) f_2(s) s^\beta \, ds, \quad t \in I.$$

Here, $R_\alpha(t, s)$ *is the resolvent kernel given by (1.3.12) in Theorem 1.3.3.*

The proof uses the *superposition principle* for solutions of linear VIEs with the same kernel but different non-homogeneous terms, and the result of Theorem 1.3.5. The details are left as an exercise (see Exercise 2.4.3).

Remark 2.1.5 The statement of Theorem 2.1.4 yields a regularity result for the solution of the VIE (2.1.11) corresponding to $f_1(t) \equiv 0$. This raises the question as to whether we can choose β and smooth f_2 so that the solution of (2.1.11) is smooth on I, provided K is smooth. Exercise 2.4.4 addresses this question. ◇

If the weakly singular VIE (2.1.3) has a kernel singularity of the form $\log(t - s)$, corresponding to $\alpha = 0$ in (2.1.5), the regularity of the solution is described in the following theorem (due to Brunner, Pedas & Vainikko, 1999); it is the analogue of Theorem 2.1.2(a).

Theorem 2.1.6 *If* $f \in C^m(I)$ *and* $K \in C^m(D)$, *with* $K(t, t) \neq 0$ *on* I, *the solution of the VIE*

$$u(t) = f(t) + \int_0^t \log(t - s) K(t, s) u(s) \, ds, \quad t \in I, \tag{2.1.12}$$

exhibits the regularity

$$u \in C^m(0, T] \cap C(I), \quad \text{with} \quad u'(t) \sim C_0 \log(t) \ \text{as} \ t \to 0^+,$$

for some non-zero constant C_0.

Proof Since the above regularity result can be obtained as a special case of Theorem 2.1.7, we leave its direct proof as an exercise. □

The following general regularity result for the linear VIE

$$u(t) = f(t) + \int_0^t K(t, s) u(s) \, ds, \quad t \in I, \tag{2.1.13}$$

was established in Brunner, Pedas & Vainikko (1999). It includes VIEs with kernels that either contain general weak singularities (including logarithmic ones) or are bounded but have *unbounded derivatives* as $s \to t$, and with non-homogeneous terms f that may be non-smooth.

Theorem 2.1.7 *Assume:*

(K) *The kernel $K = K(t, s)$ in (2.1.13) is $m \geq 1$ times continuously differentiable with respect to t and s in $D^- := \{(t, s) : 0 \leq s < t \leq T\}$, and there exist a real number $v \in (-\infty, 1]$ and a positive constant C_K so that, for $0 \leq s < t \leq T$ and for non-negative integers i, j, ℓ with $i + j + \ell \leq m$, the following estimate holds:*

$$\left| \left(\frac{\partial}{\partial t} \right)^i \left(\frac{\partial}{\partial t} + \frac{\partial}{\partial s} \right)^j K(t, s) \right| \leq C_K \begin{cases} 1 & \text{if } v + i < 0, \\ 1 + |\log(t - s)| & \text{if } v + i = 0, \\ (t - s)^{-v-i} & \text{if } v + i > 0. \end{cases}$$
$$(2.1.14)$$

(F) *$f \in C^{m,v}(0, T]$; that is, $f \in C^m(0, T]$ and there exists a constant $C_f > 0$ so that the estimate*

$$|f^{(\ell)}(t)| \leq C_f \begin{cases} 1 & \text{if } \ell < 1 - v, \\ 1 + |\log(t)| & \text{if } \ell = 1 - v, \\ t^{1-v-\ell} & \text{if } \ell > 1 - v, \end{cases}$$

is true for $\ell = 0, 1, \ldots, m$ and $t \in (0, T]$.

Then the solution u of the VIE (2.1.13) lies in the space $C^{m,v}(I)$.

Its proof is a particular case of the analogous theorem (cf. Theorem 3.1.24) for non-linear VIEs due to Brunner, Pedas & Vainikko (1999). We leave it to the reader to adapt that proof to the linear case (2.1.13).

Remark 2.1.8 Assumption (K) means that $K(t, s)$ may possess a weak (integrable) singularity as $s \to t$ (if $i = j = \ell = 0$ and $0 < v \leq 1$). If $v < 0$, then $K(t, s)$ is *bounded* on D but its derivatives may have a singularity when $s = t$. We will discuss a particular case in Section 2.1.3.   ◇

Example 2.1.9

(a) If $K(t, s) = (t - s)^{-\beta} H(t, s)$ $(\beta < 1)$ for some smooth $H(t, s)$, then assumption (K) holds with $v = \beta$.

(b) If $K(t, s) = \log(t - s)H(t, s)$ for some smooth $H(t, s)$, then assumption (K) holds with $\nu = 0$.

Remark 2.1.10 Results on the regularity of the solutions of general *non-linear* VIEs with weakly singular, or other types of bounded but non-smooth kernels involving both algebraic and logarithmic terms were established in Brunner, Pedas & Vainikko (1999); we will discuss these results in Section 3.1.5. ◇

2.1.3 Bounded but Non-Smooth Kernels

A typical example of a VIE with bounded but non-smooth kernel is

$$u(t) = f(t) + \int_0^t (t - s)^{\nu+\alpha-1} K(t, s) u(s)\, ds, \quad t \in I, \qquad (2.1.15)$$

where $0 < \alpha < 1$, $\nu \in \mathbb{N}$ $(\nu \geq 1)$ and smooth K. A look at the proof of Theorem 1.3.5 reveals that the iterated kernels associated with the given kernel $(t - s)^{\nu+\alpha-1} K(t, s)$ all inherit the non-smooth factor $(t - s)^{\nu+\alpha-1}$. The proof of Theorem 2.1.2(a) is therefore readily adapted to show that the corresponding resolvent kernel has the form

$$R_{\nu,\alpha}(t, s) = (t - s)^{\nu+\alpha-1} Q_{\nu,\alpha}(t, s), \quad (t, s) \in D,$$

with

$$Q_{\nu,\alpha}(t, s) := \sum_{n=1}^{\infty} (t - s)^{(n-1)(\nu+\alpha)} \Phi_{\nu,\alpha,n}(t, s)$$

and $\Phi_{\nu,\alpha,n} \in C^m(D)$ $(n \geq 1)$. The resulting regularity result for the solution of (2.1.15) is summarised in the following theorem; its proof (which we leave to the reader) can be carried out along the lines of the proof of Theorem 1.3.5.

Theorem 2.1.11 *Assume that $f \in C^m(I)$ and $K \in C^m(D)$, with $K(t, t) \neq 0$ for $t \in I$. Then for any $m \geq 1$ the solution of (2.1.15) with $\nu \geq 1$ lies in the space $C^{\nu-1,1-\alpha}(I)$, and*

$$u^{(\nu)}(t) \sim C_{\nu,\alpha} t^{\alpha-1} \quad as \quad t \to 0^+,$$

with $C_{\nu,\alpha}$ denoting some non-zero constant.

Remark 2.1.12 If we consider the VIE (2.1.15) with $\nu = 0$ we recover the regularity result described in Theorem 2.1.2. ◇

Example 2.1.13 The resolvent kernel $R_{\nu,\alpha}(t,s) = r_{\nu,\alpha}(t-s)$ associated with the VIE

$$u(t) = f(t) + \lambda \int_0^t \frac{(t-s)^{\nu+\alpha-1}}{\Gamma(\nu-\alpha)} u(s)\,ds$$

with $\nu \in \mathbb{N}$ ($\nu \geq 1$) and $0 < \alpha < 1$, has the form

$$r_{\nu,\alpha}(t-s) = \sum_{n=1}^\infty \frac{\lambda^n}{\Gamma(n(\nu+\alpha))}(t-s)^{n(\nu+\alpha)-1}$$

$$= (t-s)^{\nu+\alpha-1} \sum_{n=1}^\infty \frac{\lambda^n}{\Gamma(n(\nu+\alpha))}(t-s)^{(n-1)(\nu+\alpha)}$$

(recall also Theorem 1.3.2). In analogy to Theorem 1.3.2 we may write

$$r_{\nu,\alpha}(t-s) = \frac{d}{dt} E_{\nu+\alpha}\big(\lambda(t-s)^{\nu+\alpha}\big), \quad (t,s) \in D.$$

For $f(t) = u_0$ the solution of the above VIE is thus given by

$$u(t) = \Big(1 + \int_0^t r_{\nu,\alpha}(t-s)\,ds\Big)u_0 = E_{\nu+\alpha}(\lambda t^{\nu+\alpha})u_0, \quad t \in I,$$

whose regularity agrees with the result of Theorem 2.1.11.

2.1.4 Kernels with Boundary Singularities

A representative example of a second-kind VIE whose kernel contains both *diagonal* and *boundary* singularities is

$$u(t) = f(t) + (\mathcal{V}_{\alpha,\beta}u)(t), \quad t \in I, \tag{2.1.16}$$

with

$$(\mathcal{V}_{\alpha,\beta}u)(t) := \int_0^t (t-s)^{\alpha-1}s^{-\beta}K(t,s)u(s)\,ds \quad (0<\alpha<1,\ 0<\beta<\alpha). \tag{2.1.17}$$

To avoid degeneracy we assume that $K \in C(D)$ satisfies $K(t,t) \neq 0$ and $K(t,0) \neq 0$ for $t \in I$.

The following result (and its proof) on the regularity of the solution of (2.1.16) can be found in Kolk & Pedas (2013), pp. 241–242 (see also Pedas & Vainikko, 2006). Its modification for $\beta \geq \alpha$ is obvious.

Theorem 2.1.14 *Assume that the kernel in (2.1.17) is subject to the conditions stated above. Then the VIE (2.1.16) possesses a unique solution u in the Hölder space $C^{\alpha+\beta}(I)$ whenever $f \in C(I)$. Moreover, if $f \in C^m(I)$ and $K \in$*

$C^m(D)$, *the solution u lies in* $C^{m,\alpha+\beta}(I]$, *and there exists a non-zero constant* $C_{\alpha,\beta}$ *so that* $u'(t) \sim C_{\alpha,\beta} t^{\alpha+\beta-1}$ *as* $t \to 0^+$.

2.1.5 Kernel Singularities of the Form $(t^2 - s^2)^{-1/2}$

We first consider two examples of the weakly singular second-kind VIE

$$u(t) = f(t) + \int_0^t (t^2 - s^2)^{-1/2} K(t,s)u(s)\,ds, \quad t \in I, \qquad (2.1.18)$$

where $K \in C(D)$. While they are seemingly quite similar, their underlying Volterra integral operators possess fundamentally different properties.

As we shall see in Section 7.1 (cf. Example 7.1.11), if the kernel in (2.1.18) has the form $K(t,s) = \kappa s$, with $\kappa \neq 0$ (implying that $K(0,0) = 0$), the corresponding Volterra integral operator is compact and has the spectrum $\{0\}$. This means that for any $f \in C(I)$ and every κ, (2.1.18) has a unique solution $u \in C(I)$. This is in general no longer true if we have $K(t,s) = \kappa \neq 0$ (and hence $K(0,0) \neq 0$): in this case, the corresponding Volterra integral operator is not compact and has an uncountable spectrum (see Example 7.1.21 with $\gamma = 2$); hence (2.1.18) may not have a unique solution. The next theorem reveals that if (2.1.18) possesses a unique continuous solution it inherits the regularity of f and K.

Theorem 2.1.15 *Assume that* $f \in C^m(I)$ *and* $K \in C^m(D)$ *for some* $m \geq 0$. *Then the following statements hold:*

(a) *If* $K(0,0) = 0$ *the cordial VIE (2.1.18) possess a unique continuous solution u.*

(b) *If* $K(0,0) \neq 0$ *there is a unique continuous solution only if* 1 *is not an eigenvalue of the cordial Volterra integral operator underlying (2.1.18),*

$$(\mathcal{V}_\varphi u)(t) := \int_0^t \left(t^\gamma - s^\gamma\right)^{-1/\gamma} K(t,s)u(s)\,ds$$

$$= \int_0^t t^{-1}\varphi(s/t)K(t,s)u(s)\,ds, \qquad (2.1.19)$$

with $\gamma = 2$ *and core* $\varphi(z) := (1 - z^\gamma)^{-1/\gamma}$. *If (2.1.18) possesses a unique solution u, then it lies in* $C^m(I)$.

Proof Since Theorem 2.1.15 is a special case of Theorem 7.2.5 we will illustrate the validity of assertion (c) for the case where $K(0,0) = 0$ by looking at (2.1.18) with $K(t,s) = \lambda s$ ($\lambda \neq 0$),

$$u(t) = f(t) + \lambda \int_0^t (t^2 - s^2)^{-1/2} s\, u(s)\, ds, \quad t \in I. \tag{2.1.20}$$

As Lemma 2.1.16 below shows, the unique (continuous) solution can in fact be obtained explicitly, by using Picard iteration for (2.1.18) (see also Becker, 2011 for an equivalent expression of the resolvent in terms of the error function, using the relationship $E_{1/2}(z^{1/2}) = (1 + \mathrm{erf}(z^{1/2}))e^z$; cf. Section 1.3.1). ☐

Lemma 2.1.16 *The resolvent kernel of the kernel* $\lambda(t^2 - s^2)^{-1/2} s$ *is*

$$\begin{aligned}
R_{2,1/2}(t,s) &= \lambda s (t^2 - s^2)^{-1/2} + \frac{\pi \lambda^2 s}{2} E_{1/2}\!\left(\tfrac{\lambda}{2}\left(\pi(t^2 - s^2)\right)^{1/2}\right) \\
&= \lambda s (t^2 - s^2)^{-1/2}\Big(1 + \frac{\pi \lambda}{2}\big((t^2 - s^2)^{1/2} \\
&\quad + \frac{\tfrac{\lambda}{2}\pi(t^2 - s^2)}{\Gamma(3/2)} + \cdots \big)\Big),
\end{aligned}$$

where $E_{1/2}(z)$ *is the Mittag-Leffler function.*

The asserted regularity of u now follows since the solution of the VIE (2.1.20) is given by

$$u(t) = f(t) + \int_0^t R_{2,1/2}(t,s) f(s)\, ds, \quad t \in I,$$

where

$$\begin{aligned}
\int_0^t (t^2 - s^2)^{-1/2} f(s)\, ds &= \sum_{j=1}^{d-1} \frac{t^{2j-1}}{2j-1} f^{(j-1)}(t) \\
&\quad + \frac{1}{2d-1} \int_0^t (t^2 - s^2)^{(2d-1)/2} f^{(d)}(s)\, ds. \qquad \square
\end{aligned}$$

Remark 2.1.17 The regularity of solutions of the related VIEs

$$u(t) = f(t) + \int_0^t p(t,s) K(t,s) u(s)\, ds, \quad t \in I, \tag{2.1.21}$$

and

$$u(t) = f(t) + \int_0^t q(t,s) K(t,s) u(s)\, ds, \quad t \in I, \tag{2.1.22}$$

where

$$p(t,s) := \pi^{-1/2} (\log(t/s))^{-1/2} \left(\frac{s}{t}\right)^{\mu} s^{-1} \tag{2.1.23}$$

and

$$q(t,s) := \left(\frac{s}{t}\right)^{\mu} s^{-1},$$

with $\mu > 0$, was analysed in Han (1994). As shown in Diogo, McKee & Tang (1991), the above VIEs are closely related: (2.1.22) can be obtained from (2.1.21) by a simple change of variables. A typical result states that the VIE (2.1.21) with $p(t, s)$ as in (2.1.23) and $\mu > 1$, and with $f \in C^m(I)$, $K \in C^m(D)$ ($m \geq 1$) has a unique solution $u \in C^m(I)$. In other words, for such a weakly singular VIE the solution inherits the regularity of the data on the *closed* interval I. We shall return to VIEs corresponding to the above (and related) Volterra integral operators (which are non-compact for certain values of $\mu > 0$) in Chapter 7. ◇

2.2 VIEs of the First Kind

2.2.1 VIEs with Smooth Kernels

The regularity properties of solutions of first-kind VIEs with smooth kernels are similar to the ones for VIEs of the second kind with smooth kernels. As we have already mentioned in Section 1.2.1, Volterra was led to his theory of VIEs of the second kind when he studied the solvability (*inversione*) of the first-kind integral equation

$$\int_0^t H(t, s) u(s) \, ds = g(t), \quad t \in I = [0, T]. \tag{2.2.1}$$

If H and g satisfy the conditions stated in Theorem 1.4.1, this integral equation is equivalent to the second-kind VIE

$$u(t) = f(t) + \int_0^t K(t, s) u(s) ds, \quad t \in I, \tag{2.2.2}$$

where the functions $f \in C(I)$ and $K \in C(D)$ describing this equation are

$$f(t) := g'(t)/H(t, t) \quad \text{and} \quad K(t, s) := -\big(\partial H(t, s)/\partial t\big)/H(t, t).$$

This equivalence allows us to derive the following regularity result for the solution of (2.2.1).

Theorem 2.2.1 *Let $m \geq 0$ and assume that the functions in the first-kind VIE (2.2.1) are such that*

(i) $g \in C^{m+1}(I)$, with $g(0) = 0$;
(ii) $H \in C^{m+1}(D)$, with $H(t, t) \neq 0$ for all $t \in I$.

Then its unique solution lies in the space $C^m(I)$.

Proof The assumptions (i) and (ii) imply that f and K defining the second-kind VIE (2.2.2) satisfy the conditions of Theorem 2.1.1, namely $f \in C^m(I)$ and $K \in C^m(D)$. Hence, the assertion on the regularity of the solution of the first-kind VIE (2.1.1) follows from Theorem 2.1.1. □

Remark 2.2.2 The regularity of solutions of first-kind Volterra integral equations with *convolution kernels*,

$$\int_0^t h(t-s)u(s)ds = g(t), \quad t \in I,$$

with $h \in C^{m+1}(I)$ and $h(0) \neq 0$, follows readily from Theorem 2.1.2. We leave the statement of the analogous theorem and its proof to the reader. ◇

2.2.2 VIEs with Weakly Singular Kernels

It was shown in Section 1.5.1 that for $g \in C^1(I)$ and $\alpha \in (0, 1)$, the solution of the Abel integral equation

$$\frac{1}{\Gamma(\alpha)} \int_0^t (t-s)^{\alpha-1} u(s)\, ds = g(t), \quad t \in I \ (0 < \alpha < 1), \qquad (2.2.3)$$

has the form

$$u(t) = \frac{g(0)}{\Gamma(1-\alpha)} t^{-\alpha} + \int_0^t \frac{(t-s)^{-\alpha}}{\Gamma(1-\alpha)} g'(s)\, ds, \quad 0 < t \leq T. \qquad (2.2.4)$$

For smoother g, e.g. $g \in C^{r+1}(I)$ with $g^{(j)}(0) = 0$ $(j = 0, \dots, r)$, the solution (2.2.4) becomes

$$u(t) = \int_0^t \frac{(t-s)^{-\alpha}}{\Gamma(1-\alpha)} g^{(r+1)}(s)\, ds, \quad t \in I \qquad (2.2.5)$$

(recall Theorem 1.5.6). This shows that for $r \geq 0$, the derivative $u^{(r+1)}(t)$ behaves like $C_\alpha t^{-\alpha}$ near $t = 0^+$.

The above regularity result remains true for the general weakly singular VIE of the first kind,

$$\int_0^t (t-s)^{\alpha-1} H(t,s) u(s)\, ds = g(t), \quad t \in I \ (\alpha \in (0,1)). \qquad (2.2.6)$$

This is a consequence of the equivalence of (2.2.6) and the first-kind VIE

$$\int_0^t H_\alpha(t,s) u(s)\, ds = \int_0^t (t-s)^{-\alpha} g(s)\, ds =: G_\alpha(t), \quad t \in I, \qquad (2.2.7)$$

with smooth kernel $H_\alpha(t, s)$ and non-smooth right-hand side $G_\alpha(t)$ (recall (1.5.8) in Section 1.5.2). Thus, according to Theorem 2.2.1 the regularity of

u is governed by the regularity of G_α. We make this precise in the following theorem.

Theorem 2.2.3 *Assume that the given functions in (2.2.6) satisfy*

(i) $g \in C^{m+1}(I)$;
(ii) $H \in C^{m+1}(D)$, with $H(t, t) \neq 0$ for $t \in I$;
(iii) $g^{(\nu)}(0) = 0$ ($\nu = 0, 1, \dots, r$) for some $r < m$.

If $0 < \alpha < 1$ the unique solution of (2.2.6) lies in the Hölder space $C^{r, 1-\alpha}(I)$. More precisely, it possesses the regularity

$$u \in C^q(I) \cap C^m(0, T], \quad with \quad u^{(r+1)}(t) \sim C_\alpha t^{-\alpha}, \quad as \quad t \to 0^+,$$

where C_α denotes some non-zero constant. For $r = 0$ the solution of (2.2.6) has a representation similar to (2.1.8) in Theorem 2.1.2, with $1-\alpha$ replacing α.

Proof Using assumption (i) and integration by parts in the expression for the function G_α in (2.2.7) we obtain

$$G_\alpha(t) = \sum_{\nu=0}^{m} \frac{g^{(\nu)}(0) t^{\nu+1-\alpha}}{(1-\alpha)_{\nu+1}} + \frac{1}{(1-\alpha)_{m+1}}$$

$$\times \int_0^t (t-s)^{m+1-\alpha} g^{(m+1)}(s) \, ds, \quad 0 < t \le T,$$

where, for given $n \in \mathbb{N}$ ($n \ge 1$) and $\beta \in \mathbb{R}$, $(\beta)_n := \beta(\beta+1) \cdots (\beta+n-1)$ denotes the *Pochhammer symbol*. Thus, as we already saw in Section 1.5.2 (cf. the proof of Theorem 1.5.6), differentiation of (2.2.7) yields the second-kind VIE

$$H_\alpha(t, t) u(t) + \int_0^t \frac{\partial H_\alpha(t, s)}{\partial t} u(s) \, ds = \frac{d}{dt} G_\alpha(t), \quad 0 < t \le T, \qquad (2.2.8)$$

with

$$\frac{d}{dt} G_\alpha(t) = \sum_{\nu=0}^{m} \frac{g^{(\nu)}(0) t^{\nu-\alpha}}{(1-\alpha)_\nu} + \frac{1}{(1-\alpha)_m} \int_0^t (t-s)^{m-\alpha} g^{(m+1)}(s) \, ds.$$

The asserted regularity result now follows from assumption (iii). □

Remark 2.2.4 When $r = 0$, the derivative $u'(t)$ of the solution of (2.2.6) near zero behaves like $C_\alpha t^{-\alpha}$, compared to $u'(t) \sim C_\alpha t^{\alpha-1}$ when u is the solution of the *second-kind* VIE (2.2.1) with the same weakly singular kernel (recall Theorem 2.1.2). ◇

In the previous theorem we assumed that the function g in (2.2.6) is smooth. The next theorem – the analogue to Theorem 2.1.4 – deals with the case of non-smooth g in the VIE

$$\int_0^t (t-s)^{\alpha-1} H(t,s) u(s)\, ds = g(t), \quad t \in I \ (0 < \alpha < 1), \tag{2.2.9}$$

with arbitrarily smooth kernel H satisfying $H(t,t) \neq 0$ for $t \in I$, and with $g(t) = t^\beta \tilde{g}(t)$ ($\beta > 0$, $\beta \notin \mathbb{N}$). The result in Theorem 2.2.5 below is similar to a more general result due to Atkinson (1974, where the singular part of the kernel is given by $(t^\gamma - s^\gamma)^{\alpha-1}$ with $\gamma > 0$; see also Theorem 2.2.6 in the following section). A concise proof for the case $\gamma = 1$ can also be found in Lubich (1987).

Theorem 2.2.5 *Assume that the function g in (2.2.9) has the form $g(t) = t^\beta \tilde{g}(t)$, with $\beta > 0$ ($\beta \notin \mathbb{N}$) and arbitrarily smooth $\tilde{g}$. Then the solution u of (2.2.9) has the form*

$$u(t) = t^{\beta-\alpha}\big(a_0 + t a_1(t)\big), \quad t \in (0, T],$$

with smooth a_1. We have $a_0 = 0$ if, and only if, $\tilde{g}(0) = 0$. If this condition does not hold and if $\beta > 0$ ($\beta \notin \mathbb{N}$), the solution has in the same regularity as the function $t^{1-\alpha+\beta}$.

Proof We use the second-kind VIE (2.2.8) derived in the proof of Theorem 2.2.3, where $G_\alpha(t)$ is now replaced by

$$G_{\alpha,\beta}(t) := \int_0^t (t-s)^{-\alpha} s^\beta \tilde{g}(s)\, ds.$$

By assumption, $\beta > 0$ and $\tilde{g}$ is arbitrarily smooth. The regularity of the solution of the VIE (2.2.9) then follows from Theorem 1.2.3 where in equation (1.2.11) the resolvent kernel $R(t,s)$ is arbitrarily smooth (by the assumption on H) but where the above non-smooth $G_{\alpha,\beta}(t)$ assumes the role of $G_\alpha(t)$. It is readily verified that $(d/dt)G_{\alpha,\beta}(t)$ possesses the same regularity as $t^{\beta-\alpha}$. $\quad\square$

2.2.3 Other Types of Kernel Singularities

In applications (compare Gorenflo & Vessella, 1990; see also Chapter 9) one encounters weakly singular first-kind VIEs corresponding to the cordial Volterra integral operator $\mathcal{V}_\varphi$ in (2.1.19) with core $\varphi(z) = (1-z^\gamma)^{-1/\gamma}$,

$$\int_0^t (t^\gamma - s^\gamma)^{-1/\gamma} H(t,s) u(s)\, ds = \int_0^t t^{-1} \varphi(s/t) H(t,s) u(s)\, ds$$

$$= g(t), \quad t \in I, \qquad (2.2.10)$$

where $\gamma = 2$ (recall the discussion in Section 1.5.3). The regularity result contained in the following theorem (due to Atkinson (1974)) is the analogue of Theorem 2.2.4 (where we had $\gamma = 1$). See also Vainikko (2014, Section 7) where a somewhat more general form of (2.2.10) is studied. We will come back to such (cordial) first-kind VIEs in Section 7.2.2.

Theorem 2.2.6 *Assume that the kernel H in (2.2.10) is in $C^{m+1}(D)$ for some $m \geq 0$, with $H(t,t) \neq 0$ ($t \in I$). If g has the form $g(t) = t^\beta \tilde{g}(t)$, where $\gamma(1 - \alpha) + \beta > 0$ and $\tilde{g} \in C^{m+1}(I)$, then the solution $u \in C(0,T]$ of*

$$\int_0^t (t^\gamma - s^\gamma)^{\alpha - 1} H(t,s) u(s)\, ds = g(t)$$

has the form

$$u(t) = t^{\beta - \alpha} \big(a_0 + t w(t) \big), \quad t \in (0,T],$$

with $w \in C^m(I)$. We have $a_0 = 0$ if, and only if, $\tilde{g}(0) = 0$.

Proof Instead of reproducing the proof of Atkinson (1974) we defer the proof of Theorem 2.2.6 to Section 7.2.2, where we shall also study the regularity of the solution of cordial VIEs of the form (2.2.10) when $g \in C^{m+1}(I)$. □

2.2.4 The Generalised Abel Integral Equation

The weakly singular first-kind VIE

$$c(x) \int_0^x \frac{(x-s)^{\alpha - 1}}{\Gamma(\alpha)} u(s)\, ds + d(x) \int_x^1 \frac{(s-x)^{\alpha - 1}}{\Gamma(\alpha)} u(s)\, ds = g(x), \quad x \in [0,1]$$

$$(2.2.11)$$

(with $0 < \alpha < 1$) is known as the *generalised Abel integral equation* (cf. Gakhov, 1966, Sakaljuk, 1960, 1965, von Wolfersdorf, 1965; also Gorenflo & Vessella, 1991, p. 61 and Appendix 6.A). The given real-valued functions c and d are assumed to be (Hölder) continuous on $[0,1]$ and satisfy $|c(x)| + |d(x)| \neq 0$ for all $x \in [0,1]$.

The generalised Abel integral equation (2.2.11) is closely related to a boundary-value problem for *Tricomi's equation*,

$$y^m \Psi_{xx} + \Psi_{yy} = 0 \quad (x \in \mathbb{R}, \ y > 0), \tag{2.2.12}$$

$$\Psi(x, 0) = u_0(x) \ \text{if} \ 0 < x < 1,$$

$$\Psi_y(x, 0) = 0 \ x < 0 \ \text{or} \ x > 1$$

(compare, for example, Chang & Lundgren, 1959 and the review paper by Chakrabarti, 2008). The reduction of the above boundary-value problem to an integral equation of the form (2.2.11) is described in detail in von Wolfersdorf (1965, pp. 172–177); it is based on a certain Ansatz for the solution of the boundary-value problem (2.2.12).

The somewhat more general equation, with 'inner coefficients' c and d,

$$\int_0^x c(s) \frac{(x-s)^{\alpha-1}}{\Gamma(\alpha)} u(s) \, ds + \int_x^1 d(s) \frac{(s-x)^{\alpha-1}}{\Gamma(\alpha)} u(s) \, ds = g(x) \ x \in [0, 1], \tag{2.2.13}$$

(with $0 < \alpha < 1$) was studied in Sakaljuk (1965) under the assumption that c and d are Hölder continuous on $[0, 1]$.

The VIE (2.2.11) contains as special cases the weakly singular first-kind *Fredholm* integral equation

$$\int_0^1 \frac{|x-s|^{\alpha-1}}{\Gamma(\alpha)} u(s) \, ds = g(x), \quad x \in [0, 1] \tag{2.2.14}$$

(when $c(x) = d(x) \equiv 1$; it was studied by Carleman in 1922); *Abel's* integral equation (when $c(x) \equiv 1$ and $d(x) \equiv 0$) studied in Section 2.2.2; and the *adjoint* Abel integral equation,

$$\int_x^1 \frac{(s-x)^{\alpha-1}}{\Gamma(\alpha)} u(s) \, ds = g(x), \quad x \in [0, 1], \tag{2.2.15}$$

which corresponds to $c(x) \equiv 0$ and $d(x) \equiv 1$ (see also Exercise 2.4.10).

As Theorem 2.2.3 and Exercise 2.4.10 show, under appropriate conditions on g the solutions of Abel's integral equation and its adjoint version (2.2.15) lie in the Hölder space $C^{1-\alpha}[0, 1]$: at $t = 0^+$ and $t = 1^-$, they behave like $u'(t) \sim C_0 t^{-\alpha}$ and $u'(t) \sim C_1 (1-t)^{-\alpha}$, respectively. Thus, we expect the solution of the general equation (2.2.11) to lie in some Hölder space and to exhibit similar singularities in its derivative at *both endpoints* of the given interval $[0, 1]$.

Sakaljuk (1960) and von Wolfersdorf (1965; see also Gorenflo & Vessella, 1991) proved that if the function g in (2.2.11) is of the form

$$g(x) = x^{-\beta}(1-x)^{-\beta} \tilde{g}(x),$$

where $\tilde{g}$ is in the Hölder space $C^\epsilon[0, 1]$ with $\epsilon > \alpha$, and c, $d \in C^\beta[0, 1]$ ($\beta > \alpha$), then the corrsponding solution exhibits singularities at both endpoints of $(0, 1)$; that is,

$$x^{-\mu}(1 - x^{-\nu})\tilde{u}(x), \ x \in (0, 1)(\mu, \nu < 1),$$

where $\tilde{u}$ lies in some Hölder space $C^\rho[0, 1]$.

It was shown by von Wolfersdorf (1965) and Sakaljuk (1965; see also Gakhov, 1966, Samko, 1968 and Chakrabarti, 2008) that the generalised Abel integral equations (2.2.11) and (2.2.13) admit closed-form solutions. For example, if $c(x) = -d(x) \equiv 1$ in (2.2.11), its solution has the form

$$
\begin{aligned}
u(x) = {} & C_0\big(x(1 - x)\big)^{(1-\alpha)/2-1} \\
& + \frac{1}{2\pi} \tan\big(\pi(1 - \alpha)/2\big) \frac{d}{dx}\left(\int_0^1 |x - s|^{-\alpha} g(s)\, ds\right) \\
& + C_1 \frac{d}{dx}\left(\int_0^1 K(x, s)g(s)\, ds\right), \ x \in (0, 1),
\end{aligned}
$$

with constants C_0 and

$$C_1 := -\frac{\sin(\pi(1 - \alpha)/2)}{\alpha\pi} \frac{1}{\Gamma^2((1 - \alpha)/2)}.$$

The kernel $K(x, s)$ is of the form

$$K(x, s) := x^{-\alpha}(1 - x)^{-\alpha} F\big((1 + \alpha)/2, \alpha; 1 + \alpha; \frac{s - x}{x(s - 1)}\big),$$

where $F(\cdot, \cdot; \cdot; \cdot)$ denotes the hypergeometric function defined by

$$F(a, b; c; z) := \frac{\Gamma(c)}{\Gamma(a)\Gamma(b)} \sum_{n=0}^{\infty} \frac{\Gamma(a + n)\Gamma(b + n)}{\Gamma(c + n)} \frac{z^n}{n!}.$$

We observe that for $C_0 \neq 0$, the solution is unbounded at the endpoints of the interval $[0, 1]$. For solutions satisfying the constraint $\int_0^1 u(s)\, ds = 0$, we have $C_0 = 0$ (von Wolfersdorf, 1965, p. 168).

2.3 Linear Volterra Functional Integral Equations

2.3.1 Introduction

A Volterra functional integral equation (VFIE) contains at least one term corresponding to a Volterra integral operator that depends on a *delay function* (or *lag function*)

$$\theta(t) = t - \tau(t), \quad \text{with } delay \quad \tau(t) \ge 0 \ (t \in I := [t_0, T]).$$

Our subsequent discussion will focus on VFIEs with the following two (fundamentally different) types of delays.

Definition 2.3.1 A delay $\tau(t)$ is called a *vanishing delay* (with respect to a given interval $[t_0, T]$) if

$$\tau(t_0) = 0 \quad \text{and} \quad \tau(t) > 0 \quad \text{for} \quad t_0 < t \le T.$$

A delay $\tau(t)$ is said to be a *non-vanishing delay* with respect to the interval $[t_0, T]$ if

$$\tau(t) \ge \tau_0 > 0 \quad \text{for all} \quad t \in [t_0, T].$$

A *constant delay* occurs if $\tau(t) = \tau_0 > 0$ on $[t_0, T]$.   ◇

When studying VFIEs with *vanishing delays* we will assume that the delay function θ possesses the following properties:

(V1) $\theta(t) = t - \tau(t)$, with $\tau \in C^d[t_0, T]$ for some $d \ge 0$;
(V2) $\theta(t_0) = 0$ and $\theta(t) < t$ for $t > t_0$;
(V3) θ is strictly increasing on $[t_0, T]$.

In analogy to (V1)–(V3), the delay function θ associated with a *non-vanishing* delay τ is assumed to be such that

(N1) $\theta(t) = t - \tau(t)$, with $\tau \in C^d[t_0, T]$ for some $d \ge 0$;
(N2) $\tau(t) \ge \tau_0 > 0$ for $t \in [t_0, T]$;
(N3) θ is strictly increasing on $[t_0, T]$.

Remark 2.3.2 The analyses in the following sections will reveal that the conditions (V3) and (N3) (strict monotonicity of θ) are introduced mainly for technical reasons. The reader is encouraged to consult Section 2.1 in Bellen and Zennaro (2003) for many illuminating examples and remarks on the complications that can arise when (V3) or (N3) do not hold (compare also Brunner & Maset, 2009).   ◇

For smooth data, *non-vanishing delays* will in general lead to lower regularity in the solution of a VFIE at certain points. The following definition, together with illuminating examples, can be found in Bellen & Zennaro (2003, Section 2.2).

Definition 2.3.3 The points $\{\xi_\mu : \mu \geq 0\}$ generated by the recursion

$$\theta(\xi_{\mu+1}) = \xi_{\mu+1} - \tau(\xi_{\mu+1}) = \xi_\mu, \quad \mu = 0, 1, \ldots; \quad \xi_0 := t_0, \qquad (2.3.1)$$

are called the *primary discontinuity points* (or *breaking points*) associated with the delay function $\theta(t) = t - \tau(t)$. ◇

As the name indicates, at these points the solution of a VFIE will in general exhibit a lower degree of regularity than that of the given functions (see Tables 2.1 and 2.2 in Sections 2.3.4 and 2.3.5). We note that additional, *secondary* discontinuity points may arise if the given initial function ϕ is only piecewise continuous; that is, if it contains one ore more finite jump discontinuities. We will not pursue this problem in the subsequent presentation (but see Exercise 2.4.18).

Remark 2.3.4 There exists a more refined classification of the primary discontinuity points $\{\xi_\mu\}$ that is related to the degree of regularity at such a point. More precisely (Bellen & Zennaro, 2003, pp. 21–24), ξ_μ is said to be a primary discontinuity point of order k if $u^{(j)}(\xi_\mu)$ exists for $j = 0, \ldots, k$ and $u^{(k)}$ is (Lipschitz) continuous at ξ_μ. We will not use this distinction in the following discussion. ◇

Example 2.3.5 If the delay $\tau(t) = \tau_0 > 0$ is constant on the given interval $[t_0, T]$, the primary discontinuity points induced by $\theta(t) = t - \tau_0$ are

$$\xi_\mu = t_0 + \mu\tau_0, \quad \mu = 0, 1, \ldots .$$

Example 2.3.6 Consider the interval $[t_0, T]$ with $t_0 > 0$. The delay function

$$\theta(t) = qt = t - (1 - q)t, \quad \text{with} \quad 0 < q < 1$$

now corresponds to the *non-vanishing* proportional delay $\tau(t) = (1 - q)t$. The corresponding primary discontinuity points are therefore given by

$$\xi_\mu = q^{-\mu}t_0, \quad \mu = 0, 1, \ldots .$$

We note that when $t_0 > 0$ is replaced by $t_0 = 0$, there are no longer any primary discontinuity points.

In the following sections we will study the general VFIE

$$u(t) = f(t) + (\mathcal{V}_\alpha u)(t) + (\mathcal{V}_{\theta,\alpha} u)(t), \quad t \in I, \qquad (2.3.2)$$

and its special cases

$$u(t) = f(t) + (\mathcal{V}_{\theta,\alpha} u)(t), \quad t \in I \qquad (2.3.3)$$

and

$$u(t) = f(t) + (\mathcal{W}_{\theta,\alpha}u)(t), \quad t \in I. \tag{2.3.4}$$

The underlying Volterra integral operators are

$$(\mathcal{V}_\alpha u)(t) := \int_0^t (t-s)^{\alpha-1} K^{(0)}(t,s)u(s)\,ds,$$

$$(\mathcal{V}_{\theta,\alpha}u)(t) := \int_0^{\theta(t)} (t-s)^{\alpha-1} K^{(1)}(t,s)u(s)\,ds,$$

$$(\mathcal{W}_{\theta,\alpha}u)(t) := \int_{\theta(t)}^t (t-s)^{\alpha-1} K(t,s)u(s)\,ds,$$

with $0 < \alpha \le 1$ and continuous kernels $K^{(0)}$, $K^{(1)}$ and K. When $\alpha = 1$ we will employ the notation $\mathcal{V} := \mathcal{V}_1$, $\mathcal{V}_\theta := \mathcal{V}_{\theta,1}$, $\mathcal{W}_\theta := \mathcal{W}_{\theta,1}$. The delay function θ corresponds either to a vanishing delay (Sections 2.3.2 and 2.3.3), or to a non-vanishing delay (Section 2.3.4 and 2.3.5). Unless stated otherwise, we will from now on assume that $t_0 = 0$; hence, I will again denote the interval $[0, T]$.

Remark 2.3.7

  (i) If the delay function θ corresponds to a *vanishing* delay τ, the operators $\mathcal{V}_{\theta,\alpha}$ and $\mathcal{W}_{\theta,\alpha}$ are well defined for all $u \in C(I)$.

 (ii) In the case of a *non-vanishing* delay, the integrals $(\mathcal{V}_{\theta,\alpha}u)(t)$ and $(\mathcal{W}_{\theta,\alpha}u)(t)$ are well defined only if the values of u on the interval $[\theta_0, 0]$ (where $\theta_0 := \inf\{\theta(t) : t \in I\}$) have been prescribed. In the VFIEs (2.3.2)–(2.3.4) these values will correspond to given *initial conditions* on the interval $[\theta_0, 0]$ (see Section 2.3.4).

(iii) In applications (see Section 9.2.2 for examples), VFIEs of the second kind occasionally occur in a somewhat different form, namely,

$$u(t) = f(t) + (\mathcal{V}_\alpha u)(t) + \int_0^t (t-s)^{\alpha-1}$$

$$\times\, K(t,s)u(\theta(s))\,ds, \quad t \in I \ (0 < \alpha \le 1) \tag{2.3.5}$$

(see Exercise 2.4.21).   ◇

2.3.2 Second-Kind VFIEs with Vanishing Delays

The second-kind delay VFIE

$$u(t) = f(t) + \int_0^{\theta(t)} K(t,s)u(s)\,ds, \quad t \in I = [0, T], \tag{2.3.6}$$

with the linear *proportional delay* function $\theta(t) = qt$ $(0 < q < 1)$ was first studied by Andreoli in 1913 and 1914; see also Chambers (1990), where the following theorem can be found.

Theorem 2.3.8 *Let f and K satisfy $f \in C(I)$ and $K \in C(D_\theta)$, where $D_\theta :=$ $\{(t, s) : 0 \le s \le \theta(t), t \in I\}$. Then for any $\theta(t) = qt$ with $q \in (0, 1)$ the delay integral equation (2.3.6) possesses a unique solution $u \in C(I)$. This solution is given by*

$$u(t) = f(t) + \sum_{n=1}^{\infty} \int_0^{q^n t} K_n(t, s) f(s)\, ds$$

$$= f(t) + \int_0^t \left(\sum_{n=1}^{\infty} q^n K_n(t, q^n s) f(q^n s) \right) ds, \quad t \in I. \quad (2.3.7)$$

The iterated kernels $K_n(t, s)$ $(= K_n(t, s; q))$ are obtained recursively by

$$K_{n+1}(t, s) := \int_{q^{-n} s}^{qt} K(t, v) K_n(v, s)\, dv, \quad (t, s) \in D_\theta^{(n+1)} \ (n \ge 1), \quad (2.3.8)$$

with $K_1(t, s) := K(t, s)$, $D_\theta^{(k)} := \{(t, s) : 0 \le s \le q^k t, t \in I\}$ $(k \ge 0)$ and $D_\theta^{(0)} := D_\theta$.

Remark 2.3.9

(i) The solution representation (2.3.7) reveals that, owing to the coupling of the values of the iterated kernels with those of the given function f, the solution of the second-kind VFIE (2.3.6) cannot be written in the form (1.2.11): there is no resolvent kernel associated with the given kernel $K(t, s)$ in (2.3.6) when $0 < q < 1$. For $q = 1$, the solution representation of course reduces to (1.2.11).

(ii) In Cerha (1976) it is asserted that the solution of a VFIE whose delay function θ occurs in the integrand,

$$u(t) = f(t) + \int_0^t K(t, s) u(\theta(s))\, ds, \quad t \in I,$$

and is such that $0 \le \theta(t) \le t$, admits the variation-of-constants formula

$$u(t) = f(t) + \int_0^t R(t, s) f(\theta(s))\, ds, \quad t \in I,$$

where the resolvent kernel R satisfies the resolvent equations

$$R(t, s) = K(t, s) + \int_s^t K(t, v) R(\theta(v), s)\, dv, \quad (t, s) \in D,$$

and

$$R(t, s) = K(t, s) + \int_s^t R(t, v) K(\theta(v), s) dv, \quad (t, s) \in D$$

similar to (1.2.9) and (1.2.10). However, these assertions appear to be false for delay functions like $\theta(t) = qt$ $(0 < q < 1)$, owing to an error that occurs when interchanging the order of integration in the proof of Theorem 9 (p. 116) in that paper. Thus, for vanishing delays the solution of the above VFIE does not admit a variation-of-constant formula, in analogy to (i). $\diamond$

Proof Picard iteration applied to (2.3.6) yields, for $u_0(t) := f(t)$,

$$u_1(t) := f(t) + \int_0^{qt} K(t, s) f(s) \, ds,$$

and hence, setting $K_1(t, s) := K(t, s)$,

$$u_2(t) := f(t) + \int_0^{qt} K(t, s) \left(f(s) + \int_0^{qs} K(s, v) f(v) dv \right) ds$$

$$= f(t) + \int_0^{qt} K(t, s) f(s) \, ds$$

$$+ \int_0^{qt} \left(\int_{q^{-1}v}^{qt} K(t, s) K_1(s, v) ds \right) f(v) \, dv$$

$$=: f(t) + \int_0^{qt} K_1(t, s) f(s) \, ds + \int_0^{qt} K_2(t, s) f(s) \, ds,$$

where

$$K_2(t, s) := \int_{q^{-1}s}^{qt} K(t, v) K_1(v, s) \, dv.$$

Using a simple induction argument one readily verifies that the iterated kernels $K_n(t, s)$ of the given kernel $K(t, s)$ are generated recursively by

$$K_{n+1}(t, s) = \int_{q^{-n}s}^{qt} K(t, v) K_n(v, s) \, dv, \quad (t, s) \in D_\theta^{(n+1)} \quad (n \geq 1)$$

(see also Chambers, 1990). We leave the detailed steps of the proof to the reader but state the uniform bounds for the iterated kernels. $\square$

Lemma 2.3.10 *Uniform bounds on $I = [0, T]$ for the iterated kernels $K_n(t, s)$ defined in Theorem 2.3.1 are given by*

$$|K_n(t, s)| \leq \frac{q^{n(n-1)/2}}{(n-1)!} T^{n-1} \bar{K}_\theta^n, \quad (t, s) \in D_\theta^{(n)} \quad (n \geq 1),$$

where we have set $\bar{K}_\theta := \max\{|K(t, s)| : (t, s) \in D_\theta\}$.

The proof of the *uniqueness* of the solution of (2.3.6) is left to the reader. It employs the analogue of Gronwall's lemma (Lemma 1.2.17) for the integral inequality

$$z(t) \leq f(t) + \int_0^{qt} k(s)z(s)\,ds, \quad t \in I \ (0 < q < 1)$$

(see Exercise 2.4.13). $\qquad\qquad\qquad\qquad\qquad\qquad\qquad\qquad\qquad\qquad\square$

The existence, uniqueness and regularity properties hold also for the more general linear delay VIE with proportional delay,

$$u(t) = f(t) + (\mathcal{V}u)(t) + (\mathcal{V}_\theta u)(t), \quad t \in I, \tag{2.3.9}$$

corresponding to the Volterra integral operators

$$(\mathcal{V}u)(t) := \int_0^t K^{(0)}(t,s)u(s)\,ds, \qquad (\mathcal{V}_\theta u)(t) := \int_0^{\theta(t)} K^{(1)}(t,s)u(s)\,ds,$$

with $\theta(t) := qt$ ($0 < q < 1$) and continuous kernels $K^{(0)}$ and $K^{(1)}$.

Theorem 2.3.11 *Assume that the kernels of the linear Volterra operators $\mathcal{V}$ and $\mathcal{V}_\theta$ in (2.3.9) (corresponding to (2.3.2) with $\alpha = 1$) possess the regularity $K^{(0)} \in C^d(D)$ and $K^{(1)} \in C^d(D_\theta)$, for some $d \geq 0$. Then the VFIE (2.3.9) with $\theta(t) = qt$ ($0 < q < 1$) has a unique solution $u \in C^d(I)$ for any f with $f \in C^d(I)$.*

Proof Theorem 2.3.8 shows that the iterated kernels $K_n(t,s)$ associated with the kernel K of the special delay integral equation (2.3.6) inherit the regularity of the given kernel $K(t,s)$. Since the additional (non-delay) term $(\mathcal{V}u)(t)$ in the general linear VFIE (2.3.9) will not induce a lower regularity in the corresponding Picard iterates, the assertion of Theorem 2.3.11 follows from the uniform convergence of the Picard iterates on I, for any $q \in (0,1)$. $\qquad\square$

For *non-linear* delay functions θ possessing the properties (V1), (V2) and (V3) (cf. Section 2.3.1),

(V1) $\theta(t) = t - \tau(t)$, with $\tau \in C^d(I)$ for some $d \geq 0$;
(V2) $\theta(0) = 0$ and $\theta(t) < t$ for $t > 0$;
(V3) θ is strictly increasing on I;

the result of Theorem 2.3.11 on the existence, uniqueness and regularity of the solutions to VFIEs (2.3.6) remains valid. We leave the proof to the reader.

The above regularity results for solutions of VFIEs with vanishing delays were established for smooth kernels. If the Volterra integral operators $\mathcal{V}_\alpha$ and

$\mathcal{V}_{\theta,\alpha}$ in the VFIE (2.3.2) contain the weak singularity $(t - s)^{\alpha-1}$ $(0 < \alpha < 1)$, the solution corresponding to smooth data f and K will no longer be smooth, and its regularity, not surprisingly, will be the same as in the case when $\mathcal{V}_{\theta,\alpha} = 0$ (cf. Theorem 2.2.1). In order to understand this, consider the special case

$$u(t) = f(t) + \int_0^{\theta(t)} (t - s)^{\alpha-1} K(t, s) u(s) \, ds, \quad t \in I = [0, T], \quad (2.3.10)$$

with $\theta(t) = qt$ $(0 < q < 1)$ and $0 < \alpha < 1$. It was shown by Bai (2011) that Picard iteration for (2.3.10) leads again to a solution representation analogous to the one in Theorem 2.3.8. We make this precise in the following theorem.

Theorem 2.3.12 *Let $\theta(t) = qt$ $(0 < q < 1)$ and assume that $K \in C(D_\theta)$. Then for any $f \in C(I)$ and $\alpha \in (0, 1)$, the weakly singular VFIE (2.3.10) possesses a unique solution $u \in C(I)$. This solution has the form*

$$u(t) = f(t) + \sum_{n=1}^{\infty} \int_0^{q^n t} H_{\alpha,n}(t, s; q) f(s) \, ds, \quad t \in I, \quad (2.3.11)$$

where the iterated kernels $H_{\alpha,n}(t, s; q)$ are defined recursively by

$$H_{\alpha,n}(t, s; q) := \int_{q^{-n}s}^{qt} H_{\alpha,1}(t, v) H_{\alpha,n}(v, s) \, dv$$

on $D_\theta^{(n)} := \{(t, s) : 0 \leq s \leq q^n t, \ t \in I\}$, with

$$K_{\alpha,1}(t, s; q) := H_{\alpha,1}(t, s) := (t - s)^{\alpha-1} K(t, s).$$

If $f \in C^m(I)$ and $K \in C^m(D_\theta)$, then for any $m \geq 1$ the solution u lies in the Hölder space $C^\alpha(I)$: its regularity is described by

$$u \in C^m(0, T] \cap C[0, T] \quad and \quad u'(t) \sim C_\alpha t^{\alpha-1} \quad as \ t \to 0^+.$$

Proof The existence of a unique solution $u \in C(I)$ of (2.3.10) can be established along the lines of the proof of Theorem 1.3.5 (which concerns the VIE (2.3.10) with $\theta(t) = t$). The proof makes use of the analogues of the generalisation of (Gronwall) Lemma 1.3.13 to VFIEs with weakly singular and vanishing delays (see Exercise 2.4.13). Since the regularity of the kernel $K_{\alpha,1}(t, s; q) = (t - s)^{\alpha-1} K(t, s)$ on D_θ is inherited by the iterated kernels $K_{\alpha,n}(t, s; q)$ $(n \geq 2)$, the Picard iteration process then yields the regularity result $u \in C^m(0, T] \cap C[0, T]$, under the assumption that $f \in C^m(I)$. $\qquad\square$

Remark 2.3.13

(i) The regularity result in Theorem 2.3.12 remains valid for the weakly singular VFIE

$$u(t) = f(t) + \int_{\theta(t)}^{t} (t-s)^{\alpha-1} K(t,s) u(s) \, ds, \quad t \in I,$$

with $\theta(t) = qt$ ($0 < q < 1$) and $0 < \alpha < 1$; see Bai (2011).
(ii) The VFIE

$$u(t) = f(t) + b(t)u(\theta(t)) + \int_{\theta(t)}^{t} (t-s)^{\alpha-1}$$

$$\times K(t,s)u(s) \, ds, \quad t \in I \quad (0 < \alpha \le 1) \tag{2.3.12}$$

with vanishing delay θ is more general than the VFIE (2.3.2), due to the presence of the delay term $b(t)u(\theta(t))$. The special case corresponding to $\theta(t) = qt$ ($0 < q < 1$) and $\alpha = 1$,

$$u(t) = f(t) + b(t)u(qt) + \int_{0}^{t} K(t,s)u(s) \, ds, \quad t \in I, \tag{2.3.13}$$

was studied by Volterra in his book of 1913 (cf. pp. 85–88); compare also (2.3.15) below. For more general vanishing delays, but still with $\alpha = 1$, the existence and uniqueness properties of the solutions of (2.3.12) and (2.3.13) was analysed in Denisov & Lorenzi (1997) and in Pukhnacheva (1990; compare also Xie, Zhang & Brunner, 2011). For $0 < \alpha < 1$ analogous results remain to be established. ◇

2.3.3 First-Kind VFIEs with Vanishing Delays

In 1897, the year following the publication of his *Nota I* on the standard first-kind VIE

$$(\mathcal{V}u)(t) := \int_{0}^{t} H(t,s)u(s) \, ds, \quad t \in I = [0, T],$$

Volterra turned his attention to the first-kind VFIE

$$(\mathcal{W}_{\theta}u)(t) := \int_{\theta(t)}^{t} H(t,s)u(s) \, ds = g(t), \quad t \in I. \tag{2.3.14}$$

Here, the delay function is $\theta(t) = qt = t - (1-q)t$, with $0 < q < 1$. It corresponds to the vanishing delay $\tau(t) = (1-q)t$. Under assumptions similar

to those in Theorem 1.3.2 (see Theorem 2.3.14 below), (2.3.14) is equivalent
to the second-kind VFIE

$$H(t,t)u(t) = g'(t) + qH(t,qt)u(qt) - \int_{qt}^{t} \frac{\partial H(t,s)}{\partial t} u(s)\,ds, \quad t \in I,$$

or, if $H(t,t) \neq 0$ on I, to

$$u(t) = g_1(t) + H_q(t)u(qt) + \int_{qt}^{t} H_1(t,s)u(s)\,ds, \quad t \in I, \qquad (2.3.15)$$

where

$$g_1(t) := g'(t)/H(t,t), \quad H_q(t) := qH(t,qt)/H(t,t),$$
$$H_1(t,s) := -\big(\partial H(t,s)/\partial t\big)/H(t,t).$$

Volterra then applies Picard iteration to (2.3.15): owing to the presence of the
term $H_q(t)u(qt)$ and the lower limit of integration qt, this iteration process
becomes much more complex. The following theorem is due to Volterra (1897;
see also Lalesco, 1908, p. 156, and Volterra, 1913 pp. 95–97). We use the
notation

$$\bar{D}_\theta := \{(t,s) : \theta(t) \leq s \leq t \ (t \in I)\}.$$

Theorem 2.3.14 *Assume that the given functions in (2.3.14) satisfy*

(i) $\theta(t) = qt, \ 0 < q < 1;$
(ii) $H \in C(\bar{D}_\theta), \ \partial H/\partial t \in C(\bar{D}_\theta),$ and $H(t,t) \neq 0$ for $I;$
(iii) $g \in C^1(I),$ with $g(0) = 0.$

Then for any $q \in (0,1)$, the VFIE (2.3.14) possesses a unique solution $u \in C(I)$.

Proof   The reader is invited either to look at Volterra's ingenious arguments in
Volterra (1897) or to carry out the proof independently, using Picard iteration
for (2.3.15).   □

Remark 2.3.15

(i) In his book of 1913 (pp. 95–96), Volterra also studied the first-kind
functional integral equation

$$\int_{pt}^{qt} K(t,s)u(s)\,ds = g(t), \quad t \in I \ (0 < p < q < 1).$$

Such VFIEs are the subject of Exercise 2.4.23. We note that analogous first-kind VFIEs with *variable* lower and upper limits of integration are studied in Denisov & Korovin (1992).

(ii) If the kernel $H(t, s)$ in (2.3.14) is of the form $H(t, s) = h(s)$, (2.3.15) reduces to the *functional equation*

$$u(t) = g_1(t) + h_q(t)u(qt), \quad t \in I, \tag{2.3.16}$$

where

$$g_1(t) := \frac{g'(t)}{h(t)} \quad \text{and} \quad h_q(t) := \frac{qh(qt)}{h(t)},$$

and hence we have $u(0) = \dfrac{g'(0)}{(1 - q)h(0)}$. Existence and uniqueness results for such functional equations can be found in Xie, Zhang & Brunner (2011). ◇

An extension of Theorem 2.3.14 to the first-kind VFIE

$$\int_{\theta(t)}^{t} H(t, s)u(s)\, ds = g(t), \quad t \in I, \tag{2.3.17}$$

with more general (non-linear) delay functions $\theta(t)$ satisying $\theta(0) = 0$ was established in Denisov & Korovin (1992); see also Theorem 3.3.1 in the book by Apartsyn (2003).

Theorem 2.3.16 *Assume that the given functions in (2.3.17) have the following properties:*

(i) $H \in C(\bar{D}_\theta)$, $H_t := \partial H/\partial t \in C(\bar{D}_\theta)$, $H(t, t) \neq 0 \ (t \in I)$;

(ii) $g \in C^1(I)$, $g(0) = 0$;

(iii) $\theta \in C^1(I)$, $\theta(0) = 0$, $\theta(t) < t \ (t \in (0, T])$, $\theta'(0) < 1$, $\theta'(t) > 0 \ (t \in (0, T])$.

Then the first-kind VFIE (2.3.17) possesses a unique continuous solution u on I.

Proof The idea on which the proof is based consists in introducing a sequence of points $\{\xi_m\}$ in $[0, T]$ by setting $\xi_m := \theta^m(T) \ (m = M, M - 1, \ldots, 0)$ for some appropriate M defined below. It follows from the assumptions (i)–(iii) that the given VFIE is equivalent to the second-kind VFIE

$$u(t) = \frac{g'(t)}{H(t, t)} + \theta'(t)\frac{H(t, \theta(t))}{H(t, t)}u(\theta(t)) - \int_0^t \frac{H_t(t, s)}{H(t, t)}u(s)\, ds, \quad t \in I.$$

It is governed by the linear operator

$$(\tilde{\mathcal{V}}u)(t) := \theta'(t)\frac{H(t,\theta(t))}{H(t,t)}u(\theta(t)) - \int_{\theta(t)}^{t}\frac{H_t(t,s)}{H(t,t)}u(s)\,ds\,.$$

By assumption (iii) there exists an integer $M > 0$ such that $\tilde{\mathcal{V}} : C[0,\xi_M] \to C[0,\xi_M]$ is a contraction mapping. This implies that on this initial interval the VFIE (2.3.17) possesses a unique solution $u_M \in C[0,\xi_M]$. For $t \in [\xi_M,\xi_{M-1}]$ the VFIE (2.3.17) can be written as

$$u(t) = g_M(t) - \int_{\xi_M}^{t}\frac{H_t(t,s)}{H(t,t)}u(s)\,ds,$$

where

$$g_M(t) := \frac{g'(t)}{H(t,t)} + \theta'(t)\frac{H(t,\theta(t))}{H(t,t)}u_M(\theta(t)) - \int_{\theta(t)}^{\xi_M}\frac{K_t(t,s)}{K(t,t)}u_M(s)\,ds.$$

This is a classical VIE of the second kind with continuous data which has a continuous solution u_{M-1} on $[\xi_M,\xi_{M-1}]$ that satisfies $u_{M-1}(\xi_M) = u_M(\xi_M)$. This process can obviously be extended to the subsequent intervals $[\xi_{m-1},\xi_{m-2}]$ $(m = M, \ldots, 2)$, where $\xi_0 = T$. $\qquad\square$

Remark 2.3.17 The case where in (iii) the assumption on the delay function in (2.3.17) has now the property $\theta'(0) = 1$ was analysed by Denisov & Lorenzi (1995). This case was already treated, albeit in a somewhat sketchy way, by Lalesco (1911). $\qquad\diamond$

2.3.4 Second-Kind VFIEs with Non-Vanishing Delays

We have seen in the previous section that if the given functions in second-kind VFIEs with *vanishing delay* are smooth, the solution of the VFIE inherits this smoothness on the entire interval I. This is no longer true for VFIEs with *non-vanishing delays*. For such equations there will in general exist points – the so-called *primary discontinuity points* described in Definition 2.3.2 – in the interval I at which the degree of smoothness is, at least initally, much lower than that of the given data. This is of course a well-known phenomenon in delay differential equations with non-vanishing delays (cf. Bellen & Zennaro, 2003).

Throughout this chapter the delay function θ will be subject to conditions that are the analogues of (V1)–(V3) for vanishing delays. For the convenience of the reader we recall these from Section 2.3.1:

(N1) $\theta(t) = t - \tau(t)$, $\tau \in C^d(I)$ for some $d \geq 0$;
(N2) $\tau(t) \geq \tau_0 > 0$ for $t \in I$;
(N3) θ is strictly increasing on I.

Moreover, we will assume for ease of exposition that the interval $I = [0, T]$ is such that $T = \xi_{M+1}$ for some $M \geq 1$.

We now analyse the nature of the solution to the general linear VFIE

$$u(t) = f(t) + (\mathcal{V}_\alpha u)(t) + (\mathcal{V}_{\theta,\alpha} u)(t), \quad t \in (0, T], \qquad (2.3.18)$$

where $0 < \alpha \leq 1$ and with the delay function θ being subject to the assumptions (N1)–(N3). The Volterra integral operators in (2.3.18) are defined as in Section 2.3.1,

$$(\mathcal{V}_\alpha u)(t) := \int_0^t K_\alpha^{(0)}(t, s) u(s) \, ds, \quad (\mathcal{V}_{\theta,\alpha} u)(t) := \int_0^{\theta(t)} K_\alpha^{(1)}(t, s) u(s) \, ds,$$

with kernels $K_\alpha^{(i)}(t, s) := (t - s)^{\alpha - 1} K^{(i)}(t, s)$ $(i = 0, 1; \ 0 < \alpha \leq 1)$. Later on we shall also consider the VFIE

$$u(t) = f(t) + (\mathcal{W}_{\theta,\alpha} u)(t), \quad t \in I,$$

which is described by the Volterra integral operator

$$(\mathcal{W}_{\theta,\alpha} u)(t) := \int_{\theta(t)}^t (t - s)^{\alpha - 1} K(t, s) u(s) \, ds.$$

This VFIE may be viewed as a particular case of (2.3.18) when $K^{(1)}(t, s) = -K^{(0)}(t, s)$.

Since $\theta(0) < 0$, these VFIEs are defined only if they are complemented by an *initial condition*,

$$u(t) = \phi(t), \quad t \in [\theta(0), 0].$$

We point out that the VFIE (2.3.18) is considered on the *left-open* interval $(0, T]$: we shall see below (Theorem 2.3.21) that solutions to second-kind VFIEs with non-vanishing delays typically possess a finite (jump) discontinuity at $t = 0$, while for first-order delay differential equations the solution u is continuous at this point, with the discontinuity occurring in u' (cf. Bellen & Zennaro, 2003, Section 2.1).

In Section 2.3.1 we pointed out that a non-vanishing delay $\tau(t)$ gives rise to *primary discontinuity points* $\{\xi_\mu\}$ for the solution u of (2.3.18): they are determined by the recursion (2.3.1). Condition (N2) ensures that these discontinuity points have the uniform separation property

$$\xi_\mu - \xi_{\mu-1} = \tau(\xi_\mu) \geq \tau_0 > 0 \quad \text{for all } \mu \geq 1;$$

that is, no clustering of the ξ_μ can occur as μ increases.

We will first study the case $\alpha = 1$, using the notation $\mathcal{V}_\theta := \mathcal{V}_{\theta,1}$ and $\mathcal{W}_\alpha := \mathcal{W}_{\theta,1}$.

Theorem 2.3.18 *Assume that the given functions in (2.3.18) with $\alpha = 1$ are continuous on their respective domains and that the delay function θ is subject to the conditions (N1)–(N3). Then for any initial function $\phi \in C[\theta(0), 0]$ there exists a unique (bounded) $u \in C(0, T]$ that solves the VFIE (2.3.18) and coincides with ϕ on $[\theta(0), 0]$. In general, this solution has a finite (jump) discontinuity at $t = 0$; that is, we have*

$$\lim_{t \to 0^+} u(t) \neq \lim_{t \to 0^-} u(t) = \phi(0).$$

The solution has no jump discontinuity at $t = 0$ only for initial functions ϕ satisfying the condition

$$f(0) - \int_{\theta(0)}^{0} K^{(1)}(0, s)\phi(s)\, ds = \phi(0),$$

where $K^{(1)}(t, s)$ is the kernel of the Volterra operator $\mathcal{V}_\theta$.

Proof For $t \in I^{(\mu)} := [\xi_\mu, \xi_{\mu+1}]$ $(\mu = 0, 1, \ldots, M)$ the VFIE (2.3.18) $(\alpha = 1)$ may be written as a local VIE of the second kind,

$$u(t) = f_\mu(t) + \int_{\xi_\mu}^{t} K^{(0)}(t, s)u(s)\, ds, \tag{2.3.19}$$

with $f_\mu(t) := f(t) + \Phi_\mu(t)$ and

$$\Phi_\mu(t) := \int_{0}^{\xi_\mu} K^{(0)}(t, s)u(s)\, ds + \int_{0}^{\theta(t)} K^{(1)}(t, s)u(s)\, ds.$$

If $\mu = 0$ (and hence $\xi_0 = 0$) this function is known:

$$\Phi_0(t) = -\int_{\theta(t)}^{0} K^{(1)}(t, s)\phi(s)\, ds,$$

and it follows from our assumptions that $\Phi_0 \in C(I^{(0)})$. The classical Volterra theory of Section 1.2.1 tells us that for each $\mu \geq 0$ (so that $I^{(\mu)} \subset I$) the integral equation (2.3.19) possesses a unique continuous solution on $I^{(\mu)}$.

As for its regularity, we first observe that for $\mu = 0$ (and $\xi_0 = 0$),

$$\lim_{t \to 0^+} u(t) = f(0) + \Phi_0(0) = f(0) - \int_{\theta(0)}^{0} K^{(1)}(0, s)\phi(s)\, ds$$

which, for arbitrary (continuous) data f, $K^{(1)}$, ϕ, will not coincide with the value $\phi(0)$. If $\mu \geq 1$ we see that

$$u(\xi_\mu^-) = f(\xi_\mu) + \int_0^{\xi_\mu} K^{(0)}(\xi_\mu, s)u(s)\,ds + \int_0^{\theta(\xi_\mu)} K^{(1)}(\xi_\mu, s)u(s)\,ds$$

and

$$u(\xi_\mu^+) = f(\xi_\mu) + \int_0^{\xi_\mu} K^{(0)}(\xi_\mu, s)u(s)\,ds + \int_0^{\theta(\xi_\mu)} K^{(1)}(\xi_\mu, s)u(s)\,ds.$$

Hence,

$$u(\xi_\mu^+) - u(\xi_\mu^-) = 0,$$

whenever f, $K^{(0)}$, $K^{(1)}$ and θ are continuous functions. This completes the proof of Theorem 2.3.18. □

We have seen in Section 1.2.1 that the solution of a linear Volterra integral equation of the second kind can be expressed in term of the resolvent kernel and the non-homogeneous terms f (recall Theorem 1.2.3). As the above proof implicitly shows, an analogous (local) representation can be derived for the solution of the VFIE (2.3.18) (with $\alpha = 1$) on each interval $I^{(\mu)}$, since by (N2) the delay $\tau = \tau(t)$ in $\theta(t) = t - \tau(t)$ does not vanish in I. Suppose, for ease of notation and without loss of generality, that T defining the interval $I = [0, T]$ is such that $\xi_{M+1} = T$ (or, alternatively, $T \in (\xi_M, \xi_{M+1}]$) for some $M \geq 1$.

Theorem 2.3.19 *Suppose that (N1)–(N3) and the assumptions of Theorem 2.3.18 hold, and set*

$$f_0(t) := f(t) - \int_{\theta(t)}^0 K^{(1)}(t, s)\phi(s)\,ds, \quad t \in I^{(0)}.$$

Then for $t \in I^{(\mu)} := [\xi_\mu, \xi_{\mu+1}]$ ($\mu \geq 1$) the unique solution u of (2.3.18) with $\alpha = 1$ corresponding to the initial function ϕ can be expressed by the variation-of-constants formula

$$u(t) = f(t) + \int_{\xi_\mu}^t R^{(0)}(t, s)f(s)\,ds + F_\mu(t) + \Phi_\mu(t), \tag{2.3.20}$$

where

$$F_\mu(t) := \int_0^{\xi_1} R_{\mu,0}(t, s)f_0(s)\,ds + \sum_{v=1}^{\mu-1} \int_{\xi_v}^{\xi_{v+1}} R_{\mu,v}(t, s)f(s)\,ds,$$

$$\Phi_\mu(t) := \int_0^{\theta^\mu(t)} Q_{\mu,0}(t, s)f_0(s)\,ds + \sum_{v=1}^{\mu-1} \int_{\xi_v}^{\theta^{\mu-v}(t)} Q_{\mu,v}(t, s)f(s)\,ds.$$

On the initial interval $(\xi_0, \xi_1]$ (with $\xi_0 = 0$) the solution u is given by

$$u(t) = f_0(t) + \int_0^t R^{(0)}(t, s) f_0(s) \, ds. \qquad (2.3.21)$$

Here, $R^{(0)}$ is the resolvent kernel associated with the given kernel $K^{(0)}$ of the Volterra integral operator $\mathcal{V}$, $R_{\mu,\nu}$ and $Q_{\mu,\nu}$ denote functions which are continuous on their respective domains and depend on $K^{(0)}$, $K^{(1)}$, $R^{(0)}$ and θ, and $\theta^k := \underbrace{\theta \circ \cdots \circ \theta}_{k \text{ times}}$.

Proof The solution of the local integral equation (2.3.19),

$$u(t) = f_\mu(t) + \int_{\xi_\mu}^t K^{(0)}(t, s) u(s) \, ds, \quad t \in I^{(\mu)},$$

is given by

$$u(t) = f_\mu(t) + \int_{\xi_\mu}^t R^{(0)}(t, s) f_\mu(s) \, ds, \quad t \in I^{(\mu)}, \qquad (2.3.22)$$

with $R^{(0)}$ defined by the resolvent equation

$$R^{(0)}(t, s) = K^{(0)}(t, s) + \int_s^t R^{(0)}(t, v) K^{(0)}(v, s) \, dv, \quad (t, s) \in D^{(\mu)},$$

where $D^{(\mu)} := \{(t, s) : \xi_\mu \le s \le t \le \xi_{\mu+1}\}$.

The expression (2.3.21) for the solution on the interval $I^{(0)}$ ($\mu = 0$) thus follows immediately. On $I^{(1)}$ we thus have

$$
\begin{aligned}
f_1(t) &= f(t) + \int_0^{\xi_1} K^{(0)}(t, s) u(s) \, ds + \int_0^{\theta(t)} K^{(1)}(t, s) u(s) \, ds \\
&= f(t) + \int_0^{\xi_1} K^{(0)}(t, s) f_0(s) \, ds + \int_0^{\theta(t)} K^{(1)}(t, s) f_0(s) \, ds \\
&\quad + \int_0^{\xi_1} \left(\int_v^{\xi_1} K^{(0)}(t, s) R^{(0)}(s, v) \, ds \right) f_0(v) \, dv \\
&\quad + \int_0^{\theta(t)} \left(\int_v^{\theta(t)} K^{(0)}(t, s) R^{(0)}(s, v) \, ds \right) f_0(v) \, dv,
\end{aligned}
$$

and hence

$$
\begin{aligned}
f_1(t) &= f(t) + \int_0^{\xi_1} \left(K^{(0)}(t, s) + \int_s^{\xi_1} K^{(0)}(t, v) R^{(0)}(v, s) \, dv \right) f_0(s) \, ds \\
&\quad + \int_0^{\theta(t)} \left(K^{(1)}(t, s) + \int_s^{\theta(t)} K^{(1)}(t, v) R^{(0)}(v, s) \, dv \right) f_0(s) \, ds
\end{aligned}
$$

$$=: f(t) + \int_0^{\xi_1} Q_{1,1}^{(1)}(t, s) f_0(s) \, ds + \int_0^{\theta(t)} Q_{1,0}^{(1)}(t, s) f_0(s) \, ds,$$

with obvious meaning of the (continuous) functions $Q_{1,0}^{(1)}$ and $Q_{1,1}^{(1)}$.

The solution of (2.3.19) (with $\mu = 1$) can therefore (upon some trivial algebraic manipulations) be written as

$$u(t) = f(t) + \int_0^t R^{(0)}(t, s) f(s) \, ds + \int_0^{\xi_1} \left(Q_{1,1}^{(1)}(t, s) + \hat{Q}_{1,1}^{(1)}(t, s) \right) f_0(s) \, ds$$

$$+ \int_0^{\theta(t)} \left(Q_{1,0}^{(1)}(t, s) + \hat{Q}_{1,0}^{(1)}(t, s) \right) f_0(s) \, ds.$$

This yields (2.3.22) for $\mu = 1$, by setting

$$R_{1,0}(t, s) := Q_{1,1}^{(1)}(t, s) + \hat{Q}_{1,1}^{(1)}(t, s), \qquad Q_{1,0}(t, s) := Q_{1,0}^{(1)}(t, s) + \hat{Q}_{1,0}^{(1)}(t, s).$$

Clearly, the functions describing this expression for u are continuous on their respective domains. The proof is now concluded by a simple (but notationwise tedious) induction argument. ☐

The proof reveals that in the solution representation (2.3.20) the integrals over $[\xi_\mu, \xi_{\mu+1}]$ with $\mu \geq 1$ will contribute terms involving only $f(t)$, while the integrals over $[\xi_0, \xi_1]$ and $[\xi_0, \theta^\mu(t)]$ contain the complete initial function $f_0(t)$.

Remark 2.3.20 The structure of the above resolvent representation (2.3.20) of the solution u clearly reveals the interaction between the classical lag term $F_\mu(t)$ (governed by the classical Volterra operator $\mathcal{V}$) and the delay term $\Phi_\mu(t)$ (which reflects the effect of the non-vanishing delay function θ). ◇

The result of Theorem 2.3.19 and its proof lead to the following result on the *regularity* of solutions of (2.3.18) with $\alpha = 1$.

Theorem 2.3.21 *Assume that the delay function θ is subject to the conditions (N1)–(N3) and that the given functions in the VFIE (2.3.18) all possess continuous derivatives of at least order $m \geq 1$ on their respective domains. Then:*

(a) *The unique solution u of (2.3.18) lies in $C^m(\xi_\mu, \xi_{\mu+1}]$ for each $\mu = 0, 1, \ldots, M$ and is bounded on $Z_M := \{\xi_\mu : \mu = 0, 1, \ldots, M\}$, and hence on I.*

(b) *At $t = \xi_\mu$ ($\mu = 1, \ldots, \min\{m, M\}$) we have*

$$\lim_{t \to \xi_\mu^-} u^{(\mu-1)}(t) = \lim_{t \to \xi_\mu^+} u^{(\mu-1)}(t),$$

Table 2.1 *Regularity and smoothing of solutions to VFIEs with*
smooth kernels

VFIE with C^m-data	Regularity at $t = \xi_\mu$ $(\mu = 0, 1, \ldots, M; \; \xi_{M+1} = T)$
• $u(t) = f(t) + (\mathcal{V}_\theta u)(t)$	$C^{\mu-1}$ (finite jump at $t = 0$)
• $u(t) = f(t) + (\mathcal{W}_\theta f)(u)$	$C^{\mu-1}$ (finite jump at $t = 0$)
• $u(t) = f(t) + b(t)u(\theta(t)) + (\mathcal{V}_\theta u)(t)$	C^{-1} (finite jump at $t = 0$; no smoothing at $t = \xi_\mu$)

while the μ-th derivative of u is in general not continuous at ξ_μ. Moreover,
if $\min\{m, M\} = m < M$, the solution also lies in $C^m[\xi_{m+1}, T]$.

Proof   This is left as an exercise.   □

We summarise the regularity and smoothing properties of solutions to various types of VFIEs in Table 2.1; the proofs of some of these results can be found Zhang & Brunner (1999).

We now consider the VFIE (2.3.18) with $0 < \alpha < 1$, assuming that the delay function θ is again subject to the conditions (N1)–(N3). How does a non-vanishing delay affect the regularity of the solution of the *weakly singular* second-kind VFIE

$$u(t) = g(t) + (\mathcal{V}_\alpha u)(t) + (\mathcal{V}_{\theta,\alpha} u)(t) \quad (0 < \alpha < 1) \tag{2.3.23}$$

and

$$u(t) = g(t) + (\mathcal{W}_{\theta,\alpha} u)(t),$$

with initial condition $u(t) = \phi(t)$ ($t \leq 0$), when all the given functions are arbitrarily smooth on their respective domains?

We summarise a number of relevant regularity results in Table 2.2 below; they extend those described in Table 2.1. The proofs are left to the reader.

2.3.5 First-Kind VFIEs with Non-Vanishing Delays

Consider the linear first-kind Volterra functional integral equation

$$(\mathcal{V}_\alpha u)(t) + (\mathcal{V}_{\theta,\alpha} u)(t) = g(t), \quad t \in (0, T] \quad (0 < \alpha \leq 1), \tag{2.3.24}$$

with initial condition $u(t) = \phi(t)$, $t \in [\theta(0), 0]$. The delay function θ is subject to the assumptions (N1)–(N3), and as before we will assume that T

Table 2.2 *Regularity and smoothing of solutions to weakly singular VFIEs*

VFIE with C^m-data $(0 < \alpha < 1)$	Regularity on $I^{(\mu)} = (\xi_\mu, \xi_{\mu+1}]$ $(\mu = 0, 1, \ldots, M)$
• $u(t) = g(t) + (\mathcal{V}_{\theta,\alpha}u)(t)$	$\begin{cases} C^{\mu,\alpha} & \text{if } \mu = 0, 1, \ldots, \min\{m, M\} \\ C^m & \text{if } \mu > \min\{m, M\} \end{cases}$ (finite jump at $t = 0$)
• $u(t) = g(t) + (\mathcal{W}_{\theta,\alpha}u)(t)$	$\begin{cases} C^{\mu,\alpha} & \text{if } \mu = 0, 1, \ldots, \min\{m, M\} \\ C^m & \text{if } \mu > \min\{m, M\} \end{cases}$ (finite jump at $t = 0$)
• $u(t) = b(t)u(\theta(t)) + (\mathcal{V}_{\theta,\alpha}u)(t)$	C^α (finite jump at $t = 0$; no smoothing at $t = \xi_\mu$ $(\mu \geq 1)$)
• $u(t) = b(t)y(\theta(t)) + (\mathcal{W}_{\theta,\alpha}u)(t)$	C^α (finite jump at $t = 0$; no smoothing at $t = \xi_\mu$ $(\mu \geq 1)$)

coincides with one of the primary discontinuity points; that is, $T = \xi_{M+1}$ for some $M \geq 1$. The Volterra integral operators are the ones introduced in (2.3.18); they correspond to the kernels $K_\alpha^{(i)}(t,s) := (t-s)^{\alpha-1}K^{(i)}(t,s)$ $(i = 0, 1; \ 0 < \alpha \leq 1)$, respectively.

The notation introduced in the previous section allows us to write (2.3.24) in the local form

$$\int_{\xi_\mu}^t K_\alpha^{(0)}(t,s)u(s)\,ds = g_\mu(t), \quad t \in (\xi_\mu, \xi_{\mu+1}], \tag{2.3.25}$$

with

$$g_\mu(t) := g(t) - \int_0^{\xi_\mu} K_\alpha^{(0)}(t,s)u(s)\,ds - \int_0^{\theta(t)} K_\alpha^{(1)}(t,s)u(s)\,ds \tag{2.3.26}$$

$(\mu \geq 1)$. On the initial interval $(0, \xi_1]$ $(\mu = 0)$ this reduces to

$$g_0(t) = g(t) + \int_{\theta(t)}^0 K_\alpha^{(1)}(t,s)\phi(s)\,ds. \tag{2.3.27}$$

According to Theorem 1.4.1 $(\alpha = 1)$ and Theorem 1.5.7 $(0 < \alpha < 1)$ the first-kind VIE (2.3.25) corresponding to $\mu = 0$ possesses a continuous solution on the *closed* interval $[0, \xi_1]$ only if $g_0(0) = 0$. For arbitrarily given (non-trivial) $K^{(0)}$, g, ϕ and θ this will clearly not be true: in general we have

$$g_0(0) = g(0) + \int_{\theta(0)}^0 K_\alpha^{(0)}(0,s)\phi(s)\,ds \neq 0.$$

This means that the solution u of the local first-kind VFIE (2.4.25) with $\mu = 0$ will be *unbounded* at $t = 0^+$:

$$\lim_{t \to 0^-} u(t) = \phi(t_0) \neq \lim_{t \to 0^+} u(t) = \pm\infty.$$

For u to be *bounded* at $t = 0^+$ the given functions must be such that

$$\int_{\theta(0)}^0 K_\alpha^{(1)}(0, s)\phi(s)\,ds = -g(0) \qquad (2.3.28)$$

holds.

If (2.3.28) is true, does the solution $u(t)$ of (2.3.24) become smoother at the points ξ_μ for increasing values of μ? The answer is contained in the following theorem, together with the regularity property of the solution to the VFIE (2.3.24), with $\alpha = 1$, when the given functions are arbitrarily smooth.

Theorem 2.3.22 *Assume that the given functions in the first-kind VFIE (2.3.24) (with $\alpha = 1$) satisfy*

(i) $K^{(0)} \in C(D)$ *and* $\partial K^{(0)}/\partial t \in C(D)$, *with* $K^{(0)}(t, t) \neq 0$ *for all* $t \in I :=$ $[0, T]$;
(ii) $K^{(1)} \in C(D_\theta)$ *and* $\partial K^{(1)}/\partial t \in C(D_\theta)$;
(iii) $g \in C^1(I)$;
(iv) θ *is subject to the conditions (N1)–(N3), with* $d \geq 1$ *in (N1).*

Then for any initial function $\phi \in C[\theta(0), 0]$ *the VFIE (2.3.24) possesses a unique solution u satisfying $u \in C(\xi_\mu, \xi_{\mu+1}]$ ($\mu = 0, 1, \ldots, M$). This solution u remains bounded on $[0, T]$ if, and only if, (2.3.28) holds. If u has a finite jump discontinuity at $t = \xi_0 = 0$, it is discontinuous at all subsequent points ξ_μ ($\mu = 1, \ldots, M$). If the given functions in (2.3.24) (with $\alpha = 1$) are in C^{m+1}, then $u \in C^m(\xi_\mu, \xi_{\mu+1}]$ ($\mu = 0, \ldots, M$).*

We leave the proof of this theorem to the reader but illustrate the underlying key ideas by means of the following representative example.

Example 2.3.23 For $\alpha = 1$ and kernels $K^{(0)}(t, s) \equiv 1$, $K^{(1)}(t, s) \equiv \lambda_1 \neq 0$, (2.3.24) reads

$$\int_0^t u(s)\,ds + \lambda_1 \int_0^{\theta(t)} u(s)\,ds = g(t), \quad t \in (0, T], \qquad (2.3.29)$$

with $u(t) = \phi(t)$ when $t \in [\theta(0), 0]$. Note that for $\lambda_1 = -1$ the VFIE (2.2.29) reduces to

$$\int_{\theta(t)}^{t} u(s) \, ds = g(t), \quad t > 0.$$

Assume that $g \in C^1(I)$ and $\theta(t) = t - \tau(t)$ with $\tau(t) > 0$ and $\tau \in C^1(I)$. For ease of exposition we choose $\phi(t) = \phi_0$ to be i a constant. On the interval $(\xi_\mu, \xi_{\mu+1}]$ the local form of the VFIE (2.3.29) is

$$\int_{\xi_\mu}^{t} u(s) \, ds = g_\mu(t), \quad t \in (\xi_\mu, \xi_{\mu+1}], \tag{2.3.30}$$

where

$$g_\mu(t) := g(t) - \int_0^{\xi_\mu} u(s) \, ds - \lambda_1 \int_0^{\theta(t)} u(s) \, ds \quad (\mu \geq 0). \tag{2.3.31}$$

For $\mu = 0$ we have

$$g_0(t) := g(t) + \lambda_1 \int_{\theta(t)}^{0} \phi(s) \, ds = g(t) - \lambda_1 \phi_0 \theta(t),$$

and thus the solution of (2.3.29) on $(0, \xi_1]$ is

$$u(t) = g_0'(t) = g'(t) - \lambda_1 \phi_0 \theta'(t). \tag{2.3.32}$$

It is right-continuous (hence bounded) at $t = 0^+$ only if $g(t)$ is such that $g_0(0) = 0$; that is, if for given λ_1, θ and ϕ_0 the value $g(0)$ is

$$g(0) = \lambda_1 \phi_0 \theta(0) = -\lambda_1 \phi_0 \tau(0). \tag{2.3.33}$$

For arbitrary data θ, λ_1, ϕ_0 the equation

$$u(0^+) - u(0) = g'(0) - \lambda_1 \phi_0 \theta'(0)$$

then implies that u has a finite jump discontinuity at $t = \xi_0 = 0$.

Let now $\mu = 1$ and $t \in (\xi_1, \xi_2]$. Using (2.3.32) and (2.3.31) we find that $g_1(\xi_1) = g(0)$. Thus, if (2.3.33) holds, the solution is right-continuous (and hence bounded) at $t = \xi_1$. An analogous analysis shows that under (2.3.33) the solution of (2.3.29) exhibits this discontinuity property at all subsequent primary discontinuity points $\xi_\mu \in I$.

Remark 2.3.24 To readers familiar with *neutral* delay differential equations (NDDEs) (cf. Hale & Verduyn Lunel, 1993, Section 9.1, or Bellen & Zennaro, 2003, pp. 5–6) the non-smoothing result in Theorem 2.3.26 will not be

surprising, since for sufficiently smooth data the VFIE (2.3.29) is equivalent to
the linear NDDE equation

$$u'(t) + a_1(t)u(\theta(t)) + a_2(t)u'(\theta(t)) = g''(t),$$

with $a_1(t) := \lambda_1 \theta''(t)$ and $a_2(t) := \lambda_1(\theta'(t))^2$. For such NDDEs there is no
smoothing in the solution.   ⋄

We now turn to first-kind VFIEs with non-vanishing delays and *weakly singular kernels* and employ the following example to obtain insight into the
typical regularity properties of their solutions.

Example 2.3.25 Consider the weakly singular VFIE of the first kind, with
delay function $\theta(t) = t - \tau$ ($\tau > 0$),

$$\int_{\theta(t)}^{t} \frac{(t-s)^{\alpha-1}}{\Gamma(\alpha)} u(s)\, ds = g(t), \quad t \in (0, T] \ (0 < \alpha < 1), \qquad (2.3.34)$$

with $u(t) = \phi_0$ ($t \in [\theta(0), 0]$) and $g \in C^1(I)$. Here, the primary discontinuity
points are $\xi_\mu = \mu\tau$, and we will assume that T is such that $T = \xi_{M+1}$ for
some $M \geq 2$. The analogue of the local version (2.3.30) is

$$\int_{\xi_\mu}^{t} \frac{(t-s)^{\alpha-1}}{\Gamma(\alpha)} u(s)\, ds = g_\mu(t), \quad t \in (\xi_\mu, \xi_{\mu+1}] \ (\mu = 0, 1, \ldots, M),$$
$$(2.3.35)$$

where

$$g_\mu(t) := g(t) - \int_{\theta(t)}^{\xi_\mu} \frac{(t-s)^{\alpha-1}}{\Gamma(\alpha)} u(s)\, ds. \qquad (2.3.36)$$

According to Theorem 1.5.1 the local first-kind VIE (2.3.35) has the solution

$$u(t) = \frac{g_\mu(\xi_\mu)}{\Gamma(1-\alpha)}(t-\xi_\mu)^{-\alpha} + \int_{\xi_\mu}^{t} \frac{(t-s)^{-\alpha}}{\Gamma(1-\alpha)} g'_\mu(s)\, ds, \quad t \in (\xi_\mu, \xi_{\mu+1}].$$
$$(2.3.37)$$

This reveals that for $g \in C^1(I)$ the solution is unbounded at the primary discontinuity point $t = \xi_\mu$ if $g_\mu(\xi_\mu) \neq 0$. In order to analyse the regularity of
the solution of (2.3.34) at these points ξ_μ, consider first the interval $(0, \xi_1]$. On
this interval the function $g_0(t)$ is given by

$$g_0(t) = g(t) + \frac{\phi_0}{\alpha\Gamma(\alpha)}(t^\alpha - \tau^\alpha),$$

and hence we have $g_0(0) = g(0) - \phi_0\tau^\alpha/\Gamma(1+\alpha)$. We thus see that, in analogy
to the solution of (2.3.29), the solution of (2.3.34) is bounded at $t = 0^+$ only
if, for given α, τ and ϕ_0, g is such that

$$g(0) = \frac{\phi_0 \tau^\alpha}{\Gamma(1+\alpha)} \tag{2.3.38}$$

(the analogue of (2.3.33) with $\lambda_1 = -1$) holds. Assuming that $g \in C^1(I)$, it follows from (2.3.37) that on $(0, \xi_1]$ the solution can be written in the form

$$
\begin{aligned}
u(t) &= \left(\frac{g(0)}{\Gamma(1-\alpha)} - \frac{\phi_0 \tau^\alpha}{\alpha \Gamma(\alpha) \Gamma(1-\alpha)} \right) t^{-\alpha} + \int_0^t \frac{(t-s)^{-\alpha}}{\Gamma(1-\alpha)} g'(s)\, ds + \phi_0 \\
&= \frac{g(0)}{\Gamma(1-\alpha)} t^{-\alpha} + \int_0^t \frac{(t-s)^{-\alpha}}{\Gamma(1-\alpha)} g'(s)\, ds \\
&\quad + \left(1 - \frac{\tau^\alpha t^{-\alpha}}{\Gamma(1+\alpha) \Gamma(1-\alpha)} \right) \phi_0.
\end{aligned} \tag{2.3.39}
$$

The first two terms in (2.3.39) represent the solution of the classical Abel integral equation on the interval $(\xi_0, \xi_1] = (0, \tau]$, while the remaining terms reflect the effect of the initial function associated with the constant delay τ.

A straightforward calculation shows that for any $g(t)$ (regardless of whether or not the boundedness condition (2.3.38) is satisfied) there holds $g_0(\xi_1) = 0$; hence the solution of the first-kind VFIE (2.3.34) is always bounded at ξ_1 (but in general exhibits a finite jump discontinuity). An analogous calculation for the subsequent intervals $(\xi_\mu, \xi_{\mu+1}]$ yields that at all points ξ_μ with $\mu \geq 2$ the solution is always bounded and right-continuous. (This, incidentally, generalises the observation in Ito & Turi (1991, p. 1715) that for their specific example $(g(t) \equiv 1, \phi_0 = 0)$ the above boundedness result is true.) The following theorem summarises the above observations.

Theorem 2.3.26 *Assume that* $\theta(t) = t - \tau$ $(\tau > 0)$, $\phi(t) = \phi_0$, *and* $g \in C^1(I)$. *Then the solution* $u(t)$ *of the first-kind weakly singular VFIE (2.3.34) is continuous on each left-open interval* $(\xi_\mu, \xi_{\mu+1}]$ $(\mu = 0, \dots, M)$. *It remains bounded at all the primary discontinuity points* $t = \xi_\mu$ *in* I *only if* $g(t)$ *is such that (2.3.38) holds.*

2.4 Exercises and Research Problems

Exercise 2.4.1 In Theorem 2.1.1, asssume that $f \in C^q(I)$ for some $q \geq 1$, while $K \in C^m(D)$ $(m \geq 0)$. What can be said about the regularity of the solution of the corresponding VIE (1.2.2)?

Exercise 2.4.2 Suppose that the kernel $K(t, s)$ in the weakly singular VIE (2.2.1) does not satify the condition $K(t, t) \neq 0$ on I but vanishes for all $t \in I$.

State and prove the corresponding analogue of the regularity result (2.1.7) of Theorem 2.1.2.

Exercise 2.4.3 Prove Theorem 2.1.4.

Exercise 2.4.4 Consider the VIE

$$u(t) = t^\beta f_2(t) + \int_0^t (t-s)^{\alpha-1} u(s)\, ds \ \ t \in I,$$

with $0 < \alpha < 1$ and $\beta > 0$ ($\beta \notin \mathbb{N}$). Are there functions f_2 in $C^m(I)$ for which the corresponding solution u is also in $C^m(I)$?

Exercise 2.4.5 Analyse the regularity of the solution of the VIE

$$u(t) = f(t) + \int_0^t \log(t-s) s^{-\beta} u(s)\, ds,$$

when $f \in C^m(I)$ and $0 < \beta < 1$.

Exercise 2.4.6 (*Research Problem*)
Analyse the existence and the regularity of the solution of

$$u(t) = f(t) + \int_{-t}^t |t-s|^{\alpha-1} K(t,s) u(s)\, ds, \ \ t \in \tilde{I} := [-T, T] \ (0 < \alpha < 1),$$

assuming that $f \in C^d(I)$ and $K \in C^d(\tilde{D})$ for some $d \in \mathbb{N}$ (see Example 1.2.4 of Section 1.2.4 for the notation).

Exercise 2.4.7 If $H \in C^{m+1}(D)$, with $H(t,t) = 1$ for $t \in I$, describe the functions g for which the weakly singular first-kind VIE

$$\int_0^t (t-s)^{\alpha-1} H(t,s) u(s)\, ds = g(t), \ \ t \in I \ (0 < \alpha < 1), \qquad (2.4.1)$$

possesses a unique solution $u \in C^m(I)$.

Exercise 2.4.8 Prove that under the assumptions (a), (b) and (c) (with $q = 0$) in Theorem 2.2.3 the solution of the weakly singular first-kind VIE (2.4.1) has a representation analogous to the one in (2.1.8) (Theorem 2.1.2) for a weakly singular second-kind VIE, except that in that expansion α is now replaced by $1 - \alpha$.

Exercise 2.4.9 (*Research Problem*)
(This continues Exercise 1.7.16.) Extend the regularity result of Theorem 2.2.3 to the system of first-kind VIEs

$$\int_0^t \mathbf{H}_\alpha(t,s)\mathbf{u}(s)\,ds = \mathbf{g}(t), \quad t \in I,$$

whose weakly singular matrix kernel has the form

$$\mathbf{H}_\alpha(t,s) := \left[\begin{matrix} (t-s)^{\alpha_{ij}-1}H_{ij}(t,s) \\ (i,j=1,\ldots,M) \end{matrix} \right].$$

Here, $M \geq 2$ and $0 < \alpha_{ij} \leq 1$, with $\alpha_{ij} < 1$ for at least one index pair (i,j).

Exercise 2.4.10 Analyse, under assumptions analogous to the ones in Theorem 2.2.3, the regularity of the solution to the *adjoint* weakly singular VIE

$$\int_t^T (s-t)^{\alpha-1}K(s,t)u(s)\,ds = g(t), \quad t \in [0,T].$$

Exercise 2.4.11 Is the generalised Abel integral equation

$$\frac{c(x)}{\Gamma(\alpha)}\int_0^x (x-s)^{\alpha-1}u(s)\,ds + \frac{d(x)}{\Gamma(\alpha)}\int_x^1 (s-x)^{(\alpha-1)}u(s)\,ds = g(x), \quad x \in [0,1],$$

with $0 < \alpha < 1$ and $|c(x)| + |d(x)| \neq 0$ on $[0,1]$, an ill-posed problem?

Exercise 2.4.12 Prove the uniqueness parts of Theorems 2.3.8 and 2.3.12.

Exercise 2.4.13 Extend the generalised Gronwall Lemmas (cf. Lemma 1.2.17) and (1.3.8) to the Volterra functional inequalities

$$z(t) \leq f(t) + M\int_0^{qt} \frac{(t-s)^{\alpha-1}}{\Gamma(\alpha)}z(s)\,ds, \quad t \in [0,T]$$

and

$$z(t) \leq f(t) + M\int_{qt}^t \frac{(t-s)^{\alpha-1}}{\Gamma(\alpha)}z(s)\,ds, \quad t \in [0,T],$$

where $M > 0$, $0 < q < 1$ and $0 < \alpha \leq 1$.

Exercise 2.4.14 Discuss the solvability of the VFIE

$$u(t) = f(t) + \int_{pt}^{qt} K(t,s)u(s)\,ds, \quad t \in [0,T] \quad (0 < p < q < 1)$$

in the space $C[0,T]$, assuming that $f \in C[0,T]$ and $K \in C(D)$.

Exercise 2.4.15 Provide the details in the proof of Theorem 2.3.21.

Exercise 2.4.16 Does the general VFIE (2.3.13),

$$u(t) = f(t) + b(t)u(qt) + \int_0^{qt} K(t,s)u(s)\,ds, \quad t \in [0,T],$$

with continuous f, K and $q \in (0,1)$, have a continuous solution u for any $b \in C[0,T]$?

Exercise 2.4.17 Analyse the regularity properties of the solution of the VFIE

$$u(x) = f(x) + \frac{\gamma}{\theta(x)} \int_0^{\theta(x)} u(s)\,ds, \quad x \in [0,1],$$

where, for given $a > 1$ and $0 < \gamma < 1$,

$$\theta(x) := \begin{cases} ax & \text{if } 0 \le x \le a^{-1} \\ 1 & \text{if } a^{-1} \le x \le 1 \end{cases}$$

(Piila & Pitkäranta, 1999; see also Section 9.2.7), assuming that the given function f is arbitrarily smooth.

Exercise 2.4.18 Discuss the regularity of the solution of the following VFIE:

$$u(t) = f(t) + \int_{t/2}^t K(t,s)u(s)\,ds, \quad t \in [1,10],$$

$$\phi(t) = \begin{cases} 1 & \text{if } 1/2 \le t < 3/4, \\ 1/2 & \text{if } 3/4 \le t \le 1, \end{cases}$$

with $f \in C^m[1,10]$, $K \in C^m(D)$ ($D = \{(t,s) : 1 \le s \le t \le 10\}$.

Exercise 2.4.19 (*Research Problem*)

(a) Consider the VFIE

$$u(t) = f(t) + \int_0^{\theta(t)} K(t,s)u(s)\,ds, \quad t \in [0,T],$$

with continuous f and K and the vanishing, *non-monotonic* delay function

$$\theta(t) = \gamma t e^{-t} \quad (0 < \gamma < 1).$$

Does this VFIE possess a unique solution $u \in C[0,T]$ for any $T > 1$?

(b) A VFIE with a much more general non-monotonic delay function is

$$u(t) = 1 + \int_{\theta(t)}^t K(t,s)u(s)\,ds, \quad t \in [0,T],$$

with $K \in C(D)$ and $\theta(t) = q_1 t + (q_2 - q_1) t \sin^2(\omega t)$ $(0 < q_1 < q_2 < 1, \omega \geq 1)$. Does this VFIE possesses a unique continuous solution for any (e.g. very large) ω? Is the solution highly oscillatory when $\omega \gg 1$? (Compare also Exercise 4.5.14.)

Exercise 2.4.20 Consider the VFIE

$$u(t) = f(t) + \int_{\theta(t)}^{t} K(t, s) u(s) \, ds, \quad t \in (0, T],$$

with $\theta(t) = t - \tau_0$ $(\tau_0 > 0)$ and $u(t) = \phi(t)$ $(t \in [-\tau_0, 0])$. Assuming that $f \in C^d(I)$ and $K \in C^d(D)$ for some $d \geq 0$, is it possible to find an initial function ϕ for which the solution is in $C^d(I)$?

Exercise 2.4.21 Discuss the existence and the regularity of solutions to the VFIE

$$u(t) = f(t) + \int_{0}^{t} \left(K^{(0)}(t, s) u(s) + K^{(1)}(t, s) u(\theta(s)) \right) ds, \quad t \in [0, T],$$

both for vanishing and for non-vanishing delays $\tau(t)$ in $\theta(t) = t - \tau(t)$ (assuming that f and K are abitrarily smooth functions).

Exercise 2.4.22 Let the functions $\theta_i(t)$ $(i = 1, 2)$ be continuously differentiable. Show that for smooth $f(t, s)$ there holds

$$\frac{d}{dt} \left(\int_{\theta_1(t)}^{\theta_2(t)} f(t, s) \, ds \right) = \int_{\theta_1(t)}^{\theta_2(t)} \frac{\partial f(t, s)}{\partial t} \, ds + f(\theta(t)) \frac{d\theta_2(t)}{dt} - f(\theta_1(t)) \frac{d\theta_1(t)}{dt}$$

(Leibniz's theorem for the differentiation of an integral).

Exercise 2.4.23 In his 1913 book (pp. 85–88) Volterra briefly mentions the first-kind VFIE

$$\int_{pt}^{qt} K(t, s) u(s) \, ds = g(t), \quad t \in [0, T] \; (0 < p < q < 1).$$

If $K \in C^1(D)$ and $g \in C^1[0, T]$, with $g(0) = 0$, does this VFIE possess a unique solution $u \in C[0, T]$?

Exercise 2.4.24 (*Research Problem*)
In Nota II of 1896 Volterra established the existence and uniqueness of a continuous solution for the first-kind VIE

$$\int_{qt}^{t} (t - s)^{\alpha - 1} K(t, s) u(s) \, ds = g(t), \quad t \in [0, T] \; (0 < \alpha < 1),$$

when $q = 0$ (cf. Theorem 1.5.7). Extend his result to the above VFIE with $0 < q < 1$.

Exercise 2.4.25

(a) Extend the analysis in Example 2.3.4 to the VFIE

$$\int_0^t \frac{(t-s)^{\alpha-1}}{\Gamma(\alpha)} u(s)\, ds + \lambda_1 \int_0^{\theta(t)} \frac{(t-s)^{\alpha-1}}{\Gamma(\alpha)} u(s)\, ds = g(t),$$

where $\theta(t) = t - \tau$ $(\tau > 0)$ and $\lambda_1 \neq 0$. Does the solution experience *smoothing* at ξ_μ as μ increases?

(b) Do the same for the VFIE (2.3.34) with variable delay $\tau(t) \geq \tau_0 > 0$ and non-constant initial function $\phi \in C^1[\theta(0), 0]$.

2.5 Notes

Section 2.1: *VIEs of the Second Kind*
The regularity of solutions to VIEs with algebraic kernel singularities was analysed in Lubich (1983a) and Brunner (1983); earlier contributions are due to Tychonoff (1938) and by Miller & Feldstein (1971). The paper by Han (1994) deals with weakly singular VIEs that are related to *cordial* VIEs (cf. Chapter 7). The most comprehensive analysis of the regularity of solutions of VIEs with general (diagonal) kernel singularities is arguably the one in Brunner, Pedas & Vainikko (1999). In addition, see Kolk & Pedas (2013) on equations with diagonal *and* boundary singularities. The regularity of weakly singular *Fredholm* integral equations is studied in Richter (1976), Schneider (1979) [Hölder continuity of solutions] and Graham (1982); see also Vainikko (1993) and Pedas & Vainikko (2006).

Section 2.2: *VIEs of the First Kind*
Regularity for weakly singular VIEs of the first kind can be found in Atkinson (1974) and Lubich (1987); see also Gorenflo & Vessella (1991) and Brunner (2004, Section 6.1).

The fundamental paper by von Wolfersdorf (1965) discusses the regularity properties of solutions to the *generalised Abel integral equation* in Hölder spaces; it also includes comprehensive references on their history (going back to Carleman's 1922 paper). A brief summary of the results in this paper can be found in Gorenflo & Vessella (1991, pp. 121–123). Gakhov's book of 1966 contains both a discussion of the theory of this equation, as well as of the underlying physical motivation for studying it. A selection of related later papers is Samko (1968), Linz & Noble (1971), Mandal, Chakrabarti & Mandal (1996) and Chakrabarti (2008, which describes an approach different from Gakhov's 1966 book but does not seem to be aware of von Wolfers-dorf paper of 1965). Systems of generalised Abel integral equations are studied in Lowengrub & Walton (1979; see also for additional references).

Section 2.3: *Linear VIEs with Delay Arguments*
Second-kind VFIEs with *vanishing* (proportional) *delays* were first studied in Andreoli (1914); Chambers (1990) contains a comprehensive analysis of solutions to such VFIEs (including non-linear ones). See also Brunner (2004, Ch. 5) and Brunner (2004a) on more general VFIEs of this type. The analysis in Volterra (1897) on first-kind VFIEs with proportional delays was extended by Lalesco (1908, 1911; compare also Volterra, 1913, pp. 92–101 and Fenyö & Stolle, 1984, pp. 324–327). A particular case of the VFIE (2.3.14) ($\alpha = 1$) was studied by Picard (1907). Further generalisations were obtained in the early 1990s, by Denisov & Korovin (1992) and Denisov & Lorenzi (1995). The most authoritative source on such first-kind VFIEs is Apartsyn (2003; cf. Chapter 3). In Apartsyn (2014) first-kind VIEs of the form

$$\sum_{i=1}^{r} \int_{a_i(t)}^{a_{i-1}(t)} K_i(t,s)u(s)\,ds = g(t), \quad t \in I,$$

with delays $0 \leq a_r(t) \leq \cdots \leq a_0(t) \equiv 1$ are studied. Chapter 2 in Vogel (1965) contains an illuminating survey of the historical development of VFIEs. Bownds & Cushing (1976) analyses the existence, uniqueness and extendibility of solutions of systems of VFIEs with variable delays.

Gronwall-type inequalities for Volterra functional inequalities with non-negative delay functions $\theta(t) \leq t$ have been studied in a number of papers; see, for example, Lipovan (2006) and its references.

As in DDEs with variable *non-vanishing delays*, a key problem in the analysis of analogous VFIEs (especially when the delay is state-dependent) is the identification (and numerical computation) of the (primary) discontinuity points $\{\xi_\mu\}$. The book by Bellen & Zennaro (2003) contains a complete characterisation of more general ('k-level') discontinuity points. See also Neves & Feldstein (1976) for the original pioneering study, and as well as Willé & Baker (1992), Feldstein, Neves & Thompson (2006), and Guglielmi & Hairer (2008).

Chapter 2 in Apartsyn (2003) contains a detailed analysis of the solutions of first-kind VFIEs with non-vanishing delays. The papers by Kappel & Zhang (1986) and Ito & Turi (1991) are excellent sources for results on weakly singular first-kind VFIEs with constant delay $\tau > 0$ (they also contain numerous references on e.g. related papers by Burns et al.).

Numerical Analysis of VFIEs
The 2003 monograph by Bellen & Zennaro contains a comprehensive introduction to the numerical analysis of delay differential equations (DDEs), with focus on various types of Runge–Kutta methods. See also the *Acta Numerica* review paper by Bellen, Maset, Zennaro & Guglielmi (2009). Numerical methods for VFIEs are studied in e.g. Brunner (2004, Chapters 4, 5, 6) and in Brunner (2004a: this *Acta Numerica* review

paper contains an extensive list of references on the numerical analysis of VFIEs). In addition, see also Xie, Zhang & Brunner (2011).

Brunner & Maset (2009, 2010) present a general mathematical framework for analysing Runge–Kutta and collocation methods for DDEs with very general (e.g. non-monotone or state-dependent) delays. This framework can potentially be extended to encompass the analysis of the existence and the regularity of solutions to state-dependent VFIEs, and to their numerical analysis.

3

Non-Linear Volterra Integral Equations

Summary

In the first part of this chapter we present an introduction to the theory of non-linear Volterra integral equations. Its focus is on results on the existence and uniqueness of solutions, maximal and minimal solutions, VIEs with multiple solutions, comparison theorems, and on non-linear VIEs whose solutions develop singularities (finite-time blow-up; finite-time quenching). The second part is concerned with non-linear VIEs of non-standard forms; they include auto-convolution equations, implicit VIEs, and non-linear Volterra functional integral equations with time-dependent and state-dependent delays.

3.1 Non-Linear Second-Kind VIEs

3.1.1 General Existence Theorems

A standard non-linear VIE of the second kind has the form

$$u(t) = f(t) + \int_0^t k(t, s, u(s)) \, ds, \quad t \in I := [0, T], \qquad (3.1.1)$$

where f and k are given functions. The underlying non-linear Volterra integral operator

$$(\mathcal{U}u)(t) := \int_0^t k(t, s, u(s)) \, ds$$

is usually referred to as the *Volterra–Urysohn* integral operator.

If $k = k(t, s, u)$ in (3.1.1) is of the form $k(t, s, u) = K(t, s)G(s, u)$, the corresponding VIE

$$u(t) = f(t) + \int_0^t K(t, s)G(s, u(s)) \, ds, \quad t \in I, \qquad (3.1.2)$$

103

is called a VIE of *Hammerstein type* (cf. Hammerstein, 1930) or, in short, a *Volterra–Hammerstein* integral equation (VHIE). It corresponds to the *Volterra–Hammerstein* integral operator

$$(\mathcal{H}u)(t) := \int_0^t K(t, s)G(s, u(s))\, ds\,.$$

We shall study these particular non-linear VIEs in Section 3.1.2.

In Section 3.4 we shall encounter various classes of more general, *non-standard* non-linear VIEs, for example,

$$u(t) = f(t) + \int_0^t K(t, s)G(u(t), u(s))\, ds$$

and

$$u(t) = f(t) + \int_0^t K(t, s)G(u(t - s))G(u(s))\, ds,$$

as well as *implicit* VIEs,

$$W\big(t, u(t)\big) = f(t) + \int_0^t K(t, s)u(s)\, ds.$$

For the equation (3.1.1) the Picard iteration process introduced in Section 1.2 assumes the form

$$u_n(t) := f(t) + \int_0^t k(t, s, u_{n-1}(s))\, ds, \quad t \in I \quad (n \geq 1), \tag{3.1.3}$$

with $u_0(t) := f(t)$. The following (local) existence theorem (Miller, 1971) generalises the classical result for

$$u(t) = u_0 + \int_0^t G(s, u(s))\, ds, \quad t \in I,$$

which is the integrated form of the initial-value problem $u'(t) = G(t, u(t))$, $u(0) = u_0$. Here, the kernel $k(t, s, u) = G(s, u(s))$ does not depend on t, and hence this non-linear VIE is a very special case of (3.1.2): it is equivalent to an initial-value problem for a non-linear differential equation.

In order to state the general existence result for (3.1.1) we introduce the set

$$\Omega_B := \{(t, s, u) : (t, s) \in D, \ u \in \mathbb{R} \text{ and } |u - f(t)| \leq B\} \tag{3.1.4}$$

(where $D := \{(t, s) : \ 0 \leq t \leq T < \infty\}$), and define the constant

$$M_B := \max\{|k(t, s, u)| : \ (t, s, u) \in \Omega_B\}.$$

Here, $B > 0$ is a given real number. We also remind the reader of the following definition.

Definition 3.1.1 Let $k = k(t, s, u)$ be defined for $0 \leq s \leq t < \infty$ and $u \in \mathbb{R}$.

(i) If, for any given positive constants A and B, there exists a constant $L \geq 0$ such that

$$|k(t, s, u_1) - k(t, s, u_2)| \leq L|u_1 - u_2| \text{ for all}$$
$$0 \leq s \leq t \leq T \text{ and } u_1, u_2 \leq B,$$

then k is said to be *locally Lipschitz continuous* in u.
(ii) If the Lipschitz constant L does not depend on B, then k is called *globally Lipschitz continuous* in u. ◇

Theorem 3.1.2 *Assume that $f = f(t)$ and $k = k(t, s, u)$ are such that*

(i) $f \in C(I)$ and $k \in C(\Omega_B)$;
(ii) k satisfies a local Lipschitz condition with respect to u.

Then the following statements are true:

(a) For all $n \geq 1$ the Picard iterates $u_n(t)$ defined by (3.1.3) exist and are continuous on the interval $I_0 := [0, \delta_0]$, where $\delta_0 := \min\{T, B/M_B\}$. On I_0 they converge uniformly to a solution $u \in C(I_0)$ of the non-linear VIE (3.1.1).
(b) This solution u is the unique continuous solution of (3.1.1) on I_0.

Proof To establish uniqueness suppose that (3.1.1) possesses two continuous solutions u and w on the interval I_0. Hence, by (ii),

$$|u(t) - w(t)| \leq \int_0^t |k(t, s, u(s)) - k(t, s, w(s))| \, ds$$
$$\leq L_B \int_0^t |u(s) - w(s)| \, ds, \quad t \in I_0.$$

The continuity of $|u - w|$ and Gronwall's lemma (cf. Lemma 1.2.17) imply that

$$|u(t) - w(t)| \leq 0 \cdot \exp(L_B t) = 0, \quad t \in I_0.$$

Hence, $\|u - w\|_{0,\infty} := \max_{t \in I_0} |u(t) - w(t)| = 0$, meaning that the two solutions are identical on I_0.

To prove the existence of a local continuous solution on I_0 we first show that the Picard iterates defined in (3.1.3) satisfy

$$|u_n(t) - f(t)| \leq M_B t \leq B \quad \text{for all } t \in I_0.$$

This assertion is certainly true for $n = 0$. Assuming that it holds for all integers up to some $n > 1$, we have $(t, s, u_n(s)) \in \Omega_B$ when $t \in I_0$. Hence, $k(t, s, u_n(s))$ is well defined and

$$|k(t, s, u_n(s))| \le M_B \quad \text{for } (t, s) \in D.$$

It follows from (3.1.3) that

$$|u_{n+1}(t) - f(t)| \le \int_0^t |k(t, s, u_n(s))| \, ds \le M_B t \le B, \quad t \in I_0.$$

Thus, $u_{n+1}(t)$ is defined on I_0, and the continuity of f and k on I and Ω_B, respectively, implies that $u_{n+1} \in C(I_0)$. We now prove that the sequence $\{u_n(t)\}$ of the Picard iterates is a Cauchy sequence on I_0. To this end, let $z_n(t) := u_{n+1}(t) - u_n(t)$. It is readily verified that

$$|z_n(t)| \le \frac{M_B L_B^n t^{n+1}}{(n+1)!}, \quad t \in I_0 \quad (n \ge 0).$$

Since

$$u_{n+m}(t) - u_n(t) = \sum_{j=0}^{m-1} \left(u_{n+j+1}(t) - u_{n+j}(t) \right),$$

we see that, for all $t \in I_0$ and all $m \ge 1$,

$$|u_{n+m}(t) - u_n(t)| \le \sum_{j=0}^{m-1} |z_{n+j}(t)| \le M_B \sum_{j=0}^{m-1} \frac{L_B^{n+j} t^{n+j+1}}{(n+j+1)!}$$

$$= M_B \sum_{j=n+1}^{m+n} \frac{L_B^{j-1} t^j}{j!}.$$

Thus the sequence of (continuous) Picard iterates $\{u_n(t)\}_{n \ge 0}$ converges uniformly on I_0 to a continuous function $u(t)$. Using the local Lipschitz condition for $k(t, s, u)$ with respect to u (assumption (ii)), we obtain

$$\left| \int_0^t [k(t, s, u_n(s)) - k(t, s, u(s))] \, ds \right|$$

$$\le L_B \int_0^t |u_n(s) - u(s)| \, ds \longrightarrow 0, \quad t \in I_0,$$

as $n \to \infty$. This allows us to carry out the final step in the existence proof, namely to show that u solves the non-linear integral equation (3.1.1) on I_0: owing to the continuity of $k(t, s, u)$ we have

$$u(t) = \lim_{n \to \infty} u_n(t) = f(t) + \lim_{n \to \infty} \int_0^t k(t, s, u_{n-1}(s)) \, ds$$

$$= f(t) + \int_0^t k(t, s, \lim_{n \to \infty} u_{n-1}(s)) \, ds$$

$$= f(t) + \int_0^t k(t, s, u(s)) \, ds, \quad t \in I_0.$$

The proof of Theorem 3.1.2 is now complete. $\qquad\qquad\qquad\qquad\square$

Does the solution exist (and remain continuous) beyond $t = \delta_0$? Setting $w(t) := u(t + \delta_0)$, the VIE (3.1.1) can be written in 'shifted' form, namely,

$$w(t) = f_0(t) + \int_0^t k_0(t, s, w(s)) ds, \quad t \geq 0,$$

where

$$f_0(t) = f(t + \delta_0) + \int_0^{\delta_0} k(t + \delta_0, s, u(s)) \, ds$$

and

$$k_0(t, s, w) := k(t + \delta_0, s + \delta_0, w), \quad t + \delta_0 \leq T.$$

Since f_0 and k_0 satisfy the hypotheses of Theorem 3.1.2, we may deduce the existence of a (unique) continuous solution w on some interval $[0, \delta_1] \subset I$, with $\delta_1 > 0$, and this implies that the solution u of the original VIE (3.1.1) has been *continued* continuously to the interval $[\delta_0, \delta_1]$.

The above continuation argument can obviously be repeated, to generate the continuation of the solution u to some interval $[0, \delta^*)$, where $\delta^* := \sum_{j=1}^{\infty} \delta_j$. If $\delta^* = \infty$, the non-linear VIE possesses a *global* solution on $[0, \infty)$; if $\delta^* < \infty$ the solution does not exist globally.

Definition 3.1.3 A continuous solution $u(t)$ of (3.1.1) that is defined on an interval $[0, \delta^*]$, where $\delta^* > 0$ is such that either $\delta^* = \infty$, or $\delta^* < \infty$ with $\lim_{t \to \delta^*} \sup |u(t)| = \infty$, is called a *maximally defined* solution. The interval $[0, \delta^*)$ is said to be the maximal interval for $u(t)$. $\qquad\qquad\diamond$

A closer look at the proof of Theorem 3.1.2 reveals that it can be adapted to cover non-linear VIEs with *weakly singular kernels*,

$$u(t) = f(t) + \int_0^t k_\alpha(t - s) k(t, s, u(s)) \, ds, \quad t \in I, \qquad (3.1.5)$$

with $k_\alpha(t - s)$ as in (1.3.3):

$$k_\alpha(t - s) := \begin{cases} (t - s)^{\alpha-1} & \text{if } 0 < \alpha < 1, \\ \log(t - s) & \text{if } \alpha = 0. \end{cases}$$

Since $\int_0^t |k_\alpha(t - s)|\, ds$ is finite for $t \leq T < \infty$ and $0 \leq \alpha < 1$, the set $\Omega = \Omega(\alpha)$ and the constant $M_B(\alpha)$ now depend on α via this integral. Hence this also holds for the existence interval $I_0 = I_0(\alpha) = [0, \delta_0(\alpha)]$. We leave the details of this proof to the reader (see also Exercise 3.7.1) but summarise, for the sake of completeness, the conclusion in the following analogue of Theorem 3.1.2.

Theorem 3.1.4 *Let the assumptions stated in Theorem 3.1.2 hold and set*

$$\delta_0(\alpha) := \min\{T, \ B/M_B(\alpha)\} \qquad and \qquad I_0(\alpha) := [0, \delta_0(\alpha)].$$

Then the VIE (3.1.4) possesses a unique solution $u \in C(I_0(\alpha))$ for any $\alpha \in [0, 1)$.

Remark 3.1.5 The Hammerstein version of the VIE (3.1.5) (corresponding to the kernel $(t - s)^{\alpha-1}G(s, u)$ $(0 < \alpha < 1)$) was studied in Dinghas (1958). Reinermann & Stallbohm (1971) used *Edelstein's fixed-point theorem* to analyse the general case; compare also Gorenflo & Vessella (1991, pp. 132–139). $\diamond$

The solution of a non-linear VIE cannot always be continued onto an arbitrarily large interval because it may develop a singularity as t approaches some finite value. This singularity may occur in the form of *finite-time blow-up* or *finite-time quenching* (that is, the derivative blows up in finite time). The following two examples illustrate this.

Example 3.1.6 *A VIE with a blow-up solution*
The VIE

$$u(t) = u_0 + \int_0^t \left(\lambda u(s) + \epsilon u^p(s)\right) ds, \ \ t \geq 0, \tag{3.1.6}$$

is equivalent to the initial-value problem

$$u'(t) = \lambda u(t) + \epsilon u^p(t), \ \ t \geq 0, \ \text{ with } \ u(0) = u_0.$$

The following result shows that when $\lambda \leq 0$, $\epsilon > 0$, $p > 1$ and $u_0 > 0$ the solution of (3.1.6) may *blow up* in *finite time*; that is, there exists a positive $T_b < \infty$ so that $\lim_{t \to T_b^-} u(t) = +\infty$. Since the above differential equation is a *Bernoulli* differential equation the solution of the VIE (3.1.6) can be found explicitly, and hence the proof of Theorem 3.1.7 is elementary.

Theorem 3.1.7 *Assume that* $\lambda \leq 0$, $\epsilon > 0$, $p > 1$ *and* $u_0 > 0$. *Then the (unique) continuous solution of (3.1.6) is formally given by*

$$u(t) = \left(u_0^{1-p} e^{-\lambda(p-1)t} - (\epsilon/\lambda)(1 - e^{-\lambda(p-1)t} \right)^{1/(1-p)}, \quad t \geq 0. \quad (3.1.7)$$

For given λ and p it blows up in finite time T_b if, and only if, the initial value u_0 is sufficiently large, namely,

$$u_0 > \left(-\lambda/\epsilon \right)^{1/(p-1)}.$$

The corresponding blow-up time is

$$T_b = \frac{1}{\lambda(p-1)} \ln \left(1 + \frac{\lambda}{\epsilon y_0^{p-1}} \right).$$

If $u_0 < \left(-\lambda/\epsilon \right)^{1/(p-1)}$, the solution exists for all $t > 0$.

We will describe the general theory of finite-time blow-up for Volterra–Hammerstein integral equations in Section 3.2.

Example 3.1.8 *A VIE with a quenching solution*
If the kernel $k(t, s, u)$ of the non-linear VIE (3.1.1) is singular for certain values of u, the continuous solution may exist on some finite closed interval $[0, T_q]$ but its *derivative* blows up as t tends to T_q^-; that is,

$$\sup\{|u(t)| : t \in [0, T_q)\} =: u_q < \infty \quad \text{and} \quad \lim_{t \to T_q^-} |u'(t)| = \infty.$$

Such a solution is said to *quench* at $t = T_q$, and T_q is called the *quenching time*. We shall discuss this behaviour in detail in Section 3.3. Here, we illustrate it by means of the VIE

$$u(t) = u_0 + \int_0^t \frac{1-s}{1-u(s)} \, ds, \quad t \geq 0, \quad u_0 \in (0, 1) \quad (3.1.8)$$

(cf. Yang & Brunner, 2014). This VIE is equivalent to the initial-value problem

$$u'(t) = (1-t)/(1-u(t)), \quad u(0) = u_0.$$

It is readily seen that the solution of (3.1.8) is formally given by

$$u(t) = 1 - \left((1-u_0)^2 + t^2 - 2t \right)^{1/2}, \quad t \geq 0.$$

Its derivative,

$$u'(t) = (t-1)\big((1-u_0)^2 + t^2 - 2t\big)^{-1/2},$$

becomes infinite as t tends to $T_q = 1 - \big(1 - (1-u_0)^2\big)^{1/2}$, while $u(t)$ has obviously a finite value at $t = T_q$. However, the solution ceases to exist for $t > T_q$.

We leave it to the reader to show that for $u_0 \leq 0$ the solution of the non-linear VIE (3.1.8) exists for all $t > 0$ (see also Example 3.3.1 for the case $u_0 = 0$). Section 3.3 will present the analysis of quenching in solutions to non-linear VIEs of Hammerstein type.

An alternative approach to establishing the local existence of solutions to non-linear VIEs is based on the *Carathéodory method* of successive approximation (described in Nohel, 1962; see also Miller, 1971, Ch. II) used originally to establish the existence of solutions for non-linear ordinary differential equations. For given fixed $\gamma > 0$ and each $n \geq 1$, the Carathéodory iterates for the non-linear VIE (3.1.1) are defined by

$$u_n(t) := \begin{cases} f(t) & \text{for } 0 \leq t \leq \gamma/n, \\ f(t) + \displaystyle\int_0^{t-\gamma/n} k(t-\gamma/n, s, u_n(s))\, ds & \text{for } \gamma/n \leq t \leq \gamma. \end{cases}$$

$$(3.1.9)$$

In contrast to the *Picard* iterate (3.1.3), the *Carathéodory* iterate $u_n(t)$ is, for each n, defined *recursively* on the interval $[0, \gamma]$. Note also that it is independent of $u_{n-1}(t)$ and previous iterates. Moreover, the sequence $\{u_n(t)\}$ of Carathéodory iterates (3.1.9) is uniformly bounded on $[0, \gamma]$, and it is equicontinuous at each $t \in [0, \gamma]$. It can also be shown, using the Ascoli Lemma (cf. Miller, 1971, pp. 36–41) that $\{u_n(t)\}$ contains a subsequence that converges to a continuous solution of the VIE (3.1.1). We summarise this in the following theorem (Miller, 1971, p. 36).

Theorem 3.1.9 *Assume that the given functions in (3.1.1) satisfy the conditions stated in Theorem 3.1.2. Then there exists a $\gamma > 0$ and a subsequence of the Carathéodory iterates $\{u_n(t)\}$ that converges uniformly on $[0, \gamma]$ to a solution $u \in C[0, \gamma]$. This solution is unique on $[0, \gamma]$.*

Remark 3.1.10 See Exercise 3.7.21 for a typical application of Carathédory iteration to a non-linear VIE in which the kernel $k(t, s, u)$ does not satisfy a Lipschitz condition with respect to u. ◇

3.1.2 VIEs of Hammerstein Type

Definition 3.1.11 A non-linear Volterra integral operator $\mathcal{H}_\alpha : C(I) \to C(I)$ of the form

$$(\mathcal{H}_\alpha u)(t) := \int_0^t k_\alpha(t - s)K(t, s)G(s, u(s)) \, ds, \quad t \in I, \qquad (3.1.10)$$

with $k_\alpha(t - s)$ as in (3.1.5), is called a *Volterra–Hammerstein operator*. The non-linearity G is usually a smooth function (but see also Section 3.4.3), and the kernel $K_\alpha(t, s) := (t - s)^{\alpha-1} K(t, s)$ $(0 < \alpha \le 1)$ may be weakly singular. Accordingly, we will refer to the corresponding second-kind VIE

$$u(t) = f(t) + (\mathcal{H}_\alpha u)(t), \quad t \in I, \qquad (3.1.11)$$

as a *Volterra–Hammerstein* integral equation (VHIE) of the second kind. ◇

The analogue of Theorem 3.1.2 for the VHIE (3.1.11) with $\alpha = 1$ is stated in the following theorem; its proof is readily obtained by adapting the proof of Theorem 3.1.2. A similar result is true for (3.1.11) with weakly singular kernel singularities $(t - s)^{\alpha-1}$ $(0 < \alpha < 1)$ or $\log(t - s)$ (cf. Exercise 3.7.2).

Theorem 3.1.12 *Assume that $K \in C(D)$ and G is continuous on $I \times \Omega_B$ (where Ω_B now denotes the set $\{(s, u) : |u - f(s)| \le B\}$). If in addition G satisfies the (local) Lipschitz condition with respect to u,*

$$|G(s, u) - G(s, v)| \le L_B|u - v| \text{ for all } (s, u), (s, v) \in \Omega_B,$$

then the VHIE (3.1.11) possesses a unique solution $u \in C(I_0)$, where $I_0 := [0, \delta_0]$ and $\delta_0 := \min\{T, B/M_B\}$, with $M_B := \max\{|G(s, u)| : (s, u) \in \Omega_B\}$.

It is often advantageous (for example, in the numerical analysis of VHIEs; see Brunner, 1991) to rewrite the VHIE (3.1.11) in an implicit form which contains a *linear* Volterra integral operator. It is based on the Nemytzkiĭ operator (frequently transliterated as *Niemytzki operator*; *Mathematical Reviews* uses both versions), or *substitution* operator, $\mathcal{N} : C(I) \to C(I)$ defined by

$$(\mathcal{N}u)(t) := G(t, u(t)), \quad t \in I.$$

Setting $z(t) := (\mathcal{N}u)(t)$, the Hammerstein integral operator $\mathcal{H}_\alpha$ is the composition $\mathcal{V}_\alpha\mathcal{N} := \mathcal{V}_\alpha \circ \mathcal{N}$ of the linear Volterra integral operator

$$(\mathcal{V}_\alpha u)(t) := \int_0^t K_\alpha(t,s)u(s)\,ds \quad (0 < \alpha \le 1)$$

with kernel $K_\alpha(t,s) := (t-s)^{\alpha-1}K(t,s)$, and the Nemytziĭ operator $\mathcal{N}$. Hence the VHIE (3.1.11) can be written as

$$u(t) = f(t) + (\mathcal{V}_\alpha\mathcal{N}u)(t), \quad t \in I. \tag{3.1.12}$$

It can be transformed into an *implicitly linear* integral equation for z,

$$z(t) = G\big(t, f(t) + (\mathcal{V}_\alpha z)(t)\big) = \mathcal{N}\big(f + \mathcal{V}_\alpha z\big)(t), \quad t \in I. \tag{3.1.13}$$

If (3.1.13) has a unique solution $z \in C(I)$, then the (unique) solution $u \in C(I)$ of the original VIE (3.1.12) is given by

$$u(t) = f(t) + (\mathcal{V}_\alpha z)(t), \quad t \in I.$$

It is shown in Krasnosel'skiĭ & Zabreĭko (1984, pp. 143–146) that, under suitable assumptions on the (smooth) non-linearity G, there is a one-to-one correspondence between (continuous) solutions of (3.1.13) and (3.1.12); hence, if (3.1.13) possesses a unique solution $z \in C(I)$ then (3.1.12) has a unique solution $u \in C(I)$, and vice versa (compare also Corduneanu (1991, pp. 86–87, p. 153). General results on the existence of solutions to Hammerstein integral equations can, for example, be found in Dolph & Minty (1964) and in Brezis & Browder (1975).

Remark 3.1.13 Note that while the Volterra integral operator $\mathcal{V}_\alpha : C(I) \to C(I)$ $(0 \le \alpha \le 1)$ with continuous kernel $K(t,s)$ is *compact*, this is not true for the Nemytzkiĭ operator $\mathcal{N} : C(I) \to C(I)$ (unless G does not depend on u). However, the compositions $\mathcal{V}_\alpha\mathcal{N}$ and $\mathcal{N}\mathcal{V}_\alpha$ are *compact* operators on $C(I)$ (cf. Appendix, Theorem A.2.16(f)). ◇

3.1.3 Maximal Solutions and a Comparison Theorem

If the (scalar) non-linear VIE

$$u(t) = f(t) + \int_0^t k(t,s,u(s))\,ds, \quad t \ge 0, \tag{3.1.14}$$

does not possess a unique continuous local solution (for example, if k does not satisfy assumption (iii) in Theorem 3.1.2), it may have a maximal and a minimal solution on some interval $[0, \delta]$. The following definition (cf. Miller, 1971, pp. 116–120), Gripenberg, Londen & Staffans, 1990, Section 13.4, or Corduneanu, 1991, pp. 99–102) makes the meaning of these concepts precise.

Definition 3.1.14 Suppose that there exists a continuous solution $\bar{u}$ of the VIE (3.1.14) on an interval $[0, \delta]$ such that any other continuous solution $u(t)$ of (3.1.14) satisfies $u(t) \leq \bar{u}(t)$, as long as $u(t)$ exists and $t \in [0, \delta]$. Then $\bar{u}(t)$ is called a *maximal solution* of (3.1.14) on $[0, \delta]$.

Analogously, if $\underline{u}(t)$ is a continuous solution so that for any other continuous solution of (3.1.14) we have $u(t) \geq \underline{u}(t)$, as long as $u(t)$ exists and $t \in [0, \delta)$, then $\underline{u}(t)$ is called a *minimal solution* of (3.1.14) on $[0.\delta]$.   ◇

It is well known that scalar ordinary differential equations always have maximum and minimum solutions. This is *not* true for non-linear VIEs of the second kind with continuous $f = f(t)$ and $k = k(t, s, u)$: while there is always at least one (local) continuous solution (Nohel, 1962), a maximal solution may not exist, as the following example reveals. The VIE

$$u(t) = -\int_{-1}^{t} t\big(u(s)\big)^{1/2} ds, \ t \in [-1, 1],$$

where $a^{1/2} := |a|^{1/2}\mathrm{sign}\, a$ for any $a \in \mathbb{R}$, has the one-parameter family of (continuous) solutions

$$u(t; \gamma) := \begin{cases} -\dfrac{t}{4}\Big(\displaystyle\int_{-\gamma}^{t} s^{1/2}ds\Big)^{2} & \text{if } t \in [-\gamma, \gamma], \\ 0 & \text{otherwise}, \end{cases}$$

with $0 \leq \gamma \leq 1$ (Gollwitzer & Hager, 1970). Let $u_1(t) := u(t; 1)$ and $u_2(t) := -u_1(t)$. Then any (continuous) solution $u(t)$ satisfies

$$u_2(t) \leq u(t) \leq u_1(t) \text{ if } t \in [-1, 0] \quad \text{and}$$
$$u_1(t) \leq u(t) \leq u_2(t) \text{ if } t \in [0, 1].$$

However, this VIE possesses no maximal solution (see also the geometrical illustration in Gollwitzer & Hager, 1970, p. 302, and Exercise 3.7.5 at the end of this chapter).

The following two theorems describe sufficient conditions for the existence of a maximal solution (Miller, 1971 or Gripenberg, Londen & Staffans, 1990). We leave it as an exercise to show that if (3.1.14) possesses a maximal solution $\bar{u}(t)$ [minimal solution $\underline{u}(t)$] then $\bar{u}(t)$ [$\underline{u}(t)$] must be unique.

Theorem 3.1.15 *Assume that* $f = f(t)$ *and* $k = k(t, s, u)$ *are continuous for* $0 \leq s \leq t < \infty$ *and* $\{(t, s) : 0 \leq s \leq t < \infty\} \times \mathbb{R}$, *respectively, and* k *is non-decreasing in* u *for each fixed pair* (t, s) *(that is,* $u \leq v$ *implies* $k(t, s, u) \leq k(t, s, v)$ *for each pair* (t, s)*). Then the VIE (3.1.14) possesses a (non-continuable) maximal solution* $\bar{u}(t)$ *and a (non-continuable) minimal solution on an interval* $[0, \delta]$.

Proof   The proof is based on introducing the sequence of VIEs

$$u_n(t) = f(t) + \phi_n + \int_0^t k(t, s, u_n(s)) \, ds, \quad t \geq 0 \ (n \in \mathbb{N}),$$

with $\phi_n > 0$ and $\lim_{n \to \infty} \phi_n = 0$. Assuming that this VIE has a continuous solution u_n on $[0, \delta]$, one then shows that $u_n(t) \geq u(t)$ for any solution of (3.1.14), using the assumption that $k(t, s, u)$ is non-decreasing in u. The details of the proof can be found in Miller (1971, pp. 116–122) or in Gripenberg, Londen & Staffans (1990, pp. 403–408).   □

The conclusions in the above theorem remain true for non-linear VIEs (3.1.14) with more general kernel functions $k(t, s, u)$ (e.g. functions containing integrable singularities). Theorem 3.1.7 is due to Miller (1971, where it is stated for systems of non-linear VIEs; cf. pp. 116–120 for its proof); compare also the related Exercise 3.7.6 below.

Theorem 3.1.16 *Assume that* f *and* k *in (3.1.14) have the following properties:*

(i)   *f is continuous and bounded on* $\mathbb{R}^+$;

(ii) *$k = k(t, s, u)$ is continuous in* u *for each fixed* (t, s) *with* $0 \leq s \leq t < \infty$, *and* $k(t, s, u) = 0$ *whenever* $s > t$;

(iii) *For each* $\kappa > 0$ *and each bounded subset* J *of* $\mathbb{R}$ *there exists an integrable function* $h = h(t, s) \geq 0$ *for which*

$$|k(t, s, u)| \leq h(t, s) \quad (0 \leq s \leq t \leq \kappa, \ u \in J)$$

and

$$\sup \left\{ \int_0^t h(t, s) \, ds : 0 \leq t \leq \kappa \right\} < \infty$$

hold.

(iv) *For each compact subinterval* $[a, b]$ *of* $\mathbb{R}^+$, *each bounded set* J *of* $\mathbb{R}$, *and each* $t_0 \in \mathbb{R}^+$ *there holds*

$$\lim_{t \to t_0} \left(\sup \left\{ \int_a^b |k(t, s, z(s)) - k(t_0, s, z(s))| ds : z \in C[a, b] \right\} \right) = 0.$$

In addition, suppose that k is such that for any given $\kappa > 0$ and any bounded interval J of $\mathbb{R}$,

$$\lim_{\varepsilon \to 0} \int_t^{t+\varepsilon} |k(t + \varepsilon, s, z(s)| \, ds = 0$$

is true uniformly in (t, z) for $0 \le t \le \kappa$ and $z \in C(0, \kappa+1]$, with $z(t) \in J$. Then the assertion in Theorem 3.1.15 remains true.

The concept of a maximal solution is often a useful tool in the analysis of non-linear VIEs. The following comparison theorem shows such an application; we will see a further one in the proof of Theorem 3.1.20.

Consider the non-linear VIE (3.1.14) and the related VIE

$$U(t) = F(t) + \int_0^t K(t, s, U(s)) \, ds, \quad t \ge 0, \qquad (3.1.15)$$

where the majorising functions F and K satisfy

$$F(t) \ge |f(t)| \ (t \ge 0) \quad \text{and}$$
$$K(t, s, |u|) \ge |k(t, s, u)| \ (0 \le s \le t < \infty, \ u \in \mathbb{R}). \qquad (3.1.16)$$

We will refer to the VIE (3.1.15) as the *comparison VIE* associated with the VIE (3.1.14).

The following comparison theorem (which can be found, for example, in Miller (1971, pp. 120–122) or in Corduneanu (1991, pp. 107–108) shows that, under certain (continuity) assumptions on f, k, F and G, we have $|u(t)| \le U(t)$ on some interval $[0, \bar{\delta}^*)$.

Theorem 3.1.17 *Let the assumptions of Theorem 3.1.15 hold for f, k and F, K. Suppose in addition that the maximal solution $\bar{U}(t)$ of the comparison VIE (3.1.15) exists on the maximal interval $[0, \delta^*)$. Then any continuous solution $u(t)$ of the VIE (3.1.14) can be continued to the right as long as $\bar{U}(t)$ exists, and it satisfies $|u(t)| \le \bar{U}(t)$ for $t \in [0, \delta^*)$.*

Proof Assume that $u(t)$ is defined on $[0, \delta_0^*)$ for some $\delta_0^* < \delta^*$, and choose an $\varepsilon_0 > 0$ such that for $0 < \varepsilon \le \varepsilon_0$, any solution $U_\varepsilon(t)$ of

$$U_\varepsilon(t) = F(t) + \varepsilon + \int_0^t k(t, s, U_\varepsilon(s)) \, ds$$

exists on the entire interval $[0, \delta_0^*]$. It then follows from $|u(0)| = |f(0)| < F(0) + \varepsilon = U_\varepsilon(0)$ and the continuity assumptions that $|u(t)| < U_\varepsilon(t)$ on some interval $[0, t_0)$. If $t_0 \leq \delta_0^*$ and if we assume that $|u(t_0)| = U_\varepsilon(t_0)$, we obtain

$$
\begin{aligned}
0 &= U_\varepsilon(t_0) - |u(t_0)| = F(t_0) + \varepsilon + \int_0^{t_0} k(t_0, s, U_\varepsilon(s))\, ds \\
&\quad - \left| f(t_0) + \int_0^{t_0} k(t_0, s, u(s))\, ds \right| \\
&\geq \varepsilon + \left(F(t_0) - |f(t_0)| + \int_0^{t_0} \big(k(t_0, s, U_\varepsilon(s)) \right. \\
&\quad \left. - |k(t_0, s, u(s))| \big)\, ds \right) \geq \varepsilon > 0.
\end{aligned}
$$

This contradiction implies that we must have $|u(t)| < U_\varepsilon(t)$ for $t \in [0, \delta_0^*]$. The proof is completed by recalling the continuity assumptions on the given functions and the definition of the maximal solution $\bar{U}(t)$ (cf. Theorem 3.1.15). $\qquad\square$

3.1.4 VIEs with Multiple Solutions

We have seen in Section 1.2 that a linear homogeneous VIE

$$
u(t) = \int_0^t (t - s)^{\alpha - 1} K(t, s) u(s)\, ds, \quad t \geq 0,
$$

with $\alpha > 0$ and $K \in C(D)$ can have only the trivial solution $u(t) \equiv 0$. This is in general no longer true for a non-linear homogeneous VIE like the Hammerstein equation

$$
u(t) = \int_0^t k(t - s) G(u(s))\, ds, \quad t \geq 0, \tag{3.1.17}
$$

with continuous G satisfying $G(0) = 0$: in addition to the trivial solution $u(t) \equiv 0$ there may exist a continuous *non-trivial* solution.

Example 3.1.18 It is readily verified (see also Bushell & Okrasiński, 1990) that for $\alpha > 0$ and $0 < p < 1$, the VIE

$$
u(t) = \int_0^t (t - s)^{\alpha - 1} u^p(s)\, ds, \quad t \geq 0 \tag{3.1.18}
$$

(which corresponds to $k(z) = z^{\alpha - 1}$ and $G(u) = u^p$ in (3.1.17)) possesses, in addition to the trivial solution $u(t) \equiv 0$, the non-trivial solution

$$
u(t) = \big(B(\alpha, 1 + \alpha/(1 - p)) \big)^{1/(1-p)} t^{\alpha/(1-p)}, \quad t \geq 0,
$$

where $B(\cdot, \cdot)$ is the Euler beta function.

It was shown by Gripenberg (1990) that for certain functions G and convolution kernels k the VIE (3.1.17) possesses only the trivial solution, while there exist other kernels for which there is also a (continuous) non-trivial solution. This is made precise in the following theorem where the given real-valued functions k and G describing the VIE (3.1.17) belong respectively to the sets

$$\mathcal{K}^+ := \{k \in L^1(\mathbb{R}^+) : k(z) \geq 0, \ \int_0^t k(z)\,dz > 0 \text{ for } t > 0\}$$

and

$$\mathcal{G} := \{G \in C(\mathbb{R}) : \ G(u) = 0 \text{ for } u \leq 0\}.$$

Theorem 3.1.19 *For every kernel $k \in \mathcal{K}^+$ there exists a $G \in \mathcal{G}$ so that the trivial solution of the VIE (3.1.17) is unique, and for every non-linearity $G \in \mathcal{G}$ there exists a $k \in \mathcal{K}^+$ so that the same statement is true.*

Conversely, for every $k \in \mathcal{K}^+$ there exists a $G \in \mathcal{G}$ so that (3.1.17) possesses a continuous solution u on some interval $[0, T_0]$ $(T_0 > 0)$ so that $u(t) > 0$ when $t \in (0, T_0]$, and for every $G \in \mathcal{G}$ there exists a $k \in \mathcal{K}^+$ for which the same statement holds.

Instead of providing a proof for the above result (we refer the reader to Gripenberg, 1990) we will study in some detail the existence of continuous non-trivial solutions of a particular class of VIEs that arise as mathematical models in numerous applications (see Section 9.2). It corresponds to (3.1.17) with kernel $k(t - s) = (t - s)^{\alpha-1}$,

$$u(t) = \int_0^t (t - s)^{\alpha-1} G(u(s))\,ds, \ t \geq 0, \tag{3.1.19}$$

where $\alpha > 0$. This is described in the next theorem (due to Mydlarczyk, 1991, but see also Gripenberg, 1981a and Okrasiński, 1991b): it illustrates the subtle interplay between the non-linearity $G(u)$ and the convolution kernel in the VIE (3.1.19). Related results can be found in Gripenberg (1990), Bushell & Okrasiński (1990), Okrasiński (1990) and Mydlarczyk (1994).

Theorem 3.1.20 *Let $G : \mathbb{R} \to \mathbb{R}$ be a continuous, non-decreasing function with the property that $G(0) =$ and $G(u) > 0$ for $u > 0$. Then for any $\alpha > 0$ the VIE (3.1.19) possesses a continuous non-trivial solution if, and only if,*

$$\int_0^\delta \frac{du}{u\big(H(u)\big)^{1/\alpha}} < \infty \ (\delta > 0), \tag{3.1.20}$$

with $H(u) := G(u)/u$, holds.

Proof Since the proof is rather technical, we will outline the principal steps that lead to the above result. Thus, following Mydlarczyk (1991), assume that in (3.1.17) we have $k : \mathbb{R}^+ \to \mathbb{R}^+$, with k locally integrable on $(0, \infty)$ (i.e. $k \in L^1_{loc}(\mathbb{R}^+)$), $K(t) := \int_0^t k(z)dz > 0$ for all $t > 0$, and G satisfies the above assumptions. For an arbitrarily given number $M > 0$ we can construct a strictly increasing function $\Phi : [0, M] \to \mathbb{R}^+$ such that $\Phi(t) \leq t/G(t)$ for $t \in (0, M]$, with $\Phi(0) := 0$. It is then easy to show that for any non-negative continuous solution $u(t)$ of (3.1.17) there holds $\phi(t) := \Phi^{-1}(K(t))$ ($t \in [0, \delta]$), where $\delta > 0$ is such that $K(\delta) < \Phi(M)$.

We now define, for a continuous function $w : \mathbb{R}^+ \to \mathbb{R}^+$, the operator $\mathcal{K} : C[0, \delta] \to C[0, \delta]$ by

$$(\mathcal{K}w)(t) := \int_0^t k(t - s)G(w(s))\, ds, \ \ t \in [0, \delta].$$

The sequence $\{\phi_n(t)\}_{n \geq 0}$ generated by setting $\phi_0(t) := \phi(t)$, $\phi_n(t) := (\mathcal{K}\phi_{n-1})(t)$ ($n \geq 1$) converges uniformly (on $[0, \delta]$) to the maximal solution $\bar{u}(t)$ of (3.1.17). The key step now consists in verifying that (3.1.17) possesses a non-trivial (non-negative) solution if, and only if, we can find a non-trivial continuous function Ψ for which $\Psi(t) \leq (\mathcal{K}\Psi)(t)$ for all $t \in [0, \delta]$. In the case of (3.1.19) (where we have $k(z) = z^{\alpha-1}$) the role of Ψ is taken by the inverse function of

$$P(t) := c \int_0^t s^{-1}\big(s/G(s)\big)^{1/\alpha}ds, \ \ t \in (0, \delta),$$

where the constant c can be chosen as $c = \big(2^{-1/\alpha} - 4^{-1/\alpha}\big)^{-1}\alpha^{1/\alpha-1}$ (cf. Mydlarczyk, 1991, p. 88). □

Example 3.1.21 For any $0 < p < 1$ the function $G(u) = u^p$ clearly satisfies the assumptions in Theorem 3.1.9. It follows from $H(u) = G(u)/u = u^{p-1}$ that for $\delta > 0$ the integral in (3.1.20) has the value

$$\int_0^\delta \frac{du}{u\big(H(u)\big)^{1/\alpha}} = \int_0^\delta \frac{du}{u^{1+(p-1)/\alpha}} = \frac{1}{(1-p)/\alpha}\delta^{(1-p)/\alpha},$$

which is finite for $\delta > 0$. Thus we conclude that (3.1.19) with $G(u) = u^p$ ($0 < p < 1$) has a continuous non-trivial solution for all $\alpha > 0$. This confirms the insight gained in Example 3.1.18, but without having to resort to the explicit solution of (3.1.18).

Remark 3.1.22 What happens if in Example 3.1.21 we have $G(u) = u^p$ with $p > 1$? We shall see in Section 3.2.3 that in addition to the trivial solution there may exist a continuous (positive) non-trivial solution $u(t)$, but not for all $t > 0$: this solution *blows up* in finite time (Mydlarczyk, 1994, 1999). ⋄

3.1.5 Regularity Results

We have seen in Section 2.1 (cf. Theorem 2.1.2) that a weak singularity in a linear VIE with otherwise smooth data f and K induces a singularity in the derivative of the solution u at $t = 0$. The same is true for analogous non-linear VIEs. This was shown in Miller & Feldstein (1971) for VIEs of Hammerstein form with weakly singular convolution kernels. The first regularity result for general non-linear VIEs of the form

$$u(t) = f(t) + \int_0^t (t-s)^{\alpha-1} k(t, s, u(s)) \, ds, \quad t \in I \ (0 < \alpha < 1), \quad (3.1.21)$$

is due to Lubich (1983a, pp. 88–90). It reveals that for real-valued functions f and k that are smooth (or even analytic) on their respective domains, the corresponding solutions do not have these properties at $t = 0$: their expansions near $t = 0$ contain mixed powers of t and t^α, in analogy to (2.1.8) in Theorem 2.1.2 for the linear version of (3.1.21).

Theorem 3.1.23 *Assume that $f = f(t)$ and $k = k(t, s, u)$ are real and analytic at $t = 0$ and $(t, s, u) = (0, 0, 0)$, respectively. Then there exists a $\delta_0 > 0$ so that the solution u of (3.1.21) is real and analytic in $(0, \delta_0)$. Moreover, there exists a function $U = U(v, w)$ that is real and analytic at $(0, 0)$ and is such that $u(t) = U(t, t^\alpha)$.*

While Theorem 3.1.23 is concerned only with VIEs whose kernels contain weak singularities of algebraic type, the next theorem (Brunner, Pedas & Vainikko, 1999) presents a general regularity result for non-linear VIEs

$$u(t) = f(t) + \int_0^t k(t, s, u(s)) \, ds, \quad t \in I, \quad (3.1.22)$$

whose kernel functions either contain very general weak singularities (including logarithmic ones) or are bounded but have *unbounded derivatives* as $s \to t$. As in the analogous Theorem 2.1.7 we use the notation $D^- := \{(t, s) : 0 \le s < t \le T\}$.

Theorem 3.1.24 *Assume:*
(K) The kernel function $k = k(t, s, u)$ in (3.1.22) is $m \ge 1$ times continuously differentiable with respect to t, s and u on $D^- \times \mathbb{R}$, and there exists a real number $v \in (-\infty, 1]$ so that, for $(t, s) \in D_-$ and for

non-negative integers i, j, ℓ *with* $i + j + \ell \leq m$, *the following estimates hold:*

$$\left| \left(\frac{\partial}{\partial t} \right)^i \left(\frac{\partial}{\partial t} + \frac{\partial}{\partial s} \right)^j \left(\frac{\partial}{\partial u} \right)^\ell k(t, s, u) \right|$$

$$\leq C_k(|u|) \begin{cases} 1 & \text{if } v + i < 0, \\ 1 + |\log(t - s)| & \text{if } v + i = 0, \\ (t - s)^{-v-i} & \text{if } v + i > 0, \end{cases}$$

and

$$\left| \left(\frac{\partial}{\partial t} \right)^i \left(\frac{\partial}{\partial t} + \frac{\partial}{\partial s} \right)^j \left(\frac{\partial}{\partial u} \right)^\ell k(t, s, u) \right.$$

$$\left. - \left(\frac{\partial}{\partial t} \right)^i \left(\frac{\partial}{\partial t} + \frac{\partial}{\partial s} \right)^j \left(\frac{\partial}{\partial u} \right)^\ell k(t, s, u) \right|$$

$$\leq C_f(\max\{|u_1|, |u_2|\})|u_1 - u_2| \begin{cases} 1 & \text{if } v + i < 0, \\ 1 + |\log(t - s)| & \text{if } v + i = 0, \\ (t - s)^{-v-i} & \text{if } v + i > 0, \end{cases}$$

with functions b_i : $[0, \infty) \rightarrow [0, \infty)$ $(i = 1, 2)$ *that are monotone increasing.*

(F) $f \in C^{m,v}(0, T]$; *that is,* $f \in C^m(0, T]$ *and the estimate*

$$|f^{(\ell)}(t)| \leq C \begin{cases} 1 & \text{if } \quad \ell < 1 - v, \\ 1 + |\log(t)| & \text{if } \quad \ell = 1 - v, \\ t^{1-v-\ell} & \text{if } \quad \ell > 1 - v \end{cases}$$

$(\ell = 0, 1, \ldots, m)$ *holds for* $t \in (0, T]$.
(V) The VIE (3.1.22) possesses a unique solution $u \in L^\infty(0, T)$.

Then the solution u *of the non-linear VIE (3.1.22) lies in the space* $C^{m,v}(I)$.

Proof For kernels described in Example 3.1.25(a) the regularity of the solution of (3.1.22) was first analysed in Lubich (1983), assuming that f and k are analytic functions. For more general kernels described in assumption (K) (which include the kernels in Example 3.1.25(b)) the proof is based on extending the kernel function $k(t, s, u)$ in the VIE (3.1.22) to (t, s) with $t \leq s \leq T$, using a reflection technique of Brunner, Pedas & Vainikko (1999). The regularity of the resulting non-linear Fredholm integral equations can then be studied by resorting to Vainikko (1993;

see also Pedas & Vainikko, 2006 and its references to related earlier work). □

Example 3.1.25

(a) If $k(t, s, u) = (t - s)^{-\beta} h(t, s, u)$ ($\beta < 1$) for some smooth h, the assumption (K) holds with $\nu = \beta$.
(b) If $k(t, s, u) = \log(t - s) h(t, s, u)$ with some smooth h, the assumption (K) is satisfied for $\nu = 0$.

Remark 3.1.26

(i) The assumptions (K) and (F) guarantee the existence and uniqueness of a continuous solution to (3.1.22) on some (possibly small) interval $[0, \delta] \subset I$. Instead of assuming a global Lipschitz condition for $k(t, s, u)$, we here use the assumption (V) on the existence of a unique continuous solution on I.
(ii) The assumption (K) says that $k(t, s, u)$ may possess a weak (integrable) singularity as $s \to t$ (if $i = j = \ell = 0$ and $0 < \nu \le 1$). If $\nu < 0$, then $k(t, s, u)$ is *bounded*, for fixed u, on $D_- \times \mathbb{R}$ but its derivatives may have a singularity at $s = t$.   ◇

3.2 Solutions with Finite-Time Blow-Up

3.2.1 Introduction

Example 3.1.6 and Theorem 3.1.7 showed that the solution of a non-linear VIE like

$$u(t) = f(t) + \int_0^t k(t - s) G(s, u(s)) \, ds, \quad t \ge 0, \qquad (3.2.1)$$

may exist globally (that is, for all $t \ge 0$), or exist only on some *finite* interval $[0, T_b)$ and become unbounded (*blows up*) as t approaches some finite (blow-up) time $T_b > 0$.

Definition 3.2.1 A continuous solution u of the non-linear VIE (3.2.1) is said to *blow up in finite time* if there exists a positive $T_b < \infty$ so that $u(t)$ is defined and continuous on $[0, T_b)$, but

$$\lim_{t \to T_b^-} |u(t)| = \infty. \qquad (3.2.2)$$

We will refer to T_b as the *blow-up time* of u.   ◇

In this section we will consider the following problems:

- For which (continuous or weakly singular) convolution kernels $k(t - s)$ and non-linearities $G(s, u)$ does the solution of (3.2.1) blow up in finite time?
- If the non-linearity is of the form $G(s, u) = r(s)u^p$, where r is a positive and non-decreasing function, does the solution of (3.2.1) blow up in finite time for all $p > 1$?
- Since the blow-up time T_b can in general not be found explicitly, is it possible to derive realistic lower and upper bounds for T_b? Also, how does the solution behave near the blow-up time T_b (that is, what is its *blow-up profile*)?

In Example 3.1.6 we studied a VIE of the form (3.2.1) whose kernel $k(t - s)$ did not depend on t, and we were able to determine the solution of the VIE, and thus its blow-up time, explicitly. The next example shows that while even for still very simple non-constant but linear convolution kernels the blow-up analysis becomes much more complex, the blow-up time can still be found explicitly.

Example 3.2.2 Consider the VIE

$$u(t) = u_0 + \lambda \int_0^t (t - s)u^p(s)\, ds, \quad t \geq 0, \tag{3.2.3}$$

with $u_0 > 0$, $\lambda > 0$ and $p > 1$. Since (3.2.3) is equivalent to the initial-value problem

$$u''(t) = \lambda u^p(t), \quad u(0) = u_0, \quad u'(0) = 0,$$

it is readily seen (see also Miller, 1971, pp. 46–48) that $u'(t)$ satisfies

$$u'(t) = \left(\frac{2\lambda}{p+1}\right)^{1/2} \left(u^{p+1}(t) - u_0^{p+1}\right)^{-1/2}, \quad t \geq 0.$$

This initial-value problem can be solved by resorting to separation of variables, and we thus find that the solution $u(t)$ of (3.2.3) is defined implicitly by the equation

$$\left(\frac{2\lambda}{p+1}\right)^{1/2} t = \int_{u_0}^{u(t)} \left(v^{p+1} - u_0^{p+1}\right)^{-1/2} dv$$

$$= u_0^{-(p-1)/2} \int_1^{u(t)/u_0} \left(s^{p+1} - 1\right)^{-1/2} ds, \quad t > 0.$$

This reveals that $u(t)$ must tend to infinity as t approaches the *finite* value $T_b > 0$ given by the solution of the equation

$$\left(\frac{2\lambda}{2+\beta}\right)^{1/2} T_b = u_0^{-1/2} \int_1^\infty \left(s^{2+\beta} - 1\right)^{-1/2} ds.$$

Thus, the *blow-up time* T_b is

$$T_b = \left(\frac{p+1}{2\lambda}\right)^{1/2} u_0^{-(p-1)/2} \int_1^\infty \left(s^{p+1} - 1\right)^{-1/2} ds.$$

Remark 3.2.3 We observe that under the stated assumptions on λ and p the solution of (3.2.3) blows up for any initial value $u_0 > 0$. ◇

Before presenting the blow-up analysis for general Hammerstein-type VIEs in Section 3.2.2 we will look at a class of VIEs that arise in applications (see Example 3.2.7 below) and whose special structure of the non-linearity $G(s, u)$ permits the derivation of explicit lower and upper bounds for the blow-up time. This kernel non-linearity is of the form

$$G(s, u) = r(s)F\big(u + h(s)\big), \tag{3.2.4}$$

where the given functions r, F, h are assumed to be smooth on their respective domains and satisfy

$$r(t) > 0, \quad r'(t) \geq 0 \quad \text{for } t \geq 0; \quad F^{(j)}(z) > 0 \ (j = 0, 1) \text{ for } z > 0;$$
$$\tag{3.2.5}$$

and

$$0 < h_0 \leq h(z) \leq h_\infty < \infty, \quad h'(z) \geq 0 \text{ for } z \geq 0. \tag{3.2.6}$$

The convolution kernel k is subject to the hypotheses

$$k(z) \geq 0 \quad \text{and} \quad k'(z) < 0 \quad (z > 0) \tag{3.2.7}$$

(which are typical for *memory kernels* reflecting the diffusive nature of the underlying physical problems).

In Theorem 3.2.4 (due to Roberts, Lasseigne & Olmstead, 1993) the following quantities will play important roles. Denoting by M^* the smallest root of the equation

$$M/F(M + h_\infty) = 1/F'(M + h_\infty), \tag{3.2.8}$$

we set

$$\kappa := \int_{h_0}^\infty \frac{dz}{F(z)} < \infty \ (h_0 > 0) \tag{3.2.9}$$

and

$$J(t) := \int_0^t k(t-s)r(s)\,ds, \quad t \geq 0. \tag{3.2.10}$$

Observe that the values of M^* and κ do not depend on the kernel $k(t-s)$. Moreover, the inverse function $J^{-1}(t)$ is well defined because of the assumptions (3.2.5) and (3.2.7) on r and k.

Theorem 3.2.4 *Assume that the given functions in the VIE (3.2.1) are subject to the assumptions (3.2.5)–(3.2.7), and that the values $M^*/F(M^*+h_\infty)$ and κ lie in the range of $J(t)$. Then the VIE (3.2.1) with kernel non-linearity (3.2.4) possesses a unique continuous, positive and strictly increasing solution $u(t)$ that ceases to exist at some $T_b \in [t^*, t^{**}]$, where*

$$t^* := J^{-1}\big(M^*/F(M^*+h_\infty)\big) > 0 \quad and \quad t^{**} := J^{-1}(\kappa). \tag{3.2.11}$$

Proof We will first show that a unique continuous solution $u(t) \geq 0$ exists for all $t \in [0, t^*)$ for some $t^* > 0$ and, if the solution continues to exist beyond this interval, there is a $T_b \geq t^*$ so that it can no longer exist at $t \geq T_b$.

Lemma 3.2.5 *Under the assumptions (3.2.4)–(3.2.7), any continuous solution of (3.2.1) is positive and strictly increasing for all $t > 0$ for which it exists.*

It is straightforward to verify that, under the assumptions (3.2.4)–(3.2.7), while the solution $u(t)$ satisfies $u(0) = 0$, it is positive on the remaining part of its interval of existence, $(0, \delta)$. In order to show that it is strictly increasing on this interval, we differentiate the VIE (3.2.1) and then use integration by parts for evaluating the resulting integral term. This leads to the equation

$$u'(t) = k(t)r(0)F\big(u(0) + h(0)\big) + \int_0^t k(t-s)\Big(r'(s)F\big(u(s) + h(s)\big)$$
$$+ r(s)F'\big(u(s) + h(s)\big)\big(u'(s) + h'(s)\big)\Big)ds.$$

Suppose that there exists a point $t = t_0 \in (0, \delta)$ for which $u'(t_0) = 0$. This yields an immediate contradiction in the above equation: while its left-hand side is zero, the assumptions (3.2.5)–(3.2.7) and $u(0) = 0$ imply that the integral on the right-hand side is positive. $\qquad\square$

Lemma 3.2.6 *Assume that (3.2.5)–(3.2.7) hold and let M^* be be the smallest (positive) solution of the equation (3.2.8). Then there exists a unique continuous solution of (3.2.1) which satisfies*

$$0 \leq u(t) \leq M \quad (0 \leq t < t^*) \quad for\ any\ M < M^*.$$

To establish this result, let $\mathcal{H} : C(I) \to C(I)$ be the Volterra–Hammerstein integral operator given by

$$(\mathcal{H}u)(t) := \int_0^t k(t-s)r(s)F\big(u(s)+h(s)\big)ds,$$

and assume that u is a continuous solution such that $0 \leq u(t) \leq M$ ($0 \leq t \leq t_0$) for some $t_0 > 0$. We will show that u is unique for $M > 0$ if $M < M^*$, where M^* is the smallest real number satisfying the equation $M^*\big(F(M^* + h_\infty)\big)^{-1} = \big(F'(M^* + h_\infty)\big)^{-1}$. It follows from the definition and from (3.2.5)–(3.2.7) that $(\mathcal{H}u)(t) \leq F(M + h_\infty)J(t)$, $t \in [t, t_0]$. Hence, if $J(t) \leq M\big(F(M+h_\infty)\big)^{-1}$, the operator $\mathcal{H}$ maps the set of continuous functions u with $0 \leq u(t) \leq M$ ($t \in [0, t_0]$) into itself. Moreover, $\mathcal{H}$ is contractive: since for any continuous u_1 and u_2 in this set of functions we have

$$|(\mathcal{H}u_1)(t) - (\mathcal{H}u_2)(t)|$$
$$\leq \int_0^t k(t-s)r(s)ds \cdot \big(F'(M + h_\infty)\big) \sup\big\{|u_1(t) - u_2(t)| : t \in [0, t_0]\big\}$$
$$= J(t)F'(M + h_\infty) \sup\big\{|u_1(t) - u_2(t)| : t \in [0, t_0]\big\},$$

we obtain a contraction mapping whenever $J(t) < \big(F'(M + h_\infty)\big)^{-1}$ for $t \in [0, t_0]$. Observing that $J(t)$ is strictly increasing for $t \geq 0$, we see that the largest possible value $J(t^*)$ of $J(t)$ is obtained for the smallest M^* that satisfies

$$J(t^*) = M^*\big(F(M^* + h_\infty)\big)^{-1} = \big(F'(M^* + h_\infty)\big)^{-1}.$$

The existence of a unique continuous solution on $[0, t^*]$ then follows.   □

The first part of assertion (3.2.11) in Theorem 3.2.4 now follows if $J(t^*)$ is in the range of J. The final step in the proof of Theorem 3.2.4 consists in verifying that there exists a t^{**} with $t^* \leq t^{**} < \infty$ such that the solution of (3.2.1) ceases to exist at this point. Thus, supposing that there is a continuous solution u on $[0, t_1]$ for some $t_1 > 0$, we have

$$u(t) = (\mathcal{H}u)(t) \geq J_1(t) := \int_0^t k(t_1 - s)r(s)F\big(u(s)+h(s)\big)\,ds, \ t \in [0, t_1].$$

Since the integrand in $J_1(t)$ does not depend on t, and F is strictly increasing, the derivative of $J_1(t)$ satisfies

$$J_1'(t) = k(t_1 - t)r)t)F\big(u(t)+h(t)\big) \geq k_1(t_1-t)r(t)F\big((J_1(t)+h_0\big).$$

Using separation of variables in integrating the above differential inequality we obtain

$$\int_{h_0}^{J_1(t_1)+h_0} \frac{dz}{F(z} \geq \int_0^{t_1} k(t_1-s)r(s)\,ds = J(t_1).$$

It now follows from the continuity of u on $[0, t_1]$ that $J_1(t_1) < \infty$, and hence there must hold $J(t_1) < \int_0^\infty dz/F(z) = \kappa$. The hypothesis that the value of $J(t_1)$ can be increased to $J(t^{**})$ for some $t^{**} > t_1$ with $J(t^{**}) = \kappa$ leads to a contradiction. □

The inverse function J^{-1} is well defined because J is strictly increasing. Thus, the above insights can be combined to yield

$$0 < J^{-1}\big(M^*/F(M^* + h_\infty)\big) = t^* \leq T_b \leq t^{**} = J^{-1}(\kappa) < \infty,$$

as asserted in (3.2.11). □

Example 3.2.7 The study of the blow-up behaviour of solutions to (3.2.1) with non-linearity (3.2.4) was motivated by the VIE

$$u(t) = \frac{\gamma}{\Gamma(\alpha)} \int_0^t (t-s)^{\alpha-1}(1+s)^q [u(s)+1]^p\,ds, \quad t \geq 0, \qquad (3.2.12)$$

where $\gamma > 0$, $0 < \alpha < 1$, $q \geq 0$ and $p > 1$ denote certain material parameters. The case $\alpha = 1/2$ arises in the mathematical modelling of shear band formation in steel when subjected to very high strain rates (see Example 9.2.11 in Section 9.2.4 for details and references).

Roberts, Lasseigne & Olmstead, 1993, pp. 539–540 established upper and lower bounds for the blow-up time of the solution of (3.2.12) in the case where $\alpha = 1/2$. The following theorem generalises their bounds to any $\alpha \in (0, 1)$. See also Mydlarczyk (1991, Theorem 2.5) for similar results when $h(t) \equiv 0$ in (3.2.4).

Theorem 3.2.8 *For every $0 < \alpha < 1$ the solution of the VIE (3.2.12) always blows up in finite time, and the blow-up time T_b is bounded by $t^* \leq T_b \leq t^{**}$, with*

$$t^* := \left(\frac{\alpha\Gamma(\alpha)}{\gamma}\right)^{1/\alpha} (p-1)^{-1/\alpha}(1-p^{-1})^{-p/\alpha},$$

$$t^{**} := \left(\frac{\alpha\Gamma(\alpha)}{\gamma}\right)^{1/\alpha} (p-1)^{-1/\alpha}.$$

Proof It is easy to verify that for the given functions $k(t-s) = \gamma/\Gamma(\alpha)(t-s)^{\alpha-1}$, $r(s) = 1$, $F(z) = z^p$ and $h(s) = 1$ in (3.2.12) we obtain, using (3.2.8)–(3.2.10),

$$M^* = \frac{1}{p-1}, \quad \kappa = \frac{1}{p-1}, \quad J(t) = \frac{\gamma}{\alpha\Gamma(\alpha)}t^\alpha, \quad \text{and}$$

$$J^{-1}(t) = \left(\frac{\alpha\Gamma(\alpha)}{\gamma}\right)^{1/\alpha} t^{1/\alpha}.$$

The asserted lower and upper bounds t^* and t^{**} for the blow-up time T_b then follow from the last inequality in the proof of Theorem 3.2.1. □

Example 3.2.9 For $\alpha = 1/2$ (with $\Gamma(1/2) = \pi^{1/2}$) we find the bounds

$$t^* := \frac{\pi}{4\gamma^2}(p-1)^{-2}(1-p^{-1})^{2p}, \quad t^{**} := \frac{\pi}{4\gamma^2}(p-1)^{-2},$$

as originally shown in Roberts, Lasseigne & Olmstead (1993).

Remark 3.2.10 The results of Roberts, Lasseigne & Olmstead (1993) were extended by Mydlarczyk (1999) to VIEs (3.2.1) in which the kernel linearity (3.2.4) is such that $h(t) \equiv 0$. He established lower and upper bounds, implicit in the blow-up time T_b, for the maximal solutions of such VIEs with $k(z) = z^{\alpha-1}$ ($0 < \alpha < 1$ or $\alpha > 1$); cf. Theorems 2.5 and 2.6 in that paper. ◇

3.2.2 Blow-Up Theory for General Hammerstein VIEs

We will now study the blow-up theory for the more general Volterra–Hammerstein integral equation

$$u(t) = f(t) + \int_0^t k(t-s)G(s, u(s))\,ds, \quad t \ge 0, \tag{3.2.13}$$

under the assumption that the given functions are subject to the following hypotheses:

(F) f is continuous, positive and non-decreasing on $[0, \infty)$;

(K) $k \in L^1_{loc}(\mathbb{R}^+)$, and $K(z) := \int_0^z k(s)\,ds > 0$ is a non-decreasing function;

(G1) $G(s, 0) \equiv 0$ for $s \ge 0$, and $G(s_1, u_1) > G(s_0, u_0)$ for all $(s_i, u_i) > (0, 0)$ with $s_1 \ge s_0$ and $u_1 > u_0$;

(G2) $\displaystyle\lim_{u \to \infty} \frac{G(0, u)}{u} = \infty$.

In this section we will first study the global existence of a solution of (3.2.13) and then derive necessary and sufficient conditions for finite-time blow-up, following Brunner & Yang (2013). In the subsequent analysis the functions

$$F(t, u) := \int_0^t k(t - s)G(s, u)\, ds - u \quad (t \in (0, \infty),\ u \in \mathbb{R}^+)$$

and

$$F_{min}(t) := \inf\left\{ F(t, u) : u \in \mathbb{R}^+ \right\} \quad (t \in (0, \infty)) \qquad (3.2.14)$$

will be needed.

We first summarise some basic properties of the solution of (3.2.13) and the above functions.

Lemma 3.2.11

(a) *If (F), (K) and (G1) hold, the solution u of (3.2.13) is non-decreasing, with $u(t) > f(0)$ for all t in its interval of existence.*

(b) *If (K), (G1) and (G2) hold, then $F_{min}(t)$ exists, and F_{min} is non-positive and non-decreasing, with $\lim_{t \to 0^+} F(t) = -\infty$.*   □

The proof is left as an exercise.

Next we show that if the solution exists globally it will, under certain conditions, become infinite as $t \to \infty$.

Theorem 3.2.12 *Assume (F), (K), (G1) and (G2), and suppose that the solution of (3.2.13) exists for all $t \in [0, \infty)$. If there is a $\bar{t} \in (0, \infty)$ for which $f(\bar{t}) + F_{min}(\bar{t}) > 0$, then $\lim_{t \to \infty} u(t) = \infty$.*

Proof Suppose the contrary, namely that $\lim_{t \to \infty} u(t) = u_\infty < \infty$ exists. It then follows from Lemma 3.2.11(a) that for any $\varepsilon > 0$ with $f(\bar{t}) - \varepsilon + F_{min}(\bar{t}) > 0$, there is a $T_\varepsilon > \bar{t}$ such that $u_\infty - \varepsilon + F_{min}(\bar{t}) > 0$ whenever $t > T_\varepsilon$. Hence, using the assumptions (K) and (G1) we find that for such t there holds

$$u_\infty > u(t) = f(t) + \int_0^t k(t - s)G(s, u(s))\, ds$$

$$\geq f(t) + \int_0^{T_\varepsilon} k(t - s)G(s, u(s))\, ds + \int_{T_\varepsilon}^t k(t - s)G(s, u(s))\, ds$$

$$\geq f(t) + \int_0^{t - T_\varepsilon} k(s)G(t - T_\varepsilon - s, u_\infty - \varepsilon)\, ds.$$

On the other hand, it follows from Lemma 3.2.11(b) that $f(t) - \varepsilon + F(t - t_\varepsilon, u_\infty - \varepsilon) < 0$. This contradicts the hypothesis that $f(\bar{t}) - \varepsilon + F_{min}(\bar{t}) > 0$.   □

The following corollary is an obvious consequence of Theorem 3.2.12.

Corollary 3.2.13 *Suppose that (F), (K), (G1) and (G2) are true and that the solution of (3.2.13) exists on* $[0, \infty)$. *Then either* $\lim_{t \to \infty} u(t) = \infty$ *if* $\lim_{t \to \infty} f(t) = \infty$, *or*

$$\lim_{t \to \infty} \int_0^t k(t - s) G(s, u) \, ds = \infty \quad \text{for all } u \in (0, \infty). \tag{3.2.15}$$

Proof For $t > 0$, let $\bar{u}(t) := \sup \{ u : F(t, u) \le 0 \}$. Condition (G2) then implies that $\bar{u}(t) \in [0, \infty)$. We claim that, by (3.2.15), $\liminf \{ \bar{u}(t) : t \in (0, \infty) \} = 0$. If not, then there exists a $u_L > 0$ and a sequence $\{t_n\}$ with $\lim_{n \to \infty} t_n = \infty$ so that for all n, $u_L \le \bar{u}(t_n) \le 2u_L$. But it follows from (G1) that

$$1 = \left(\bar{u}(t_n) \right)^{-1} \int_0^{t_n} k(t_n - s) G\big(s, \bar{u}(t_n)\big) ds \ge (2u_L)^{-1} \int_0^{t_n} k(t_n - s) G(s, u_L) \, ds,$$

which contradicts (3.2.15). The proof is now complete since, by (3.2.14), $F_{min}(t) \le -\bar{u}(t)$. $\qquad \square$

The next theorem describes sufficient conditions for finite-time blow-up.

Theorem 3.2.14 *Let (F), (K), (G1) and (G2) hold, and assume in addition that*

(i) $k(z) = z^{\alpha - 1} k_1(z)$, *with* $\alpha > 0$ *and* $k_1(z) \ge 0$ *such that there exists a* $\delta > 0$ *for which*

$$\kappa_1 := \inf \{ k_1(z) : z \in [0, \delta] \} > 0,$$

and

(ii) *there is a* $\bar{t} > 0$ *such that* $f(\bar{t}) + F_{min}(\bar{t}) > 0$ *and*

$$\int_{u_0}^\infty \frac{du}{u \left(\bar{H}(u) \right)^{1/\alpha}} < \infty \quad \text{for } u_0 > 0, \tag{3.2.16}$$

with $\bar{H}(u) := G(\bar{t}, u)/u$.

Then the solution of (3.2.13) blows up in finite time T_b.

Proof Suppose that (F), (K), (G1) and (G2) hold and that the solution u of (3.2.13) exists for all $t \in [0, \infty)$. It is then not difficult to see that if there is a $\bar{t} > 0$ for which the first part of assumption (ii) is true, we can find, for any given $\rho > 1$, a non-negative, strictly increasing sequence $\{t_n\}$ such that $\lim_{n \to \infty} t_n = \infty$, $\lim_{n \to \infty} (t_{n+1} - t_n) = 0$, and $u(t_n) = \rho^n$ for all sufficiently large n (depending on $f(0)$ and ρ). If the kernel is given by

$k(z) - z^{\alpha-1}k_1(z)$ $(\alpha > 0)$, we can choose $\rho = r^\alpha$ for some $r > 1$. Hence, for any $\delta > 0$ there is an integer $N = N(\delta)$ so that $h_n := t_{n+1} - t_n \le \delta$ and $t_n \ge \bar{t}$ for all $n \ge N$. Using (F), (K) and (G1) we are then led to the estimate

$$u(t_{n+1}) = r^{(n+1)\alpha} = f(t_{n+1}) + \int_0^{t_{n+1}} k(t_{n+1} - s)G(s, u(s))\, ds$$

$$\ge G(t_n, u(t_n))K(t_{n+1} - t_n) \ge G(\bar{t}, u(t_n))K(t_{n+1} - t_n).$$

Owing to the definition of κ_1, there follows

$$\frac{\kappa_1}{\alpha} h_n^\alpha \le \frac{r^{(n+1)\alpha}}{G(\bar{t}, r^{n\alpha})} \quad (n \ge N),$$

and therefore we obtain

$$t_{n+1} - t_n = h_n \le \left(\frac{\alpha}{\kappa_1}\right)^{1/\alpha} \frac{r^{n+1}}{\left(G(\bar{t}, r^{n\alpha})\right)^{1/\alpha}} \le \left(\frac{\alpha}{\kappa_1}\right)^{1/\alpha} \frac{r^{n+1} - r^n}{\left(G(\bar{t}, r^{n\alpha})\right)^{1/\alpha}}$$

$$\le \frac{r^2}{r-1}\left(\frac{\alpha}{\kappa_1}\right)^{1/\alpha} \int_{r^{n-1}}^{r^n} \frac{ds}{\left(G(\bar{t}, s^\alpha)\right)^{1/\alpha}}$$

$$= \frac{r^2}{\alpha(r-1)}\left(\alpha/\kappa_1\right)^{1/\alpha} \int_{u(t_{n-1})}^{u(t_n)} \frac{du}{\left(\bar{H}(u)\right)^{1/\alpha}} < \infty \quad (n \ge N)$$

(by assumption (3.2.16)). Since $t_{n+1} = t_n + h_n < \infty$ for all $n \ge N$, this contradicts the above statement that $\lim_{n \to \infty} t_n = \infty$. $\square$

The solution of (3.2.13) does not always blow up in finite time: it may exist globally on $[0, \infty)$. This is made precise in the following corollary (whose proof is left as an exercise).

Corollary 3.2.15 *Assume (F), (K), (G1) and (G2). If for all $t \in \mathbb{R}^+$ we have*

$$F_{min}(t) \le 0 \quad and \quad f(t) + F_{min}(t) \le 0, \tag{3.2.17}$$

the solution u of (3.2.13) exists globally on $(0, \infty)$.

We now turn to describing necessary conditions for finite-time blow-up: it will be seen that the condition (3.2.16) is also necessary for finite-time blow-up of the solution of (3.2.13) (Brunner & Yang, 2013).

Theorem 3.2.16 *Suppose that, in addition to (F), (K), (G1) and (G2), there exists a $\bar{t} \in (0, \infty)$ such that $f(\bar{t}) + F_{min}(\bar{t}) > 0$. If the convolution kernel*

$k(z)$ *is of the form* $k(z) = z^{\alpha-1}$ *with* $\alpha > 0$ *(that is, if* $k_1(z) \equiv 1$*) and if there is a* $u_0 > 0$ *for which*

$$\int_{u_0}^{\infty} \frac{du}{u\left(\bar{H}(u)\right)^{1/\alpha}} = \infty, \qquad (3.2.18)$$

then the solution of (3.1.13) does not blow up in finite time: it exists for all $t \in (0, \infty)$.

Proof Since many of the ideas used in the proof are very similar to the ones employed in the proof of Theorem 3.2.14, we will only give a broad outline of the proof of Theorem 3.2.16 and leave the technical details to the reader (they can also be found in Brunner & Yang, 2013, pp. 497–499).

Starting with the hypothesis that there exists a finite $T_b > 0$ such that $\lim_{t \to T_b^-} u(t) = \infty$, we choose an increasing sequence of points $t_n > 0$ with $t_n \to T_b$ as $n \to$, with the property that $u(t_n) = r^{n\alpha}$ for

$$r := \begin{cases} \left(2(1 + f(T_b))\right)^{1/\alpha} & \text{if } 0 < \alpha < 1 \\ \left(2(2^{\alpha} + f(T_b))\right)^{1/\alpha} & \text{if } \alpha \geq 1. \end{cases}$$

Thus, we have $h_n := t_{n+1} - t_n \to 0$, as $n \to \infty$, and so there is an integer $N > 0$ for which $h_n < \min\{1, t_n\}$ for all $n \geq N$. It can then be verified that

$$u(t_{n+1}) \leq \left(2^{\alpha} + 2\alpha^{-1}\right)h_n^{\alpha} G\left(T_b, u(t_{n+1})\right),$$

both for $0 < \alpha < 1$ and for $\alpha \geq 1$. The next step, similar to the one used in the proof of Theorem 3.2.14, consists in showing that

$$h_n \geq \frac{1}{r-1}C_{\alpha}^{-1/\alpha} \int_{r^{n+1}}^{r^{n+2}} \frac{ds}{\left(G(T_b, s^{\alpha})\right)^{1/\alpha}},$$

where $C_{\alpha} := 2^{\alpha} + 2\alpha^{-1}$. Recalling the assumption (3.2.18), we see that $t_{n+1} = t_n + h_n \to \infty$ as $n \to \infty$. This is a contradiction to the hypothesis that $T_b < \infty$. $\qquad\square$

The proofs of Theorems 3.2.14 and 3.2.16 suggest necessary and sufficient conditions under which the solution of (3.2.13) with convolution kernel $k(z) = z^{\alpha-1}k_1(z)$ ($\alpha > 1$) blows up in finite time (Brunner & Yang, 2013, pp. 499–500). We leave the details of the proof of the next theorem as an exercise.

Theorem 3.2.17 *Let the following assumptions hold:*

(i) The given functions in (3.2.13) satisfy (F), (K), (G1) and (G2);
(ii) the kernel $k(z) = z^{\alpha-1}k_1(z)$ ($\alpha > 0$) is subject to the hypothesis (i) in Theorem 3.2.14, and k_1 is bounded on any finite interval.

Then the solution of (3.2.13) blows up in finite time if, and only if, there exists a $\bar{t} \in (0, \infty)$ so that

$$f(\bar{t}) + F_{min}(\bar{t}) > 0$$

(where $F_{min}(t)$ is as in (3.2.14)) and

$$\int_{u_0}^{\infty} \frac{du}{u\big(\hat{H}(u)\big)^{1/\alpha}} < \infty \text{ for all } u_0 > 0$$

are true.

Remark 3.2.18 We shall see in Section 3.6.2 that analogous blow-up results are true, under very similar conditions as (F), (K), (G1) and (G2), for VFIEs with delay functions $\theta(t) = t - \tau(t)$ where $\tau(t) \geq 0$.   ◇

When the solution of a differential equation or a VIE blows up in finite time, one would like to know not only good estimates for the blow-up time T_b but also how the solution behaves as t approaches T_b^-; that is, the *blow-up profile* near T_b. Of the few existing results we mention the one in Roberts & Olmstead (1996) for the VIE (3.2.13) whith $k(z) = z^{\alpha-1}$ ($0 < \alpha < 1$) and non-linearity $G(s, u) = r(s)F(u + h(s))$. They showed that if $F(u) = u^p$ ($p > 1$), the solution u behaves like

$$u(t) \sim \gamma\big(T_b - t\big)^{-\alpha/(p-1)}, \quad \text{as} \quad t \to T_b^-,$$

with the constant $\gamma > 0$ depending on α, p, T_b, and r.

Mydlarczyk (1999) proved similar, but slightly more general results (see, for example, Theorem 2.6 in his paper). However, the derivation of the blow-up profile for the VIE (3.2.13) with general convolution kernels $k(z)$ and general non-linearities $G(s, u)$ satisfying (K), (G1) and (G2) remains open.

3.2.3 Multiple Solutions – Revisited

In Section 3.1.4 (cf. Theorem 3.1.19 and Example 3.1.21) we studied the existence of a non-trivial solution of the homogeneous Hammerstein VIE

$$u(t) = \int_0^t (t - s)^{\alpha-1} u^p(s) \, ds, \quad t > 0, \tag{3.2.19}$$

with $\alpha > 0$ and $0 < p < 1$. The analysis led to the question as to whether a non-trivial (positive) solution also exists when $p > 1$ and how such a solution behaves. The following two theorems (Mydlarczyk, 1991, 1994, 1999) are concerned with the more general VIE

$$u(t) = \int_0^t (t-s)^{\alpha-1} r(s) G(u(s)) \, ds, \ t > 0 \ (\alpha > 0) \qquad (3.2.20)$$

(which corresponds to (3.2.1) with $h(s) \equiv 0$ in the non-linearity (3.2.4)). For r and G satisfying $r(0) = G(0) = 0$ we obviously have the trivial solution $u(t) \equiv 0$. While the following theorem furnishes an answer to the above question, Theorem 3.2.13 gives a necessary and sufficient condition for the interval of existence of the (non-trivial) maximal solution of (3.2.20) to be finite.

Theorem 3.2.19 *Assume that*

(i) r is continuous and non-decreasing for $t \geq 0$, with $r(0) = 0$ and $r(t) > 0$ for $t > 0$;

(ii) G is continuous and non-decreasing for $u \geq 0$, with $G(0) = 0$ and $G(u) > 0$ for $u > 0$.

Then the following statements are true:

(a) For $\alpha > 0$ the VIE (3.2.20) possesses a continuous maximal solution $u(t) \geq 0$ which is non-decreasing for $t \geq 0$.

(b) If this maximal solution is positive for $t > 0$, it is the only solution with this property.

(c) The maximal solution satisfies $u(t) > 0$ for $t > 0$ if, and only if,

$$\int_0^\delta \frac{du}{u \big(H(u) \big)^{1/\alpha}} < \infty \ (\delta > 0), \qquad (3.2.21)$$

where $H(u) := G(u)/u$.

Proof We leave the proof of the assertions (a) and (b) to the reader but sketch proof of (c). In order to establish the necessity part, asssume that $u(t) \not\equiv 0$ is a solution of (3.2.20). Then there exists a $\delta > 0$ so that, setting $\bar{r} := \sup\{r(s) : s \in [0, \delta]\}$, we have

$$u(t) \leq \bar{r} \int_0^t (t-s)^{\alpha-1} G(u(s)) \, ds, \ t \in [0, \delta].$$

It is readily seen that this inequality can hold for non-trivial $u(t) \geq 0$ if, and only if, the given $\alpha > 0$ and the non-linearity $G(u)$ are subject to the condition (3.2.21). The proof of the sufficiency part is based on the integral operator

$$(\mathcal{H}_\alpha u)(t) := \int_0^t (t-s)^{\alpha-1} r(s) G(u(s))\, ds, \ \ t \ge 0,$$

and the properties of the non-negative, non-decreasing functions $v(t)$ satisfying $v(t) \le (\mathcal{H}_\alpha v)(t)$ for $t \ge 0$. It can then be shown that the solution of (3.2.20) is obtained as the limit $u(t) = \lim_{n\to\infty} (\mathcal{H}_\alpha^n v)(t)$ when $t \in (0, \delta]$. The details may be found in Mydlarczyk, 1999, pp. 151–152). $\square$

The next theorem reveals that the condition (3.2.21) in that theorem is, under the assumptions stated in that theorem, necessary and sufficient for the maximal solution of (3.2.20) to blow up in finite time. Its proof can be found in Mydlarczyk (1999, pp. 152–154).

Theorem 3.2.20 *Let r and G be subject to the assumptions (i) and (ii) in Theorem 3.2.19. Then the maximal solution of (3.2.20) blows up in finite time if, and only if,*

$$\int_\delta^\infty \frac{du}{u\big(H(u)\big)^{1/\alpha}} < \infty \ \ for \ \delta > 0. \tag{3.2.22}$$

3.3 Quenching of Solutions

3.3.1 Quenching in Differential Equations

In Example 3.1.8 we encountered the initial-value problem

$$u'(t) = \frac{1-t}{1-u(t)}, \ \ t \ge 0; \ \ u(0) = u_0 \in [0, 1),$$

whose solution $u(t)$ quenches in finite time T_q; that is, there exists a $T_q < \infty$ so that

$$\lim_{t\to T_q^-} u(t) = 1 \ \ \text{and} \ \ \lim_{t\to T_q^-} u'(t) = \infty.$$

Since $u(t)$ could be found explicitly, we were able to derive the explicit expression

$$T_q = 1 - \big(1 - (1-u_0)^2\big)^{1/2}$$

for the quenching time T_q.

Example 3.3.1 The above problem is closely related to an initial-boundary-value problem for a semilinear parabolic differential equation which was used by Kawarada (1975) to introduce the notion of the quenching of solutions:

$$u_t = u_{xx} + f(u), \quad t > 0, \quad 0 < x < L,$$
$$u(t, 0) = u(t, L) = 0, \quad t > 0,$$
$$u(0, x) = u_0(x), \quad 0 \le x \le L,$$

with $f(u) = 1/(1 - u)$. This term becomes *singular* when $u = u(t, x)$ reaches the value $u = 1$. Kawarada proved that if $L > 2\sqrt{2}$ then there exists a finite time $T_q > 0$ so that

$$\lim_{t \to T_q^-} u(t, L/2) = 1 \quad \text{and} \quad \lim_{t \to T_q} u_t(t, L/2) = \infty :$$

the solution of the above problem *quenches* at T_q. It was later shown by other authors (cf. the references in Yang & Brunner, 2014) that there exists an $L_0 < 2\sqrt{2}$ so that the solution u cannot quench in finite time and hence exists for all $t > 0$.

For more general problems it will not be possible to obtain an explicit expression for the quenching time T_q, and hence we will now describe how the quenching time (or the existence of a global solution) can be determined without knowledge of the solution $u(t)$. This will be done for the initial-value problem

$$u'(t) = r(t)F(u(t)), \quad t \ge 0, \quad \text{with } u(0) = u_0 \in [0, b). \quad (3.3.1)$$

Here, r is assumed to be continuous and non-negative, and $b > 0$ is such that $F \in C^1[0, b)$ satisfies $F(u) \to \infty$ as $u \to b^-$ (that is, $F(u)$ has a singularity at $u = b$).

The initial-value problem (3.3.1) is equivalent to a VIE with *constant* forcing function $f(t) \equiv u_0$ and *constant* convolution kernel $k(t - s) = 1$:

$$u(t) = u_0 + \int_0^t r(s)F(u(s))\,ds, \quad t \ge 0. \quad (3.3.2)$$

In order to state our first quenching result (cf. Yang & Brunner, 2014) we introduce, in analogy to (3.2.9) and (3.2.10), the functions

$$H(u) := \int_0^u \frac{dz}{F(z)}\,dz, \quad u \ge 0, \quad (3.3.3)$$

$$J(t) := \int_0^t r(s)\,ds, \quad t \ge 0. \quad (3.3.4)$$

Theorem 3.3.2 *Assume that r and F are subject to the hypotheses stated above, with $F(u) > 0$ for $u \in [0, b)$. If $u_0 \in [0, b)$ the following statements hold:*

(a) *The solution of (3.3.2) exists for all $t \in [0, \infty)$ if*

$$J(t) < H(b) - H(u_0) \quad \text{for all} \quad t \geq 0.$$

(b) *The solution u of (3.3.2) quenches in finite time T_q if*

$$T_q := \sup\{t \geq 0: \ J(t) < H(b) - H(u_0)\} < \infty \quad \text{and} \quad r(T_q) \neq 0.$$

Proof It follows from (3.3.1) (by using separation of variables) that the solution $u(t)$ satisfies

$$H(u(t)) - H(u_0) = J(t), \quad t \geq 0. \tag{3.3.5}$$

Moreover, since by the assumptions on F the function $H(u)$ is well defined and is strictly increasing for $u \in [0, b)$, its inverse H^{-1} exists. Thus, using (3.3.3) it follows from (3.3.5) that

$$u(t) = H^{-1}(H(u_0) + J(t)).$$

This reveals that $u(t)$ exists whenever we have $J(t) < H(b) - H(u_0)$ for all $t \geq 0$. Otherwise,

$$\lim_{t \to T_q^-} u(t) = \lim_{t \to T_q^-} H^{-1}(H(u_0) + J(t)) = b,$$

provided that

$$T_q := \sup_{t \geq 0}\{J(t) < H(b) - H(u_0)\} < \infty.$$

If, in addition, $r(T_q) \neq 0$, then $\lim_{t \to T_q^-} u'(t) = \infty$. $\qquad\square$

Example 3.3.3 In the VIE introduced in Example 3.1.8,

$$u(t) = u_0 + \int_0^t r(s)F(u(s))\,ds, \quad t \geq 0,$$

we have $r(s) = 1 - s$ and $F(u) = (1-u)^{-1}$ (which is singular for $u = b = 1$). It follows that

$$J(t) = \frac{1}{2}(1 - (1-t)^2) \quad \text{and} \quad H(u) = \frac{1}{2}(1 - (1-u)^2),$$

and hence

$$H(b) - H(u_0) = (1/2)(1 - (1 - (1 - u_0)^2)) = (1/2)(1 - u_0)^2.$$

Using Theorem 3.3.2(b) we readily see that the solution $u(t)$ quenches in finite time T_q for any $u_0 \in (0, 1)$, with T_q given by the root of the equation

$$J(t) - (H(1) - H(u_0)) = \frac{1}{2}(1 - (1-t)^2 - (1 - u_0)^2).$$

This root is $T_q = 1 - \left(1 - (1 - u_0)^2\right)^{1/2}$, thus confirming the result found in Example 3.1.8. Since $r(T_q) = 1 - T_q = \left(1 - (1 - u_0)^2\right)^{1/2}$ we observe that $r(T_q) \neq 0$ when $u_0 \in (0, 1)$, but $r(T_q) = 0$ for $u_0 = 0$.

Example 3.3.4 (Yang & Brunner, 2014)
The VIE

$$u(t) = u_0 + a \int_0^t e^{\beta s} \left(1 - u(s)\right)^{-\gamma} ds, \quad t \geq 0, \tag{3.3.6}$$

is equivalent to the initial-value problem

$$u'(t) = ae^{\beta t} \left(1 - u(t)\right)^{-\gamma}, \quad u(0) = u_0.$$

We will assume that $u_0 \in [0, 1)$, $a > 0$, $\beta \in \mathbb{R}$ and $\gamma > 0$. Since the corresponding functions $G(u)$ and $J(t)$ are

$$H(u) = (1 + \gamma)^{-1}\left(1 - (1 - u)^{1+\gamma}\right) \quad \text{and} \quad J(t) = a\beta^{-1}\left(e^{\beta t} - 1\right),$$

we see that for $\beta \geq 0$ there holds $\lim_{t \to \infty} J(t) = \infty$. As we have $b = 1$ it follows that

$$H(b) - H(u_0) = \frac{1}{1 + \gamma}\left(1 - u_0\right)^{1+\gamma},$$

and thus, by Theorem 3.3.2(b), the solution of (3.3.6) must quench in finite time, namely at

$$T_q = \frac{1}{\beta} \ln \left(1 + \frac{\beta}{a(1 + \gamma)}(1 - u_0)^{1+\gamma}\right),$$

for any $u_0 \in [0, 1)$. Quenching still occurs when $\beta = 0$: the quenching time is now seen to be $T_q = (1 - u_0)^{1-\gamma}/(a(1 + \gamma))$. If $\beta < 0$ then $I(t) = (-a/\beta)(1 - e^{\beta t})$ is strictly increasing but remains bounded by $-a/\beta$. The solution of (3.3.6) can therefore quench only for sufficiently large $u_0 \in (0, 1)$; that is, for u_0 satisfying

$$a(1 + \gamma) + \beta(1 - u_0)^{1+\gamma} > 0.$$

For smaller u_0 the solution exists for all $t > 0$.

The first studies of quenching in VIEs were motivated by the analysis of solutions to certain initial-boundary-value problems for semilinear parabolic partial differential equations (cf. Kirk & Roberts, 2002, 2003, Roberts, 2007, and the list of references in Yang & Brunner, 2014).

The quenching of a solution can happen if the non-linear term in the PDE contains a singularity, or if the PDE is linear but is governed by certain non-linear boundary conditions that depend on the unknown solution (see the 1992

review paper by Levine and its list of references). As a representative example
we consider the linear diffusion equation with non-linear boundary conditions
(Levine, 1983), to reveal the connection with quenching of solutions to certain
VIEs (Kirk & Roberts, 2002).

Example 3.3.5 Let $v = v(t, x)$ be the solution of the initial-boundary-value
problem with Neumann boundary conditions,

$$
\begin{aligned}
v_t &= v_{xx}, \quad t > 0, \quad x \in (0, 1); \\
v(0, x) &= v_0(x) \in [0, 1), \quad x \in (0, 1); \\
v_x(t, 1) &= \lambda G(v(t, 1)), \quad t > 0; \\
v_x(t, 0) &= 0, \quad t > 0.
\end{aligned}
\tag{3.3.7}
$$

Here, $\lambda \in \mathbb{R}$ is a non-zero parameter. It can be shown that at the right boundary
$x = 1$, the solution $v(t, 1) =: u(t)$ satisfies a VIE of Hammerstein form,

$$
u(t) = f(t) + \lambda \int_0^t k(t - s) G(u(s)) \, ds, \quad t > 0,
\tag{3.3.8}
$$

where G is the non-linear function in the right boundary condition (3.3.7).
The function f and the convolution kernel k depend on the *Green's function*
$\Gamma_N(t, x; \tau, \xi)$ for the above Neumann problem.

3.3.2 Quenching in VIEs of Hammerstein Type

We now turn to the more general Hammerstein VIE

$$
u(t) = f(t) + \int_0^t k(t - s) r(s) F(u(s)) \, ds, \quad t \geq 0;
\tag{3.3.9}
$$

that is, to a VIE which is not reducible to initial-value problems for an ordinary
differential equation or which is not necessarily related to a non-linear initial-
boundary-value problem for a parabolic PDE (i.e. whose convolution kernels
k are not given in terms of some Green's function, as in Example 3.3.4).

In the context of quenching in VIEs we assume that the given data are such
that the solution $u(t)$ is continuously differentiable in its interval of existence.
If the kernel non-linearity $F(u)$ in the VIE (3.3.9) becomes singular for a certain value of $u = b$ ($b > 0$), the solution may be continuous and exist on some
interval $[0, T_q)$, have a finite limit as $t \to T_q^-$, but its derivative $u'(t)$ may blow
up at $t = T_q$. If this occurs, the solution of (3.3.9) is said to *quench* at T_q. We
make this now precise in the following formal definition.

Definition 3.3.6 The solution of the VIE (3.3.9) *quenches* in finite time $T_q > 0$, called the *quenching time*, if it exists in $[0, T_q)$, lies in $C^1[0, T_q)$ and satisfies

$$\lim_{t \to T_q} |u(t)| < \infty, \quad \text{with} \quad \lim_{t \to T_q^-} |u'(t)| = \infty. \qquad (3.3.10)$$

◇

We shall of course have to present conditions under which the solution $u(t)$ of (3.3.9) actually *attains* this singular valye b. This problem will be addressed in Theorem 3.3.14, Corollary 3.3.15, and Theorem 3.3.16.

In analogy to the assumptions stated at the beginning of Section 3.2.2 we assume that the given functions in (3.3.9) are subject to the following conditions:

(F) $f \in C^1(\mathbb{R}^+)$ is non-negative, with $f(0) \in [0, b)$;
(K) $k \in C^1(0, \infty)$ is locally integrable, and the integral $K(t) :=$ $\int_0^t k(s)\,ds$ is a strictly positive and non-decreasing function on $(0, \infty)$;
(R) $r \in C^1(\mathbb{R}^+)$ is non-negative and does not vanish identically;
(G) F is an increasing positive function on some interval $[0, b)$ and satisfies $\lim_{u \to b^-} F(u) = \infty$.

The following results on quenching and non-quenching of solutions to the VIE (3.3.9) are due to Yang & Brunner (2014). Before presenting a number of general results we will, in order to exhibit the basic ideas, first again consider VIEs that are equivalent to initial-value problems for ODEs (recall also Example 3.1.8); that is, we consider the VIE (3.3.9) with *constant* f and *constant* kernel k. We again require the functions we introduced for the quenching analysis of (3.3.2),

$$H(u) := \int_0^u \frac{dz}{F(z)}\,dz \quad \text{and} \quad J(t) := \int_0^t r(s)\,ds. \qquad (3.3.11)$$

Theorem 3.3.7 *Let $f(t) = u_0 \in [0, b)$ and $k(z) = \kappa > 0$ be constant, and assume that (R) and (G) hold. Then the following statements are true for the VIE (3.3.9) with the above data:*

(a) It possesses a global solution $u \in C^1[0, \infty)$ if

$$J(t) < H(b) - H(u_0) \quad \text{for all } t \geq 0, \qquad (3.3.12)$$

where $J(t)$ and $H(u)$ are defined in (3.3.11).

(b) The solution of (3.3.9) quenches in finite time if

$$T_q := \sup_{t \geq 0} \left\{ J(t) < H(b) - H(u_0) \right\} < \infty \quad \text{and} \quad r(T_q) \neq 0.$$

Proof If f and k in (3.2.9) are assumed to be constant, the corresponding VIE is equivalent to the initial-value problem

$$u'(t) = r(t)F(u(t)) \ (t \geq 0), \ u(0) = u_0.$$

Using separation of variables this leads to

$$\int_{u_0}^{u} \frac{dz}{F(z)} = \int_0^t r(s) \, ds = J(t),$$

and hence to $H(u(t)) - H(u_0) = J(t)$, $t \geq 0$, where $H(u)$ and $J(t)$ are defined in (3.3.11). Since $H(u)$ is continuous and strictly increasing in the interval $[0, b)$ it follows that the inverse H^{-1} is well defined, and hence we may write

$$u(t) = H^{-1}\big(H(u_0) + J(t)\big), \ t \geq 0.$$

This reveals that the solution of (3.2.9) exists whenever

$$H(u_0) + J(t) < H(b) \quad \text{or} \quad J(t) < H(b) - H(u_0) \quad \text{for all } t \geq 0.$$

Otherwise, there exists a $T_q := \sup \left\{ t \geq 0 : J(t) < H(b) - H(u) \right\}$ so that

$$\lim_{t \to T_q^-} u(t) = \lim_{t \to T_q^-} H^{-1}\big(H(u_0) + J(t)\big) = b.$$

Moreover, since the derivative of H^{-1} with respect to t contains the factor $J'(t) = r(t)$, it follows that $\lim_{t \to T_q^-} u(t) = \infty$ only if $r(T_q) \neq 0$. $\qquad\square$

Remark 3.3.8 In Theorem 3.3.7(b) we assumed that $r(T_q) \neq 0$. The following example shows that if $r(T_q) = 0$, the solution $u(t)$ of the VIE (3.3.9) can attain the value b in finite time T_q but does not quench at $t = T_q$. $\qquad\diamond$

Example 3.3.9 We saw in Example 3.1.8 that the solution of

$$u(t) = u_0 + \int_0^t \frac{1 - s}{1 - u(s)} \, ds, \ t \geq 0 \ (u(0) = u_0 \in [0, 1)),$$

attains the value $u(b) = u(1) = 1$ as $t \to T_q = 1$ and quenches for any $u_0 \in (0, 1)$. However, for $u_0 = 0$ we have

$$\lim_{t \to 1^-} u'(t) = \lim_{t \to 1^-} (t - s)[(1 - t)^2]^{1/2} = 1 < \infty :$$

the solution does not quench at $t = 1$.

We will now establish a number of results on the quenching of solutions of the Hammerstein VIE (3.3.9) with *non-constant f* and *non-constant* (and possibly weakly singular) convolution kernel k. We first extend Theorem 3.3.2 to the VIE (3.3.9) with a non-constant k: it gives a sufficient condition for its solution to quench in finite time.

Theorem 3.3.10 *Assume, in addition to (F), (K), (R) and (G), that $r(s) > 0$ for all $s > 0$, and f and the convolution kernel k are increasing functions, with $k(0) > 0$. Then the solution of the VIE (3.3.9) quenches in finite time if*

$$H(b) - H(f(0)) \leq k(0) \int_0^t r(s)\, ds \quad \text{for some } t > 0.$$

Here, $H(u)$ is defined in (3.3.11).

Proof Since f and k are continuously differentiable, it follows from (3.2.9) that

$$u'(t) = f'(t) + k(0)r(t)F(u(t)) + \int_0^t k'(t-s)r(s)F(u(s))\, ds.$$

Thus, it follows from the assumptions stated above that

$$u'(t) \geq k(0)r(t)F(u(t)), \quad t \geq 0.$$

The proof can then be concluded along the lines of the proof of Theorem 3.3.7. □

The next theorem presents sufficient conditions for quenching of the solution of (3.3.9) corresponding to a particular class of convolution kernels (which includes weakly singular kernels). Its proof is left to the reader (see also Yang & Brunner, 2014, Theorem 3.6).

Theorem 3.3.11 *Assume that the following assumptions hold in addition to the conditions (F), (R) and (G):*

(i) $u(t)$ attains the singular point b in finite time T_q, and $r(T_q) \neq 0$;
(ii) the convolution kernel k in (3.3.9) has the form $k(z) = z^{\alpha-1}k_1(z)$ ($\alpha > 1$) with non-decreasing and strictly positive $k_1 \in C^1(\mathbb{R})$.

Then the solution of (3.3.9) quenches in finite time T_q if there exists a constant $\gamma > 0$ such that

$$F(u) \geq \gamma(b-v)^{1-\alpha} \quad \text{for} \quad u \in (0, b].$$

If f is non-decreasing, then the above condition can be weakened to

$$\int_u^b (b-v)^{\alpha-2} F(v)\,dv = \infty \quad \text{for all} \quad u \in [0,b).$$

For such f this condition is also necessary for u to quench in finite time.

Assuming that u attains the singular point b of F, we obtain the following sufficient condition for finite-time quenching of the solution of (3.3.9) (Yang & Brunner, 2014, Theorem 3.3).

Theorem 3.3.12 *Let the hypotheses (F), (K), (R) and (G) hold, and assume that*

(i) the solution $u(t)$ of (3.3.9) attains the value b in finite time T_q;
(ii) r is non-decreasing and such that $r(T_q) \neq 0$.

Then $u(t)$ quenches at $t = T_q$ if

$$k_{inf} := \inf\{k(t) : z \in (0, z_0)\} > 0 \quad \text{for some} \quad z_0 > 0. \tag{3.3.13}$$

Proof We proceed along the lines of the proof of Theorem 3.3.10, except that now the value $k(0)$ of the kernel $k(z) = z^{\alpha-1}k_1(z)$ ($\alpha > 0$) may not be defined. Hence, instead of using the equation in the second line in the proof of Theorem 3.3.10, we employ as our starting point the analogous equation

$$u'(t) = f'(t) + k(t)r(0)G(u(0)) + \int_0^t k(t-s)r'(s)G(u(s))\,ds$$
$$+ \int_0^t k(t-s)r(s)G'(u(s))\,ds.$$

Assuming, without loss of generality, that $t_0 < T_q$ ($t_0 > 0$) and $r(T_q - t_0) > 0$, we see that $f'(t)$, $k(t)r(0)G(u(0))$ and

$$\int_0^{T_q-t_0} k(t-s)r'(s)G(u(s))\,ds$$

are bounded for $t \in (T_q - t_0, T_q)$, with

$$\int_0^t k(t-s)r'(s)G(u(s))\,ds \geq 0.$$

On the other hand, it is easy to show that

$$\int_{T_q-t_0}^t k(t-s)r(s)G'(u(s))\,ds \geq k_{inf}r(T_q - t_0)$$
$$\times \left(G(u(t)) - G(u(t_q - t_0))\right) \to \infty,$$

as $t \to T_q^+$. This contradiction completes the proof. $\qquad\square$

Remark 3.3.13 The condition (3.3.13) is satisfied for VIEs (3.3.9) with weakly singular kernels $k(z) = z^{\alpha-1}$ $(0 < \alpha < 1)$ and $k(z) = \log(z)$. In the former case we have, for $z_0 = 1$, $k_{inf} = 1 > 0$, and $K(z) = \int_0^t k(z)\,dz = t^\alpha/\alpha$ is non-decreasing. ◇

Under which conditions on the given functions in (3.3.9) does the solution $u(t)$ *attain* the value b at which the non-linearity $G(u)$ becomes singular? The following theorems (Yang & Brunner, 2014) provide answers to this question.

Theorem 3.3.14 *Assume that the functions describing the VIE (3.3.9) are subject to the conditions (F), (K), (R) and (G).*

(a) *If there exists a $\bar{t} > 0$ for which*

$$f(\bar{t}) + \int_0^{\bar{t}} k(\bar{t} - s)r(s)F(f(s))\,ds = b,$$

then the solution of (3.3.9) attains the singular value b of $F(u)$ in finite time.

(b) *If there is a $\bar{b} \in (0, b)$ such that*

$$f(t) + F(\bar{b})J(t) < \bar{b} \quad \text{for all} \quad t \ge 0,$$

then the solution of (3.3.9) exists globally, with $u(t) \le \bar{b}$ for $t \ge 0$.

Proof (a) Since $u(0) = f(0) < b$, assume that $u(t) \in [0, b)$ for $t \in [0, \bar{t}]$. It follows from (F), (K), (R) and (G) that this leads to a contradiction:

$$b > u(\bar{t}) = f(\bar{t}) + \int_0^{\bar{t}} k(\bar{t} - s)r(s)F(u(s))\,ds$$

$$\ge f(\bar{t}) + \int_0^{\bar{t}} k(\bar{t} - s)r(s)F(f(s))\,ds = b.$$

(b) The hypothesis that $t^* := \inf\{t > 0 : u(t) > \bar{b}\} < \infty$ leads again to a contradiction, because we obtain

$$\bar{b} = f(t^*) + \int_0^{t^*} k(t^* - s)r(s)F(u(s))\,ds \le f(t^*) + F(\bar{b})J(t^*) < \bar{b}.$$

□

Corollary 3.3.15 *Under the assumptions of Theorem 3.3.11 the solution of (3.3.9) attains the value b in finite time if $J(\bar{t})F(0) \ge b$ for some finite $\bar{t} > 0$.*

A sharper result can be obtained if both f and r are non-decreasing functions.

Theorem 3.3.16 *Assume that the assumptions stated in Theorem 3.3.14 hold. If f and r are non-decreasing functions on $\mathbb{R}^+$, then the solution $u(t)$ of (3.3.9) attains the singular value b of $F(u)$ if, and only if, there exists a $\bar{t} \in [0, \infty)$ such that $f(\bar{t}) + Q_{inf}(\bar{t}) > 0$. Here,*

$$Q_{inf}(t) := \inf\{Q(t, u) : u \in [0, b)\} \quad and \quad Q(t, u) := J(t)F(u) - u.$$

Proof We will prove the necessity part of Theorem 3.3.14 and leave the proof of the sufficiency part to the reader. The assumption that $f(t) + Q_{inf}(t) \leq 0$ for all $t \geq 0$ leads to $Q_{inf}(t) \leq 0$, and hence to

$$u_Q(t) := \sup\{u \in [0, b) : Q(t, u) = Q_{inf}(t)\} < b \quad \text{for all } t > 0.$$

Since $Q_{inf}(0) = -b$ and $u_Q(0) = b > u(0)$, suppose that

$$t := \inf\{s \geq 0 : u(s) > u_Q(s)\} < \infty,$$

where we define $\inf\{\emptyset\} := \infty$. It follows from the assumptions (F), (K), (R) and (G) that we arrive at a contradiction since we obtain

$$u(t) = f(t) + \int_0^t k(t - s)r(s)F(u(s))\,ds < f(t)$$
$$+ \int_0^t k(t - s)r(s)F(u(t))\,ds$$
$$\leq -Q_{inf}(t) + J(t)F(u_Q(t)) = u_Q(t).$$

Hence we have shown that $u(t) \leq u_Q(t) < b$ for all $t > 0$. $\qquad\square$

Remark 3.3.17 While the problem of establishing lower and upper bounds for the quenching time T_q (analogous to the ones for the blow-up time in Theorem 3.2.4) for the general Hammerstein VIE (3.3.9) is still open, some results exist for special cases of (3.3.9); see Kirk & Roberts (2002). ◇

3.4 Other Types of Non-Linear VIEs

The kernel function in a general standard non-linear VIE (3.1.1) is of the form $k(t, s, u(s))$. However, in certain mathematical modelling processes (see, for example, Section 9.2) the non-linearity will also depend on u at time t and hence has the form $k(t, s, u(t), u(s))$ or $k(t, s, u(t - s), u(s))$. In view of their importance in applications we will restrict our analysis to Hammerstein equations with kernel non-linearities $G(u(t), u(s))$ and $G(u(t-s))G(u(s))$. We will study these two cases in Sections 3.4.1 and 3.4.2, respectively.

3.4.1 Non-Standard Second-Kind VIEs

The existence and uniqueness of a solution for the non-linear VIE

$$u(t) = f(t) + \int_0^t k(t, s, u(t), u(s)) \, ds, \quad t \in I = [0, T],$$

was studied in Guan, Zhang & Zou (2012). We will refer to such a non-linear VIE as a *non-standard* VIE of the second kind. In this section we carry out the analysis of the existence of solutions for the non-standard Volterra–Hammerstein version,

$$u(t) = f(t) + \int_0^t K_\alpha(t, s) G(u(t), u(s)) \, ds, \quad t \in I, \qquad (3.4.1)$$

where $K_\alpha(t, s) := (t - s)^{\alpha-1} K(t, s)$ $(0 < \alpha \leq 1)$ and $K \in C(D)$; the non-linearity $G = G(u, v)$ is assumed to be a smooth (or at least locally Lipschitz continuous; cf. Definition 3.1.1) in both variables. Its basic version (which we write down for the sake of later comparison, in Section 3.4.2, with an auto-convolution VIE) has the non-linearity $G(u, v) = uv$; it is

$$u(t) = f(t) + \lambda \int_0^t (t - s)^{\alpha-1} u(t) u(s) \, ds, \quad t \in I \; (0 < \alpha \leq 1), \qquad (3.4.2)$$

where λ is a real constant.

The following theorem is the adaptation of a more general result in Guan, Zhang & Zou (2012) (cf. Theorem 2.2, where $\alpha = 1$) to the VIE (3.4.1) with $\alpha = 1$. Their proof can readily be extended to include the case where $0 < \alpha < 1$ (see also Theorem 3.1.4).

Theorem 3.4.1 *Assume that $f \in C(I)$, $K \in C(D)$, and let $G = G(u, v)$ satisfy a local Lipschitz condition in both u and v. Then there exists a $\delta_0 = \delta_0(\alpha) > 0$ such that for any $\alpha \in (0, 1]$ the non-standard Hammerstein VIE (3.4.1) possesses a unique continuous solution on the interval $I_0 := [0, \delta_0]$.*

Proof The existence and uniqueness of a local continuous solution of (3.4.1) can be established by employing the approach used in the proof of Lemma 3.4.7 in Section 3.4.2 below: there is no essential difference between the Hammerstein VIE (3.4.1) and the auto-convolution VIE (3.4.5) on small initial intervals $[0, \delta_0]$. Thus, we will not duplicate that argument here. $\square$

Remark 3.4.2 The reader is invited to study the *continuation* of the solution of (3.4.1) beyond the interval $[0, \delta_0)$. In particular: can the blow-up analysis of Brunner & Yang (2013) be extended to the non-standard Volterra–Hammerstein equation (3.4.1) with $f(t) \geq 0$ and non-linearities like $G(u, v) = u^p v^q$ with $p > 1$ and $q > 1$? To the best of my knowledge, this problem has not yet been investigated. ◇

3.4.2 VIEs of Auto-Convolution Type

A non-linear VIE of the form

$$u(t) = f(t) + \lambda \int_0^t u(t - s)u(s)\,ds, \quad t \in I \ (\lambda \in \mathbb{R}), \tag{3.4.3}$$

is called an *auto-convolution* VIE (AVIE) of the second kind. We will refer to (3.4.3) as the *basic* auto-convolution VIE. It is a 'quadratic' VIE (the terminology is motivated by the fact that the application of the Laplace transform yields a quadratic algebraic equation for the image of $u(t)$; see below). As we shall see below, the nature of this VIE differs significantly from that of the non-standard VIE (3.4.2) because the integral term is now the convolution of the unknown solution with itself.

Example 3.4.3 (von Wolfersdorf, 2000)
Assume that $a \neq 0$ and $f \in C[0, \infty)$ is bounded. We use the Laplace transform (cf. Doetsch, 1973 or Schiff, 1999) to establish the existence of a continuous and bounded solution of the basic auto-convolution VIE

$$au(t) = f(t) + \int_0^t u(t - s)u(s)\,ds, \quad t \geq 0, \tag{3.4.4}$$

with $a \neq 0$. The application of the Laplace transform $\mathcal{L}$ to (3.4.4) shows that $U(\zeta) := (\mathcal{L}u)(t)$ satisfies the quadratic equation

$$aU(\zeta) - U^2(\zeta) = (\mathcal{L}f)(t) =: F(\zeta),$$

where by assumption on f its Laplace transform is such that $F(\zeta) \to 0$ as $\text{Re}\,\zeta \to \infty$. Of its two solutions

$$U_{1,2}(\zeta) = \frac{a}{2} \pm \frac{a}{2}\left(1 - \frac{4}{a^2}F(\zeta)\right)^{1/2},$$

we are interested in the one that satisfies $U(\zeta) \to 0$ as $\text{Re}\,\zeta \to \infty$. Thus, the desired solution of (3.4.4) is given by the inverse Laplace transform of

$$U(\zeta) = U_2(\zeta) = \frac{a}{2} - \frac{a}{2}\left(1 - \frac{4}{a^2}F(\zeta)\right)^{1/2}.$$

We summarise the result in the following theorem (von Wolfersdorf, 2000, pp. 8–9).

Theorem 3.4.4 *Assume that f is continuous and bounded on $[0, \infty)$. Then the auto-convolution VIE (3.4.4) possesses a unique continuous and bounded solution u on $[0, \infty)$.*

Remark 3.4.5

(i) If $a = 0$ the VIE (3.4.4) becomes an auto-convolution VIE of the *first kind*. These ill-conditioned problems are studied in e.g. Gorenflo & Hofmann (1994), Fleischer & Hofmann (1996, 1997), and von Wolfersdorf (2000, p. 9).

(ii) Abstract AVIEs of the form

$$u(t) + (G_0 u)(t) + \int_0^t (G_1 u)(t - s)(G_2 u)(s) \, ds, \quad t \in I,$$

where G_i $(i = 0, 1, 2)$ are non-linear operators on a Banach space (e.g. $C(I)$ or $L^p(0, T)$), were studied by von Wolfersdorf (1995; see also von Wolfersdorf, 2000 for additional references). ⋄

In the following we will refer to the VIE

$$u(t) = f(t) + \int_0^t K(t, s) G(u(t - s)) G(u(s)) \, ds, \quad t \in I, \tag{3.4.5}$$

with non-constant kernel $K(t, s)$ and non-linear $G(u)$ as the *generalised auto-convolution* equation of the second kind. An important special case corresponds to $G(u) = u^\beta$ $(\beta > 0)$ and $K(t, s) = \lambda$:

$$u(t) = f(t) + \lambda \int_0^t u^\beta(t - s) u^\beta(s) \, ds, \quad t \in I. \tag{3.4.6}$$

Theorem 3.4.6 *The generalised auto-convolution VIE (3.4.5) with $G(u) = u$ and $K \in C(D)$ possesses a unique solution $u \in C(I)$ for every $f \in C(I)$.*

Proof The proof consists of two steps. We first establish the local existence and uniqueness of a continuous solution on some small interval $[0, \delta_0]$. It will then be seen that on the remaining interval $[\delta_0, T]$ the VIE (3.4.5) becomes a *linear* VIE to which we may apply the classical theory of Section 1.2.

Lemma 3.4.7 *Assume that* $f \in C(I)$, $K \in C(D)$ *and* $G(u) = u$. *Then there exists a unique* $u \in C[0, \delta_0]$ *satisfying the VIE (3.4.5) on* $[0, \delta_0]$, *where*

$$\delta_0 := \min\left\{\frac{1}{4\bar{K}(\bar{F}+1)^2}, T\right\}, \quad \bar{F} := \max_{t \in I}|f(t)|, \quad \bar{K} := \max_{(t,s) \in D}|K(t,s)|.$$

We will prove the above assertion by resorting to Picard iteration for (3.4.5),

$$u_{n+1}(t) := f(t) + \int_0^t K(t,s)u_n(t-s)u_n(s)\,ds, \quad t > 0 \ (n \geq 0), \quad (3.4.7)$$

with $u_0(t) := f(t)$, and show that

$$|u_n(t) - f(t)| \leq 1, \quad t \in [0, \delta_0] \ (n \geq 0). \quad (3.4.8)$$

Since for $n = 0$ the estimate (3.4.8) is trivial, a straightforward induction step reveals that (3.4.8) holds for any $n \geq 1$:

$$|u_{n+1}(t) - f(t)| = \left|\int_0^t K(t,s)u_n(t-s)u_n(s)\,ds\right| \leq \bar{K}\left(\bar{F}+1\right)^2 t \leq 1,$$

with each u_{n+1} defined on $[0, \delta_0]$. It follows from the continuity of f and K that $u_{n+1} \in C[0, \delta_0]$. We now show that the sequence $\{u_n(t)\}$ is a Cauchy sequence on $[0, \delta_0]$. Setting $y_n(t) := u_{n+1}(t) - u_n(t)$, we obtain

$$|y_0(t)| = |u_1(t) - u_0(t)| = \left|\int_0^t K(t,s)f(t-s)f(s)\,ds\right|$$

$$\leq \bar{K}\,\bar{F}^2 t \leq \bar{K}\,(\bar{F}+1)^2 t.$$

Hence, if

$$|y_n(t)| \leq \frac{\bar{K}^{n+1}(\bar{F}+1)^{n+2}2^n t^{n+1}}{(n+1)!}, \quad t \in [0, \delta_0],$$

an induction argument shows that the above estimate holds for all n and $t \in [0, \delta_0]$:

$$|y_{n+1}(t)| = \left|\int_0^t K(t,s)\left(u_{n+1}(t-s)u_{n+1}(s) - u_n(t-s)u_n(s)\right)ds\right|$$

$$\leq \bar{K}\int_0^t \left(|u_{n+1}(t-s)||u_{n+1}(s) - u_n(s)|\right.$$

$$\left. + |u_{n+1}(t-s) - u_n(t-s)||u_n(s)|\right)ds$$

$$\leq 2\bar{K}(\bar{F}+1)\int_0^t |u_{n+1}(s) - u_n(s)|\,ds$$

$$\leq 2\bar{K}(\bar{F}+1)\int_0^t \frac{\bar{K}^{n+1}(\bar{F}+1)^{n+2}2^n s^{n+1}}{(n+1)!}\,ds$$

$$= \frac{\bar{K}^{n+2}(\bar{F}+1)^{n+3}2^{n+1}t^{n+2}}{(n+2)!}.$$

This implies that $\lim_{n\to\infty} u_n(t) =: u(t)$ uniformly on $[0, \delta_0]$, with $u \in C[0, \delta_0]$.
Since

$$\left| \int_0^t K(t,s)\big(u_n(t-s)u_n(s) - u(t-s)u(s)\big) ds \right|$$

$$\leq \int_0^t |K(t,s)|\big(|u_n(t-s)||u_n(s) - u(s)|$$

$$+|u_n(t-s) - u(t-s)||u(s)|\big) ds$$

$$\leq 2\bar{K}(\bar{F}+1) \int_0^t |u_n(s) - u(s)| ds \to 0, \quad t \in [0, \delta_0],$$

as $n \to \infty$, we see that

$$u(t) = \lim_{n\to\infty} u_n(t) = f(t) + \lim_{n\to\infty} \int_0^t K(t,s)u_{n-1}(t-s)u_{n-1}(s) ds$$

$$= f(t) + \int_0^t K(t,s)u(t-s)u(s) ds, \quad t \in [0, \delta_0];$$

that is, there exists a $u \in C[0, \delta_0]$ which is a solution of the VIE (3.4.5) (with $G(u) = u$) on $[0, \delta_0]$.

In order to establish the uniqueness of the solution, suppose that (3.4.5) possesses two continuous solutions z_1 and z_2 on $[0, \delta_0]$. Then

$$|z_1(t) - z_2(t)| = \left| \int_0^t K(t,s)\big(z_1(t-s)z_1(s) - z_2(t-s)z_2(s)\big) ds \right|$$

$$\leq \bar{K} \int_0^t \big(|z_1(t-s)||z_1(s) - z_2(s)| + |z_1(t-s)$$

$$-z_2(t-s)||z_2(s)|\big) ds$$

$$\leq 2\bar{K}(\bar{F}+1) \int_0^t |z_1(s) - z_2(s)| ds, \quad t \in [0, \delta_0].$$

It follows from the classical Gronwall lemma (cf. Lemma 1.2.17) that

$$|z_1(t) - z_2(t)| \leq 0 \cdot \exp\big(2\bar{K}(\bar{F}+1)t\big) = 0, \quad t \in [0, \delta_0].$$

The continuity of z_1 and z_2 then implies that $z_1(t) = z_2(t)$ for all $t \in [0, \delta_0]$. We denote the unique solution of (3.4.5) on this interval by $w_1(t)$.

In order to extend the solution to the entire interval $[0, T]$ consider first the interval $[\delta_0, 2\delta_0]$, assuming that $2\delta_0 \leq T$. On this interval (3.4.5) may be written as

$$
\begin{aligned}
u(t) &= f(t) + \int_0^{\delta_0} K(t, s) u(t-s) u(s)\, ds + \int_{\delta_0}^t K(t, s) u(t-s) u(s)\, ds \\
&= f(t) + \int_{t-\delta_0}^t K(t, t-s) u(t-s) u(s)\, ds + \int_{\delta_0}^t K(t, s) u(t-s) u(s)\, ds \\
&= f(t) + \int_{t-\delta_0}^{\delta_0} K(t, t-s) u(t-s) u(s)\, ds \\
&\quad + \int_{\delta_0}^t \big(K(t, t-s) + K(t, s) \big) u(t-s) u(s)\, ds,
\end{aligned}
$$

or as

$$
u(t) = f_1(t) + \int_{\delta_0}^t \big(K(t, t-s) + K(t, s) \big) w_1(t-s) u(s)\, ds, \quad (3.4.9)
$$

where

$$
f_1(t) := f(t) + \int_{t-\delta_0}^{\delta_0} K(t, t-s) w_1(t-s) w_1(s)\, ds, \quad t \in [\delta_0, 2\delta_0].
$$

We observe that the VIE (3.4.9) is *linear* in u. Thus it follows from the classical Volterra theory (cf. Ch. 1) that (3.4.9) has a unique continuous solution on $[\delta_0, 2\delta_0]$. This in turn implies that (3.4.4) possesses a unique continuous solution on the interval $[0, 2\delta_0]$. We will denote this solution by $w_2(t)$.

Since $T < \infty$ we can continue this process. Suppose that $k\delta_0 < T$ and that, for $t \in [0, k\delta_0]$, the unique solution $w_k(t)$ is in $C[0, k\delta_0]$. Then for $t \in [k\delta_0, (k+1)\delta_0]$, (3.4.5) assumes the form

$$
\begin{aligned}
u(t) &= f(t) + \int_0^{k\delta_0} K(t, s) u(t-s) u(s)\, ds + \int_{k\delta_0}^t K(t, s) u(t-s) u(s)\, ds \\
&= f_k(t) + \int_{k\delta_0}^t K\big((t, t-s) + K(t, s) \big) w_k(t-s) u(s)\, ds, \quad (3.4.10)
\end{aligned}
$$

with

$$
f_k(t) := g(t) + \int_{t-k\delta_0}^{k\delta_0} K(t, t-s) w_k(t-s) w_k(s)\, ds.
$$

By an adaptation of the above argument it can be seen that this VIE possesses a unique continuous solution on $[0, (k+1)\delta_0]$. Since $\delta_0 > 0$ is fixed, the existence and uniqueness of a continuous solution can be extended to the entire interval $I = [0, T]$ in a finite number of analogous steps.   □

As the proof of the following corollary shows, the regularity of the solution to the generalised auto-convolution VIE

$$u(t) = f(t) + \int_0^t K(t,s)u(t-s)u(s)\,ds, \quad t \in I, \tag{3.4.11}$$

can be deduced from the regularity of the Picard iterates $u_n(t)$ for (3.3.7) and from Theorem 2.1.1.

Corollary 3.4.8 *If* $f \in C^m(I)$ *and* $K \in C^m(D)$, *the generalised auto-convolution VIE (3.4.5) with* $G(u) = u$ *possesses a unique solution* $u \in C^m(I)$.

Remark 3.4.9

(i) The extension of Theorem 3.4.6 to the AVIE (3.4.5) with $K(t,s)$ replaced by $(t-s)^{\alpha-1}K(t,s)$ (with $0 < \alpha < 1$) is left to the reader. The existence of a unique continuous solution on some small interval $[0, \delta_0]$ can be established in a straightforward way, namely by replacing $\bar{K}$ by $\bar{K}T^\alpha/\alpha$ in the proof of Lemma 3.4.7.

(ii) More general AVIEs of the form

$$u(t) = f(t) + \int_0^t K_0(t,s)u(s)\,ds + \int_0^t K_1(t,s)G_1\big(u(t-s)\big)G_2(u(s))\,ds$$

are analysed in von Wolfersdorf (2000, pp. 10–13).

(iii) Auto-convolution VIEs of the *first kind*,

$$\int_0^t u(t-s)u(s)\,ds = g(t), \quad t \in I = [0, T],$$

arise in many applications (see, for example, the references in Dai & Lamm, 2008). As shown in Gorenflo & Hofmann (1994), Fleischer & Hofmann (1996, 1997), Fleischer, Gorenflo & Hofmann (1999), and Dai & Lamm (2008), such equations are severely ill-posed. $\diamond$

The non-linear VIE

$$u(t) = f(t) + \lambda \int_t^T u^\beta(s-t)u^\beta(s)\,ds, \quad t \in I \ (\beta > 0), \tag{3.4.12}$$

is seemingly closely related to the auto-convolution VIE (3.4.6): it is formally the adjoint equation of (3.4.6). However, as we shall see below, the analysis of the existence of a continuous solution is fundamentally different from the one for (3.4.6).

The basic example corresponding to $\beta = 1$,

$$u(t) = f(t) + \lambda \int_t^T u(s-t)u(s)\,ds, \ t \in I \ (\lambda \geq 0), \tag{3.4.13}$$

was studied in Pimbley (1967), Ramalho (1976, and Nussbaum (1980, 1981), and in the seminal paper by Nussbaum & Baxter (1981). A closer look at (3.4.13) shows that the argument we employed to establish the existence of a solution of (3.4.4) on $I = [0, T]$ (cf. the proof of Theorem 3.4.4) no longer works since on a small initial interval $[0, \delta]$ (or, alternatively, $[T - \delta, T]$) the solution values on *both* of these intervals are involved, and hence the VIE (3.4.13) no longer reduces to a linear one on $[0, T - \delta]$.

Since a detailed treatment of the general adjoint AVIE (3.4.12) is beyond the scope of this book, we just cite a number of representative results on the existence and non-existence of continuous solutions of the basic equation (3.4.13). They are due to Pimbley (1967, and were extended by Ramalho, 1976; see also Nussbaum & Baxter, 1981); they reveal that even for this special VIE the existence of a (continuous) solution is no longer guaranteed for every value of $\lambda > 0$.

Theorem 3.4.10 *Assume that $f(t) = 1$ and $T = 1$. Then:*

(a) For every $\lambda \in (0, 1/2)$ the VIE (3.4.13) possesses exactly two continuous solutions on $[0, 1]$.
(b) If $\lambda = 1/2$ the VIE (3.4.13) has a unique continuous solution on $[0, 1]$.
(c) The solutions described in (a) and (b) are positive, monotone decreasing and lie in $C^2[0, 1]$, with $u''(t) \geq 0$.
(d) The VIE (3.4.13) has no real solution if $\lambda > 1/2$.

Proof We will sketch the proofs of (c) and (d) and leave the verification of (a) and (b) to the reader. In order to analyse the regularity of the solutions when $\lambda \in (0, 1/2]$ we formally differentiate the VIE (3.4.13) and use integration by parts. This leads to

$$u'(t) = -\lambda u(0)u(t) - \lambda \int_t^1 u'(s-t)u(s)\,ds$$

$$= -\lambda u(1-t) + \lambda \int_t^1 u(s-t)u'(s)\,ds,$$

where we have used $u(1) = f(1)$. If we set $f_1(t) := -\lambda f(1-t)$, the above equation becomes a VIE for $w_1(t) := u'(t)$:

$$w_1(t) = f_1(t) + \lambda \int_t^1 f(s-t)w_1(s)\,ds, \ t \in [0, 1].$$

The existence of a continuous solution $w_1(t)$ can then be established by resorting to Picard iteration. We note in passing that this process can be continued, to show that $u \in C^m[0, 1]$ for any $m \geq 2$.

Assertion (d) is proved by defining $J := \int_0^1 u(s)ds$ and using the symmetry of the product $u(v - s)u(s)$. More precisely, we have

$$J = \int_0^1 \left(1 + \lambda \int_s^1 u(v - s)u(v)\, dv \right) ds$$

$$= 1 + \lambda \int_0^1 \left(\int_s^1 u(v - s)u(v) dv \right) ds$$

$$= 1 + \lambda \int_0^1 \left(\int_0^v u(s)ds \right) u(v)\, dv = 1 + \frac{\lambda}{2} J^2.$$

This quadratic equation for J has no real roots when $\lambda > 1/2$. If $\lambda = 1/2$ there is exactly one real root, while there are two distinct real roots for $0 < \lambda < 1/2$ (compare also Exercise 3.7.20). □

3.4.3 Implicit VIEs

A non-linear VIE of the form

$$W\big(t, z(t)\big) = f(t) + \int_0^t k\big(t, s, z(s)\big)\, ds, \quad t \in I,$$

where $W : I \times \mathbb{R} \to \mathbb{R}$ is a given function that is non-linear in z, is called an *implicit* VIE.

A particular example of an implicit VIE arises in the analysis of the existence of a continuous solution to the non-linear first-kind VIE

$$\int_0^t h\big(t, s, z(s)\big)\, ds = g(t), \quad t \in I.$$

Under appropriate smoothness assumptions on h and g (cf. Section 3.5) it is equivalent to the *implicit* VIE

$$W\big(t, z(t)\big) = g'(t) - \int_0^t \frac{\partial h\big(t, s, z(s)\big)}{\partial t}\, ds,$$

where $W\big(t, z(t)\big) := k(t, t, z(t))$.

Before studying the existence of solutions of general implicit VIEs of the form

$$W\big(z(t)\big) = f(t) + \int_0^t K(t, s)z(s)\, ds, \quad t \in I, \tag{3.4.14}$$

we look at a special class of such equations that frequently arise in applications (see Example 9.2.10 in Chapter 9), namely

$$z^\beta(t) = f(t) + \int_0^t K(t, s) z(s) \, ds, \quad t \in I, \tag{3.4.15}$$

with either $0 < \beta < 1$ or $\beta > 1$. Setting $u(t) := z^\beta(t)$, we see that the VIE (3.4.15) is equivalent to the non-linear second-kind Hammerstein VIE

$$u(t) = f(t) + \int_0^t K(t, s) u^{1/\beta}(s) \, ds, \quad t \in I. \tag{3.4.16}$$

Case I: If $0 < \beta < 1$, the non-linearity $G(u) = u^{1/\beta}$ in (3.4.16) is locally Lipschitz continuous since $1/\beta > 1$. We may thus resort to Theorem 3.1.4 to establish the existence of a (local) continuous positive solution for the Hammerstein VIE (3.1.4), provided $f \in C(I)$ and $K \in C(D)$ are non-negative on their respective domains I and D. The degree of regularity of the solution u depends on the particuar value of β, as shown in the following theorem.

Theorem 3.4.11 *Assume that the non-negative functions f and K in (3.4.16) satisfy $f \in C^m(I)$ and $K \in C^m(D)$ for some $m \geq 1$. If $0 < \beta < 1$, the solution of (3.4.16) is in C^m on its interval of existence, and the solution z of (3.4.15) has the same property.*

Proof Using Picard iteration for (3.4.16) with $u_0(t) = f(t)$, we see that the iterates u_n are in C^m since $f^{1/\beta} \in C^m(I)$ for all $\beta \in (0, 1)$. The regularity assertion for u is a consequence of the uniform convergence of the sequence $\{u_n(t)\}$ on its interval of definition. Moreover, as $1/\beta > 1$, $u \in C^m(I)$ implies that $z = u^{1/\beta}$ has the same regularity property. $\square$

Case II: When $\beta > 1$, the non-linearity $G(u) = u^{1/\beta}$ does no longer satisfy a local Lipschitz condition, and hence the Picard iteration method cannot be used to establish the existence of a continuous (local) solution of (3.4.16). (See, however, Exercise 3.7.21.) Moreover, the non-negativity of f no longer guarantees the existence of a unique continuous solution, as the following example shows.

Example 3.4.12 If f and K in the VIE (3.4.16) are C^1-functions, the VIE is equivalent to the initial-value problem

$$u'(t) = f'(t) + K(t, t) u^{1/\beta}(t) + \int_0^t \frac{\partial K(t, s)}{\partial t} u^{1/\beta}(s) \, ds, \quad u(0) = f(0).$$

For $f(t) = a_0 + a_1 t$ $(a_i \geq 0)$ and $K(t, s) = \lambda > 0$, this becomes $u'(t) = a_1 + \lambda u^{1/\beta}(t)$, with $u(0) = a_0$. If $\beta = 2$ and $a_1 = 0$ this well-known problem possesses the (smooth) solution $u(t) = \frac{1}{4}(t - c)_+^2$ for any $c \in I$ (here, $(t - c)_+^k$ denotes the truncated power function). Note that the function f is not strictly positive on I, in contrast to the assumption in the next theorem.

The following theorem is due to Bushell (1976, Theorem 2.3); it also encompasses VIEs (3.4.16) with $-1 < \beta < 0$.

Theorem 3.4.13 *Assume that $f \in C(I)$ is positive on I and $K \in C(D)$ satisfies $K(t, s) \geq 0$ on D. Then for each p with $0 < |p| < 1$ the VIE*

$$u(t) = f(t) + \int_0^t K(t, s) u^p(s)\, ds, \quad t \in I, \tag{3.4.17}$$

possesses a unique positive solution $u \in C(I)$.

Proof We resort to the following lemma (Bushell, 1976, Theorem 1.2) concerning certain properties of positive operators on a cone in the Banach space $C(I)$ of continuous functions with the norm $\| \cdot \|_\infty$. In order to state this result we introduce, for $g \in C(I)$ with $g(t) > 0$ on $I_0 := (0, T]$, the Banach space

$$C_g := \{u : u \in C(I),\ \|u\|_g := \sup_{t \in I_0} [\|u(t)\|/g(t)] < \infty\}.$$

In addition, we need the cones

$$\mathcal{C} := \{u \in C(I) : u(t) \geq 0\ (t \in I)\} \quad \text{and}$$
$$\mathcal{C}_0 := \{u \in C(I) : \inf_{t \in I_0} [u(t)/g(t)] \geq 0\}.$$

$\square$

Lemma 3.4.14 *Assume that $A : \mathcal{C} \to \mathcal{C}$ is a monotone, increasing mapping which is homogeneous of degree $p \in (0, 1)$ (that is, $A(\gamma u) = \gamma^p A(u)$ for any scalar γ), and let $B : \mathcal{C}_0 \to \mathcal{C}_0$ be defined by $Bu := f + Au$. If A is a continuous and uniformly bounded map on $\mathcal{C}_0 \cap \{u : \|u\|_\infty = 1\}$, then the equations*

$$u = Bu = f + Au \quad \text{and} \quad w = B(1/w) = f + A(1/w)$$

possess unique solutions in $\mathcal{C}_0$.

The obvious identification of the operators A and B of Lemma 3.4.14 with the ones defining the VIE (3.4.17) leads to the assertion in Theorem 3.4.13. We note in passing that Lipovan (2009, where Bushell's paper is not mentioned) contains a concise proof for (3.4.17) with convolution kernel and with $0 < p < 1$. □

Remark 3.4.15 The existence of L^2 solutions for VIEs involving powers of linear Volterra integral operators,

$$u(t) = \sum_{j=1}^{r} a_j \left(f_j(t) + \int_0^t K_j(t, s) u(s) \, ds \right)^j, \quad t \in I \ (r \geq 2), \quad (3.4.18)$$

with $a_j \in \mathbb{R}$, $f_j \in L^\infty(I)$ and $K_j \in L^\infty(D)$, was studied by Sloss & Blyth (1994; see also Exercise 3.7.18 at the end of this chapter when $a_j \in \mathbb{R}$, $f_j \in C(I)$, $K_j \in C(D)$). Their analysis was motivated by the special case

$$u(t) = \left(f_2(t) + \int_0^t K_2(t, s) u(s) \, ds \right)^2, \quad t \in I \quad (3.4.19)$$

corresponding to $r = 2$, $a_1 = 0$, $a_2 = 1$. It is related (via $u(t) := y''(t)$) to the initial-value problem $y''(t) = 2y(t)$, $y(0) = 1$, $y'(0) = 0$, whose solution is a Weierstrass elliptic function. The VIE (3.4.19) can be written as an *implicit* VIE (3.4.15) with $\beta = 1/2$ (see Theorem 3.4.13). The existence proof for an L^2 solution in Sloss & Blyth (1994) is readily adapted to establish the existence of a solutions $u \in C(I)$ of these non-standard VIEs. ◇

Consider now the general implicit VIE (3.4.14) and assume that $f \in C(I)$, $f(t) \geq 0$ and $K \in C(D)$, $K(t, s) \geq 0$. Let $W : \mathbb{R}^+ \to \mathbb{R}$ be strictly increasing and continuous on $\mathbb{R}^+$, $W(0) = 0$, and assume that W' exists on $\mathbb{R}^+$. Since the inverse mapping W^{-1} is defined, (3.4.14) can be written either as

$$z(t) = W^{-1} \left(f(t) + \int_0^t (t - s)^{\alpha - 1} K(t, s) z(s) \, ds \right), \quad t \in I,$$

or as

$$u(t) = f(t) + \int_0^t (t - s)^{\alpha - 1} K(t, s) W^{-1} \big(u(s) \big) \, ds, \quad t \in I, \quad (3.4.20)$$

where we have set $u(t) := W\big(z(t)\big)$. The non-linear VIE (3.4.20) is of Hammerstein type (3.1.11), with $W^{-1}(u)$ assuming the role of $G(s, u)$. Thus, the existence an uniqueness of a (local) continuous solution u of (3.4.20) can be established by resorting to Theorem 3.1.12. On the other hand, the equation preceding (3.4.20) is a particular case of the *implicitly linear* VIE (3.1.13) where now $G(t, u) = G(u)$ does not depend on t.

Remark 3.4.16

(i) Motivated by a non-linear VIE arising as a mathematical model of water percolation from a cylindrical reservoir (see also Section 9.2.7). Okrasiński (1986) presented a detailed analysis of the existence of positive solutions of

$$W(z(t)) = f(t) + \int_0^t k(t-s)z(s)\, ds, \ \ t \geq 0,$$

where $W : \mathbb{R}^+ \to \mathbb{R}^+$ is continuous, stricly increasing and concave, with $W(0) = 0$, $W \in C^1(0, \infty)$ and $W'(u) \to \infty$ as $u \to 0^+$.

(ii) An analysis of the more general version of the implicit VIE (3.4.20), namely

$$u(t) = F\Big(t, u(\theta(t)), \int_0^t k(t-s)G\big(s, z(s)\big)\, ds\Big), \ \ t \geq 0,$$

where $\theta(t) = t - \tau$ ($\tau > 0$) is continuous and bounded on $\mathbb{R}^+$, can be found, for example, in Karakostas (1981). ◇

3.5 Non-Linear First-Kind VIEs

We consider the general non-linear VIE of the first kind,

$$\int_0^t (t-s)^{\alpha-1} h(t, s, u(s))\, ds = g(t), \ \ t \in I, \tag{3.5.1}$$

with $0 < \alpha \leq 1$ and appropriate (differentiable) functions h and g.

If $\alpha = 1$, we may differentiate both sides of (3.5.1), to obtain the equivalent *implicit* VIE *implicit* second-kind VIE

$$W_1\big(t, u(t)\big) = g'(t) - \int_0^t \frac{\partial h(t, s, u(s))}{\partial t}\, ds, \ \ t \in I, \tag{3.5.2}$$

where we have set $W_1(t, u) := h(t, t, u)$.

The following result is due to Gorenflo & Pfeiffer (1991). Related versions can be found in Branca (1978: $\alpha = 1/2$), Dixon, Mckee & Jeltsch (1986: $\alpha = 1$) and, for $0 < \alpha < 1$, in Gorenflo & Vessella (1991, Section 5.2), Ang & Gorenflo (1991) and in Deimling (1995).

Theorem 3.5.1 *Assume that the given functions in the general non-linear first-kind VIE (3.5.1) with $\alpha = 1$, satisfy*

(i) $h \in C(D \times \mathbb{R})$, $\partial h/\partial t \in C(D \times \mathbb{R})$;

(ii) *there exists a constant* $\gamma > 0$ *so that* $\left| \dfrac{\partial h(t, t, u)}{\partial u} \right| \geq \gamma > 0$ *for all* $t \in I$

   and $u \in \mathbb{R}$;

(iii) *there exists a constant* $L > 0$ *so that*

$$\left| \frac{\partial h(t, s, u)}{\partial t} - \frac{\partial h(t, s, v)}{\partial t} \right| \leq L |u - v|$$

   for all $(t, s) \in D$ *and* $u, v \in \mathbb{R}$;

(iv) $g \in C^1(I)$, *with* $g(0) = 0$.

Then the first-kind VIE (3.5.1) possesses a unique solution $u \in C(I)$ *for all* $\alpha \in (0, 1]$. *Moreover, the value* $u(0)$ *is given implicitly by*

$$h(0, 0, u(0)) = \alpha \lim_{t \to 0^+} t^{-\alpha} g(t).$$

Proof If $0 < \alpha < 1$, we apply to (3.5.1) Volterra's transformation discussed in Section 1.5.2. This yields, in analogy to (1.5.8), the non-linear first-kind VIE

$$\int_0^t h_\alpha(t, s, u(s)) \, ds = G_\alpha(t), \quad t \in I, \tag{3.5.3}$$

where

$$G_\alpha(t) := \int_0^t (t - s)^{-\alpha} g(s) \, ds.$$

The bounded (and differentiable) kernel function in (3.5.3) is given by

$$h_\alpha(t, s, u) := \int_0^1 (1 - w)^{-\alpha} w^{\alpha - 1} h(s + (t - s)w, s, u(s)) \, dw;$$

it inherits the regularity of $h(t, s, u)$. As we have seen before, it follows from the assumption (iv) that $G_\alpha \in C^1(I)$ and $G_\alpha(0) = 0$, differentiation of (3.5.3) yields the analogue of the *implicit* VIE (3.5.2), namely

$$W_\alpha(t, u(t)) = G'_\alpha(t) - \int_0^t \frac{\partial h_\alpha(t, s, u(s))}{\partial t} \, ds, \quad t \in I, \tag{3.5.4}$$

with $W_\alpha(t, u(t)) := h_\alpha(t, t, u(t))$. The assumptions (ii) and (iii) imply that we can resort to the Implicit Function Theorem (see Appendix A.3.2) to establish the existence and uniqueness of a local continuous solution u of the implicit VIE (3.5.2), and hence of the implicit VIE (3.5.4) for any $\alpha \in (0, 1)$. We leave the details to the reader. $\qquad \square$

Remark 3.5.2

(i) Since the assumption (iii) of a *global* Lipschitz condition in Theorem 3.5.1 is often too restrictive (for example, when $k(t, s, u) = K(t, s)u^2$), it can be replaced by a *local* one, in analogy to the one in Theorem 3.1.2. We leave the derivation of the corresponding analogue of Theorem 3.5.1 (in particular the analysis of the maximal interval of existence) as an exercise.

(ii) Assumption (ii) in Theorem 3.5.1 can be replaced by the weaker (accretivity) condition,

$$\big(h(t, t, u_1) - h(t, t, u_2)\big)(u_1 - u_2) \geq \gamma(u_1 - u_2)^2 \quad (t \in I, \ u_i \in \mathbb{R});$$

see Ang & Gorenflo (1991) and Deimling (1995). ◇

3.6 Non-Linear Volterra Functional Integral Equations

3.6.1 State-Independent Delays

The existence and uniqueness of (local) solutions to non-linear Volterra functional integral equations (VFIEs) with state-independent delay function $\theta(t) = t - \tau(t) \leq t$, for example,

$$u(t) = f(t) + \int_0^{\theta(t)} k(t, s, u(s)) \, ds, \ t \in I = [0, T], \tag{3.6.1}$$

and

$$u(t) = f(t) + \int_{\theta(t)}^t k(t, s, u(s)) \, ds, \ t \in I, \tag{3.6.2}$$

can be established by resorting to Theorem 3.1.2. We will illustrate this by first looking at the above VFIEs with *vanishing delays*; that is, with delay functions that are subject to the assumptions we introduced in Section 2.3.1:

(V1) $\theta(t) = t - \tau(t)$, with $\tau \in C^d(I)$ for some $d \geq 0$;
(V2) $\theta(0) = 0$ and $\theta(t) < t$ for $t > 0$;
(V3) θ is strictly increasing on I.

The basic example is $\theta(t) = qt = t - (1 - q)t$ $(0 < q < 1)$, corresponding to the proportional delay $\tau(t) = (1 - q)t$. In this case a proof of the existence of a unique (local) solution $u \in C[0, \delta_0]$ for the VFIE (3.6.1) was given by Chambers (1990, pp. 30–34). This result can also be established by a straightforward adaptation of the proof of Theorem 3.1.2; in analogy to that theorem (cf. (a)) this proof yields an explicit expression for δ_0 in terms of the upper bound of $|k(t, s, u)|$ and q. A similar approach can be used for analysing the existence of a continuous solution to the more general VFIE (3.6.2) on some small interval $[0, \delta_0]$. We leave the details of these proofs as an exercise.

Turning now to the non-linear VFIEs (3.6.1) and (3.6.2) with *non-vanishing* delays $\tau(t)$ we assume as in Section 2.3.1 that the delay function θ is such that

(N1) $\theta(t) = t - \tau(t)$, with $\tau \in C^d(I)$ for some $d \geq 0$;
(N2) $\tau(t) \geq \tau_0 > 0$ for $t \in I$; and
(N3) θ is strictly increasing on I;

hold. Let $\{\xi_\mu\}$ denote the primary discontinuity points (recall Definition 2.3.3) induced by $\theta(t)$, and set $I_\mu := [\xi_\mu, \xi_{\mu+1}]$ (if $\mu \geq 1$), with $I_0 := (0, \xi_1]$. The assumptions (N2) and (N3) imply that the primary discontinuity points do not cluster as t increases. Hence we can again employ the method of steps to establish the (local) existence of a unique continuous solution for (3.6.1) and (3.6.2). For the VFIE (3.6.1) the argument is trivial since for any given initial function $u(t) = \phi(t)$ ($t \leq 0$) the solution $u(t)$ can be found explicitly on each subinterval I_μ ($\mu \geq 0$).

For the VFIE (3.6.2) and $t \in I_\mu$ (and hence $\theta(t) \in I_{\mu-1}$), the VFIE can be written in the local form

$$u(t) = f_\mu(t) + \int_{\xi_\mu}^t k(t, s, u(s))\, ds,$$

with

$$f_\mu(t) := f(t) + \int_{\theta(t)}^{\xi_\mu} k(t, s, u(s))\, ds.$$

Each of these equations is a non-linear VIE on I_μ, and therefore we can resort to Theorem 3.1.2 to continue our analysis. If $\mu = 0$ the solution either exists only on some proper subinterval $[0, \delta_0] \subset [0, \xi_1)$, or it exists on $[0, \xi_1]$. In the latter case we apply Theorem 3.1.2 to the non-linear VIE

$$u(t) = f_0(t) + \int_{\xi_1}^t k(t, s, u(s))\, ds, \quad t \in [0, \xi_1],$$

and continue the preceding process. It follows that the solution of the VFIE (3.6.2) can be extended to some (maximal) interval $[0, T_f]$ ($T_f < T$), or it exists on the given interval $[0, T]$. We refrain from stating the formal existence and uniqueness theorem.

3.6.2 Blow-Up Theory for Non-Linear VFIEs

How does the blow-up dynamics change if we replace the Hammerstein VIE

$$u(t) = f(t) + \int_0^t k(t - s)G(s, u(s))\, ds, \ t \in I,$$

by the analogous Hammerstein VFIE

$$u(t) = f(t) + \int_{\theta(t)}^t k(t - s)G(s, u(s))\, ds, \ t \in I, \tag{3.6.3}$$

with a delay function θ that corresponds either to a *vanishing delay* satisfying the conditions (V1), (V2) and (V3); or to a *non-vanishing delay* subject to the conditions (N1), (N2) and (N3) (see Section 3.6.1).

In this section we will focus on the VFIE (3.6.3) with *vanishing delays* (for analogous blow-up results on VFIEs with *non-vanishing delays* we refer the reader to Yang & Brunner, 2013a, pp. 772–776)).

In analogy to the non-delay case analysed in Section 3.2.2 we will assume that the given functions are subject to the following conditions:

(F) f is continuous, positive and non-decreasing on $[0, \infty)$;
(K) $k \in L^1_{loc}(\mathbb{R}^+)$, $k(z) > 0$ $(z \in \mathbb{R}^+)$, with $K(z) := \int_0^z k(s)\, ds > 0$;
(G1) $G(s, 0) \equiv 0$ for $s \geq 0$, and $G(s_1, u_1) > G(s_0, u_0)$ for all $(s_i, u_i) > (0, 0)$ with $s_1 \geq s_0$ and $u_1 > u_0$;
(G2) $\displaystyle\lim_{u \to \infty} \frac{G(0, u)}{u} = \infty$.

The following theorems reveal that the blow-up results for VFIEs of the form (3.6.3) with vanishing delays are essentially identical with the ones for VIEs without delay arguments. We start with a theorem (the analogue of Theorem 3.2.14) that describes *sufficient* conditions for finite-time blow-up in the VFIE with vanishing delay.

Theorem 3.6.1 *Let (F), (K), (G1) and (G2) hold, and assume in addition that*

(i) $k(z) = z^{\alpha-1}k_1(z)$, *with $\alpha > 0$ and $k_1(z) \geq 0$ such that there exists $\delta > 0$ for which*

$$\kappa_1 := \inf\{k_1(z) : z \in [0, \delta]\} > 0;$$

(ii) *there is a $\bar{t} > 0$ such that $f(\bar{t}) + F_{min}(\bar{t}) > 0$, and for which*

$$\int_{u_0}^\infty \frac{du}{u\big(\bar{H}(u)\big)^{1/\alpha}} < \infty, \tag{3.6.4}$$

with $\bar{H}(u) := G(\bar{t}, u)/u$, is true for all $u_0 > 0$.

(iii) The delay function $\theta(t) = t - \tau(t)$ with vanishing delay $\tau(t)$ is subject to the conditions (V1)–(V3).

Then the solution of (3.6.3) blows up in finite time T_b.

Proof Since the delay function satisfies $0 < \theta(t) < t$ when $t > 0$, the proof is essentially identical to the one of Theorem 3.2.14. The reader interested in the details may find them in Yang & Brunner (2013a, pp. 269–272). □

The next theorem (the analogue of Theorem 3.2.16) is concerned with *necessary* conditions for finite-time blow-up in (3.6.3). Its proof is a straightforward adaptation of the proof of Theorem 3.2.16.

Theorem 3.6.2 *Suppose that, in addition to (F), (K), (G1) and (G2), there exists a $\bar{t} \in (0, \infty)$ such that $f(\bar{t}) + F_{min}(\bar{t}) > 0$. If the convolution kernel $k(z)$ is of the form $k(z) = z^{\alpha-1}$ with $\alpha > 0$ (that is, if $k_1(z) \equiv 1$) and if there is a $u_0 > 0$ for which*

$$\int_{u_0}^{\infty} \frac{du}{u\left(\bar{H}(u)\right)^{1/\alpha}} = \infty, \qquad (3.6.5)$$

then the solution of (3.6.3) does not blow up in finite time: it exists for all $t \in (0, \infty)$.

The *necessary* and *sufficient* conditions for the solution of the VFIE (3.6.3) to blow up in finite time are, not surprisingly after having seen the above results, identical to the ones for VIEs without delay. Theorem 3.6.3 is the analogue of Theorem 3.2.17.

Theorem 3.6.3 *Assume that, in addition to (F), (K), (G1) and (G2), the assumption (ii) of Theorem 3.6.1 and the assumptions in Theorem 3.6.2 hold. Then the solution of the VFIE (3.6.3) blows up in finite time if, and only if, the solution of the VIE (3.2.13) (that is, the VFIE (3.6.3) with $\theta(t) \equiv 0$) blows up in finite time.*

Remark 3.6.4 It was shown in Yang & Brunner (2013a) that the previous theorem is not valid for VFIEs with *non-vanishing delays*. This is due to the subtle interplay between the non-homogeneous term $f(t)$ and the initial function $\phi(t)$ in the VFIE (3.6.2). This markedly different dynamical behaviour is illustrated by the following examples; the reader is invited to prove the assertions in (i), (ii), and (iii) stated below. Thus, consider the VFIE

$$u(t) = f(t) + \int_{t-1}^{t} e^{-(t-s)^2} u^2(s) \, ds, \quad t \in (0, T],$$

with the initial condition $u(t) = \phi(t)$ ($t \in [-1, 0]$).

(i) If $f(t) \equiv 1/3$ and $\phi(t) \equiv 1$, the solution of this VFIE does not blow up in finite time, but the solution of the corresponding VIE (where the lower limit of integration $t - 1$ is replaced by 0) does blow up in finite time.

(ii) If $f(t) \equiv 1/4$ and $\phi(t) \equiv 2$, the reverse of (i) is true: while the solution of the VFIE blows up in finite time, the solution of the VIE does not.

(iii) If $f(t) \equiv 1$ and $\phi(t) \equiv 2$, the solutions of the VFIE and the VIE both blow up in finite time.   ◇

3.6.3 State-Dependent Delays

If the delay function θ in a VFIE depends on the unknown solution $u(t)$, that is, if

$$\theta = \theta(t, u(t)) = t - \tau(t, u(t)) \le t, \quad t \ge 0,$$

the VFIE

$$u(t) = f(t) + \int_0^t k\big(t, s, u(s), u(\theta(s, u(s)))\big) \, ds, \quad t \ge 0, \tag{3.6.6}$$

is a VFIE with *state-dependent delay* $\tau(t, u(t))$. The Hammerstein analogue of the state-dependent VFIE (3.6.6) is

$$u(t) = f(t) + \int_0^t K(t, s) G\big(s, u(s), u(\theta(s, u(s)))\big) \, ds, \quad t \ge 0. \tag{3.6.7}$$

In most applications the kernel $K(t, s)$ in (3.6.7) is given by a *memory kernel*; that is, it is of the form $K(t, s) = k(t - s)$, with positive and non-increasing k.

An alternative form arises in certain mathematical models of population dynamics: here, the state-dependent delay occurs in the lower limit of integration of the integral operator, and the VFIE has the form

$$u(t) = f(t) + \int_{\theta(t, u(t))}^{t} k(t - s) G(s, u(s)) \, ds, \quad t \ge 0. \tag{3.6.8}$$

Typically, the delay $\tau(u)$ in $\theta(u) = t - \tau(u)$ is a function that is positive and locally Lipschitz continuous, with $\tau(u) \le \bar{\tau}$ for some constant $\bar{\tau}$. A particular case of the VFIE (3.6.8), namely

$$u(t) = \int_{t-\tau(u(t))}^{t} G(u(s)) \, ds, \quad t > 0, \tag{3.6.9}$$

was introduced and analysed in Bélair (1991; see also Example 9.2.5 in Chapter 9 of this book for more details).

Since the delay function $\theta(t, u(t)) = t - \tau(t, u(t))$ in (3.6.6), (3.6.7) and (3.6.8) now depends on the unknown solution $u(t)$, the primary discontinuity points $\{\xi_\mu\}$ are not known a priori. (See, however, the papers by Neves & Feldstein, 1976, Willé & Baker, 1992, and Guglielmi & Hairer, 2008 on the classification and the numerical computation of these points ξ_μ.) While there are some results for state-dependent VFIEs (3.6.6) (see e.g. Cahlon, 1992), the theory on the existence and regularity of solutions to state-dependent VFIEs of the form (3.6.8) is still little understood. A possible approach to dealing with this problem is sketched in Brunner & Maset (2010): it is based on employing so-called time transformations to reduce a VFIE (3.6.7) or (3.6.8) to an equivalent VFIE whose delay function is no longer state-dependent and for which the (transformed) primary discontinuity points can thus be determined a priori.

3.7 Exercises and Research Problems

Exercise 3.7.1 Provide the detailed proof of Theorem 3.1.4.

Exercise 3.7.2 Prove Theorem 3.1.12 directly, without resorting to the proof of Theorem 3.1.2. Then formulate and prove the analogous result for the weakly singular Hammerstein VIE

$$u(t) = f(t) + \int_0^t k_\alpha(t - s)G(s, u(s))\, ds \quad (0 \le \alpha < 1),$$

where $k_\alpha(t - s) := (t - s)^{\alpha-1}$ if $0 < \alpha \le 1$, and $k_0(t, s) := \log(t - s)$.

Exercise 3.7.3 Suppose that the given functions in the non-linear VIE (3.1.1) are subject to the assumptions in Theorem 3.1.16. Show that if $u(t)$ is a bounded continuous solution of (3.1.13) on $[0, \delta)$, then $u(t)$ can be continued (as a continuous solution) to $[0, \bar{\delta})$ for some $\bar{\delta} > 0$.

Exercise 3.7.4

(a) Show that if the VIE (3.1.11) has a maximal solution $u^*(t)$ on $[0, \delta^*]$, it must be unique.
(b) Verify that the assumptions in Theorem 3.1.16 are satisfied when the kernel function $k(t, s, u)$ in (3.1.14) is of the form $k(t, s, u) = (t - s)^{\alpha-1}k_0(t, s, u)$ $(0 < \alpha < 1)$, with k_0 continuous on $\{0 \le s \le t < \infty\} \times \mathbb{R}$.

Exercise 3.7.5 Prove that the VIE

$$u(t) = \int_0^t t\big(u(s)\big)^{1/2} ds, \ \ t \in [-1, 1],$$

possesses a maximal solution. (Recall the result by Gollwitzer & Hager, 1970 described in Section 3.1.3: if the above integral is preceded by the minus sign then a maximal solution does not exist.)

Exercise 3.7.6 Suppose that $G \in C(\mathbb{R}^+ \times \mathbb{R})$. Show that the assumptions (ii)–(iv) and the one after (iv) in Theorem 3.1.16 are satisfied when the kernel of the non-linear VIE (3.1.13) is of the form $k_\alpha(t - s)G(s, u)$, with $k_\alpha(z) := z^{\alpha-1}$ $(0 < \alpha < 1)$ and $k_0(z) := \log(z)$.

Exercise 3.7.7 Consider the VIE

$$u(t) = \frac{\gamma}{\Gamma(\alpha)} \int_0^t (t - s)^{\alpha-1} e^{u(s)+\beta} ds,$$

with $0 < \alpha \le 1$ and $\beta > 0$. Does its solution blow up in finite time T_b ? If so, find lower and upper bounds for T_b.

Exercise 3.7.8 Consider the VIE

$$u(t) = f(t) + \int_0^t k(t - s)r(s)u^p(s) ds, \ \ t > 0 \ \ (p > 1),$$

where the kernel $k(z)$ satisfies (3.2.7) but $r(t)$ is such that instead of assumption (3.2.5) we have $r(t) > 0$ and $r'(t) < 0$ $(t \ge 0)$ (instead of $r'(t) \ge 0$). Does the solution still cease to exist at some finite $T_b > 0$ (cf. Theorem 3.2.4).

Exercise 3.7.9 (*Research Problem*)
In Brunner & Yang (2013) finite-time blow-up for VIEs

$$u(t) = f(t) + \int_0^t k(t - s)G(s, u(s)) ds, \ \ t \ge 0,$$

was analysed for kernels $k(z)$ that were either smooth or had the form $z^{-\beta}k_1(z)$, with $\beta \in (0, 1)$ and smooth $k_1(z)$. Establish analogous blow-up results for the above VIE for kernels of the form $k(z) = \log(z)k_1(z)$.

Exercise 3.7.10 (*Research Problem*)
Analyse the finite-time blow-up of solutions to the non-linearly perturbed VIE

$$u(t) = f(t) + \int_0^t \big(k_0(t - s)u(s) + k_1(t - s)G(u(s))\big) ds, \ \ t \ge 0,$$

where f and the kernels $k_i(t - s)$ satisfy the assumptions (F) and (K) in Section 3.2.2, or the assumptions of Theorem 6.3.1 (implying that their resolvent kernels are in $L^1(\mathbb{R}^+)$).

Exercise 3.7.11 In his 1975 paper Kawarada studied the quenching of solutions of the initial-value problem

$$\frac{du}{dt} = \frac{1}{1 - u}, \quad t > 0; \quad u(0) = u_0 \neq 1.$$

Does the solution quench in finite time if $u_0 > 1$? What happens if $u_0 < 0$?

Exercise 3.7.12 In Example 3.1.8 discussed a quenching problem for a VIE. Prove that for any $u_0 \leq 0$ the solution of the VIE does not quench in finite time but exists for all $t > 0$.

Exercise 3.7.13 Assume that the functions r and G in the VIE (3.3.2) satisfy the hypotheses of Theorem 3.3.1, except that there exists a $u^* \in (u_0, b)$ for which $G(u^*) = 0$. Show that for such G the solution of (3.3.2) never quenches.

Exercise 3.7.14 (*Research Problem*)
Extend the quenching theory of Section 3.3.2 to VFIEs

$$u(t) = f(t) + \int_{\theta(t)}^{t} k(t - s)G(s, u(s))\, ds, \quad t \geq 0,$$

where the delay function θ corresponds to a vanishing delay $\tau(t) = qt$ $(0 < q < 1)$ (cf. Section 3.2.2).

Exercise 3.7.15 Show that the unique continuous and bounded solution of the auto-convolution VIE

$$2u(t) = \sin(t) + \int_{0}^{t} u(t - s)u(s)\, ds, \quad t \geq 0,$$

is $u(t) = J_1(t)$ (where J_1 is the Bessel function of order one).

Exercise 3.7.16 Discuss the existence and the regularity properties o f real solutions $u \in C(I)$ of the basic auto-convolution VIE (3.4.3) when $f(t) = 1$ and $\lambda < 0$.

Exercise 3.7.17 (*Research Problem*)
Consider the general auto-convolution VIE

$$u(t) = f(t) + \int_0^t k(t-s)u^\beta(t-s)u^\beta(s)\,ds, \quad t \geq 0, \qquad (3.7.1)$$

with $\beta > 1$ ($\beta \neq 1$). The (real-valued) function $f \in C[0, \infty)$ is non-negative, and the positive convolution kernel k is either continuous or locally integrable ($k(z) = z^{\alpha-1}$ with $0 < \alpha < 1$ is a representative example for the latter case). Can the solution of (3.7.1) blow up in finite time; that is, does there exist, for certain values of β and certain functions f and k, a $T_b < \infty$ so that $u(t)$ is defined (and continuous) on $[0, T_b)$ but

$$\lim_{t \to T_b^-} u(t) = +\infty ?$$

Exercise 3.7.18 Suppose that we have $a_j \in \mathbb{R}$, $f_j \in C(I)$, $K_j \in C(D)$ ($j = 1, 2$) in the Sloss-Blyth VIE (3.4.19) with $r = 2$. Analyse the existence of a unique solution $u \in C(I)$ of the slightly more general VIE

$$u(t) = \sum_{j=1}^2 a_j\left(f_j(t) + \int_0^t (t-s)^{\alpha-1}K_j(t,s)u(s)\,ds\right)^j, \quad t \in I \ (0 < \alpha \leq 1).$$

Exercise 3.7.19 (*Research Problem*)
Analyse the existence and regularity of solutions to the implicit VIE

$$u^\beta(t) = f(t) + \int_0^t K(t,s)u(t-s)u(s)\,ds, \quad t \geq 0,$$

with $\beta > 0$ ($\beta \neq 1$). (*Hint:* Exploit the connection of the above VIE with the one in (3.7.1).)

Exercise 3.7.20 (Pimbley, 1967, p. 111) Suppose that $u(t)$ is any continuous, real-valued solution of the VIE

$$u(t) = 1 + \lambda \int_t^1 u(t-s)u(s)\,ds, \quad t \in [0, 1] \ (\lambda \in \mathbb{R}).$$

Show that $J := \int_0^1 u(s)\,ds$ satisfies the quadratic equation $\lambda J^2 - 2J + 2 = 0$. What is the implication of this result for the existence of (real) solutions of the above VIE?

Exercise 3.7.21 Consider the non-linear VIE

$$u(t) = f(t) + \int_0^t K(t,s)u^p(s)\,ds, \quad t \in [0,1] \ (0 < p < 1), \qquad (3.7.2)$$

where f and K are continuous functions satisfying $f(t) > 0$ and $K(t,s) \geq 0$. Since $G(u) = u^p$ is not Lipschitz continuous, use Carathéodory successive approximation (recall Theorem 3.1.9) to prove that the above VIE has a (local) positive solution, provided $f(t) > 0$ and $K(t,s) \geq 0$. As an example, show that the implicit VIE (3.7.2) with $f(t) = 1$, $K(t,s) = (t-s)^{1/2}$ and $p = 1/2$ possesses a unique positive solution on $[0,1]$.

Exercise 3.7.22 Analyse the regularity of the solution of

$$u(t) = f(t) + \int_0^t K(t,s)u^{1/\beta}(s)\,ds, \quad t \in [0,T],$$

assuming that $\beta > 1$, $f \in C^m(I)$, $K \in C^m(D)$ $(m \geq 1)$, with $f(t) > 0$ and $K(t,s) \geq 1$ (i.e. establish the counterpart of Theorem 3.4.11).

Exercise 3.7.23 Consider the Hammerstein version of the non-linear first-kind VIE (3.5.1),

$$\int_0^t (t-s)^{\alpha-1}H(t,s)G(s,u(s))\,ds = g(t), \quad t \in I \ (0 < \alpha < 1).$$

State and prove the analogue of Theorem 3.5.1 for this VIE.

Exercise 3.7.24 Consider the non-linear VFIEs

$$u(t) = f(t) + \int_0^{\theta(t)} K(t,s)G(s,u(s))\,ds, \quad t \geq 1,$$

and

$$u(t) = f(t) + \int_0^t K(t,s)G(s,u(s),u(\theta(s)))\,ds, \quad t \geq 0,$$

where $f \in C(I)$, $K \in C(D)$, $G \in C(I \times \mathbb{R})$, $G = G(s,u)$ is smooth with respect to u, and $\theta(t) = qt$ $(0 < q < 1)$. Show that these VFIEs possess (local) unique continuous solutions $u(t)$ on some interval $[0,\delta] \subset I$, with $\delta > 0$. What can be said about their regularity properties if the given functions have continuous derivatives of order $d \geq 1$?

Exercise 3.7.25

$$u(t) = f(t) + \int_0^{\theta(t)} k(t-s)G(u(s))\,ds, \quad t \geq 0,$$

with vanishing delay $\theta(t) = qt$ $(0 < q < 1)$. If f, k and G satisfy the assumptions of Theorem 3.2.14, can the solution of the above VFIE blow up in finite time?

Exercise 3.7.26 Assume that in the VFIE

$$u(t) = u_0 + \int_0^{\theta(t)} r(s)G(u(s))\,ds, \quad t \geq 0,$$

the given functions r and G satisfy the hypotheses stated in Theorem 3.3.2, and $\theta(t)$ corresponds to a vanishing delay (that is, the conditions (V1)–(V3) of Section 3.6.1 hold). Can the solution of this VIE quench in finite time?

Exercise 3.7.27 Suppose that the given functions f and k describing the non-linear VFIEs (3.6.1) and (3.6.2) with proportional delay $\tau(t) = (1 - q)t$ $(0 < q < 1)$ are subject to conditions analogous to the ones in the classical existence theorem (Theorem 3.1.2) for non-linear VIEs. Prove that under these conditions both VFIEs possess unique continuous solutions on some interval $[0, \delta_0]$, and determine δ_0.

Exercise 3.7.28 (*Research Problem*)
In Brunner & Maset (2010) the so-called *time transformation* was used to transform a general state-dependent DDE into one with prescribed delay that is not state-dependent. Extend this approach to state-dependent VFIEs of the forms (3.6.7) and (3.6.8).

3.8 Notes

Section 3.1: *Non-Linear Second-Kind VIEs*
The theory of non-linear Volterra integral equations of the second kind has its origin in the 1908 paper by Lalesco, and the follow-up papers by Cotton (1910, on systems of VIEs of Hammerstein type) and Picone (1910); see also Galajikian (1913) for details. In his 1912 master's thesis and two corresponding papers, Galajikian (1913, 1914) extended their analysis to (systems of) more general non-linear VIEs of the form

$$u(t) = G\left(t, \int_0^t k_1(t, s, u(s))ds, \int_0^t k_2(t, s, u(s))ds\right).$$

In his 1930 paper A. Hammerstein analysed the solvability of the non-linear *Fredholm-type* integral equation

$$u(t) = f(t) + \int_0^T K(t, s)G(s, u(s))ds, \quad t \in I.$$

Hammerstein's analysis was extended by Nemytzkiĭ (1934); see also the 1975 paper by Brezis & Browder, and the books by Tricomi (1957), Krasnosel'skiĭ & Zabreĭko (1984), and Corduneanu (1991, pp. 86–88) for the theory of Hammerstein equations of Fredholm or Volterra type. The monograph by Gripenberg, Londen & Staffans (1990) offers a comprehensive general treatment of Volterra–Hammerstein integral equations (see in particular Theorems 12.1.1 and 12.1.2, and Section 11.3 for equations of convolution type). Non-linear superposition operators are studied in the monograph by Appell & Zabrejko (1990). O'Regan & Meehan (1998, Section 4.3) contains a concise discussion of general existence results (based on monotonicity arguments) for Volterra–Hammerstein integral equations (also for the unbounded interval $[0, \infty)$; cf. pp. 85–89).

Further developments, both in theory and applications, of non-linear VIEs are due to Tonelli (1928b), Sato (1951, 1953a), Mann & Wolf (1951), Roberts & Mann (1951), Iwasaki (1955: systems of non-linear VIEs), Picone (1960) and Nohel (1962). A detailed survey (including a comprehensive list of references) of the early developments of this theory can be found in Wouk (1964), Nohel (1964, 1976) and Walter (1967, dealing with non-linear VIEs in several variables). In addition, see Miller & Sell (1968, 1970), Nohel & Shea (1976 on frequency domain methods, p. 283), Herdman (1977, on maximally defined solutions) and von Wolfersdorf (2000).

Detailed analyses of the *continuation of solutions* of non-linear VIEs can be found in the books by Miller (1971, pp. 93–98 and 116–120), Burton (1983, pp. 66–89) and, especially, Gripenberg, Londen & Staffans (1991, Section 13.4). We note that the behaviour of the solution of (3.1.1) on its maximal interval of existence $[0, \delta)$, as $t \to t^-$, is also discussed in Herdman (1976, 1977).

Miller (2000) describes the pioneering role the group at the University of Wisconsin at Madison played in the development of the theory of non-linear VIEs. The classical book on non-linear VIEs is Miller (1971). Artstein (1975) contains a deep study of the topologies on the class of kernel functions k so that of the solution of (3.1.1) depends continuously on f and k (see also the related paper by Gyllenberg, 1981). The most comprehensive expositions of existence results for general (systems of) non-linear VIEs are due to Gripenberg, Londen & Staffans (1990: see in particular Chapters 11, 12 and 13) and to O'Regan & Meehan (1998). A more concise treatment can be found in Corduneanu (1991: this book starts with an excellent overview of the many contributions to the subject). The 2013 monograph by Agarwal, O'Regan & Wong is dedicated to the study of constant-sign solutions of general non-linear Fredholm and Volterra integral equations; VIEs are discussed in Chapters 9–12 and 19–20.

The theory of *maximal solutions* for the initial-value problem

$$u'(t) = F(t, u(t)), \quad t \in (a, b); \quad u(a) = u_0, \tag{3.8.1}$$

is well known (see, for example, Coppel, 1965, pp. 28–29): if F is continuous in an open set $J := (a, b) \times \Omega$, with $\Omega \subset \mathbb{R}$, a solution $\bar{u}(t)$ of the above problem is said to be a (right) *maximal solution* [*minimal solution*] on J if any solution $u(t)$ of (3.8.1) such that $u(t_0) \leq \bar{u}(t_0)$ [$u(t_0) \geq \underline{u}(t_0)$] satisfies $u(t) \leq \bar{u}(t)$ [$u(t) \geq \underline{u}(t)$] for every

$t > t_0$ for which is defined. It follows that the above initial-value problem possesses a unique maximal [minimal] solution passing through any point $(t_0, u_0) \in J$. An early study of maximal solutions (and comparison theorems) for non-linear VIEs in Sato & Iwasaki (1955). Detailed treatments of such solutions can be found in Miller (1971) and Gripenberg, Londen & Staffans (1990, Section 13.4).

The existence of *multiple solutions* for non-linear VIEs was studied in e.g. Gripenberg (1981, 1990), Bushell & Okrasiński (1989), Okrasiński (1991a, 1991b, 1992, 2000), and Castillo & Okrasiński (1993, 1994). For homogeneous VIEs of Hammerstein type with $G(0) = 0$ the existence of *non-trivial solutions* is discussed in Gripenberg (1981a, 1990), Mydlarczyk (1991, 1992), Bushell & Okrasiński (1992, 1996), Arias & Benítez (2003a, 2003b), Niedziela & Okrasiński (2006), Małolepszy & Okrasiński (2008). The papers by Mydlarczyk (2001) and Mydlarczyk & Okrasińskii (2001) are concerned with this problem for VIEs of the form

$$u(t) = \int_{-\infty}^{t} (t-s)^{\alpha-1} G(u(s)) ds \quad (t \in \mathbb{R}), \text{ with } 0 < \alpha < 1 \text{ and } \alpha >$$

1, respectively. For a certain class of non-homogeneous VIEs, a similar problem is studied (in the context of finite-time blow-up) in Roberts, Lasseigne & Olmstead (1993) and Olmstead, Roberts & Deng (1995) (an extension of the results in these two papers can be found in Mydlarczyk (1999)), and in Olmstead & Roberts (1996), Mydlarczyk, Okrasiński & Roberts (2005) and Kirk, Olmstead & Roberts (2013).

For *comparison theorems*, see Beesack (1969), Miller (1971), Cochran (1972), Gripenberg, Londen & Staffans (1990), Pachpatte (1998, Ch. 2) and Meehan & O'Regan (1999). Qun Lin's 1963 paper appears to be the first study of comparison theorems for systems of VFIEs with constant delay $\tau > 0$.

Among the papers on *variation-of-constants formulas* for VIEs are the ones by Brauer (1972), Bernfeld & Lord (1978), Friedman (1963) and Diekmann & van Gils (1981: variation-of-constants formula for non-linear VIEs with convolution kernels), and Beesack (1987); see also Brauer (1972) as well as the recent unifying treatment in Agyingi & Baker (2013).

Non-linear VIEs as *dynamical systems* are investigated in the 1970 monograph by Miller & Sell; see also Diekmann & van Gils (1984, for an illuminating discussion of invariant manifolds), Arias & Castillo (1999), Arias (2000), Arias & Benítez (2003a, 2003b) and Arias et al. (2005, 2007).

Section 3.2: *Solutions with Finite-Time Blow-Up:*
The study of finite-time blow-up in semilinear parabolic PDEs originates with Fujita's 1966 paper; see the review papers by Levine (1990), Bandle & Brunner (1998), Deng & Levine (2000). Early treatments of finite-time blow-up in VIEs are due to Miller (1971, pp. 46–51) and Herdmann (1976, 1977); see also Mann & Wolf (1951). The 1990s saw a considerable interest in VIEs with finite-time blow-up; cf. Okrasinski (1991), Mydlarczyk (1991, 1992, 1994, 1996), Bushell & Okrasiński (1992, 1996), and Mydlarczyk & Okrasiński (2003). The review papers by Roberts (1998, 2007) describe this development; see also Brunner & Yang (2013) for an extensive list of references.

Results on the blow-up time and the blow-up profile of the solutions can be found in Małolepszy & Okrasiński (2010), Olmstead (1983), Roberts & Olmstead (1996), Roberts (1997, 2007). The paper by Kadem et al. (2014) is concerned with finite-time blow-up in systems of VIEs with exponential non-linearities. The general theory of finite-time blow-up for general Hammerstein VIEs was established by Brunner & Yang (2013). The analogous analysis for non-Hammerstein VIEs remains to be carried out.

Section 3.3: *Quenching of Solutions:*
As we mentioned at the beginning of Section 3.3, the quenching of solutions to initial-boundary-value problems for semilinear parabolic partial differential equations was first studied by Kawarada (1975). Comprehensive urveys of quenching results for semilinear parabolic PDEs can be found in Levine (1983, 1992).

The analysis of quenching in VIEs has its origin in the mid-1990s; see Deng & Roberts (1997), Deng & Xu (1999), Olmstead & Roberts (2001), and Kirk & Roberts (2002). A review of such quenching results can be found in Roberts (2007). The paper by Yang & Brunner (2014) presents the general quenching theory for VIEs with various types of convolution kernels; it also contains a comprehensive list of references.

Section 3.4: *Other Types of Non-Linear VIEs*
The existence and regularity of solutions to general (non-Hammerstein) VIEs (3.4.1) with bounded kernels ($\alpha = 1$) was analysed by Guan, Zhang & Zou (2012). Non-linear VIEs whose kernels contain so-called space maps g,

$$u(t, x) = f(t, x) + \int_0^t k(t - s)u(g(t, x, v), v)\, dv,$$

were studied by Annunziato & Messina (2010) and Annunziato, Brunner & Messina (2012). Such VIEs represent an alternative and simplified formulation of the Fokker–Planck–Kolmogorov equation for piecewise deterministic processes (see Section 9.2.7).

The papers by Kosel & Wolfersdorf (1986), von Wolfersdorf & Janno (1995) and von Wolfersdorf (2000, 2008, 2011) contain detailed studies of second-kind *auto-convolution VIEs*. Von Wolfersdorf (2000, pp. 10–13) discusses the existence of solutions of the general auto-convolution VIE

$$u(t) = f(t) + \int_0^t K_0(t, s)G_0(u(s))\, ds + \int_0^t K_1(t, s)G_1(u(t - s))G_2(u(s))\, ds.$$

Auto-convolution VIEs of the *third kind* were studied by Berg & von Wolfersdorf (2005: homogeneous VIEs), Janno & von Wolfersdorf (2005: non-homogeneous VIs), von Wolfersdorf (2007), and von Wolfersdorf & Janno (2008). This work is extended to VIEs of the form

$$a(t)u(t) = f(t) + \int_0^t u(t - s)u(s)\, ds + \lambda \int_0^t (t - s)^{\alpha-1}u(s)\, ds \ (0 < \alpha < 1),$$

with $a(t) \sim ct^\alpha$ near $t = 0^+$, in von Wolfersdorf (2011). Some key papers on *first-kind* auto-convolution VIEs are Gorenflo & Hofmann (1994), Fleischer & Hofmann (1996, 1997), and Fleischer, Gorenflo & Hofmann (1999). The paper by Dai & Lamm (2008) contains a good list of references on such ill-posed problems.

Adjoint auto-convolution VIEs were studied extensively by e.g. Pimbley (1967), Ramalho (1976), and, above all, by Nussbaum (1980, 1981) and Nussbaum & Baxter (1981).

A general result on the existence of solutions to *implicit VIEs* is given in Deimling (1989, 1995). Most papers deal with implicit VIEs of the form $u^\beta(t) = f(t) + (\mathcal{V}u)(t)$, where either $0 < \beta < 1$ or $\beta > 1$. Papers representing the former case are Okrasiński (1979, 1984), Askhabov & Karapetyants (1990), Askhabov (1991) and Bushell & Okrasiński (1992); see also Kilbas & Saigo (1990), Buckwar (1997, 2000, 2005), and Lipovan (2009). The case $\beta > 1$ is considered in e.g. Okrasiński (1978, 1979, 1980), Askhabov (1991) and Askhabov & Betilgiriev (1993). The most detailed treatment (for any $\beta > 0$) is arguably the one in Karapetyants, Kilbas, Saigo & Samko (2000).

Related VIEs are studied in von Wolfersdorf (2000, p. 27) and Sloss (2002: VIEs of the form

$$Au^\beta(t) + Bu^\gamma(t) = \int_0^t K(t,s)u(s)\,ds,$$

with $\beta, \gamma \in \mathbb{R}$ and $A, B \neq 0$).

Section 3.5: *Non-Linear First-Kind VIEs*
Such VIEs were analysed by Janikowski (1962, for $0 < \alpha < 1$), Branca (1978, $\alpha = 1/2$), Dixon, Mckee & Jeltsch (1986, $\alpha = 1$), Ang & Gorenflo (1991), Gorenflo & Pfeiffer (1991), and by Deimling (1995, $0 < \alpha \le 1$). The latter paper makes no reference to the earlier studies.

Section 3.6: *Non-Linear VFIEs:*
Chambers (1990) established results on the existence and uniqueness of solutions for non-linear VFIEs of the form $u(t) = f(t) + \int_0^{qt} k(t,s,u(s))ds$ with $0 < q < 1$, as well as for multi-dimensional analogues. The finite-time blow-up of solutions to VFIEs of Hammerstein type was analysed (both for vanishing and non-vanishing delays) in Yang & Brunner (2013b).

The first discussion of a state-dependent delay differential appears to be in a paper by Poisson in 1806 (compare Walther, 2013 for details of that paper and for an illuminating discussion of that DDE). The existence and the behaviour of solutions (e.g. the regularity and the primary discontinuity points) of state-dependent DDEs are discussed, for example, in Driver (1963), Neves & Feldstein (1976), Feldstein & Neves (1984); Willé & Baker (1992, 1994), Bellen & Zennaro (2003, pp. 30–35), and in the review paper by Feldstein, Neves & Thompson (2006). See also Brunner & Maset (2010) where so-called time transformations are introduced to reduce a state-dependent DDE to a DDE with prescribed (e.g. linear) delay. The review paper by Hartung, Krisztin, Walther & Wu (2006) contains a wide-ranging survey of the theory of state-dependent VFIEs.

Certain classes of state-dependent VFIEs are studied in Kwapisz & Turo (1983), Bélair (1991), Cahlon (1992, 1993), Torrejón (1993), Gołaszewska & Turo (2010: state-dependent VFIEs with several variables), and in Lizama & Velasco (2015). See also Turo (1995) on stochastic VFIEs with state-dependent delays.

Numerical Analysis of Non-Linear VFIEs

The papers by Blom & Brunner (1987, 1991) describe the theory and the computational implementation of superconvergent collocation and iterated collocation methods for (systems of) non-linear VIEs. See also Brunner & van der Houwen (1986) on Runge-Kutta and linear multistep methods for such VIEs, and Brunner (2004) on piecewise polynomial collocation metjods. The numerical analysis and computational solution of VIEs with blow-up solutions are studied in Calabrò & Capobianco (2009) and in Yang & Brunner (2013b). Zhang, Liang & Brunner (2016) analysed the optimal order of convergence of picewise polynomial collocation methods for approximating solutions of auto-convolution VIEs of the second kind.

The 2003 monograph by Bellen and Zennaro and the 2009 review paper by Bellen, Maset, Zennaro and Guglielmi in *Acta Numerica* are the foremost references on the numerical analysis and computational solution of delay differential equations. For various aspects of the computation of primary discontinuity points (especially for state-dependent delays) we refer to Willé & Baker (1992, 1994), Feldstein, Neves & Thompson (2006), Bellen, Maset, Zennaro & Guglielmi (2009), Bellen & Guglielmi (2009: neutral DDEs), and Brunner & Maset (2010).

Collocation methods for VFIEs are discussed in Brunner (2004) and in the review article Brunner (2004a) in *Acta Numerica* (the latter contains an extensive list of references). Various numerical approaches to VFIEs with state-dependent delays are described in the papers by Cahlon (1992), Khasi, Ghoreishi & Hadizadeh (2014), and Lizima & Velasco (2015).

4

Volterra Integral Equations with Highly Oscillatory Kernels

Summary

Are the solutions of Volterra integral equations with highly oscillatory kernels also highly oscillatory? The answer to this question, for second-kind and first-kind VIEs with smooth or weakly singular kernels, will be explored in this chapter. Of particular interest will be the asymptotic behaviour of the solutions as the oscillation parameter ω tends to infinity.

4.1 Introduction

This chapter is dedicated to the study of solutions of linear VIEs described by the Volterra integral operator

$$(\mathcal{V}_{\omega,\alpha}u)(t) := \int_0^t (t-s)^{\alpha-1} K_\omega(t,s) u(s)\, ds, \ \ t \in I = [0,T] \ (0 < \alpha \le 1),$$

$$(4.1.1)$$

whose kernel

$$K_\omega(t,s) := K(t,s) e^{i\omega g(t,s)} \ \ (\omega \gg 1) \tag{4.1.2}$$

is *highly oscillatory*. It will be assumed that K and the oscillator g are real-valued, do not depend on ω and are arbitrarily smooth on D, and that g is non-negative on D. We will often refer to VIEs with highly oscillatory kernels simply as *highly oscillatory VIEs*.

In order to obtain insight into the basic properties of solutions to highly oscillatory VIEs of the second kind,

$$u(t) = f(t) + (\mathcal{V}_{\omega,\alpha}u)(t), \ \ t \in I, \tag{4.1.3}$$

and highly oscillatory VIEs of the first kind,

$$(\mathcal{V}_{\omega,\alpha}u)(t) = f(t), \ \ t \in I, \tag{4.1.4}$$

175

we mostly focus on *separable oscillators*; that is, on oscillators of the form

$$g(t, s) = g_0(t) - g_0(s), \quad (t, s) \in D, \tag{4.1.5}$$

where g_0 is smooth and strictly increasing on I, with $g_0(0) \geq 0$. The *linear oscillator*

$$g(t, s) = t - s, \quad (t, s) \in D, \tag{4.1.6}$$

corresponds to $g_0(t) = t$.

Other types of highly oscillatory VIEs will be considered in later sections. Typical examples correspond to the kernels $K(t, s) \cos(\omega(t - s))$ and $K(t, s) J_0(\omega(t - s))$ (where J_0 is the Bessel function of order zero).

4.2 VIEs of the Second Kind

4.2.1 Smooth Kernels

Consider the highly oscillatory linear VIE (4.1.3) with $\alpha = 1$. The following theorem reveals that for highly oscillatory kernels $K_\omega(t, s)$ with separable oscillators the resolvent kernel $R_\omega(t, s)$ inherits the highly oscillatory term in $K_\omega(t, s)$.

Theorem 4.2.1 *Assume that the (real-valued) functions K and g have the following properties:*

(i) K is continuous on D and does not depend on ω;
(ii) the separable oscillator $g(t, s) = g_0(t) - g_0(s)$ corresponds to smooth and strictly increasing g_0, with $g_0(0) \geq 0$, which does not depend on ω.

Then the following statements are true:

(a) The resolvent kernel associated with the kernel $K_\omega(t, s)$ in (4.1.1) (with $\alpha = 1$) is given by

$$R_\omega(t, s) = R(t, s) e^{i\omega g(t,s)}, \tag{4.2.1}$$

where the resolvent kernel $R(t, s)$ of the kernel $K(t, s)$ does not depend on ω.

(b) For any $f \in C(I)$ the highly oscillatory second-kind VIE (4.1.3) possesses a unique solution $u \in C(I)$ that has the form

$$u(t) = f(t) + \int_0^t R_\omega(t, s) f(s) \, ds, \quad t \in I. \tag{4.2.2}$$

Proof We first multiply both sides of (4.1.3) (with $\alpha = 1$) by $e^{-i\omega g_0(t)}$ and set

$$u_\omega(t) := e^{-i\omega g_0(t)} u(t) \quad \text{and} \quad f_\omega(t) := e^{-i\omega g_0(t)} f(t). \tag{4.2.3}$$

This yields a second-kind VIE for u_ω, namely

$$u_\omega(t) = f_\omega(t) + \int_0^t K(t, s) u_\omega(s)\, ds, \quad t \in I. \tag{4.2.4}$$

It follows from the classical Volterra theory of Section 1.2 that its unique solution u_ω is in $C(I)$ and has the form

$$u_\omega(t) = f_\omega(t) + \int_0^t R(t, s) f_\omega(s)\, ds, \quad t \in I,$$

where $R(t, s)$ denotes the resolvent kernel of the given kernel $K(t, s)$. The assertions (a) and (b) then follow immediately from the definitions (4.2.3) since $u(t) = e^{i\omega g_0(t)} u_\omega(t)$. $\qquad\square$

We use a simple example to illustrate the statements (a) and (b) in Theorem 4.2.1 for the case of the *linear oscillator* (4.1.6) (corresponding to $g_0(t) = t$ in (4.1.5)), and to obtain some first insight into the damping of the solutions as ω tends to ∞.

Example 4.2.2 If $K(t, s) = \lambda \neq 0$, its resolvent kernel is $R(t, s) = \lambda e^{\lambda(t-s)}$. Thus, by (4.2.1) and (4.2.2) the solution of the corresponding highly oscillatory VIE (4.1.3) with $f(t) = 1$ and $\alpha = 1$ is readily seen to be

$$u(t) = 1 - \frac{\lambda}{\lambda + i\omega} + \frac{\lambda e^{\lambda t}}{\lambda + i\omega} e^{i\omega t}, \quad t \in I.$$

It is highly oscillatory for large ω but tends to the value 1 as ω tends to ∞. (For $\omega = 0$ we have of course $u(t) = e^{\lambda t}$.) We observe that the highly oscillatory term in the expression for $u(t)$ contains the *damping factor* $(\lambda + i\omega)^{-1}$.

This raises the question as to whether *stronger damping* of the highly oscillatory terms in the corresponding solutions of (4.1.3) (with $\alpha = 1$) is possible for certain classes of smooth functions f and non-constant kernels K.

We shall find that in the case of a constant kernels $K(t, s) = \lambda \neq 0$ in (4.1.2) with $\alpha = 1$, the answer is obtained as a special case of the following general theorem on the asymptotic expansion (in terms of negative powers of $i\omega$, as $\omega \to \infty$) of the solution of (4.1.3) with $\alpha = 1$,

$$u(t) = f(t) + \int_0^t K(t, s) e^{i\omega g(t, s)} u(s)\, ds, \quad t \in I, \tag{4.2.5}$$

where $g(t, s) = t - s$. As its proof will reveal, the result of Theorem 4.2.3 is readily extended to separable oscillators $g(t, s) = g_0(t) - g_0(s)$.

Theorem 4.2.3 *Assume that $f \in C^{q+1}(I)$ and $K \in C^{q+1}(D)$ for some sufficiently large q. Then the solution of the highly oscillatory VIE (4.2.5) with $g(t, s) = t - s$ can be written as*

$$
u(t) = f(t) - \sum_{j=0}^{q} (i\omega)^{-(j+1)} \sum_{\ell=0}^{j} \binom{j}{\ell} \frac{\partial^{j-\ell} R(t, t)}{\partial s^{j-\ell}} f^{(\ell)}(t)
$$

$$
+ \sum_{j=0}^{q} (i\omega)^{-(j+1)} \sum_{\ell=0}^{j} \binom{j}{\ell} \frac{\partial^{j-\ell} R(t, 0)}{\partial s^{j-\ell}} f^{(\ell)}(0) \, e^{i\omega t} \qquad (4.2.6)
$$

$$
+ (i\omega)^{-(q+1)} \int_{0}^{t} Q_{q+1}(t, s) e^{i\omega(t-s)} \, ds,
$$

where

$$
Q_{q+1}(t, s) := \sum_{j=0}^{q+1} \binom{q+1}{j} \frac{\partial^{q+1-j} R(t, s)}{\partial s^{q+1-j}} f^{(j)}(s).
$$

Proof According to (4.2.1) with $g(t, s) = t - s$, the resolvent kernel associated with the kernel $K_\omega(t, s) = K(t, s) e^{i\omega(t-s)}$ is $R_\omega(t, s) = R(t, s) e^{i\omega(t-s)}$, with $R(t, s)$ denoting the resolvent kernel of $K(t, s)$. Hence the solution of (4.2.5) is

$$
u(t) = f(t) + \int_{0}^{t} Q_0(t, s) e^{i\omega(t-s)} \, ds, \qquad (4.2.7)
$$

where we have set $Q_0(t, s) := R(t, s) f(s)$. The assumptions on f and K imply that $Q_0 \in C^{q+1}(D)$. Set $Q_j(t, s) := \partial Q_{j-1}(t, s) / \partial s$ $(j \geq 1)$ and use repeated integration by parts in (4.2.7). This leads to

$$
u(t) = f(t) - \sum_{j=0}^{q} (i\omega)^{-(j+1)} Q_j(t, s) + \sum_{j=0}^{q} (i\omega)^{-(j+1)} Q_j(t, 0) e^{i\omega t}
$$

$$
+ (i\omega)^{-(q+1)} \int_{0}^{t} Q_{q+1}(t, s) e^{i\omega(t-s)} \, ds.
$$

Since the definition of $Q_j(t, s)$ implies that

$$
Q_j(t, s) = \frac{\partial^j Q_0(t, s)}{\partial s^j} = \sum_{\ell=0}^{j} \binom{j}{\ell} \frac{\partial^{j-\ell} R(t, s)}{\partial s^{j-\ell}} f^{(\ell)}(s),
$$

the statement (4.2.6) follows.   □

Theorem 4.2.3 yields the following generalisation of the result in Example 4.2.2.

Corollary 4.2.4 *Let* $f \in C^{q+1}(I)$ *for some* $q \in \mathbb{N}$. *Then the solution of the second-kind VIE*

$$u(t) = f(t) + \lambda \int_0^t e^{i\omega(t-s)}u(s)\,ds, \ t \in I,$$

has the form

$$u(t) = f(t) - \lambda \sum_{j=0}^q \frac{f^{(j)}(t)}{(\lambda + i\omega)^{j+1}} + \lambda \sum_{j=0}^q \frac{f^{(j)}(0)}{(\lambda + i\omega)^{j+1}} e^{\lambda t} e^{i\omega t} \quad (4.2.8)$$

$$+ \frac{\lambda}{(\lambda + i\omega)^{q+1}} \int_0^t e^{(t-s)(\lambda + i\omega)} f^{(q+1)}(s)\,ds.$$

Remark 4.2.5

(i) If $f \in C^{q+1}(I)$ ($q \geq 0$) is such that $f^{(j)}(0) = 0$ for $j = 0, 1, \ldots, r$ ($r < q$) and if the oscillator $g(t, s)$ is *separable*, the corresponding second sum in (4.2.9) reveals that the damping factor in the highly oscillatory term of the solution $u(t)$ is $(\lambda + i\omega)^{-(r+1)}$. In Example 4.2.2 we have $f(0) = 1 \neq 0$ and hence the damping factor is $(\lambda + i\omega)^{-1}$.

(ii) Some other types of highly oscillating kernels, e.g. $K_\omega(t, s) = \cos(\omega(t - s))$ and $K_\omega(t, s) = J_0(\omega(t - s)))$ (where J_0 denotes the Bessel function of order zero) will be briefly considered in Section 4.3.3 and Section 4.5 (Exercises 4.5.2 and 4.5.3). ◇

4.2.2 VIEs with Weakly Singular Kernels

The following theorem (the analogue of Theorem 4.2.1) reveals that for separable oscillators the resolvent kernel $R_{\omega,\alpha}(t, s)$ inherits both the highly oscillatory term and the singularity of the weakly singular kernel $K_{\omega,\alpha}(t, s)$ in the VIE (4.1.3).

Theorem 4.2.6 *If the functions* K *and* g *have the properties described in assumptions (i) and (ii) of Theorem 4.2.1, the following statements hold:*

(a) *For any* $\alpha \in (0, 1)$, *the resolvent kernel associated with the kernel of (4.1.3) is given by*

$$R_{\omega,\alpha}(t, s) = R_\alpha(t, s)e^{i\omega g(t,s)}, \quad (4.2.9)$$

where the resolvent kernel $R_\alpha(t,s)$ of the kernel $(t-s)^{\alpha-1}K(t,s)$ does not depend on ω. The resolvent kernel may be written as

$$R_{\omega,\alpha}(t,s) = (t-s)^{\alpha-1}Q_{\omega,\alpha}(t,s), \quad 0 \le s < t \le T, \qquad (4.2.10)$$

where

$$Q_{\omega,\alpha}(t,s) := Q_\alpha(t,s)e^{i\omega g(t,s)},$$

for some $Q_\alpha \in C(D)$ that does not depend on ω.
(b) For any $f \in C(I)$ the highly oscillatory VIE (4.1.3) possesses a unique solution $u \in C(I)$ that is given by

$$u(t) = f(t) + \int_0^t R_{\omega,\alpha}(t,s)f(s)\,ds, \quad t \in I. \qquad (4.2.11)$$

Proof Multiply both sides of (4.1.3) (with $0 < \alpha < 1$) by $e^{-i\omega g_0(t)}$ and set

$$u_\omega(t) := e^{-i\omega g_0(t)}u(t) \quad \text{and} \quad f_\omega(t) := e^{-i\omega g_0(t)}f(t). \qquad (4.2.12)$$

This yields a second-kind VIE for u_ω, namely

$$u_\omega(t) = f_\omega(t) + \int_0^t (t-s)^{\alpha-1}K(t,s)u_\omega(s)\,ds \quad (0 < \alpha \le 1). \qquad (4.2.13)$$

It follows from Theorem 1.3.5 that the unique solution $u_\omega \in C(I)$ of (4.2.13) has the form

$$u_\omega(t) = f_\omega(t) + \int_0^t R_\alpha(t,s)f_\omega(s)\,ds,$$

where $R_\alpha(t,s)$ denotes the resolvent kernel of the given kernel $(t-s)^{\alpha-1}K(t,s)$. If $0 < \alpha < 1$ the resolvent kernel inherits the weak singularity: we have

$$R_\alpha(t,s) = (t-s)^{\alpha-1}Q_\alpha(t,s)$$

for some $Q_\alpha \in C(D)$ (cf. the proof of Theorem 2.2.1). The assertions (a) and (b) then follow immediately from the definitions (4.2.12).   □

Example 4.2.7 Consider the VIE (4.1.3) with $K(t,s) = \lambda/\Gamma(\alpha)$ ($\lambda \ne 0$, $0 < \alpha < 1$) and $g(t,s) = t - s$,

$$u(t) = f(t) + \lambda \int_0^t \frac{(t-s)^{\alpha-1}}{\Gamma(\alpha)}e^{i\omega(t-s)}u(s)\,ds, \qquad (4.2.14)$$

where $f \in C(I)$. By Theorem 4.2.6(b) its unique continuous solution u is given by

$$u(t) = f(t) + \int_0^t R_\alpha(t,s)e^{i\omega(t-s)}f(s)\,ds, \quad t \in I, \qquad (4.2.15)$$

where

$$R_\alpha(t,s) = \frac{d}{dt} E_\alpha\big(\lambda(t-s)^\alpha\big) = (t-s)^{\alpha-1} \sum_{n=1}^{\infty} \frac{(\lambda\Gamma(\alpha))^n}{\Gamma(n\alpha)} (t-s)^{(n-1)\alpha}, \ (t,s) \in D$$

(cf. Corollary 1.3.8). Theorem 4.2.8 below will reveal that the solution of the weakly singular VIE (4.2.14) is highly oscillatory.

Does an asymptotic expansion, as ω tends to infinity, analogous to (4.2.6) hold for the solution of the general weakly singular VIE with highly oscillatory kernel,

$$u(t) = f(t) + \int_0^t (t-s)^{\alpha-1} K(t,s) u(s) \, ds, \ t \in I, \tag{4.2.16}$$

when $0 < \alpha < 1$? The next theorem, the analogue of Theorem 4.2.3, gives an affirmative answer to this question.

Theorem 4.2.8 *Let u be the solution of the highly oscillatory VIE (4.2.16) with $f \in C^{q+1}(I)$, $K \in C^{q+1}(D)$ and $0 < \alpha < 1$. Then there exist functions $G_{\alpha,j}^{(n)}$ and $Q_{\alpha,j}^{(n)}$ ($j = 0, 1, \ldots, q+1;\ n \geq 0$) in $C^{q+1}(D)$ so that for $\omega \to \infty$, the solution has the asymptotic expansion*

$$u(t) = f(t) - \sum_{j=0}^{q} (i\omega)^{-(j+1)} \left(\sum_{n=1}^{\infty} G_{\alpha,j}(t,0) t^{n\alpha} \right) e^{i\omega t}$$

$$+ \sum_{j=0}^{q} (i\omega)^{-(j+1)} \left(\sum_{n=1}^{\infty} Q_{\alpha,j}^{(n)}(t,t) \int_0^t (t-s)^{n\alpha-1} e^{i\omega(t-s)} \, ds \right) \tag{4.2.17}$$

$$+ (i\omega)^{-(q+1)} \sum_{n=1}^{\infty} \int_0^t (t-s)^{n\alpha-1} Q_{\alpha,q+1}^{(n)}(t,s) e^{i\omega(t-s)} \, ds,$$

where the infinite series converge absolutely and uniformly on I. The meaning of $G_{\alpha,j}$ and $G_{\alpha,j}$ is made precise in the proof below.

Proof The starting point of the proof of (4.2.17) is the solution representation (4.2.11) in Theorem 4.2.6 (see also Theorem 2.1.2),

$$u(t) = f(t) + \int_0^t (t-s)^{\alpha-1} \sum_{n=1}^{\infty} (t-s)^{(n-1)\alpha} \Phi_{\alpha,n}(t,s) e^{i\omega(t-s)} f(s) \, ds,$$

where $\Phi_{\alpha,n} \in C^{q+1}(D)$ and $\Phi_{\alpha,1}(t,s) = K(t,s)$. Owing to the presence of the singular (or non-smooth) terms $(t-s)^{\alpha-1}(t-s)^{(n-1)\alpha} = (t-s)^{n\alpha-1}$, we cannot directly use the approach employed in the proof of Theorem 4.2.3.

Instead, we proceed by subtraction of the singularity, as follows. Setting $Q_{\alpha,n}^{(0)}(t,s) := \Phi_{\alpha,n}(t,s)f(s)$, we write

$$Q_{\alpha,n}^{(0)}(t,s) = \left(Q_{\alpha,n}^{(0)}(t,s) - Q_{\alpha,n}^{(0)}(t,t)\right) + Q_{\alpha,n}^{(0)}(t,t). \tag{4.2.18}$$

Since $Q_{\alpha,n}^{(0)} \in C^{q+1}(D)$, the use of Taylor's formula with remainder term leads to

$$Q_{\alpha,n}^{(0)}(t,s) - Q_{\alpha,n}^{(0)}(t,t) = \sum_{j=1}^{q} \frac{(-1)^j}{j!} \frac{\partial^j Q_{\alpha,n}^{(0)}(t,t)}{\partial s^j}(t-s)^j + (t-s)\Psi_{q,n}^{(0)}(t,s)$$

$$= (t-s)\left(\sum_{j=1}^{q} \frac{(-1)^j}{j!} \frac{\partial^j Q_{\alpha,n}^{(0)}(t,t)}{\partial s^j}(t-s)^{j-1}\right.$$

$$\left. + \Psi_{q,n}^{(0)}(t,s)\right)$$

$$=: (t-s)G_{\alpha,n}^{(0)}(t,s).$$

Thus, on I the solution u can be expressed in the form

$$u(t) = f(t) + \sum_{n=1}^{\infty} \int_0^t (t-s)^{n\alpha-1}\left((t-s)G_{\alpha,n}^{(0)}(t,s) + Q_{\alpha,n}^{(0)}(t,t)\right)e^{i\omega(t-s)}\,ds$$

$$= f(t) + \sum_{n=1}^{\infty} \int_0^t (t-s)^{n\alpha}G_{\alpha,n}^{(0)}(t,s)\,ds$$

$$+ \sum_{n=1}^{\infty} Q_{\alpha,n}^{(0)}(t,t) \int_0^t (t-s)^{n\alpha-1}e^{i\omega(t-s)}\,ds.$$

In order to deal with the integral involving $G_{\alpha,n}^{(0)}(t,s)$ we employ integration by parts to rewrite it as

$$\int_0^t (t-s)^{n\alpha}G_{\alpha,n}^{(0)}(t,s)e^{i\omega(t-s)}\,ds = \int_0^t (t-s)^{n\alpha}G_{\alpha,n}^{(0)}(t,s)\left(-\frac{1}{i\omega}\frac{\partial e^{i\omega(t-s)}}{\partial s}\right)ds$$

$$= \frac{1}{i\omega}G_{\alpha,n}^{(0)}(t,0)e^{i\omega t}t^{n\alpha}$$

$$+ \frac{1}{i\omega}\int_0^t (t-s)^{n\alpha-1}\left(-n\alpha G_{\alpha,n}^{(0)}(t,s) + (t-s)\frac{\partial G_{\alpha,n}^{(0)}(t,s)}{\partial s}\right)e^{i\omega(t-s)}\,ds$$

$$=: \frac{1}{i\omega}G_{\alpha,n}^{(0)}(t,0)t^{n\alpha}e^{i\omega t} + \frac{1}{i\omega}\int_0^t (t-s)^{n\alpha-1}Q_{\alpha,n}^{(1)}(t,s)e^{i\omega(t-s)}\,ds.$$

The subsequent steps of the proof are now clear: as in (4.2.18) we use again singularity subtraction for $Q_{\alpha,n}^{(1)}(t,s)$ in the last integral, and conclude the proof by induction. The details are left to the reader.   □

4.2.3 Comparison with Highly Oscillatory Fredholm Integral Equations

Ursell (1969) studied the asymptotic behaviour (as $\omega \to \infty$) of the solution of the Fredholm integral equation

$$u(x) = f(x) + \lambda(\mathcal{F}_\omega u)(x), \quad x \in [-1, 1] \ (\lambda \neq 0), \tag{4.2.19}$$

with

$$(\mathcal{F}_\omega u)(t) := \int_{-1}^{1} K(x, y)e^{i\omega g(x,y)}u(y)\,dy \quad (\omega \gg 1), \tag{4.2.20}$$

and the oscillator $g(x, y) = |x - y|$. He showed that when λ^{-1} is not in the spectrum of $\mathcal{F}_\omega$, and if f and K are continuous and independent of ω, the solution $u = u(x; \omega)$ of (4.2.19) behaves like

$$u(x; \omega) - f(x) = o(1) \quad \text{as} \quad \omega \to \infty.$$

Ursell's results for the above oscillator $g(x, y)$ were refined in Brunner, Iserles & Nørsett (2010). The papers by Brunner, Iserles & Nørsett (2011) and Böttcher, Brunner, Iserles & Nørsett (2010) address the analogous problem for the highly oscillatory Fredholm integral operator (4.2.20) with the 'Fox–Li' oscillator $g(x, y) = (x - y)^2$. This oscillator arises in laser and maser engineering; see Landau (1977), Trefethen & Embree (2005, Section 60), Böttcher et al. (2010), and their references. Since these bounded linear Fredholm integral operators are compact, they have at most a countable number of (complex) eigenvalues that accumulate at the origin. However, the derivation of rigorous results on their existence and asymptotic behaviour is rather difficult since these Fredholm integral operators are complex-symmetric, but not self-adjoint. As Landau says in his 1977 paper, 'this presents a major obstacle to a theoretical understanding of the equation [(4.2.19)] – indeed, even the existence of eigenvalues is difficult to prove'.

We note in passing that, to the author's knowledge, the analysis of the asymptotic behaviour of the eigenvalues of (4.2.20) with *weakly singular kernel* $|x - y|^{-\alpha}$ $(0 < \alpha < 1)$ and oscillators $|x - y|$ or $(x - y)^2$ remains open.

As we shall see in Section 7.4, the analogous problem for highly oscillatory *cordial* (non-compact) Volterra integral operators (and corresponding second-kind VIEs) with separable oscillators is much more simple: here, the eigenvalues, as well as their asymptotic properties as $\omega \to \infty$, are now completely known (see Section 7.4.1).

4.3 VIEs of the First Kind

4.3.1 VIEs with Smooth Kernels

We shall see that the oscillation properties of solutions of highly oscillatory VIEs of the first kind,

$$\int_0^t H_{\omega,\alpha}(t,s)u(s)\,ds = f(t), \quad t \in I \ (0 < \alpha \le 1), \tag{4.3.1}$$

where

$$H_{\omega,\alpha}(t,s) := (t-s)^{\alpha-1}H(t,s)e^{i\omega g(t,s)} \ (\omega \gg 1),$$

can be rather different from the ones we derived for solutions of analogous highly oscillatory second-kind VIEs of the form (4.1.3). Before making this more precise, we look at the solution of a simple *singularly perturbed* VIE with highly oscillatory kernel linking the two kinds of VIEs (compare also the 1997 review paper by Kauthen for the case $\omega = 0$).

Example 4.3.1 Consider the highly oscillatory, singularly perturbed VIE

$$\varepsilon u(t) = f(t) + \int_0^t H(t,s)e^{i\omega g(t,s)}u(s)\,ds, \quad t \in I \ (\omega \gg 1, \ 0 < \varepsilon \ll 1). \tag{4.3.2}$$

The *reduced* VIE (corresponding to $\varepsilon = 0$; see Section 5.1.2),

$$0 = f(t) + \int_0^t H(t,s)e^{i\omega g(t,s)}u(s)\,ds, \quad t \in I, \tag{4.3.3}$$

is a first-kind VIE with highly oscillatory kernel. Suppose now that in (4.3.2) we choose $f(t) = \sin(t)$, $H(t,s) \equiv -1$, and the linear oscillator $g(t,s) = t - s$. It is readily verified that for $\varepsilon > 0$ the unique solution $u \in C(I)$ of this VIE is

$$u(t) = \gamma_\varepsilon\left(\cos(t) + \varepsilon(1 - i\omega/\varepsilon - \omega^2)\sin(t) - e^{-t/\varepsilon}e^{i\omega t}\right),$$

where $\gamma_\varepsilon := \left(1 + \varepsilon^2(1 - \omega^2) - 2i\omega\varepsilon\right)^{-1}$ and $u(0) = 0$. This reveals that when $0 < \varepsilon \ll 1$, the solution contains the *highly oscillatory boundary-layer term* ('inner solution') $-\gamma_\varepsilon e^{-t/\varepsilon}e^{i\omega t}$. On the other hand, the solution

$$u_0(t) = \cos(t) - i\omega\sin(t), \quad \text{with} \quad u_0(0) = 1 - i\omega \ne 0,$$

of the reduced first-kind VIE (4.3.3) is *non-oscillatory*.

We now show that for a certain class of first-kind VIEs with smooth highly oscillatory kernels the solutions are non-oscillatory.

Theorem 4.3.2 *The solution of the first-kind VIE*

$$\int_0^t H_\omega(t, s) u(s)\, ds = f(t), \quad t \in I, \tag{4.3.4}$$

with highly oscillatory convolution kernel

$$H_\omega(t, s) := \frac{(t - s)^{r-1}}{(r - 1)!} e^{i\omega(t-s)} \quad (r \in \mathbb{N}, \ r \geq 1), \tag{4.3.5}$$

where $f \in C^r(I)$ does not depend on ω and satisfies $f^{(j)}(0) = 0$ ($j = 0, \ldots, r - 1$), is non-oscillatory:

$$u(t) = f^{(r)}(t) + \sum_{j=1}^{r} \binom{r}{j} (-i\omega)^j f^{(r-j)}(t), \quad t \in I. \tag{4.3.6}$$

For $r = 1$ the solution reduces to

$$u(t) = f'(t) - i\omega f(t). \tag{4.3.7}$$

Proof The result is trivial when $r = 0$. If $r \geq 1$, we set $u_\omega(t) := e^{-i\omega t} u(t)$, $f_\omega(t) := e^{-i\omega t} f(t)$ and then differentiate the resulting VIE for $u_\omega(t)$ $r + 1$ times, to obtain the desired result (4.3.6). $\qquad\square$

Remark 4.3.3

(i) The solution (4.3.6) reduces to the well-known expression $u(t) = f^{(r)}(t)$ of (4.3.4) when $\omega = 0$. For $\omega > 0$ it is modified by the addition of a *non-oscillatory* 'ω-perturbation' involving lower-order derivatives of $f(t)$. This raises the obvious question as to whether this remains true for first-kind VIEs with more general (continuous) kernels, and for first-kind VIEs with weakly singular kernels.

(ii) We note that the VIE (4.3.4) with kernel (4.3.5) does not have a unique continuous solution on the closed interval $[0, T]$ when $f^{(j)}(0) = 0$ for $j = 0, \ldots, q$ ($q < r - 1$) but $f^{(r-1)}(0) \neq 0$. $\qquad\diamond$

As a first step towards answering the question in (i), we consider a first-kind VIE with linear oscillator,

$$\int_0^t h(t - s) e^{i\omega(t-s)} u(s)\, ds = f(t), \quad t \in I, \tag{4.3.8}$$

where h is a smooth, non-constant convolution kernel satisfying $h(0) \neq 0$ (in contrast to the kernel (4.3.5)) and $f \in C^1(I)$, with $f(0) = 0$. If we resort again to the substitutions (4.2.5) we see that the differentiated form of the resulting first-kind VIE

$$\int_0^t h(t-s)u_\omega(s)\,ds = f_\omega(t)$$

may be written as

$$u_\omega(t) + \int_0^t h'(t-s)u_\omega(s)\,ds = f_\omega'(t), \quad t \in I \qquad (4.3.9)$$

(where we have assumed without loss of generality that $h(0) = 1$).

Theorem 4.3.4 *Assume that the resolvent kernel associated with the kernel* $-h'(t-s)$ *in (4.3.9) is* $r_1(t-s)$. *Then the (unique) continuous solution of the first-kind VIE (4.3.8) is*

$$u(t) = f'(t) - i\omega f(t) + \int_0^t r_1(t-s)\big(f'(s) - i\omega f(s)\big)e^{i\omega(t-s)}\,ds. \quad (4.3.10)$$

Proof The proof is straightforward, since by using again the notation of (4.2.5) and the fact that the solution of (4.3.9) is given by

$$u_\omega(t) = f_\omega'(t) + \int_0^t r_1(t-s)f_\omega'(s)\,ds,$$

where $f_\omega'(t) = (-i\omega f(t) + f'(t))e^{-i\omega t}$, we obtain the desired result (4.3.10).
□

Example 4.3.5 Assume that in (4.3.8) $h(t-s) = 1+t-s$. Since the resolvent kernel $r_1(t-s)$ corresponding to $-h'(t-s)$ is $r_1(t-s) = -e^{-(t-s)}$ (cf. Example 4.2.2 with $\lambda = -1$), it is readily verified that for $f(t) = t$ the VIE

$$\int_0^t (1+t-s)e^{i\omega(t-s)}u(s)\,ds = f(t), \quad t \in I, \qquad (4.3.11)$$

possesses the unique continuous solution

$$u(t) = 1 - \frac{1}{(1-i\omega)^2} + \frac{(i\omega)^2}{1-i\omega}t + \frac{e^{-t}}{(1-i\omega)^2}e^{i\omega t}, \quad t \geq 0.$$

Hence, the solution of the first-kind VIE (4.3.11) is highly oscillatory, in contrast to (4.3.6) and (4.3.7). Moreover, the highly oscillatory term contains the damping factor $(1-i\omega)^{-2}$. This is reminiscent of the damping we observed in Example 4.2.2 for second-kind VIEs, except that now we have the stronger damping $(1-i\omega)^{-2}$.

It turns out that stronger damping of the highly oscillatory component of the solution is possible for certain classes of smooth functions f in the first-kind VIE (4.3.11), in analogy to the result of Theorem 4.2.3. We will first make this precise for the VIE (4.3.11) of Example 4.3.5.

Theorem 4.3.6 *Assume that $f \in C^{q+1}(I)$ $(q \geq 1)$, with $f(0) = 0$. Then the unique solution $u \in C(I)$ of the first-kind VIE (4.3.11) with linear oscillator $g(t, s) = t - s$ can be written in the form*

$$u(t) = f'(t) - i\omega f(t) + \sum_{j=1}^{q} \frac{(-1)^{j+1} f^{(j)}(t)}{(1 - i\omega)^j} + i\omega \sum_{j=0}^{q-1} \frac{(-1)^j f^{(j)}(t)}{(1 - i\omega)^{j+1}}$$

$$+ \sum_{j=1}^{q} \frac{(-1)^{j+1} f^{(j)}(0)}{(1 - i\omega)^j} e^{-t} e^{i\omega t} + i\omega \sum_{j=0}^{q-1} \frac{(-1)^j f^{(j)}(0)}{(1 - i\omega)^{j+1}} e^{-t} e^{i\omega t}$$

$$+ \frac{(-1)^{q+1}}{(1 - i\omega)^q} \int_0^t e^{-(t-s)} \big(f^{(q+1)}(s) - i\omega f^{(q)}(s)\big) e^{i\omega(t-s)} \, ds.$$

$$(4.3.12)$$

Proof We proceed as in the proof of Theorem 4.2.3, using integration by parts for the two highly oscillatory integrals in the solution representation (4.3.10), where the resolvent kernel is now $r_1(t-s) = -e^{-(t-s)}$. The asymptotic expansion of the second integral is the one we used in that proof; this then yields the expansion of the first integral, by replacing $f(s)$ by $f'(s)$. We leave the details to the reader (but see also the proof of Theorem 4.3.8 below). $\qquad\square$

Does the result of Theorem 4.3.6 remain valid for general (non-convolution) kernels, that is, for first-kind VIEs

$$\int_0^t H(t, s) e^{i\omega(t-s)} u(s) \, ds = f(t), \quad t \in I, \qquad (4.3.13)$$

with smooth H and f? In order to answer this question we first state the analogue of Theorem 4.3.2 for the general first-kind kind VIE (4.3.13). Its proof is an obvious generalisation of the one for Theorem 4.3.4.

Theorem 4.3.7 *Let $f \in C^1(I)$, $f(0) = 0$ and $H \in C(D)$, $\partial H/\partial t \in C^1(D)$, with $H(t, t) \neq 0$ on I, and assume without loss of generality that $H(t, t) = 1$ for $t \in I$. Then the unique solution u of (4.3.13) with regular oscillator can be written in the form*

$$u(t) = f'(t) - i\omega f(t) + \int_0^t R_1(t, s)\big(f'(s) - i\omega f(s)\big) e^{i\omega(t-s)} \, ds, \quad (4.3.14)$$

where $R_1(t, s)$ is the resolvent kernel corresponding to the kernel

$$H_1(t, s) := -\partial H(t, s)/\partial t .$$

Using the above result we now establish the asymptotic expansion, as $\omega \to \infty$, of the solution (4.3.14) of (4.3.13).

Theorem 4.3.8 *Let $f \in C^{q+2}(I)$ and $H \in C^{d+2}(D)$, with $f(0) = 0$ and $H(t, t) \neq 0$ for $t \in I$. We assume again, without loss of generality, that $H(t, t) = 1$ for $t \in I$. Under these conditions the solution of the first-kind VIE (4.3.13) has the asymptotic expansion*

$$u(t) = f'(t) - i\omega f(t) + \sum_{j=0}^{q} (i\omega)^{-j} \sum_{\ell=0}^{j} \binom{j}{\ell} \frac{\partial^{j-\ell} R_1(t, t)}{\partial s^{j-\ell}}$$

$$\times \left(f^{(\ell+1)}(t) - i\omega f^{(\ell)}(t) \right)$$

$$+ \sum_{j=0}^{q} (i\omega)^{-(j+1)} \sum_{\ell=0}^{j} \binom{j}{\ell} \frac{\partial^{j-\ell} R_1(t, 0)}{\partial s^{j-\ell}}) f^{(\ell+1)}(0) - i\omega f^{(\ell)}(0)) e^{i\omega t}$$

$$+ (i\omega)^{-(q+1)} \int_0^t Q_{q+1}(t, s) e^{i\omega(t-s)} \, ds, \tag{4.3.15}$$

where

$$Q_{q+1}(t, s) := \sum_{j=0}^{q+1} \binom{q+1}{j} \frac{\partial^{q+1-j} R_1(t, s)}{\partial s^{q+1-j}} \left(f^{(j+1)}(s) - i\omega f^{(j)}(s) \right) e^{i\omega(t-s)} ds.$$

Proof   The starting point is the solution representation (4.3.14). Setting

$$F(s) := f'(s) - i\omega f(s) \quad \text{and} \quad Q_0(t, s) := R_1(t, s) F(s),$$

and then using repeated integration by parts, we obtain

$$u(t) = F(t) + \sum_{j=0}^{q} (i\omega)^{-j)} Q_j(t, t)$$

$$+ \sum_{j=0}^{q} (i\omega)^{-(j+1)} Q_j(t, 0) e^{i\omega t} + (i\omega)^{-(q+1)} \int_0^t Q_{q+1}(t, s) e^{i\omega(t-s)} \, ds,$$

where $Q_j(t, s) := \partial Q_{j-1}(t, s)/\partial t$ $(j \geq 1)$ and

$$Q_j(t, s) = \sum_{\ell=0}^{j} \binom{j}{\ell} \frac{\partial^{j-\ell} R_1(t, s)}{\partial s^{j-\ell}} F^{(\ell)}(s).$$

The assertion now follows from the definition of the $Q_j(t, s)$ and of $F(s)$. $\square$

Remark 4.3.9 How is the result (4.3.15) in Theorem 4.3.8 related to (4.3.6) in Theorem 4.3.2? In other words: can (4.3.15) be used to characterise those kernels $H(t, s)$ in the first-kind VIE (4.3.13) for which the solution is *non-oscillatory*? We leave this as a *Research problem* (cf. Exercise 4.5.10). ◇

4.3.2 VIEs with Weakly Singular Kernels

What is the analogue of the asymptotic expansion (4.3.15) in Theorem 4.3.8 for the solution of the weakly singular first-kind VIE

$$\int_0^t (t-s)^{\alpha-1} H(t,s) e^{i\omega g(t,s)} u(s)\,ds = f(t), \quad t \in I \quad (0 < \alpha < 1), \quad (4.3.16)$$

with highly oscillatory kernel corresponding to the linear oscillator $g(t, s) = t - s$?

In order to gain some insight we first look at the form of the solution of (4.3.16) and its asymptotic expansion when the kernel has the special convolution form

$$H(t, s) = \frac{(t-s)^r}{\Gamma(r+\alpha)} \quad (r \in \mathbb{N},\ 0 < \alpha < 1).$$

Theorem 4.3.10 *If $f \in C^{r+1}(I)$ and $f^{(j)}(0) = 0$ ($j = 0, \ldots, r$), the solution of the first-kind VIE (4.3.16) with the above convolution kernel and the linear oscillator $g(t, s) = t - s$ has the form*

$$u(t) = \int_0^t \frac{(t-s)^{-\alpha}}{\Gamma(1-\alpha)} e^{i\omega(t-s)} f^{(r+1)}(s)\,ds \qquad (4.3.17)$$

$$+ i\omega \int_0^t \frac{(t-s)^{-\alpha}}{\Gamma(1-\alpha)} e^{i w(t-s)} \sum_{j=1}^{r+1} \binom{r+1}{j} (-i\omega)^j f^{(r+1-j)}(s)\,ds, \quad t \in I.$$

When $r = 0$ we have

$$u(t) = \int_0^t \frac{(t-s)^{-\alpha}}{\Gamma(1-\alpha)} e^{i\omega(t-s)} f'(s)\,ds \qquad (4.3.18)$$

$$- i\omega \int_0^t \frac{(t-s)^{-\alpha}}{\Gamma(1-\alpha)} e^{i\omega(t-s)} f(s)\,ds, \quad t \in I.$$

Proof If we multiply both sides of (4.3.16) by $e^{-i\omega t}$ and resort once more to the substitutions (4.2.12), we obtain the first-kind VIE

$$\int_0^t \frac{(t-s)^{r+\alpha-1}}{\Gamma(r+\alpha)} u_\omega(s)\,ds = f_\omega(t)$$

for $u_\omega(t) = e^{-i\omega t} u(t)$. It then follows from the classical inversion formula for Abel integral equations (cf. Section 1.5.1) and our assumptions on $f^{(j)}(0)$ $(j = 0, \ldots, r)$ that its solution is given by

$$u_\omega(t) = \int_0^t \frac{(t-s)^{-\alpha}}{\Gamma(1-\alpha)} f_\omega^{(r+1)}(s)\, ds,$$

where $f_\omega(t) = e^{-i\omega t} f(t)$. This leads to the asserted result (4.3.17).   □

Remark 4.3.11 We observe that, as for first-kind VIEs with smooth and highly oscillatory kernels (cf. Remark 4.3.3), the solution (4.3.17) of the weakly singular VIE (4.3.16) may again be viewed as an 'ω-perturbation' of the solution corresponding to $\omega = 0$ (which is the fractional derivative of order $r + 1$ of f in (4.3.16)).

Consider now the general weakly singular first-kind VIE (4.3.16) with the highly oscillatory kernel

$$(t-s)^{\alpha-1} H_\omega(t,s) := (t-s)^{\alpha-1} H(t,s) e^{i\omega(t-s)} \quad (0 < \alpha < 1).$$

By an obvious adaptation of Volterra's approach described in Section 1.5.2 we see that this equation is equivalent to the first-kind VIE

$$\int_0^t H_\alpha(t,s;\omega) u(s)\, ds = G_\alpha(t), \quad t \in I, \tag{4.3.19}$$

whose kernel

$$H_\alpha(t,s;\omega) := \int_0^1 (1-z)^{-\alpha} z^{\alpha-1} H_\omega(s+(t-s)z, s)\, dz$$

is smooth on D and where

$$G_\alpha(t) := \int_0^t (t-s)^{-\alpha} g(s)\, ds.$$

This reveals that the corresponding second-kind VIE, obtained by differentiating (4.3.19), is

$$H_\alpha(t,t;\omega) u(t) + \int_0^t \frac{\partial H_\alpha(t,s;\omega)}{\partial t} u(s)\, ds = G'_\alpha(t), \quad t \in (0,T], \tag{4.3.20}$$

where $H_\alpha(t,t;\omega) = \Gamma(1-\alpha)\Gamma(\alpha) H(t,t)$ and

$$\frac{\partial H_\alpha(t,s;\omega)}{\partial t} = \int_0^1 (1-z)^{-\alpha} z^{\alpha-1} \frac{\partial H_\omega(s+(t-s)z, s)}{\partial t}\, dz.$$

The oscillatory part of the kernel is given by

$$\frac{\partial H_\omega(t,s)}{\partial t} = \Big(\frac{\partial H(t,s)}{\partial t} + i\omega H(t,s)\Big)e^{i\omega(t-s)}. \qquad (4.3.21)$$

Although it has a somewhat more complicated structure than the highly oscillatory kernel of the weakly singular VIE (4.2.5), the proof of Theorem 4.2.8 can be adapted to deal with the second-kind VIE (4.3.20) possessing the highly oscillatory kernel (4.3.21). We leave this to the reader as a *Research problem* (cf. Exercise 4.5.11).

To end this section we briefly discuss, by means of two examples, the oscillatory behaviour of the solutions of two highly oscillatory first-kind VIEs whose kernel singularities differ from the one in (4.3.16).

Example 4.3.12 Using the result of Corollary 1.5.12 and the substitutions

$$u_\omega(t) := e^{-i\omega t}u(t), \qquad f_\omega(t) := e^{-i\omega t}f(t),$$

it is easy to see that the solution of the first-kind VIE

$$\frac{1}{\Gamma(\alpha)} \int_0^t (t^\gamma - s^\gamma)^{\alpha-1} e^{i\omega(t-s)} u(s)\,ds = f(t), \quad t \in I,$$

with $\gamma > 1$, $0 < \alpha < 1$ and $f \in C^1(I)$, $f(0) = 0$, has the form

$$u(t) = \frac{\gamma t^{\gamma-1}}{\Gamma(1-\alpha)} \int_0^t (t^\gamma - s^\gamma)^{-\alpha} e^{i\omega(t-s)} f'(s)\,ds$$

$$-i\omega \frac{\gamma t^{\gamma-1}}{\Gamma(1-\alpha)} \int_0^t (t^\gamma - s^\gamma)^{-\alpha} e^{i\omega(t-s)} f(s)\,ds.$$

For $\omega = 0$ we obtain the result already derived in Corollary 1.5.8.

Example 4.3.13 As we have seen at the end of Section 1.5.3 the Volterra integral operator

$$(\mathcal{V}_{\omega,2}u)(t) := \int_0^t (t^2-s^2)^{-1/2} e^{i\omega(t-s)} u(s)\,ds = \int_0^t t^{-1}\varphi(s/t)e^{i\omega(t-s)} u(s)\,ds,$$

with $\varphi(z) := (1-z^2)^{-1/2}$, describing the highly oscillatory VIE

$$u(t) = f(t) + \int_0^t (t^2 - s^2)^{-1/2} e^{i\omega(t-s)} u(s)\,ds, \quad t \in I \ \ (\omega \gg 1), \quad (4.3.22)$$

is *not compact* (as an operator from $C(I)$ into $C(I)$), since its spectrum $\sigma(\mathcal{V}_{\omega,2})$ is not countable. Therefore, if $1 \in \sigma(\mathcal{V}_{\omega,2})$ the existence of a unique solution $u \in C(I)$ of (4.3.22) is not guaranteed. We shall return to such *cordial VIEs* in Section 7.4.3.

4.3.3 Other Types of Oscillators

Example 4.3.14 Using Laplace transform techniques it is readily verified that the first-kind VIE

$$\int_0^t \cos\left(\omega(t-s)\right)u(s)\,ds = f(t),\ \ t \in I,$$

with $f \in C^1(I)$ and $f(0) = 0$, has the unique solution $u \in C(I)$ given by

$$u(t) = f'(t) + \omega^2 \int_0^t f(s)\,ds,\ \ t \in I.$$

We observe that, in analogy to (4.3.6) in Theorem 4.3.2, the solution is *non-oscillatory* and is again given by an 'ω-pertubation' of $f'(t)$, the solution corresponding to $\omega = 0$.

Example 4.3.15 The solution $u \in C(I)$ of the first-kind VIE

$$\int_0^t J_0\left(\omega(t-s)\right)u(s)\,ds = f(t),\ \ t \in I, \tag{4.3.23}$$

with $f \in C^1(I)$ and $f(0) = 0$, whose highly oscillatory kernel is given by the Bessel function $J_0(z)$ of order zero, is

$$u(t) = f'(t) + \omega \int_0^t \frac{J_1\left(\omega(t-s)\right)}{t-s} f(s)\,ds,\ \ t \in I,$$

where $J_1(z)$ is the Bessel function of order one.

Example 4.3.16 If the kernel of the VIE in Example 4.3.5 contains a weak singularity,

$$\int_0^t (t-s)^{-1/2} \cos\left(\omega(t-s)\right)u(s)\,ds = f(t),\ \ t \in I,$$

the unique continuous solution has the form

$$u(t) = \frac{2}{\pi\omega} \int_0^t (t-s)^{-1/2} \sin\left(\omega(t-s)\right)\left(f''(s) + \omega^2 f(s)\right)ds$$

(cf. Srivastava & Buschman, 1977, p. 81 and p. 38). It can be found by employing Laplace transform techniques: they reveal that for $\omega \neq 0$ the above expression for u is valid only under the assumptions that $f \in C^2(I)$ and $f(0) = f'(0) = 0$.

4.4 General Oscillators $e^{i\omega g(t,s)}$

As far as the author knows, the analysis of the oscillatory behaviour of solutions of second-kind VIEs

$$u(t) = f(t) + \int_0^t K(t,s) e^{i\omega g(t,s)} u(s)\, ds, \quad t \in I,$$

and analogous first-kind VIEs

$$\int_0^t H(t,s) e^{i\omega g(t,s)} u(s)\, ds, \quad t \in I,$$

with general oscillators $g(t,s)$ (especially when they possess stationary points), remains open. This is even true for 'simple-looking' non-linear oscillators like $g(t,s) = ts$ and $g(t,s) = (t-s)^2$ (which, as mentioned before, occurs in the Fox–Li equation; see Section 4.2.3 and Böttcher, Brunner, Iserles & Nørsett, 2010).

4.5 Exercises and Research Problems

Exercise 4.5.1 Analyse the oscillatory behaviour of the solution of the linear VIEs

$$u(t) = f_\omega(t) + \int_0^t K(t,s) u(s)\, ds, \quad t \in I,$$

and

$$u(t) = f_\omega(t) + \int_0^t K(t,s) e^{\omega(t-s)} u(s)\, ds, \quad t \in I,$$

where $f_\omega(t) := f(t) e^{\omega t}$. Find the asymptotic exansions of the solutions (similar to the one in Theorem 4.2.3). Here, f and K are assumed to be arbitrarily smooth.

Exercise 4.5.2 Describe the oscillatory behaviour of the solution of the VIE

$$u(t) = f(t) + \int_0^t J_0\big(\omega(t-s)\big) u(s)\, ds, \quad \omega \gg 1,$$

by deriving its asymptotic expansion as ω tends to infinity. Here, J_0 denotes the Bessel function of order zero.

Exercise 4.5.3 Consider the highly oscillatory VIE

$$u(t) = f(t) + \lambda \int_0^t k\big(\omega(t-s)\big) u(s)\, ds, \quad t \in I,$$

where either $k\big(\omega(t-s)\big) = \cos\big(\omega(t-s)\big)$ or $k\big(\omega(t-s)\big) = \sin\big(\omega(t-s)\big)$. Find the solutions and their asymptotic expansions as $\omega \to \infty$.

Exercise 4.5.4 Is the solution of the highly oscillatory VIE

$$u(t) = 1 + \int_0^t \log(t-s)e^{i\omega(t-s)}u(s)\,ds, \quad t \geq 0 \ (\omega \gg 1)$$

highly oscillatory?

Exercise 4.5.5 Use Theorem 4.2.8 and the result in Example 4.2.7 to establish the asymptotic expansion of the solution of the weakly singular VIE

$$u(t) = f(t) + \lambda \int_0^t \frac{(t-s)^{\alpha-1}}{\Gamma(\alpha)}e^{i\omega(t-s)}\,u(s)\,ds, \quad t \in I. \tag{4.5.1}$$

Exercise 4.5.6 Use Exercise 4.5.5 to derive a result, analogous to the one in Corollary 4.2.4, on the degree of damping of the solution of the weakly singular VIE (4.5.1).

Exercise 4.5.7 (*Research Problem*)
Establish the analogue of Theorem 4.2.8 for the asymptotic expansion of the solution of the VIE

$$u(t) = f(t) + \int_0^t \log(t-s)K(t,s)e^{i\omega(t-s)}u(s)\,ds, \quad t \in I,$$

assuming that $f \in C^{q+1}(I)$ and $K \in C^{q+1}(D)$.

Exercise 4.5.8 Consider the VFIEs with vanishing delays,

$$u(t) = f(t) + \int_0^{qt} K(t,s)e^{i\omega(t-s)}u(s)\,ds$$

and

$$u(t) = f(t) + \int_{qt}^t K(t,s)e^{i\omega(t-s)}u(s)\,ds,$$

where $0 < q < 1$. Assuming that f and K are smooth functions, are their solutions highly oscillatory? Also, establish the analogue of Theorem 4.2.3 on the asymptotic expansion of the solutions.

Exercise 4.5.9 (Karakostas et al., 1993, p. 588)
Consider the VFIE

$$u(t) = f(t) - \int_0^t s^2(t+\pi)^{-1}u(s-\pi)\,ds, \quad t \geq 0,$$

with $f(t) = t(t + \pi)^{-1} \cos(t)$. For which initial function ϕ is $u(t) = (t + \pi)^{-1} \sin(t)$ a solution of the above VFIE?

Exercise 4.5.10 (*Research Problem*)
Use the asymptotic expansion in Theorem 4.3.8 to obtain a characterisation of those smooth kernels $H(t, s)$ for which the solution of the highly oscillatory first-kind VIE

$$\int_0^t H(t, s) e^{i\omega(t-s)} u(s)\, ds = g(t), \quad t \in I,$$

is *non-oscillatory* (see also Theorem 4.3.2).

Exercise 4.5.11 (*Research Problem*)
Establish the analogue of the asymptotic expansion (4.3.15) (Theorem 4.3.8) for the weakly singular first-kind VIE (4.3.16), by using the equivalent second-kind VIE (4.3.20) with highly oscillatory kernel (4.3.21). Then, as in Exercise 4.5.10, characterise those kernels $K(t, s)$ for which the solution of (4.3.16) is *non-oscillatory*.

Exercise 4.5.12 Assuming that the first-kind Abel-type VIE

$$\int_0^t (t^2 - s^2)^{-1/2} e^{i\omega(t-s)} u(s)\, ds = f(t), \quad t \in [0, 1] \ (\omega \gg 1),$$

with $f \in C^1[0, 1]$, possesses a unique solution $u \in C[0, 1]$, is this solution highly oscillatory? Is the answer different if the singularity $(t^2 - s^2)^{-1/2}$ is replaced by $s(t^2 - s^2)^{-1/2}$?

Exercise 4.5.13 Derive the asymptotic expansion, as $\omega \to \infty$, of the solution of the first-kind VIE (4.3.23) in Example 4.3.15.

Exercise 4.5.14 (*Research Problem*)
Consider the linear VFIE

$$u(t) = 1 + \int_0^{\theta(t)} k(t - s) u(s)\, ds, \quad t \in [0, T],$$

with the highly oscillatory delay function

$$\theta(t) = q_1 t + (q_2 - q_1) t \sin^2(\omega t) \ (\omega \gg 1)$$

corresponding to $0 < q_1 < q_2 < 1$. If the convolution kernel is continuous (e.g. $k(z) = \lambda$, $\lambda \neq 0$), is the solution of this VFIE highly oscillatory?

Exercise 4.5.15 (*Research Problem*)
Lamm (2000) and Ring & Prix (2000) studied the sequential regularisation of linear first-kind VIEs with *noisy data* $\hat{g}(t)$. Extend their approach to analogous first-kind VIEs with highly oscillatory kernels, e.g. $K(t, s)e^{i\omega(t-s)}$ and $K(t, s)\cos(\omega(t - s))$.

4.6 Notes

Section 4.2: *VIEs of the Second Kind*
In the last ten years the analysis of solutions of integral equations with highly oscillatory kernels has received growing attention, not least as a result of the six-month programme on *Highly Oscillator Problems: Computation, Theory and Applications* at the Isaac Newton Institute (University of Cambridge, www.newton.cam.ac.uk/event/hop) in 2007. The book by Engquist et al. (2009) conveys the scope of that programme.

Among the representative papers on highly oscillatory *Fredholm integral equations* are the ones by Brunner, Iserles & Nørsett (2010, 2011), Böttcher, Brunner, Iserles & Nørsett (2010, but see also the pioneering paper by Ursell, 1969 and the one by Landau, 1977 with its references on applications). Among the papers concerned with solutions of VIEs with highly oscillatory kernels are Satco (2009), Xiang & Brunner (2013: kernels with weak singularities and Bessel function kernels), and Brunner, Ma & Xu (2015). Karakostas et al. (1993) studied the oscillatory behaviour of solutions to VFIEs of the form

$$u(t) = f(t) + \int_0^t k(t, s, u(\theta(s)))\, ds, \quad t \geq 0,$$

with $\theta(t) = t - \tau$ $(\tau > 0)$ and $u(t) = \phi(t)$ $(t \in [-\tau, 0])$. The authors discuss *strong* oscillations (the solution u is strongly oscillatory if for any $t_1 > 0$, $\inf_{t \geq t_1} u(t) < 0 < \inf_{t \geq t_1} u(t)$), and *quick* oscillations (u is quickly oscillatory if there exists a strictly increasing sequence of points $\{t_n\}$ so that $u(t_n) = 0$ $(n \geq 1)$, $\lim_{n \to \infty} t_n = \infty$, and $\lim_{n \to \infty}(t_{n+1} - t_n) = 0$.

Section 4.3: *VIEs of the First Kind*
The oscillatory behaviour of solutions of highly oscillatory *Volterra integral equations* of the first kind is analysed in Brunner (2014); see also Davies & Duncan (2004) and Wang & Xiang (2011: first-kind VIEs with Bessel function kernels).

Section 4.4: *General Oscillators $e^{i\omega g(t,s)}$*
Asymptotic expansions of integrals with general ('irregular') oscillators were studied by Iserles (2005a). As indicated before, the derivation of analogous asymptotic expansions of solutions for linear and non-linear VIEs with highly oscillatory kernels (4.1.2) with general oscillators $g(t, s)$ is waiting to be carried out.

Numerical Analysis of Highly Oscillatory VIEs

Among the fundamental papers on the numerical approximation of integrals with highly oscillatory integrands are the ones by Iserles (2005a), Iserles & Nørsett (2004, 2005), and Domínguez, Graham & Kim (2013, which deals with integrals containing weakly singular kernels). Filon-type methods for such integrals were studied by Xiang (2011). A different (collocation-based) approach to such problems is due to Levin (1997). See also Iserles (2005b) and Huybrechs & Olver (2009) and their references. A comprehensive treatment of the computation of highly oscillatory integrals can be found in the monograph by Deaño, Huybrechs & Iserles (2017).

While the papers by Xiang, Cho, Wang & Brunner (2011) and Xiang & Brunner (2014) deal with the numerical solution of certain classes of VIEs with highly oscillatory Bessel function kernels, the design and analysis of efficient numerical methods for general highly oscillatory VIEs remains essentially open.

5

Singularly Perturbed and Integral-Algebraic Volterra Equations

Summary

The solution of a singularly perturbed second-kind VIE will in general exhibit a behaviour, e.g. near the initial point of the interval of integration, that differs fundamentally from the one of the reduced VIE, the first-kind VIE corresponding to the value zero of the perturbation parameter. The first part of this chapter deals with the analysis of this behaviour. In the second part we analyse the solutions of systems of integral-algebraic Volterra equations, based on the decoupling of such a system into a subsystem of inherent VIEs of the second kind and a subsystem of (ill-posed) first-kind VIEs. Such a decoupling (analogous to the decoupling of differential-algebraic equations) provides the key for the numerical analysis (optimal order of convergence; asymptotic stability) of integral-algebraic equations. We shall also see that this decoupling theory remains open for systems of IAEs with weakly singular kernels, and for systems of non-linear IAEs.

5.1 Singularly Perturbed VIEs

Let $\mathcal{V}_\alpha : C(I) \to C(I)$ denote the Volterra integral operator

$$(\mathcal{V}_\alpha u)(t) := \int_0^t (t-s)^{\alpha-1} K(t,s) u(s)\, ds, \ \ t \in I = [0,T] \ (0 < \alpha \leq 1).$$

The second-kind VIE

$$\varepsilon u(t) = f(t) + (\mathcal{V}_\alpha u)(t), \ \ t \in I, \tag{5.1.1}$$

is called a *singularly perturbed* linear VIE if $\varepsilon > 0$ is a small *perturbation parameter*. We will refer to the first-kind VIE corresponding to $\varepsilon = 0$ in (5.1.1),

$$(\mathcal{V}_\alpha u)(t) + f(t) = 0, \ \ t \in I, \tag{5.1.2}$$

as the *reduced* VIE. For $0 < \alpha < 1$ it is equivalent to a first-kind VIE with smooth kernel,

$$\int_0^t K_\alpha(t, s) u(s) \, ds = F_\alpha(t), \quad t \in I,$$

where (in analogy to the notation introduced in Theorem 1.5.7) we have

$$K_\alpha(t, s) := \int_0^1 (1 - w)^{-\alpha} w^{\alpha - 1} K(s + (t - s)w, s) \, dw,$$

$$F_\alpha(t) := - \int_0^t (t - s)^{-\alpha} f(s) \, ds.$$

According to Theorems 1.2.3 and 1.3.5, the second-kind VIE (5.1.1) possesses a unique solution $u_\varepsilon \in C(I)$ for any $\varepsilon > 0$ and any $0 < \alpha \leq 1$, provided that $f \in C(I)$ and $K \in C(D)$. This solution satisfies $u_\varepsilon(0) = \varepsilon^{-1} f(0)$.

Under the assumptions that $\partial K / \partial t \in C^1(D)$, $K(t, t) \neq 0$ ($t \in I$), and $f \in C^1(I)$, with $f(0) = 0$, the solution u_0 of the reduced VIE (5.1.2) is also continuous on the closed interval I; at $t = 0$ it has the value

$$u_0(0) = \begin{cases} -f'(0)/K(0, 0) & \text{if } \alpha = 1, \\ \dfrac{F'_\alpha(0)}{K_\alpha(0, 0)} = 0 & \text{if } 0 < \alpha < 1 \end{cases} \tag{5.1.3}$$

(see Theorems 1.4.1 and 1.5.7). However, as $\varepsilon > 0$ approaches zero the behaviour of $u_\varepsilon(t)$ may be markedly different from that of $u_0(t)$ as $t \to 0^+$: at $t = 0^+$ the solution u_ε may exhibit a *boundary layer*. The situation becomes even more complex near $t = 0^+$ as $\varepsilon \to 0^+$ if $f \in C^1(I)$ is such that $f(0) \neq 0$, since $u_0(t)$ is then no longer continuous at $t = 0$. Thus, the behaviour of the solution of the singularly perturbed VIE (5.1.1) near $t = 0^+$ will very much depend on whether or not $f(0) = 0$.

5.1.1 Examples

Before stating our main results on the relationship between the solution $u_\varepsilon(t)$ of (5.1.1) (with $0 < \varepsilon \ll 1$ and $\alpha = 1$) and the solution $u_0(t)$ of the corresponding reduced VIE

$$0 = f(t) + \int_0^t K(t, s) u(s) \, ds, \quad t \in I, \tag{5.1.4}$$

we make the observation stated at end of the previous section more precise by looking at three simple examples.

Example 5.1.1 (Kauthen, 1997)

It is easy to verify that for $\varepsilon > 0$ the solution of

$$\varepsilon u(t) = \sin(t) - \int_0^t u(s)\,ds, \quad t \in I, \tag{5.1.5}$$

is given by

$$u_\varepsilon(t) = \frac{1}{1 + \varepsilon^2}\big(\cos(t) + \varepsilon \sin(t)\big) - \frac{1}{1 + \varepsilon}\,e^{-t/\varepsilon}, \quad t \in I. \tag{5.1.6}$$

This follows from the fact that the resolvent kernel of the kernel $k(t - s) = -1/\varepsilon$ is $\rho(t - s) = (-1/\varepsilon)e^{-(t-s)/\varepsilon}$ (recall Example 4.2.2 with $\lambda = -1/\varepsilon$). We observe that $u_\varepsilon(0) = 0$ whenever $\varepsilon > 0$.

In analogy to the terminology for singularly perturbed ordinary differential equations (see, for example, O'Malley, Jr, 1991 or Skinner, 2011) the term

$$\frac{1}{1 + \varepsilon^2}\big(\cos(t) + \varepsilon \sin(t)\big) \tag{5.1.7}$$

represents the smooth or *outer solution*, while the remaining term in (5.1.6),

$$-\frac{1}{1 + \varepsilon^2}\,e^{-t/\varepsilon}, \tag{5.1.8}$$

is the rapidly changing *inner solution*. The *reduced VIE* ((5.1.5) with $\varepsilon = 0$) possesses the solution $u_0(t) = \cos(t)$, with $u_0(0) = 1 \neq u_\varepsilon(0) = 0$. Thus it follows that for small $\varepsilon > 0$ the inner solution (5.1.8) gives rise to a *boundary layer* near $t = 0^+$. This means that as t becomes positive there is a fast transition from the value $u_\varepsilon(0) = 0$ to that of the outer solution (5.1.7). Moreover, for increasing t the outer solution (5.1.7) rapidly becomes an excellent approximation to the solution (5.1.6) of the singularly perturbed VIE (5.1.5). We note for later reference (see Theorem 5.1.4) that the solution (5.1.6) can also be written in the form

$$u_\varepsilon(t) = \frac{1}{1 + \varepsilon^2}\big(\cos(t) + \varepsilon \sin(t)\big) - \frac{1}{1 + \varepsilon^2}\sum_{n=0}^{\infty} \frac{(-1)^n}{n!}\left(\frac{t}{\varepsilon}\right)^n, \tag{5.1.9}$$

where $\tau := (t/\varepsilon)$ is the so-called *inner variable*.

Example 5.1.2 (Angell & Olmstead, 1987)

The singularly perturbed VIE

$$\varepsilon u(t) = \int_0^t (1 + t - s)\big(1 + s - u(s)\big)\,ds \quad (\varepsilon > 0),$$

which corresponds to (5.1.1) with

$$f(t) = \int_0^t (1 + t - s)(1 + s)\, ds \quad \text{and} \quad K(t, s) = -(1 + t - s),$$

possesses the solution

$$u_\varepsilon(t) = 1 + t + \frac{1}{r_1 - r_2}\left((r_2 - 1 + \varepsilon^{-1})e^{r_1 t} - (r_1 - 1 + \varepsilon^{-1})e^{r_2 t}\right), \quad (5.1.10)$$

where $r_{1,2} := \left(-1 \pm (1 - 4\varepsilon)^{1/2}\right)/(2\varepsilon)$. This follows from the equivalence of the above VIE with the singularly perturbed Volterra integro-differential equation

$$\varepsilon u'(t) = 1 + t - u(t) + \int_0^t \left(1 + s - u(s)\right) ds, \quad u(0) = 0,$$

and hence with the singularly perturbed second-order differential equation

$$\varepsilon u''(t) + u'(t) + u(t) = 2 + t, \quad \text{with } u(0) = 0, \ u'(0) = \varepsilon^{-1}.$$

The solution of the *reduced* VIE is $u_0(t) = 1 + t$, implying that for $\varepsilon > 0$ we have $0 = u_\varepsilon(0) \neq u_0(0) = 1$. We note that the solution (5.1.10) can be written as

$$u_\varepsilon(t) = (1 + t - e^{-\tau}) + (-\tau e^{-\tau})\varepsilon + \left(e^{-t} - e^{-\tau}(1 + \tau + \tau^2/2)\right)\varepsilon^2 + \mathcal{O}(\varepsilon^3),$$
$$(5.1.11)$$

where we have again used the inner variable $\tau := t/\varepsilon$.

In these two examples the non-homogeneous term $f(t)$ satisfies $f(0) = 0$, implying that the reduced (first-kind) VIE possesses a unique solution $u \in C(I)$. The next example is chosen to illustrate what happens when $f(0) \neq 0$ in (5.1.4).

Example 5.1.3 In the singularly perturbed VIE

$$\varepsilon u(t) = f(t) + \int_0^t k(t - s)u(s)\, ds, \quad t \in I,$$

let $f(t) = e^t g(t)$ and $k(t - s) = e^{t-s}$, with $g \in C^1(I)$. If $\varepsilon > 0$ the resolvent kernel corresponding to the kernel $\varepsilon^{-1}k(t - s)$ is $r(t - s) = e^{(t + \varepsilon^{-1})(t-s)}$ (see, for example, Krasnov et al. (1977, p. 22, Problem 25). Hence the solution is

$$u_\varepsilon(t) = \varepsilon^{-1}g(t)e^t + \varepsilon^{-1}\int_0^t e^{(t + \varepsilon^{-1})(t-s)}e^s g(s)\, ds, \quad t \in I \ (\varepsilon > 0),$$

with $u_\varepsilon(0) = \varepsilon^{-1}g(0)$. It is readily seen that the solution of the reduced VIE is

$$u_0(t) = -e^t g'(t) :$$

it is continuous on $(0, T]$. If $g(0) = 0$ we have $u_0 \in C(I)$ with $u_0(0) = -g'(0)$, compared to $u_\varepsilon(0) = \varepsilon^{-1}g(0) = 0$ for $\varepsilon > 0$ (see also Exercise 5.4.2). In particular, if g is of the form $g(t) = a_0 + a_1 t^p$, with $p \geq 1$, we have $g(0) = a_0$, with $g'(0) = a_1$ if $p = 1$; and $g'(0) = 0$ if $p > 1$. However, if $g(0) \neq 0$ (i.e. when $a_0 \neq 0$) the solution of the reduced VIE

$$0 = e^t g(t) + \int_0^t k(t - s)u(s)\,ds, \ t \in I,$$

is unbounded at $t = 0$ (cf. Theorem 1.4.1 and its proof).

5.1.2 VIEs with Smooth Kernels

The classical theory of singularly perturbed initial-value problems of the form

$$\varepsilon u''(t) + p(t)u'(t) + q(t, u(t)) = 0, \ t \in [0, 1] \ (0 < \varepsilon \ll 1), \quad (5.1.12)$$

with $u(0) = u_0$, $u'(0) = u_1$ and $p(t) > 0$, has been widely studied (see, for example, O'Malley, Jr, 1991). If the reduced problem

$$p(t)u'(t) + q(t, u(t)) = 0, \ u(0) = u_0,$$

possesses a solution on $[0, 1]$, the solution of (5.1.12) has, under appropriate regularity assumptions on p and q, a uniformly valid asymptotic expansion of the form

$$u_\varepsilon(t) = \sum_{n=0}^{r} \varepsilon^n \big(v_{\varepsilon,n}(t) + w_{\varepsilon,n}(\tau)\big) + \mathcal{O}(\varepsilon^{r+1}) \ (\varepsilon \to 0), \quad (5.1.13)$$

where $\tau := t/\varepsilon$. If $p \in C^1[0, 1]$ the initial-value problem (5.1.12) is equivalent to the singularly perturbed VIE

$$\varepsilon u(t) = f_\varepsilon(t) - \int_0^t \Big(\big((p(s) - (t - s)p'(s)\big)$$
$$\times u(s) + (t - s)q\big(s, u(s)\big)\Big)\,ds, \ t \in [0, 1], \quad (5.1.14)$$

with $f_\varepsilon(t) := \varepsilon(u_0 + u_1 t) + p(0)u_0 t$. Thus, (5.1.13) is also the asymptotic expansion, as ε tends to zero, of the solution of (5.1.14). We observe that the reduced equation corresponding to (5.1.14) is

$$0 = p(0)u_0 t - \int_0^t \Big(\big(p(s)-(t-s)p'(s)\big)u(s)+(t-s)q\big(s,u(s)\big)\Big)\, ds, \quad t \in [0,1].$$

We will therefore also look for an asymptotic expansion of the form (5.1.13) for the solution of the more general non-linear singularly perturbed VIE

$$\varepsilon u(t) = f(t) + \int_0^t K(t,s)G(s,u(s))\, ds, \quad t \in I \ (0 < \varepsilon \ll 1), \qquad (5.1.15)$$

and for its linear counterpart

$$\varepsilon u(t) = f(t) + \int_0^t K(t,s)u(s)\, ds, \quad t \in I \ (0 < \varepsilon \ll 1). \qquad (5.1.16)$$

We first consider the linear VIE (5.1.16). As we have seen in Section 1.4.1 (see Theorem 1.4.1), a necessary condition for the reduced VIE

$$0 = f(t) + \int_0^t K(t,s)u(s)\, ds, \quad t \in I,$$

with $K(t,t) \neq 0$ for $t \in I$, to possess a continuous (and hence bounded) solution on the *closed* interval $[0,T]$ is that $f(0) = 0$. Thus, motivated by the solution representation (5.1.9) in Example 5.1.1 (which, owing to the regularity of $\cos(t)$ and $\sin(t)$, can be written in the form (5.1.13)) we use an Ansatz (often referred to as the additive decomposition method) analogous to (5.1.13), namely

$$u_\varepsilon(t) = v_\varepsilon(t) + \phi(\varepsilon)w_\varepsilon(t/\mu(\varepsilon)), \qquad (5.1.17)$$

where v_ε and w_ε are such that (setting $\tau := t/\varepsilon$)

$$v_\varepsilon(t) = v_0(t) + o(1), \quad w_\varepsilon(\tau) = w_0(\tau) + o(1) \ \text{(as } \varepsilon \to 0).$$

Here, $\phi(\varepsilon)$ and $\mu(\varepsilon)$ represent, respectively, the *magnitude* and the *width* of the initial boundary layer. Accordingly, we look for an asymptotic expansion of $u_\varepsilon(t)$ of the form

$$u_{\varepsilon,r}(t) := \sum_{n=0}^r \varepsilon^n \big(v_{\varepsilon,n}(t) + w_{\varepsilon,n}(\tau)\big) + R_{\varepsilon,r}(t), \quad t \in I, \qquad (5.1.18)$$

with

$$R_{\varepsilon,r}(t) := \sum_{n=r+1}^\infty \varepsilon^n \big(v_{\varepsilon,n}(t) + w_{\varepsilon,n}(\tau)\big).$$

As we shall see below, $\tau := t/\varepsilon$ represents again the *inner variable*. The following theorem is due to Bijura (2002c; but see also Skinner, 1995 and Kauthen, 1997).

Theorem 5.1.4 *Assume that $f \in C^{\infty}(I)$, with $f(0) = 0$, and $K \in C^{\infty}(D)$, with $K(t, t) \leq -\kappa$ $(t \in I)$ for some constant $\kappa > 0$. Then the solution of the singularly perturbed VIE (5.1.16) has the asymptotic expansion (5.1.18), and there exist constants $C_r > 0$ and $\varepsilon_r^* > 0$ such that*

$$|R_{\varepsilon,r}(t)| = |u_{\varepsilon}(t) - u_{\varepsilon,r}(t)| \leq C_r \varepsilon^{r+1}$$

holds uniformly for $t \in I$ and $\varepsilon \in (0, \varepsilon_r^*]$.

Proof   We will sketch the main steps of the proof of the above results; its elaborate technical details (which also encompass non-linear singularly perturbed VIEs (5.1.15)) can be found in Bijura (2002c; cf. pp. 137–142). If we substitute (5.1.17) into the VIE (5.1.16) and set $\tau := t/\mu(\varepsilon)$ we obtain (using the change of the variable of integration $\sigma := s/\mu(\varepsilon)$)

$$\varepsilon v_{\varepsilon}(\mu(\varepsilon)\tau) + \varepsilon\phi(\varepsilon)w_{\varepsilon}(\tau) = f(\mu(\varepsilon)\tau)$$
$$+ \mu(\varepsilon) \int_0^{\tau} K(\mu(\varepsilon)\tau, \mu(\varepsilon)\sigma)\big(v_{\varepsilon}(\sigma)$$
$$+ \phi(\varepsilon)w_{\varepsilon}(\sigma)\big) d\sigma.$$

For fixed $\tau > 0$ and $\varepsilon \to 0$ this yields (recalling the assumptions on $v_{\varepsilon}(t)$ and $w_{\varepsilon}(t)$)

$$\varepsilon v_0(0) + \varepsilon\phi(\varepsilon)w_0(\tau) = f(0) + \mu(\varepsilon) \int_0^{\tau} K(0, 0)\big(v_0(\sigma)$$
$$+ \phi(\varepsilon)w_0(\sigma)\big) d\sigma + o(\varepsilon) + o(\mu(\varepsilon)).$$

If we compare dominant terms we see that the constants $\phi(\varepsilon)$ and $\mu(\varepsilon)$ introduced in (5.1.17) must be chosen as $\phi(\varepsilon) = 1$ and $\mu(\varepsilon) = \varepsilon$, hence justifying the form $\tau = t/\varepsilon$ of the inner variable.

By substituting

$$U_{\varepsilon,r}(t) := \sum_{n=0}^{r} \varepsilon^n \big(v_{\varepsilon,n}(t) + w_{\varepsilon,n}(\tau)\big)$$

into (5.1.16) it can be shown (Bijura, 2002c, pp. 125–129) that the functions $v_{\varepsilon,n}$ and $w_{\varepsilon,n}$ are determined recursively by the integral equations

$$v_{\varepsilon,n-1}(t) = f_n(t) + \int_0^t K(t, s)v_{\varepsilon,n}(s)\, ds + \int_0^{\infty} \tilde{K}_{n-1}(t, \sigma)\, d\sigma \quad (5.1.19)$$

(where

$$\tilde{K}_n(t, \sigma) := \sum_{j=0}^{n} \frac{1}{j!} \frac{\partial^j K(t, s)}{\partial s}\bigg|_{s=0} w_{\varepsilon,n-1}(\sigma) \,)$$

and

$$w_{\varepsilon,n}(\tau) = -\int_\tau^\infty h_n(\tau,\sigma)\,d\sigma, \tag{5.1.20}$$

with appropriate smooth functions f_n and h_n, and initial condition $\varepsilon u(0) = f(0)$. Under the assumptions on f and K stated in Theorem 5.1.4 these equations are uniquely solvable, and the resulting solution satisfies

$$v_\varepsilon(t) = v_{\varepsilon,0}(t) + o(1), \quad w_\varepsilon(\tau) = w_{\varepsilon,0}(\tau) + o(1) \ (\text{as } \tau \to \infty)$$

(ensuring the rapid decay of the initial layer). ☐

5.1.3 VIEs with Weakly Singular Kernels

In order to understand the effect a weak kernel singularity has on the asymptotic expansion, as $\varepsilon \to 0$, of the solution of the singularly perturbed VIE

$$\varepsilon u(t) = f(t) + \int_0^t (t-s)^{\alpha-1} K(t,s) u(s)\,ds, \ t \in I \ (0 < \alpha < 1), \tag{5.1.21}$$

we first consider a simple example.

Example 5.1.5 Consider the VIE

$$\varepsilon u(t) = f(t) - \int_0^t \frac{(t-s)^{\alpha-1}}{\Gamma(\alpha)} u(s)\,ds, \ t \in I \ (0 < \alpha < 1). \tag{5.1.22}$$

Writing it in the form

$$u(t) = f_\varepsilon(t) - \varepsilon^{-1}\int_0^t (t-s)^{\alpha-1} K(t,s) u(s)\,ds,$$

with $f_\varepsilon(t) := \varepsilon^{-1} f(t)$, and using Theorems 1.2.9, we see that its solution is given by

$$u_\varepsilon(t) = f_\varepsilon(0) w_\varepsilon(t) + \int_0^t w_\varepsilon(t-s) f'_\varepsilon(s)\,ds.$$

Here, $w_\varepsilon(t)$ denotes the solution of the VIE

$$w(t) = 1 - \varepsilon^{-1}\int_0^t \frac{(t-s)^{\alpha-1}}{\Gamma(\alpha)} w(s)\,ds.$$

For $f(t) = t$ (and hence $f_\varepsilon(t) = \varepsilon^{-1} t$) this becomes, by Theorem 1.3.2,

$$u_\varepsilon(t) = \varepsilon^{-1}\int_0^t E_\alpha\big(-\varepsilon^{-1}(t-s)^\alpha\big)\,ds = \varepsilon^{-1}\int_0^t \sum_{n=0}^\infty \frac{(-\varepsilon^{-1}(t-s)^\alpha)^n}{\Gamma(1+n\alpha)}\,ds$$

$$= \varepsilon^{(1-\alpha)/\alpha} \sum_{n=0}^{\infty} \frac{(-1)^n}{\Gamma(2+n\alpha)} \left(\frac{t}{\varepsilon^{1/\alpha}} \right)^{1+n\alpha}, \quad t \in I.$$

(This result was also derived by Kauthen, 1993.) As we shall see below, the above expression for the solution of (5.1.22) reveals the dependence of the *inner variable* $\tau := t/\varepsilon^{1/\alpha}$ on the value of α in the algebraic kernel singularity $(t-s)^{\alpha-1}$ in (5.1.22).

We now turn to the general singularly perturbed linear VIE (5.1.21) with weakly singular kernel and assume that $f \in C^\infty(I)$ and $K \in C^\infty(D)$. The additional assumptions $f(0) = 0$ and $K(t,t) \le -\kappa$ for $t \in I$ and some $\kappa > 0$ are necessary for the reduced VIE to possess a solution that is continuous, and hence bounded, on the closed interval $I = [0, T]$). As in Section 5.1.2 the aim is to establish an asymptotic expansion of the solution of (5.1.21) similar to (5.1.17) and (5.1.18), except that now (as we have already seen in Example 5.1.5) the inner variable τ depends on α (it is given by $\tau = t/\varepsilon^{1/\alpha}$), as do the functions

$$v_\varepsilon(t) := \sum_{n=0}^{\infty} v_{\varepsilon,n}(t)\varepsilon^n \quad \text{and} \quad w_\varepsilon(t) := \sum_{n=0}^{\infty} w_{\varepsilon,n}(\tau)\varepsilon^{n\alpha} \quad (\varepsilon \to 0^+).$$

$$(5.1.23)$$

The following theorem is due to Bijura (2002c: the special case $\alpha = 1/2$ was studied in Kauthen, 1997 and Skinner, 2000).

Theorem 5.1.6 *Assume that $f \in C^\infty(I)$, with $f(0) = 0$, and $K \in C^\infty(D)$, with $K(t,t) \le -\kappa$ ($t \in I$) for some constant $\kappa > 0$. Then the solution u_ε of (5.1.21) with $0 < \alpha < 1$ has the asymptotic expansion described by (5.1.18) and (5.1.23) with inner variable $\tau = t/\varepsilon^{1/\alpha}$; it holds uniformly for $t \in I$ and $\varepsilon \in (0, \varepsilon^*]$ with some $\varepsilon^* > 0$.*

Proof As in the proof of Theorem 5.1.4 one requires that the function $w_\varepsilon(\tau)$ satisfy $\lim_{\tau \to \infty} w_\varepsilon(\tau) = 0$ or, equivalently, $\lim_{\tau \to \infty} w_{\varepsilon,n}(\tau) = 0$ for all $n \ge 0$. Here, $\tau = t/\varepsilon^{1/\alpha}$. This means that under the assumptions on f and K stated above, the singularly perturbed VIE (5.1.21) exhibits an initial layer of width $\mathcal{O}(\varepsilon^{1/\alpha})$: the solution varies slowly for $\mathcal{O}(\varepsilon^{1/\alpha}) \le t \le T$, but changes rapidly on a small initial interval $0 \le t \le \mathcal{O}(\varepsilon^{1/\alpha})$. The proof then proceeds along the lines of the one for Theorem 5.1.4, by substituting (5.1.17), using (5.1.23), into the VIE (5.1.21 and then deriving recursively the integral equations for $v_{\varepsilon,n}$ and $w_{\varepsilon,n}$, similar to (5.1.19) and (5.1.20). The technical details can be found in Bijura (2004; but see also Kauthen, 1997 and Skinner, 2000 for the case $\alpha = 1/2$). $\qquad \Box$

5.1.4 Non-Linear VIEs

The motivation for studying non-linear singularly perturbed VIEs was provided by a nonlinear VIE of Hammerstein type representing a mathematical model for studying the time evolution of the temperature $u = u(t)$ at the surface of a conducting solid in the presence of high thermal loss (Angell & Olmstead, 1987; see also Example 9.2.7 in Section 9.2.2 of this book). This VIE is

$$\varepsilon u(t) = \pi^{-1/2} \int_0^t (t-s)^{-1/2}\big(h(s) - u^n(s)\big)\, ds \ \ t \geq 0,$$

where h represents the externally supplied heat, ε^{-1} is the radiation constant, and $n = 4$ (Stefan–Boltzmann radiation law) or $n = 1$ (thermal convection). This VIE is a particular case of the singularly perturbed Volterra–Hammerstein integral equation

$$\varepsilon u(t) = f(t) + \int_0^t k(t-s)G(s, u(s))\, ds, \ \ t \in I, \tag{5.1.24}$$

with convolution kernel k and smooth G.

The following result is due to Skinner (1995) and Bijura (2002c, who derived it for the general non-linear, singularly perturbed VIE with non-linearity $k(t, s, u)$). Its proof proceeds along the lines of the one for Theorem 1.5.1, except that the integral equations (5.1.19) and (5.1.20) assume more complicated forms (cf. Bijura, 2002c, pp. 124–129).

Theorem 5.1.7 *If f and K are subject to the assumptions stated in Theorem 1.5.1 and $G = G(s, u)$ is in $C^\infty(I \times \mathbb{R})$, the solution of the singularly perturbed Volterra–Hammerstein integral equations (1.5.25) has an asymptotic expansion analogous to (5.1.17), (5.1.18).*

Remark 5.1.8 The initial-layer properties (e.g. that the thickness of the initial layer need not always be $\mathcal{O}(\varepsilon)$, or that it may exhibit oscillatory properties) are discussed in some detail in Bijura (2006). ◇

5.2 Integral-Algebraic Equations with Smooth Kernels

5.2.1 Introduction

A system of linear integral-algebraic equations (IAEs) for the unknown function $\mathbf{u} := (u_1, \ldots, u_d)^T$ has the form

$$\mathbf{B}(t)\mathbf{u}(t) + \int_0^t \mathbf{K}(t, s)\mathbf{u}(s)\, ds = \mathbf{f}(t), \ \ t \in I = [0, T]. \tag{5.2.1}$$

Here, $d \geq 2$, and $\mathbf{B}(t)$ and $\mathbf{K}(t, s)$ are square matrices in $\mathbb{R}^{d \times d}$ $(d \geq 2)$,

$$
\mathbf{B} = \begin{bmatrix} B_{11} & B_{12} & \cdots & B_{1d} \\ B_{21} & B_{22} & \cdots & B_{2d} \\ \vdots & \vdots & & \vdots \\ B_{d1} & B_{d2} & \cdots & B_{dd} \end{bmatrix}, \quad \mathbf{K} = \begin{bmatrix} K_{11} & K_{12} & \cdots & K_{1d} \\ K_{21} & K_{22} & \cdots & K_{2d} \\ \vdots & \vdots & & \vdots \\ K_{d1} & K_{d2} & \cdots & K_{dd} \end{bmatrix}.
$$

We assume throughout this section that $\mathbf{B}(t)$ is singular for all $t \in I$ but has constant rank on I, i.e. $\det \mathbf{B}(t) = 0$ and rank $\mathbf{B}(t) = r_0 \geq 1$ for all $t \in I$, and that the *consistency relation*

$$
\mathbf{B}(0)\mathbf{u}(0) = \mathbf{f}(0) \tag{5.2.2}
$$

holds.

Example 5.2.1 The system of IAEs corresponding to $d = 2$ has the form

$$
\begin{aligned}
B_{11}(t)u_1(t) + B_{12}(t)u_2(t) &+ \int_0^t \big(K_{11}(t, s)u_1(s) \\
&+ K_{12}(t, s)u_2(s) \big)\, ds = f_1(t) \tag{5.2.3}
\end{aligned}
$$

$$
\begin{aligned}
B_{21}(t)u_1(t) + B_{22}(t)u_2(t) &+ \int_0^t \big(K_{21}(t, s)u_1(s) \\
&+ K_{22}(t, s)u_2(s) \big)\, ds = f_2(t), \tag{5.2.4}
\end{aligned}
$$

with $\det \mathbf{B}(t) = 0$, rank $[\, B_{ij}(t)\,] = 1$ $(t \in I)$, and the consistency relation

$$
\begin{bmatrix} B_{11}(0) & B_{12}(0) \\ B_{21}(0) & B_{22}(0) \end{bmatrix} \begin{bmatrix} u_1(0) \\ u_2(0) \end{bmatrix} = \begin{bmatrix} f_1(0) \\ f_2(0) \end{bmatrix}.
$$

Definition 5.2.2 If the matrix $\mathbf{B}(t) \in \mathbb{R}^{d \times d}$ of the system of IAEs (5.2.1) has the form

$$
\mathbf{B} = \begin{bmatrix} B_{11} & B_{12} & \cdots & B_{1d} \\ \vdots & \vdots & & \vdots \\ B_{d_1 1} & B_{d_1 2} & \cdots & B_{d_1 d} \\ 0 & 0 & \cdots & 0 \\ \vdots & \vdots & & \vdots \\ 0 & 0 & \cdots & 0 \end{bmatrix} \tag{5.2.5}
$$

for some d_1 with $1 \leq d_1 < d$, we will call (5.2.1) a *fully decoupled* system of IAEs.

The IAE system (5.2.5) is said to be in *semi-explicit* form if the top left $d_1 \times d_1$ submatrix of **B** has the form

$$\begin{bmatrix} B_{11} & \cdots & B_{1d_1} \\ \vdots & & \vdots \\ B_{d_1 1} & \cdots & B_{d_1 d_1} \end{bmatrix} = \mathcal{I}_{d_1} \,,$$

where $\mathcal{I}_{d_1}$ denotes the identity matrix in $\mathbb{R}^{d_1 \times d_1}$. The special form of $\mathbf{B}(t)$ implies that the IAE system (5.2.1) contains an explicit subsystem of d_2 VIEs of the first kind. ◇

Example 5.2.3 If $d = 2$ the system of IAEs (5.2.1) corresponds to (5.2.3), (5.2.4) with $B_{21}(t) = B_{22}(t) \equiv 0$; that is, it is given by

$$B_{11}(t)u_1(t) + B_{12}(t)u_2(t) + \int_0^t \Big(K_{11}(t,s)u_1(s) + K_{12}(t,s)u_2(s)\Big)ds = f_1(t)$$

$$\int_0^t \Big(K_{21}(t,s)u_1(s) + K_{22}(t,s)u_2(s)\Big)ds = f_2(t),$$

with $B_{11}(t)$ and $B_{12}(t)$ not vanishing at the same point(s) in I. It is *semi-explicit* if $B_{11}(t) = 1$ and $B_{12}(t) = 0$.

For a general (unstructured) matrix **B** the inherent subsystem of first-kind VIEs in (5.2.1) is not known a priori, and hence one is interested in a procedure that allows the 'decoupling' of (5.2.1) into the inherent system of second-kind VIEs and the complementary system of first-kind VIEs, in the sense that the solution $\mathbf{u}(t)$ is given by $\mathbf{u}(t) = [\, \mathbf{v}(t), \mathbf{w}(t)\,]^T$, with $\mathbf{v}(t)$ and $\mathbf{w}(t)$ representing respectively the solution of the two subsystems of VIEs. As in DAEs (cf. Lamour, März & Tischendorf, 2013) such a decoupling is above all motivated by the need to have a tool for the analysis of the regularity of the solution and of the optimal orders of convergence (or the asymptotic stability) of numerical schemes like collocation or Runge–Kutta methods for IAE systems (5.2.1). We shall return to this observation in Section 5.2.4 (cf. Remark 5.2.10).

As we will show below, such a decoupling depends on the tractability index μ of the IAE system (5.2.1), where the definition of μ is related to the ν-smoothing property of the Volterra integral operator

$$(\mathcal{V}\mathbf{u})(t) := \int_0^t \mathbf{K}(t,s)\mathbf{u}(s)\,ds, \quad t \in I, \tag{5.2.6}$$

defining the system of IAEs (5.2.1).

5.2.2 ν-smoothing Volterra Integral Operators

We know that the solution of a (scalar) first-kind VIE

$$(\mathcal{V}u)(t) := \int_0^t K(t, s)u(s)\, ds = g(t), \quad t \in I,$$

does not depend continuously on the given data K and g because such a VIE is an *ill-posed problem*. The ν-smoothing property introduced in Section 1.4.2 for scalar Volterra integral operators represents a measure for the degree of ill-posedness of the corresponding first-kind VIE. Before extending the concept of ν-smoothing to Volterra integral operators with matrix kernels $\mathbf{K}(t, s) \in \mathbb{R}^{d \times d}$ ($d \geq 2$), we recall the scalar version for the convenience of the reader.

Definition 5.2.4 The Volterra operator with smooth (scalar) kernel $K(t, s)$,

$$(\mathcal{V}u)(t) = \int_0^t K(t, s)u(s)\, ds, \quad t \in I,$$

is said to be ν-smoothing for some integer $\nu \geq 1$ if its kernel $K(t, s)$ has the following properties:

(a) $\quad \dfrac{\partial^j K(t, s)}{\partial t^j}\bigg|_{s=t} = 0, \quad t \in I, \quad j = 0, 1, \ldots, \nu - 2;$

(b) $\quad \dfrac{\partial^{\nu-1} K(t, s)}{\partial t^{\nu-1}}\bigg|_{s=t} = k_\nu \neq 0, \quad t \in I;$

(c) $\quad \dfrac{\partial^\nu K}{\partial t^\nu} \in C(D)$ $(D := \{(t, s) : 0 \leq s \leq t \leq T\}).$ ◇

The analogous definition for Volterra integral operators (5.2.6) corresponding to a smooth *matrix kernel* $\mathbf{K}(\cdot, \cdot) \in \mathbb{R}^{d \times d}$ with $d \geq 2$ was introduced in Liang & Brunner (2013).

Definition 5.2.5 Assume that there exist integers $\nu_{pq} \geq 1$ and

$$\nu := \max\{\nu_{pq} : 1 \leq p, q \leq d\}$$

so that the matrix kernel

$$\mathbf{K}(t, s) = \begin{bmatrix} K_{pq}(t, s) \\ (p, q = 1, \ldots, d) \end{bmatrix} \in \mathbb{R}^{d \times d}$$

of the Volterra integral operator $\mathcal{V}$ defined in (5.2.6) has the following properties:

(a) $\quad \dfrac{\partial^j K_{pq}(t, s)}{\partial t^j}\bigg|_{s=t} = 0$ for $t \in I$ and $j = 0, 1, \ldots, \nu_{pq} - 2;$

(b) $\dfrac{\partial^{\nu_{pq}-1} K_{pq}(t,s)}{\partial t^{\nu_{pq}-1}}\bigg|_{s=t} \neq 0$ for $t \in I$;

(c) $\dfrac{\partial^{\nu_{pq}} K_{pq}(t,s)}{\partial t^{\nu_{pq}}} \in C(D)$.

Then the Volterra integral operator (5.2.6) is said to be ν-*smoothing*. We set $\nu_{pq} := 0$ when $K_{pq}(t,s) \equiv 0$. A system of first-kind VIEs is called a ν-smoothing problem if the underlying Volterra integral operator is a ν-smoothing Volterra integral operator. ◇

The following examples illustrate the meaning of the above definition.

Example 5.2.6 Consider the system of IAEs (5.2.1) with $d = 2$ and

$$\mathbf{B} = \begin{bmatrix} 1 & 1 \\ 0 & 0 \end{bmatrix}, \quad \mathbf{K}(t,s) = \begin{bmatrix} 0 & 0 \\ 0 & \dfrac{(t-s)^{r-1}}{(r-1)!} \end{bmatrix} \quad (r \in \mathbb{N}, \ r \geq 1).$$

It is easy to verify that the Volterra integral operator $\mathcal{V}$ with the above kernel matrix is a ν-smoothing operator with $\nu = r$ (compare also Example 1.4.9). Written explicitly, this system of IAEs is

$$u_1(t) + u_2(t) = f_1(t) \tag{5.2.7}$$

$$\int_0^t \frac{(t-s)^{r-1}}{(r-1)!} u_2(s)\, ds = f_2(t). \tag{5.2.8}$$

If $f_2 \in C^{(r)}(I)$ and $f_2^{(j)}(0) = 0$ ($j = 0, \ldots, r-1$) it follows from equation (5.2.9) that $u_2(t) = f_2^{(r)}(t)$. Equation (5.2.8) then yields $u_1(t) = f_1(t) - f_2^{(r)}(t)$: the solution of the system IAE (5.2.8), (5.2.9) is expressed in terms of the rth derivative of f_2.

Example 5.2.7 If the matrix kernel of Example 5.2.6 is replaced by

$$\mathbf{K}(t,s) = \begin{bmatrix} t-s & -e^{t-s} \\ 1 & t-s \end{bmatrix},$$

then the corresponding Volterra integral operator is 2-smoothing (see Exercise 5.4.9).

5.2.3 The Tractability Index of a System of Linear IAEs

Motivated by the decoupling theory for systems of differential-algebraic equations (DAEs) DAE systems (see, for example, the papers by März, 1992,

2002a, 2002b and the monograph by Lamour, März & Tischendorf, 2013) we introduce the *tractability index* (Liang & Brunner, 2013). In order to do so we construct, in analogy to the approach described in the above monograph, a chain of matrices, starting with

$$\mathbf{B}_0(t) := \mathbf{B}(t), \quad \mathbf{K}_0(t) := \mathbf{K}(t,t), \quad \mathbf{K}^0(t,s) := \mathbf{K}(t,s). \tag{5.2.9}$$

For ease of notation we will now often omit the argument of the matrices and simply write $\mathbf{B}_0$ for $\mathbf{B}_0(t)$, $\mathbf{K}^0$ for $\mathbf{K}^0(t,s)$, etc.

The construction of this chain of matrices employs certain special projectors $\mathbf{P}_0$ and $\mathbf{Q}_0$ on $\mathbb{R}^d$. We briefly summarise their definitions and their basic properties (for more details see Lamour, März & Tischendorf (2013), Appendix A.1; consult also the Appendix in the present book for notation).

Definition 5.2.8

(i) A projector $Q \in \mathcal{L}(\mathbb{R}^n)$ (i.e. $Q^2 = Q$) is called a projector onto a subspace $S \subseteq \mathbb{R}^n$ if Im $Q = S$.

(ii) A projector $Q \in \mathcal{L}(\mathbb{R}^n)$ is called a projector along $S \subseteq \mathbb{R}^n$ if ker $Q = S$.   ◇

Here are some of their basic properties. Let P, $Q \in \mathcal{L}(\mathbb{R}^n)$ be projectors.

(a) Q projects onto S if, and only if, $P := \mathcal{I} - Q$ projects along S.

(b) If Q_1 and Q_2 project onto the same subspace S then $Q_2 = Q_1 Q_2$ and $Q_1 = Q_2 Q_1$ hold.

(c) If P_1 and P_2 project along the same subspace S then $P_2 = P_2 P_1$ and $P_1 = P_1 P_2$ are true.

We now describe the construction of the matrix chains mentioned above. For ease of notation we will usually omit the dependence on the variable t in what follows.

Let $\mathbf{Q}_0 \in \mathcal{L}(\mathbb{R}^d)$ denote a projector onto the kernel (null space) ker $\mathbf{B}_0$ of $\mathbf{B}_0$. The chain of matrices $\mathbf{B}_i$, $\mathbf{K}_i$, $\mathbf{P}_j$, $\mathbf{Q}_j$ is then defined recursively as follows:

(a) If $\left[\mathbf{K}_i(t,t) \right]_{pq} \neq 0$, set $\left[\mathbf{K}^{i+1}(t,s) \right]_{pq} = 0$; otherwise,

$$\left[\mathbf{K}^{i+1}(t,s) \right]_{pq} := \frac{\partial^{i+1}}{\partial t^{i+1}} \left[\mathbf{K}^i(t,s) \right]_{pq}. \tag{5.2.10}$$

We set $\mathbf{K}_{i+1}(t) := \mathbf{K}^{i+1}(t,t)$.

(b) Let $\mathbf{B}_1 := \mathbf{B}_0 + \mathbf{K}_0 \mathbf{Q}_0$ and

$$
\mathbf{B}_{i+2} := \begin{cases} \mathbf{B}_{i+1} + \displaystyle\sum_{\ell=0}^{i+1} \mathbf{K}_\ell \left(\prod_{j=0}^{i-\ell} \mathbf{P}_j \right) \mathbf{Q}_{i-\ell+1} & \text{if } 0 \le i \le \nu - 1, \\[2ex] \mathbf{B}_{i+1} + \displaystyle\sum_{\ell=0}^{\nu} \mathbf{K}_\ell \left(\prod_{j=0}^{i-\ell} \mathbf{P}_j \right) \mathbf{Q}_{i-\ell+1} & \text{if } i \ge \nu, \end{cases}
$$

$$(5.2.11)$$

where $\mathbf{Q}_j$ is a projector onto $\ker \mathbf{B}_j$ and $\mathbf{P}_j := \mathbf{I} - \mathbf{Q}_j$ (with $\mathbf{I} \in \mathcal{L}(\mathbb{R}^d)$ denoting the identity matrix).

The following definition was introduced in Liang & Brunner (2013).

Definition 5.2.9 Assume that the Volterra integral operator describing the system of IAEs (5.2.1) is a $(\nu + 1)$-smoothing operator. Then this system is said to be *index-μ tractable* if the matrices $\mathbf{B}_j(t)$ $(j = 0, 1, \ldots, \mu - 1)$ in the above matrix chain are singular and have smooth null spaces, and $\mathbf{B}_\mu(t)$ is non-singular on I. The integer μ is called the *tractability index* of the IAE system (5.2.1). ◇

Remark 5.2.10 The definition of the tractability index μ of the system of IAEs (5.2.1) does not depend on the choice of the projectors $\mathbf{Q}_i$. This is analogous to the tractability index for systems of DAEs (cf. Lamour, März & Tischendorf, 2013, Section 2.4.2). ◇

Since a system of IAEs (5.2.1) contains both the matrix $\mathbf{B}(t)$ and the matrix kernel $\mathbf{K}(t, s)$ describing the memory term, its solvability and its decoupling depend on the tractability index *and* the ν-smoothing of the Volterra integral operator $\mathcal{V}$. This is in sharp contrast to DAEs where, owing to the absence of a (non-local) memory term, it is only the tractability index that plays a role in the decoupling. The following examples illustrate the possible relationships between μ and ν (see also Exercise 5.5.9).

Example 5.2.11 Let $d = 2$ and $\mathbf{B}(t) = \begin{bmatrix} 1 & 0 \\ 0 & 0 \end{bmatrix}$.

(a) For $\mathbf{K}(t, s) = \begin{bmatrix} 0 & 0 \\ 0 & 1 \end{bmatrix}$ the operator $\mathcal{V}$ is ν-smoothing with $\nu = 1$. The tractability index of the corresponding system of IAEs is $\mu = \nu = 1$.

(b) For $\mathbf{K}(t, s) = \begin{bmatrix} 0 & t - s \\ 0 & 1 \end{bmatrix}$ we have $\nu = 2$, and the tractability index is $\mu = 1 < \nu$.

(c) For $\mathbf{K}(t, s) = \begin{bmatrix} 0 & 1 \\ t - s & 0 \end{bmatrix}$ the operator $\mathcal{V}$ is ν-smoothing with $\nu = 2$, and the tractability index is $\mu = 3 > \nu$.

Example 5.2.12 (a) The system of IAEs (5.2.1) described by $\mathbf{B}(t) = \begin{bmatrix} 1 & 1 \\ 0 & 0 \end{bmatrix}$ and $\mathbf{K}(t, s) = \begin{bmatrix} 0 & 0 \\ 0 & 1 \end{bmatrix}$ has $\nu = 1$. Since $\mathbf{B}_0 = \mathbf{B}$ and $\mathbf{K}_0 = \mathbf{K}$ we choose $\mathbf{Q}_0 = \begin{bmatrix} 1 & 0 \\ -1 & 0 \end{bmatrix}$. This leads to $\mathbf{B}_1 = \mathbf{B}_0 + \mathbf{K}_0\mathbf{Q}_0 = \begin{bmatrix} 1 & 1 \\ -1 & 0 \end{bmatrix}$. This matrix is non-singular, and hence the tractability index of this system is $\mu = 1$.

(b) For $\mathbf{B}(t) = \begin{bmatrix} 1 & 1 \\ 0 & 0 \end{bmatrix}$ and $\mathbf{K}(t, s) = \begin{bmatrix} 0 & 0 \\ 0 & t - s \end{bmatrix}$ we have ν-smoothing with $\nu = 2$ (cf. Example 5.2.6) and $\mathbf{B}_1 = \begin{bmatrix} 1 & 1 \\ 0 & 0 \end{bmatrix}$ (corrsponding to the projector $\mathbf{Q}_0 = \begin{bmatrix} 1 & 0 \\ -1 & 0 \end{bmatrix}$). It is singular, and hence $\mu \geq 2$. For the projector $\mathbf{Q}_1 = \begin{bmatrix} 0 & 0 \\ 0 & 0 \end{bmatrix}$ we obtain

$$\mathbf{B}_2 = \mathbf{B}_1 + \mathbf{K}_0\mathbf{P}_0\mathbf{Q}_1 + \mathbf{K}_1\mathbf{Q}_0 = \begin{bmatrix} 1 & 1 \\ -1 & 0 \end{bmatrix}, \quad \text{with } \det \mathbf{B}_2 \neq 0.$$

This implies that the index of tractability is $\mu = 2$.

Example 5.2.13 In (5.2.1) let $\mathbf{B}(t) = \begin{bmatrix} 1 & 0 \\ 0 & 0 \end{bmatrix}$. The Volterra integral operator corresponding to the kernel

$$\mathbf{K}(t, s) = \begin{bmatrix} t - s & e^{t-s} \\ e^{2t-s} & 0 \end{bmatrix}$$

is ν-smoothing with $\nu = 2$. The tractability index of the IAE is $\mu = 2$.

Example 5.2.14 The system of IAEs (5.2.1) with

$$\mathbf{B}(t) = \begin{bmatrix} 1 & t & t^2 \\ 1 & t+1 & t^2 \\ 2 & 2t+1 & 2t^2 \end{bmatrix}, \quad \mathbf{K}(t, s) = \begin{bmatrix} e^{t-s} & 2e^{t-s} & 0 \\ t-s+1 & 2(t-s) & 0 \\ t-s+2 & 2t-s & t-s \end{bmatrix}$$

corresponds to a 2-smoothing Volterra integral operator. The matrix $\mathbf{B}(t, s)$ has rank 2 for all $t \in [0, 1]$. Using Definition 5.2.5 we can verify that the tractability index of the above system of IAEs is $\mu = 2$.

5.2.4 The Decoupling of Index-1 IAEs

The following theorem (Liang & Brunner, 2013) describes the IAEs obtained by the decoupling of the general index-1 system of IAEs (5.2.1). The matrices $\mathbf{B}_i(t)$, $\mathbf{K}_i(t)$, $\mathbf{K}^i(t, s)$, $\mathbf{P}_i(t)$ and $\mathbf{Q}_i(t) = \mathbf{I} - \mathbf{P}_i(t)$ $(i \geq 0)$ are the ones defined in (5.2.9)–(5.2.11), and we set

$$\mathbf{C}_1(t) := \mathbf{P}_0(t)\mathbf{B}_1^{-1}(t), \quad \mathbf{D}_1(t) := \mathbf{Q}_0(t)\mathbf{B}_1^{-1}(t). \tag{5.2.12}$$

Theorem 5.2.15 *Assume that the system of IAEs (5.2.1) possesses the tractability index $\mu = 1$. If the matrices $\mathbf{B}(t)$, $\mathbf{K}(t, s)$ and the projector $\mathbf{P}_0(t)$ are $(r + 1)$-times continuously differentiable on I for some $r \in \mathbb{N}$, then the following statements are true:*

(a) The decoupled system of IAEs is given by

$$\mathbf{v}(t) + \mathbf{C}_1(t) \int_0^t \mathbf{K}(t, s)\mathbf{v}(s)\, ds + \mathbf{C}_1(t) \int_0^t \mathbf{K}(t, s)\mathbf{w}(s)\, ds = \mathbf{C}_1(t)\mathbf{f}(t),$$

$$\tag{5.2.13}$$

$$\mathbf{D}_1(t) \int_0^t \mathbf{K}(t, s)\mathbf{v}(s)\, ds + \mathbf{D}_1(t) \int_0^t \mathbf{K}(t, s) \times \mathbf{w}(s)\, ds = \mathbf{D}_1(t)\mathbf{f}(t),$$

$$\tag{5.2.14}$$

where the matrices $\mathbf{C}_1(t)$ and $\mathbf{D}_1(t)$ are defined in (5.2.12).

(b) For any $\mathbf{f} \in C^{r+1}(I)$ satisfying (5.2.2) the system of IAEs (5.2.1) has a unique solution $\big(\mathbf{v}(t), \mathbf{w}(t)\big)^T$ with $\mathbf{v}, \mathbf{w} \in C^r(I)$, and there exist matrix kernels $\mathbf{R}_i(t)$ $(i = 1, \ldots, 4)$ so that this solution can be represented in the form

$$\mathbf{v}(t) = \mathbf{C}_1(t)\mathbf{f}(t) + \int_0^t \mathbf{R}_1(t, s)\mathbf{C}_1(s)\mathbf{f}(s)ds + \int_0^t \mathbf{R}_2(t, s)\big(\mathbf{D}_1(s)\mathbf{f}(s)\big)'\, ds,$$

$$\tag{5.2.15}$$

$$\mathbf{w}(t) = -\mathbf{D}_1(t)\mathbf{K}(t, t)\mathbf{C}_1(t)\mathbf{f}(t) + \big(\mathbf{D}_1(t)\mathbf{f}(t)\big)'$$
$$+ \int_0^t \mathbf{R}_3(t, s)\mathbf{C}_1(s)\mathbf{f}(s)\, ds + \int_0^t \mathbf{R}_4(t, s)\big(\mathbf{D}_1(s)\mathbf{f}(s)\big)'\, ds.$$

$$\tag{5.2.16}$$

Proof Setting

$$\mathbf{V}_0(t) := \int_0^t \mathbf{K}(t, s)\mathbf{P}_0(s)\mathbf{u}(s)\, ds$$

and

$$\mathbf{W}_0(t) := \int_0^t \mathbf{K}(t, s)\mathbf{Q}_0(s)\mathbf{u}(s)\, ds,$$

the system of IAEs (5.2.1) assumes the form

$$\mathbf{B}(\mathbf{P}_0\mathbf{u}) + \mathbf{V}_0 + \mathbf{W}_0 = \mathbf{g},$$

which can be written as

$$(\mathbf{B} + \mathbf{K}_0\mathbf{Q}_0)(\mathbf{P}_0\mathbf{u} + \mathbf{Q}_0\mathbf{u}) + \mathbf{V}_0 + \mathbf{W}_0 - \mathbf{K}_0\mathbf{Q}_0\mathbf{u} = \mathbf{f}$$

(where, for ease of notation, we have omitted the argument t). Since (5.2.1) is index-μ tractable with $\mu = 1$, the matrix $\mathbf{B}_1(t)$ is non-singular for $t \in I$, and hence we may multiply the above equation by $\mathbf{B}_1^{-1}$. This yields

$$\mathbf{P}_0\mathbf{u} + \mathbf{Q}_0\mathbf{u} + \mathbf{B}_1^{-1}\mathbf{V}_0 + \mathbf{B}_1^{-1}\mathbf{W}_0 - \mathbf{B}_1^{-1}\mathbf{K}_0\mathbf{Q}_0\mathbf{u} = \mathbf{B}_1^{-1}\mathbf{f}. \qquad (5.2.17)$$

If we now define $\mathbf{v}(t) := \mathbf{P}_0\mathbf{u}(t)$ and $\mathbf{w}(t) := \mathbf{Q}_0\mathbf{u}(t)$, and multiply (5.2.17) respectively by $\mathbf{P}_0$ and by $\mathbf{Q}_0$ we obtain

$$\mathbf{v}(t) + \mathbf{P}_0\mathbf{B}_1^{-1}\int_0^t \mathbf{K}(t,s)\mathbf{v}(s)\,ds + \mathbf{P}_0\mathbf{B}_1^{-1}$$

$$\times \int_0^t \mathbf{K}(t,s)\mathbf{w}(s)\,ds - \mathbf{P}_0\mathbf{B}_1^{-1}\mathbf{K}_0\mathbf{w}(t) = \mathbf{P}_0\mathbf{B}_1^{-1}\mathbf{f}(t)$$

and

$$\mathbf{w}(t) + \mathbf{Q}_0\mathbf{B}_1^{-1}\int_0^t \mathbf{K}(t,s)\mathbf{v}(s)\,ds + \mathbf{Q}_0\mathbf{B}_1^{-1}$$

$$\times \int_0^t \mathbf{K}(t,s)\mathbf{w}(s)\,ds - \mathbf{Q}_0\mathbf{B}_1^{-1}\mathbf{K}_0\mathbf{w}(t) = \mathbf{Q}_0\mathbf{B}_1^{-1}\mathbf{f}(t),$$

by using the fact that

$$\mathbf{P}_0\mathbf{B}_1^{-1}\mathbf{K}_0\mathbf{w}(t) = \mathbf{P}_0\mathbf{B}_1^{-1}\mathbf{K}_0\mathbf{Q}_0\mathbf{w}(t) = \mathbf{P}_0\mathbf{B}_1^{-1}(\mathbf{B} + \mathbf{K}_0\mathbf{Q}_0)\mathbf{Q}_0\mathbf{w}(t)$$

$$= \mathbf{P}_0\mathbf{Q}_0\mathbf{w}(t) = 0$$

and

$$\mathbf{w}(t) - \mathbf{Q}_0\mathbf{B}_1^{-1}\mathbf{K}_0\mathbf{w}(t) = \mathbf{Q}_0\mathbf{w}(t) - \mathbf{Q}_0\mathbf{B}_1^{-1}\mathbf{K}_0\mathbf{Q}_0\mathbf{w}(t) = \mathbf{Q}_0\mathbf{B}_1^{-1}\mathbf{Q}_0\mathbf{w}(t) = 0.$$

Thus, recalling (5.2.12), we arrive at the asserted decoupling (5.2.13) and (5.2.14). These equations can be simplified: they become

$$\mathbf{v}(t) + \mathbf{P}_0\mathbf{B}_1^{-1}\int_0^t \mathbf{K}(t,s)\mathbf{v}(s)\,ds + \mathbf{P}_0\mathbf{B}_1^{-1}\int_0^t \mathbf{K}(t,s)\mathbf{w}(s)\,ds = \mathbf{P}_0\mathbf{B}_1^{-1}\mathbf{f}(t)$$

$$\qquad (5.2.18)$$

and

$$\mathbf{Q}_0\mathbf{B}_1^{-1}\int_0^t \mathbf{K}(t,s)\mathbf{v}(s)\,ds + \mathbf{Q}_0\mathbf{B}_1^{-1}\int_0^t \mathbf{K}(t,s)\mathbf{w}(s)\,ds = \mathbf{Q}_0\mathbf{B}_1^{-1}\mathbf{f}(t).$$

$$\qquad (5.2.19)$$

Moreover, there holds

$$\mathbf{Q}_0\mathbf{B}_1^{-1}\mathbf{K}_0\mathbf{w} = \mathbf{Q}_0\mathbf{B}_1^{-1}\mathbf{K}_0\mathbf{Q}_0\mathbf{w} = \mathbf{Q}_0\mathbf{B}_1^{-1}(\mathbf{K}_0\mathbf{Q}_0 + \mathbf{B})\mathbf{w} = \mathbf{Q}_0\mathbf{w} = \mathbf{w}.$$

Since we defined $\mathbf{P}_0\mathbf{B}_1^{-1} =: \mathbf{C}_1$ and $\mathbf{Q}_0\mathbf{B}_1^{-1} =: \mathbf{D}_1$, the equations (5.2.18) and (5.2.19) yield the decoupled system of IAEs (5.2.13) and (5.2.14). The verification of the solution representations (5.2.15) and (5.2.16) is left as an exercise. □

Remark 5.2.16

(i) As we have briefly indicated at the end of Section 5.2.1, the decoupling of the general system of linear IAEs (5.2.1) into the system (5.2.13) and (5.2.14) (that is, the explicit identification of the inherent second-kind and first-kind VIEs in (5.2.1)) is the basis for the analysis of the optimal convergence orders of piecewise polynomial collocation solutions for (5.2.1): it can be seen (cf. Liang & Brunner, 2013) that the corresponding system of collocation equations can be decoupled in the same way.

(ii) A decoupling result for an IAE system (5.2.1) with index of tractability $\mu \geq 2$ and ν-smoothing with $\nu = 1$ was given in Liang & Brunner (2013; see Section 2.3). The case $\mu = 2$ and $\nu \geq 1$ was studied in Liang & Brunner (2016). However, the decoupling analysis remains open for (5.2.1) with $\mu \geq 2$ and $\nu \geq 2$. ◊

5.3 Open Problems

The decoupling of systems of IEAs (5.2.1) with tractability index $\mu \geq 2$ can in principle be analysed by adapting the decoupling approach described in Section 5.2.3 (see Liang & Brunner, 2013 for the case $\mu \geq 2$ and $\nu = 1$, and Liang & Brunner, 2016). The added complexity in the analysis has its main roots in the more complicated interaction between the degree of ν-smoothing and the construction of the analogue of the matrix chain (5.2.10), (5.2.11) and the choice of the projectors $\mathbf{Q}_j$ and $\mathbf{P}_j$. For IAEs with higher tractability index μ the problem of decoupling remains open. Some very special (Hessenberg-type) IAEs with $d = 2$ and $d = 3$ are the subject of the papers by Shiri, Shahmorad & Hojjati (2013), Pishbin, Ghoreishi & Hadizadeh (2013), and Pishbin (2015a, 2015b).

The extension of the decoupling to systems of IAEs with *weakly singular kernels*,

$$\mathbf{B}(t)\mathbf{u}(t) + \int_0^t (t-s)^{\alpha-1}\mathbf{K}(t,s)\mathbf{u}(s)\,ds = \mathbf{f}(t), \quad t \in I \;\; (0 < \alpha < 1), \quad (5.3.1)$$

where, as in (5.2.1), $\mathbf{B}(t)$ and $\mathbf{K}(t,s)$ are (smooth) square matrices of order $d \geq 2$, with $\mathbf{B}(t)$ singular for all $t \in I$ and of constant rank $r_0 \geq 1$, and where $\mathbf{f}$ is subject to the consistency condition (5.2.2), is non-trivial and remains to be studied. This is particularly true if the kernel in (5.3.1) is replaced by a kernel $\mathbf{K}_\alpha(t,s)$ where some, but not all, elements contain a weak singularity (see, for example, the paper by Burns, Herdman & Turi, 1990 and its references, as well as Ito & Turi, 1991, for a related system of neutral Volterra integro-differential equations). The *Research Problem* 5.4.11 addresses this issue.

Finally, the definition of the tractability index and the decoupling theory of systems of *non-linear* IAEs, e.g. of Volterra–Hammerstein form

$$\mathbf{B}(t)\mathbf{u}(t) + \int_0^t \mathbf{K}(t,s)\mathbf{G}(s,\mathbf{u}(s))\,ds, \quad t \in I,$$

are waiting to be studied.

5.4 Exercises and Research Problems

Exercise 5.4.1 Analyse the behaviour of the solution of

$$\varepsilon u(t) = \cos(t) + \lambda \int_0^t u(s)\,ds, \quad t \in I = [0, T],$$

as $\varepsilon \to 0$, assuming that $\lambda \in \mathbb{R}$, $\lambda \neq 0$.

Exercise 5.4.2 Does the singularly perturbed VIE in Example 5.1.3 have an initial boundary layer for any $g \in C^1(I)$ in $f(t) = e^t g(t)$?

Exercise 5.4.3 Consider the singularly perturbed VIE

$$\varepsilon u(t) = f(t) - \int_0^t K(t,s)u(s)\,ds, \quad t \in I, \;\; (0 < \varepsilon \ll 1), \qquad (5.4.1)$$

with $f(t) = a_0 + a_1 t^p$ ($a_i \in \mathbb{R}$, $p \geq 1$), and either $K(t,s) = s$ or $K(t,s) = t$. Discuss the behaviour of the solution $u_\varepsilon(t)$ of (5.4.1) as $\varepsilon \to 0^+$, and compare this behaviour with that of the solution $u_0(t)$ of the reduced equation. Note that both kernels satisfy $K(0,0) = 0$. (See also Bijura, 2006 for a related singularly perturbed VIE.)

Exercise 5.4.4 Let $f \in C^{\infty}(I)$ and $r \in \mathbb{N}$ $(r \geq 1)$. Discuss the behaviour of the solution of the singularly perturbed VIE

$$\varepsilon u(t) = f(t) - \int_0^t \frac{(t-s)^{r-1}}{(r-1)!} u(s)\, ds,$$

as $\varepsilon \to 0$. What can be said about the boundary layer near $t = 0^+$?

Exercise 5.4.5 What is the effect on the asymptotic behaviour (as $\varepsilon \to 0$) if in (5.1.5) the kernel $K(t, s) \equiv -1$ is replaced by $-(t-s)^{\alpha-1}$, with $0 < \alpha < 1$?

Exercise 5.4.6 The singularly perturbed VIE

$$\varepsilon u(t) = \int_0^t e^{t-s} \left(u^2(s) - 1 \right) ds, \quad t \in I,$$

was studied by Angell & Olmstead (1987). Determine the asymptotic expansion of its solution, as $\varepsilon \to 0$. (The exact solution is given by the solution of an initial-value problem for a first-order non-linear ODE; it is

$$u_\varepsilon(t) = 2\varepsilon^{-1} \frac{1 - e^{\gamma t}}{\varepsilon(\gamma - 1)e^{\gamma t} + \gamma + 1}, \quad \text{with } \gamma := \varepsilon^{-1} \left(4 + \varepsilon^2 \right)^{1/2r}.)$$

Exercise 5.4.7 Analyse the behaviour of the solutions of the following systems of linearly perturbed VIEs (see also Examples 5.2.11b and 5.2.12):

$$u_1(t) + \int_0^t \left((t-s)u_1(s) + e^{t-s} u_2(s) \right) ds = f_1(s)$$

$$\varepsilon u_1(t) + \int_0^t e^{2t-s} u_1(s)\, ds = f_2(t),$$

and

$$u_1(t) + \int_0^t (t-s)u_2(s)\, ds = f_1(t)$$

$$\varepsilon u_1(t) + \int_0^t u_2(s)\, ds = f_2(t),$$

assuming that $f_i \in C(I)$ $(i = 1, 2)$.

Exercise 5.4.8 Carry out the detailed proof of Theorem 5.1.6 for the singularly perturbed VIE

$$\varepsilon u(t) = t - \lambda \int_0^t (t-s)^{\alpha-1} u(s)\, ds, \quad t \in I \ (\lambda > 0,\ 0 < \alpha < 1).$$

Exercise 5.4.9 Verify the assertions in Examples 5.2.7, 5.2.11, 5.2.13 and 5.2.14 regarding the degree of ν-smoothing and the tractability index, by constructing the matrix chains defined in (5.2.9)–(5.2.11).

Exercise 5.4.10 Show that the IAE (5.2.1) with $d = 2$, $\mathbf{B}(t) = \begin{bmatrix} 1 & 0 \\ 0 & 0 \end{bmatrix}$ and

$\mathbf{K}(t, s) = \begin{bmatrix} K_{11}(t, s) & K_{12}(t, s) \\ K_{21}(t, s) & 0 \end{bmatrix}$, with $K_{11}(t, t)$, $K_{12}(t, t)$, $K_{21}(t, t) \neq 0$

for $t \in I$, has $\nu = 1$ and tractability index $\mu = 2$. (*Hint:* Choose the projector $\begin{bmatrix} 0 & 0 \\ 0 & 1 \end{bmatrix}$.)

Exercise 5.4.11 (*Research Problem*)
In Section 5.3.2 we defined ν-smoothing and the tractability index μ for systems of IAEs (5.3.1) whose matrix kernel $\mathbf{K}_\alpha(t, s)$ contains the common singularity $(t - s)^{\alpha-1}$ (cf. Definitions 5.2.2 and 5.2.4). Extend these definitions, and the decoupling analysis, to IAEs (5.3.1) (e.g. with $d = 2$) whose matrix kernel has the elements

$$K_{ij}^{(\alpha)}(t, s) := (t - s)^{\alpha_{ij}-1} K_{ij}(t, s) \ (i, j = 1, \dots, d),$$

with $0 < \alpha_{ij} < 1$ and smooth $K_{ij}(t, s)$.

Exercise 5.4.12 (*Research Problem*)
Extend the definition of ν-smoothing to *non-linear* Volterra integral operators of Hammerstein type,

$$(\mathcal{H}\mathbf{u})(t) = \int_0^t \mathbf{K}(t, s)\mathbf{G}(\mathbf{u}(s)) \, ds$$

(with matrix kernel $\mathbf{K}(t, s)$ as (5.2.1)), and establish the theory of decoupling for the corresponding system of non-linear IAEs

$$\mathbf{B}(t)\mathbf{u}(t) + (\mathcal{H}\mathbf{u})(t) = \mathbf{f}(t), \ t \in I.$$

5.5 Notes

Section 5.1: *Singularly Perturbed VIEs*
The classical book on the singular perturbation theory of differential equations is by O'Malley (1991); Roos, Stynes & Tobiska (2008) reflect the progress and the state of the art (including the numerical analysis) of such problems. See also the books by Smith (1985) and Skinner (2011: the latter also encompasses Fredholm integral equations). Singularly perturbed *Fredholm* integral equations are studied in great detail in

the papers by Lange & Smith (1988, 1994). See also Olmstead & Angell (1989), which studies endpoint boundary layers.

The 1997 paper by Kauthen contains a review (with extensive list of references) on the theory, applications and numerical analysis of singularly perturbed VIEs. Linear problems is studied in Skinner (1975, 2000) and Shubin (2006). The most comprehensive analysis is due to Bijura (2002a, 2002b, 2006, 2010, 2012). Her 2006 paper is concerned with the thickness of initial layers (it is shown that the thickness is not always of order $\mathcal{O}(\varepsilon)$ as $\varepsilon \to 0$), while Bijura (2012) deals with systems of singularly perturbed VIEs. Non-linear problems are analysed in Skinner (1995) and Bijura (2002c, 2004, 2006).

Most of these studies were motivated by non-linear singularly perturbed VIEs arising in applications (cf. Example 9.4.9 and the Notes to Chapter 9); see, for example, Mann & Wolf (1951), Olmstead (1972), Olmstead & Handelsman (1972), Handelsman & Olmstead (1972), Angell (1985), Angell & Olmstead (1987) and the lists of references in these publications.

Linear and non-linear renewal VIEs with singular perturbation parameter $\varepsilon > 0$ were studied by Hoppensteadt (1983).

Section 5.2: *IAEs with Smooth Kernels*
The seminal monograph on the theory and the numerical analysis of differential-algebraic equations (based on the notion of the tractability index) is Lamour, März & Tischendorf (2013). It, and the underlying papers by März (e.g. 2002a, 2002b, 2004) inspired the papers by Liang & Brunner (2013, 2016) on the theory and numerical analysis of IAEs of index one and two. An excellent review of various aspects of DAEs is Rabier & Rheinboldt (2002). In addition, see the monograph by Kunkel & Mehrmann (2006); among the many topics it covers it contains, in Section 1.2, an illumination perspective of the various index concepts for DAEs.

The book by Griepentrog & März (1986) represents a major early treatise of DAEs; it also formed the basis for the projection-based approach described in the review papers by März (1992, 2000b) and refined in März (2000a, 2004a, 2004b) and Riaza & März (2008: 'simpler' construction of the matrix chain of projection matrices).

The papers by Kauthen (2000: IAEs of index 1), Hadizadeh, Ghoreishi & Pishbin (2011), Ghoreishi, Hadizadeh & Pishbin (2012), Shiri, Shahmorad & Hojjati (2013), Pishbin, Ghoreishi & Hadizadeh (2013), Shiri (2014) and Pishbin (2015a, 2015b) mostly deal with semi-explicit IAEs (usually with $d = 2$) or with IAEs of Hessenberg form. The convergence analysis is based on varying definitions of the index of the IAEs. More general IAEs are studied in Bulatov (2002), Chistyakov (2006), Bulatov & Lima (2011) and, especially, in Liang & Brunner (2013, 2016).

The terminology 'integral-algebraic equation' was coined by Gear (1990, who introduced the – to me, artificial – concept of the differentiation index for IAEs); see also Campbell & Gear (1995). The tractability index for IAEs was introduced in Liang & Brunner (2013, 2015, motivated by März's analogous concept for DAEs). Other index concepts (including the differential index and the perturbation index) were employed by Bulatov (2002), Chistyakov (2006, 2013), Bulatov & Lima (2011, who studied the IAE:

$$A(t, x)u(t, x) = \int_0^t \int_0^x K(t, x, \tau, \xi)u(\tau, \xi)\, ds\,),$$

and by Hadizadeh, Ghoreishi & Pishbin (2011), Ghoreishi, Hadizadeh & Pishbin (2012), Pishbin, Ghoreishi & Hadizadeh (2013), Shiri, Shahmorad & Hojjati (2013), Shiri (2014) and Pishbin (2015a, 2015b).

Numerical Analysis
The numerical analysis of singularly perturbed ordinary differential equations is discussed in the books by Roos, Stynes & Tobiska (2008; cf. Ch. 1) and Hairer & Wanner (2010, Ch. VI).

The books by Griepentrog & März (1986) and Kunkel & Mehrmann (2006), and the review paper März (1992) deal with various aspects of the numerical analysis and the computational solution of DAEs. The seminal monograph on this topic is the one by Lamour, März & Tischendorf (2013).

For various aspects of the numerical analysis of systems of IAEs (5.2.1) with specially structured matrix $\mathbf{B}(t)$ see the papers listed at the end of the previous paragraph. A detailed convergence analysis of piecewise collocation methods for more general IAEs can be found in Liang & Brunner (2013, 2016).

6

Qualitative Theory of Volterra Integral Equations

Summary

This chapter presents an introduction to the qualitative behaviour (integrability, asymptotic stability) of solutions to linear and Hammerstein-type non-linear Volterra integral equations with convolution kernels. Since this behaviour is governed by the qualitative behaviour of their resolvent kernels, the first part of the chapter is dedicated to results on the boundedness and integrability properties of the resolvent kernels corrsponding to convolution kernels, starting with the celebrated Paley–Wiener theorem.

6.1 Introduction

When analysing the asymptotic behaviour of resolvent kernels and solutions of linear VIEs it is customary to write the VIEs in the forms

$$u(t) + \int_0^t k(t - s)u(s)\,ds = f(t), \ t \geq 0, \tag{6.1.1}$$

and

$$u(t) + \int_0^t K(t, s)u(s)\,ds = f(t), \ t \geq 0. \tag{6.1.2}$$

We have seen in Chapter 1 (cf. Theorems 1.2.8 and 1.2.3) that for continuous or weakly singular kernels k and K and for continuous f the solutions of these VIEs on a bounded interval $I = [0, T]$ are given by

$$u(t) = f(t) - \int_0^t r(t - s)f(s)\,ds, \ t \in I, \tag{6.1.3}$$

and

$$u(t) = f(t) - \int_0^t R(t, s)f(s)\,ds, \ t \in I, \tag{6.1.4}$$

223

where r and R denote respectively the resolvent kernels of the given kernels k and K. In the present chapter we are interested in solutions of (6.1.1) and (6.1.2) on $\mathbb{R}^+ := [0, \infty)$ that lie in the space $L^1(\mathbb{R}^+)$ of functions that are integrable on $\mathbb{R}^+$. We shall show that under appropriate conditions the solution representations (6.1.3) and (6.1.4) remain valid in this more general setting.

Since the asymptotic behaviour of u on $[0, \infty)$ is then essentially governed by the asymptotic properties (e.g. the boundedness or the integrability) of r and R, as (6.1.3) and (6.1.4) reveal, the answer to the question as to whether u inherits the asymptotic behaviour of the forcing function f (e.g. $f \in L^1(\mathbb{R}^+)$, or $f(t) \to 0$ as $t \to \infty$) will depend on analogous qualitative properties of the resolvent kernels r and R.

6.2 Asymptotic Properties of Resolvent Kernels

6.2.1 VIEs with Convolution Kernels

We begin with an extension of the resolvent equations (1.2.24) and (1.2.25) for convolution VIEs with continuous kernels to the L^1 setting. It uses the fact that if $a, b \in L^1(\mathbb{R}^+)$, then their convolution $(a * b)(z) := \int_0^z a(z-s)b(s)\,ds$ is also in $L^1(\mathbb{R}^+)$. More generally, if $a \in L^1(\mathbb{R}^+)$ and $b \in L^p(\mathbb{R}^+)$ for some $p \in [1, \infty]$, their convolution lies in $L^p(\mathbb{R}^+)$ (see also Exercise 6.5.3). Also, we recall that $L^1_{loc}(\mathbb{R}^+)$ denotes the space of functions that are in $L^1(0, T)$ for any $T > 0$.

Theorem 6.2.1 *If $k \in L^1_{loc}(\mathbb{R}^+)$ there exists a solution $r \in L^1_{loc}(\mathbb{R}^+)$ that solves the resolvent equations*

$$r(z) + \int_0^z k(z-s)r(s)\,ds = k(z) \quad and \quad r(z) + \int_0^z r(z-s)k(s)\,ds = k(z)$$
(6.2.1)

for $z \in \mathbb{R}^+$. This solution depends continuously on k (with respect to the L^1 norm).

Proof To establish the uniqueness of the resolvent r on $[0, T]$ with arbitrary positive $T < \infty$, assume that there exist two solutions r_1 and r_2 of the resolvent equations (6.2.1). Thus, employing the short notation $*$ for the convolution of two L^1 functions, and observing the symmetry of the convolution integral, we obtain from

$$r_1 = k - r_1 * k \quad and \quad r_2 = k - k * r_2$$

that, on $[0, T]$,

$$r_1 = k - r_1 * k = k - r_1 * (r_2 + k * r_2) = k - (r_1 + r_1 * k) * r_2 = k - k * r_2 = r_2$$

holds. For the proof of the existence of the resolvent kernel $r \in L^1_{loc}(\mathbb{R}^+)$ we refer the reader to Gripenberg, Londen & Staffans (1991, p. 43), and we leave the proof of the continuous dependence of r on the kernel k as an exercise (Exercise 6.5.2). □

We start our discussion of the qualitative properties on $\mathbb{R}^+$ (boundedness; integrability) of the resolvent kernel r in the (scalar) VIE (6.1.1) with convolution kernel. In the celebrated *Paley–Wiener theorem* (Paley & Wiener, 1934; Gripenberg, Londen & Staffans, 1990, pp. 45–46) the key assumption is that the kernel k be integrable on $\mathbb{R}^+$.

Theorem 6.2.2 *Assume that* $k \in L^1(\mathbb{R}^+)$. *Then its resolvent kernel* r *is in* $L^1(\mathbb{R}^+)$ *if, and only if,*

$$1 + K(\zeta) \neq 0 \quad \text{for all } \zeta \in \mathbb{C} \text{ with } \operatorname{Re}\zeta \geq 0. \tag{6.2.2}$$

Here, $K(\zeta) := \int_0^\infty e^{-\zeta t} k(t) dt$ *denotes the Laplace transform of the kernel* k.

Proof We will prove the necessity of the condition (6.2.2). The key to this proof is given by Theorem 6.2.1 (in which the resolvent equations (1.2.24) and (1.2.25) were extended to the setting of $L^1_{loc}(\mathbb{R}^+)$). Assume now that $r \in L^1(\mathbb{R}^+)$ and apply the Laplace transform to the first of the resolvent equations in (6.2.1). We obtain, for ζ with $\operatorname{Re}\zeta \geq 0$ and $R(\zeta) := \int_0^\infty e^{-\zeta t} r(t) dt$,

$$R(\zeta) + K(\zeta)R(\zeta) = K(\zeta), \quad \text{or} \quad \big(1 + K(\zeta)\big)\big(1 - R(\zeta)\big) = 1.$$

Clearly this can hold only if $1 + K(\zeta) \neq 0$ for $\zeta \in \mathbb{C}$ with $\operatorname{Re}\zeta \geq 0$.

The rather more complex proof of the sufficiency part is based on Fourier transform techniques; it can be found, for example, in Miller (1971, Appendix I) or in Gripenberg, Londen & Staffans (1990, pp. 49–54). □

Remark 6.2.3 If $k(t - s)$ in (6.1.1) is a matrix kernel, say $\mathbf{k}(\cdot) \in \mathbb{R}^{d \times d}$ ($d \geq 2$), with elements in $L^1(\mathbb{R}^+)$, the condition (6.2.2) in Theorem 6.2.2 has to be replaced by

$$\det[\mathbf{I} + \hat{\mathbf{k}}(\zeta)] \neq 0 \text{ for all } \zeta \in \mathbb{C} \text{ with } \operatorname{Re}\zeta \geq 0.$$

Here, $\mathbf{I}$ denotes the identity matrix in $\mathbb{R}^{d \times d}$. ◇

There are convolution kernels k that are not in $L^1(\mathbb{R}^+)$ (an important example is the singular kernel $k(z) = (t - s)^{\alpha - 1}$ with $0 < \alpha < 1$) but whose resolvent kernels may be integrable on $\mathbb{R}^+$ if they possess one of the properties described in the following definition.

Definition 6.2.4 A function h s said to be *completely monotone* on an interval (a, b) if $f \in C^\infty(a, b)$ and $(-1)^j h^{(j)}(t) \geq 0$ is true for all $t \in (a, b)$ and all $j \in \mathbb{N}$.   ◇

Remark 6.2.5 It is easy to verify that if h is completely monotone on $(0, \infty)$, we have either $h(z) \equiv 0$, or $h(z) > 0$ for all $z \in (0, \infty)$. This follows from a result in Laplace transform theory that S. Bernstein established in 1928 (cf. Gripenberg, Londen & Staffans, 1990 pp. 143–144), which states that h is completely monotone on $(0, \infty)$ if, and only if, there exists a bounded, non-decreasing function μ such that

$$h(z) = \int_0^\infty e^{-sz}\,d\mu(s), \quad z \in [0, \infty). \tag{6.2.3}$$

(Compare also Friedman, 1963, Section 5, and Exercise 6.5.8 at the end of this chapter.)   ◇

In the remainder of this section we present a selection of representative theorems that address the question on the boundedness and the integrability of resolvent kernels when the underlying convolution kernel is not in $L^1(\mathbb{R}^+)$. Before stating these results we consider the example of the weakly singular kernel mentioned above.

Example 6.2.6 The kernel $\lambda(t - s)^{\alpha-1}/\Gamma(\alpha)$ $(0 < \alpha < 1)$ of the weakly singular VIE

$$u(t) + \lambda \int_0^t \frac{(t - s)^{\alpha-1}}{\Gamma(\alpha)} u(s)\,ds = 1, \quad t \geq 0,$$

is not in $L^1(\mathbb{R}^+)$ if $\lambda \neq 0$. As we have seen in Theorem 1.3.2 (equation (1.3.8)), the resolvent kernel of this kernel is

$$r_\alpha(t - s) = -\lambda \frac{d}{dt} E_\alpha\big(-\lambda(t - s)^\alpha\big).$$

It follows from the definition of the Mittag-Leffler function that r_α is in $L^1(\mathbb{R}^+)$ when $\lambda > 0$.

Example 6.2.7 (Londen, 1984, p. 140)
Consider the VIE (6.1.1) with convolution kernel

$$k(z) = z^{-\beta}\cos(z) \quad (0 \leq \beta < 1).$$

It is easy to verify that while $k \notin L^1(\mathbb{R}^+)$, its resolvent kernel $r(z)$ is integrable on $\mathbb{R}^+$.

The first of the theorems announced above deals with VIEs whose kernels are completely monotone. It is due to Reuter (1956; see also Gripenberg, Londen & Staffans, 1990, pp. 148–149). In the following, $f \in L^p_{loc}(\mathbb{R}^+)$ $(p \geq 1)$ means that the restriction of f to any compact subset J of $\mathbb{R}^+$ is in $L^p(J)$.

Theorem 6.2.8 *Assume that the kernel of (6.1.1) satisfies $k \in L^1_{loc}(\mathbb{R}^+)$. Then:*

(a) k is completely monotone on $(0, \infty)$ if, and only if, its resolvent kernel r is completely monotone on $(0, \infty)$, $r \in L^1(\mathbb{R}^+)$, and $\int_{\mathbb{R}^+} r(z)\, dz \leq 1$.

(b) $k \in L^1(\mathbb{R}^+)$ if, and only if, $\int_{\mathbb{R}^+} r(z)\, dz < 1$.

Proof The verification of (a) and (b) is based on properties of the Laplace transform of functions in $C(0, \infty)$ that are in $L^1_{loc}(\mathbb{R}^+)$ and are completely monotone on $\mathbb{R}^+$. Such a function k is known to have the properties that its Laplace transform $K(\zeta)$ satisfies $\lim_{\zeta \to \infty} K(\zeta) = 0$, and that $k \in L^1(\mathbb{R}^+)$ if, and only if, $\limsup_{\zeta \to 0^+} |K(\zeta)| < \infty$. The technical details of the proof may be found in Gripenberg, Londen & Staffans (1990, Sections 5.2 and 5.3); see also Miller (1971, pp. 221–224). $\qquad \square$

Remark 6.2.9 We shall see in Theorem 6.3.3 that under appropriate assumptions on f, the solution of the VIE (6.1.1) with completely monotone $L^1(\mathbb{R}^+)$ kernel is also completely monotone. $\qquad \diamond$

Example 6.2.10 We have seen in Section 1.3.1 that the Mittag-Leffler function $E_\beta(-z)$ is completely monotone for $z \in [0, \infty)$ when $0 < \beta \leq 1$) (Pollard, 1948). Thus, by Theorem 6.2.8, the resolvent kernel of Example 6.2.6 possesses the same property when $\lambda > 0$ and $\alpha \in (0, 1]$.

The second theorem, established in Shea & Wainger (1975), extends the Paley–Wiener theorem to kernels that may not be integrable on $\mathbb{R}^+$.

Theorem 6.2.11 *If $k \in L^1_{loc}(\mathbb{R}^+)$ is non-negative, non-increasing and convex on $(0, \infty)$, then its resolvent r is in $L^1(\mathbb{R}^+)$.*

Proof We refer the reader to the original paper for the proof of Theorem 6.2.11. A somewhat more refined version (containing also an upper bound for $\|r\|_{L^1}$) can be found in Gripenberg, Londen & Staffans (1990, pp. 169–176).   □

A related result where k is not assumed to be convex is stated in the next theorem.

Theorem 6.2.12 *Under the following assumptions on the convolution kernel k in (6.1.1):*

(i) $k \in L^1_{loc}(0, \infty)$, *and k is positive, continuous and non-increasing on $(0, \infty)$;*

(ii) for each $T > 0$ the function $k(t)/k(t + T)$ is non-increasing in t on $(0, \infty)$;

the resolvent kernel r corresponding to k has the following properties:

(a) r *is continuous on $(0, \infty)$, and if $k(0^+) < \infty$ it is also defined and continuous at $t = 0$;*

(b) $0 \le r(z) \le k(z)$ *for $z \in (0, \infty)$;*

(c) $\displaystyle \int_0^\infty r(z)dz = \frac{\int_0^\infty k(z)dz}{1 + \int_0^\infty k(z)dz} < 1$ *if* $\displaystyle \int_0^\infty k(z)\,dz < \infty$;

(d) $\displaystyle \int_0^\infty r(z)\,ds = 1$ *if* $\displaystyle \int_0^\infty k(z)\,dz = \infty.$

We omit the proof of this theorem (it can be found in Miller, 1968b, pp. 322–325; see also Levin, 1977); instead, we will establish the following properties of the resolvent kernel r which are a consequence of Theorem 6.2.6 (cf. Miller, 1968b).

Corollary 6.2.13 *Let the assumptions (i) and (ii) in Theorem 6.2.12 hold. Then the following statements are true:*

(a) If $\displaystyle \int_0^\infty k(s)ds = \kappa < \infty$, *then* $\displaystyle \int_0^\infty r(s)\,ds = \kappa(1 + \kappa)^{-1} < 1.$

(b) If $k \notin L^1(0, \infty)$, *then* $\displaystyle \int_0^\infty r(s)ds = 1.$

Proof Assume first that $k \in L^1(0, \infty)$. Applying the Laplace transformation to the resolvent equation

$$r(t) = k(t) - \int_0^t k(t - s)r(s)\,ds, \quad t \ge 0 \qquad (6.2.4)$$

(cf. (1.2.24)), we obtain

$$R(\zeta) = K(\zeta) - K(\zeta)R(\zeta) \ (\text{Re } \zeta \geq 0),$$

where $K(\zeta) := \int_0^\infty e^{-\zeta t} k(t)dt$. Since we have $K(0) = \kappa$ it follows from assumption (i) in Theorem 6.2.12 that

$$\int_0^\infty r(s)\,ds = R(0) = K(0)/(1 + K(0)) = \frac{\kappa}{1 + \kappa} < 1.$$

The proof of assertion (b) is left as an exercise (cf. Exercise 6.5.9). □

Remark 6.2.14 Gripenberg (1978) showed that the estimate (b) in Theorem 6.2.12 can be improved: if $k \in L^1(0, 1)$ is positive, continuous and non-increasing on $(0, \infty)$, then the resolvent equation (6.2.4) possesses a unique continuous solution r on $(0, \infty)$ that satisfies

$$0 < r(t) \leq \frac{k(t)}{1 + \|k\|_{L^1(0,t)}}, \ \ 0 < t < \infty.$$

◇

6.2.2 VIEs with General Kernels

The analysis of the qualitative properties of resolvent kernels $R(t, s)$ corresponding the (non-convolution) kernel $K(t, s)$ of the linear VIE

$$u(t) + \int_0^t K(t, s)u(s)\,ds = f(t), \ \ t \geq 0, \tag{6.2.5}$$

is much more complex than the one for VIEs with convolution kernels. In Section 8.3.1 we will briefly describe the appropriate functional analysis framework in which this analysis is best carried out: it is that of kernels and resolvents of so-called types (L^p, T) and L_{loc}^p (introduced in Miller, 1971, Chapter IV; see also Gripenberg, Londen & Staffans, 1990, Sections 9.2 and 9.3).

6.3 Asymptotic Behaviour of Solutions

6.3.1 VIEs with Convolution Kernels

As we mentioned at the beginning of this chapter, the principal motivation for discussing the integrability of the resolvent kernel r was to understand when the solution of the VIE

$$u(t) + \int_0^t k(t-s)u(s)\,ds = f(t), \quad t \geq 0, \tag{6.3.1}$$

inherits the asymptotic behaviour of the forcing function f. Two typical results are stated in the following theorems. The first of these theorem is the analogue of Theorem 1.2.8: it is concerned with the existence of solutions to (6.1.1) in $L^1_{loc}(\mathbb{R}^+)$.

Theorem 6.3.1 *If $k \in L^1_{loc}(\mathbb{R}^+)$, the VIE (6.3.1) possesses a unique solution $u \in L^1_{loc}(\mathbb{R}^+)$ for any $f \in L^1_{loc}(\mathbb{R}^+)$. This solution can be expressed in the form*

$$u(t) = f(t) - \int_0^t r(t-s)f(s)\,ds, \quad t \in \mathbb{R}^+, \tag{6.3.2}$$

where r is the resolvent kernel of k described in Theorem 6.2.1.

Proof (i) Let u be defined by the equation (6.3.2), where $f \in L^1_{loc}(\mathbb{R}^+)$ is given. Since $k \in L^1_{loc}(\mathbb{R}^+)$ it follows from (6.1.1) and (6.2.1) that

$$u + k * u = u + k * (f - r * f) = u + (k - k * r) * f = u + r * f = f.$$

This shows that u is a solution of (6.1.1) in $L^1_{loc}(\mathbb{R}^+)$.

(ii) Let $u \in L^1_{loc}(\mathbb{R}^+)$ be a solution of (6.3.1). Thus, f must be in the same space; moreover, using the second of the resolvent equations (6.2.1) we see that

$$u = f - k * u = f - (r + r * k) * u = f - r * (u + k * u) = f - r * f.$$

$\square$

The results cited in the next theorem form part of a more extensive theorem in Gripenberg, Londen & Staffans (1990, Theorem 2.4.5). That theorem also describes many other function spaces in which u inherits the asymptotic behaviour of f.

Theorem 6.3.2 *Assume that $k \in L^1(\mathbb{R}^+)$ is such that $r \in L^1(\mathbb{R}^+)$. Then the following statements are true:*

(a) *If $f \in L^1(\mathbb{R}^+)$, then the solution u of (6.3.1) is also in $L^1(\mathbb{R}^+)$.*

(b) *If f is continuous and bounded on $\mathbb{R}^+$, with $f(t) \to 0$ as $t \to \infty$, then the solution u of (6.3.1) possesses the same property.*

Proof The assertion (a) is a consequence of the solution representation (6.3.2) and the properties of the convolution of two L^1 functions mentioned at the beginning of Section 6.2.1. The proof of (b) is left as an exercise (cf. Exercise 6.5.4). □

We have seen in Section 1.3.1 that the Mittag-Leffler function $E_\beta(z)$ is completely monotone on $[0, \infty)$ for $\beta \in (0, 1)$ and $z < 0$, and that the solution of the weakly singular VIE

$$u(t) = 1 + \lambda \int_0^t \frac{(t - s)^{\alpha - 1}}{\Gamma(\alpha)} u(s) \, ds, \quad t \geq 0 \quad (0 < \alpha < 1),$$

possesses the same property. The following theorem (Friedman, 1963, pp. 400–402) describes a larger class of convolution kernels for which the solution of the VIE (6.3.1) is completely monotone.

Theorem 6.3.3 *Assume that the convolution kernel k is in $L^1(0, 1)$ and is completely monotone on $(0, \infty)$. If $f(t) \equiv 1$, the solution u of (6.3.1) is completely monotone on $(0, \infty)$.*

Proof Since k is completely monotone in every interval $(0, T)$ with $T < \infty$, it can be represented in the form (6.2.3). The basic idea of the proof then consists (i) in showing that on $[0, T]$, $k(z)$ can be approximated uniformly by

$$k_n(z) := \sum_{i=1}^{p_n} a_{n,i} e^{-\lambda_{n,i} z}$$

(with $a_{n,i} > 0$ and $0 < \lambda_{n,1} < \cdots < \lambda_{n,p_n}$), and (ii) in verifying that the solution $u_n(t)$ of the VIE (6.3.1), with $k(t - s)$ replaced by $k_n(t - s)$, is completely monotone on $(0, T)$ and converges pointwise to $u(t)$ as $p_n \to \infty$. The complete monotonicity of $u(t)$ can then be deduced from the one of $u_n(t)$. The technical details can be found in Friedman (1963, pp. 400–402). □

6.4 VIEs of Hammerstein Form

6.4.1 Non-Linear Perturbations of Linear VIEs

We now consider the non-linearly perturbed VIE

$$u(t) + \int_0^t K(t, s)\big(u(s) + G(s, u(s))\big) \, ds = f(t), \quad t \geq 0, \tag{6.4.1}$$

where the given functions $f = f(t)$, $K = K(t, s)$ and $G = G(s, u)$ are assumed to be continuous on their respective domains. The following theorem is due to Strauss (1970) and Miller (1970; see also Miller, Nohel & Wong, 1969, Miller, 1971, pp. 191–193, and Gripenberg, Londen & Staffans, 1990, pp. 312–314) for related results). The VIE (6.4.2) given below is usually called the *variation-of-constants formula* for (6.4.1). As we shall see, it plays an important role in the analysis of the asymptotic behaviour of the solution of (6.4.1).

Theorem 6.4.1 *Let* $R(t, s)$ *denote the resolvent kernel of* $K(t, s)$. *Then the VIE (6.4.1) is equivalent to the Hammerstein VIE*

$$u(t) = u_L(t) - \int_0^t R(t, s)G(s, u(s))\, ds, \quad t \geq 0. \qquad (6.4.2)$$

Here, $u_L(t)$ *is the solution of the linear VIE*

$$u(t) + \int_0^t K(t, s)u(s)\, ds = f(t), \quad t \geq 0.$$

Proof We first write the VIE (6.4.1) in the 'linear' form

$$u(t) = f(t) - F(t) - \int_0^t K(t, s)u(s)\, ds, \quad t \geq 0,$$

where

$$F(t) := \int_0^t K(t, s)G(s, u(s))\, ds\,.$$

It follows from Section 1.2.1 that its solution $u(t)$ can be formally be written in the form

$$u(t) = f(t) - F(t) + \int_0^t R(t, s)\big(f(s) - F(s)\big)\, ds,$$

and this leads to

$$
\begin{aligned}
u(t) &= f(t) - \int_0^t R(t, s)f(s)\, ds \\
&\quad - F(t) - \int_0^t R(t, s)\Big(\int_0^s K(s, v)G(v, u(v))\, dv\Big) ds \\
&= u_L(t) - F(t) - \int_0^t \Big(\int_v^t R(t, s)K(s, v)ds\Big)G(v, u(v))\, dv.
\end{aligned}
$$

Recalling the resolvent equation (1.2.10) of Section 1.2.1 we may write the inner integral as

$$\int_v^t R(t, s)K(s, v)\, ds = R(t, v) - K(t, v).$$

Hence, by the definition of $F(t)$ we obtain

$$
\begin{aligned}
u(t) &= u_L(t) - \int_0^t K(t,s)G(s,u(s))ds - \int_0^t \big(R(t,s) - K(t,s)\big)G(s,u(s))\,ds \\
&= u_L(t) - \int_0^t R(t,s)G(s,u(s))ds,
\end{aligned}
$$

as asserted in Theorem 6.4.1. □

6.4.2 Hammerstein VIEs with Convolution Kernels

We conclude this section by briefly looking at the asymptotic behaviour of the solution of the Hammerstein VIE

$$
u(t) + \int_0^t k(t-s)G(s,u(s))\,ds = f(t), \quad t \geq 0, \tag{6.4.3}
$$

which may be regarded as the natural non-linear counterpart to the linear convolution VIE (6.3.1). The following result is due to Miller (1968b, Theorem 1; see also Miller, 1971, pp. 210–212). It is closely related to a result in Friedman (1963) but is proved in a more transparent manner (see also Remark 6.4.3 below).

Theorem 6.4.2 *Assume that*

(i) $f = f(t)$ *is continuous and positive for* $t \in [0,\infty)$;
(ii) $k = k(z)$ *is continuous and positive for* $z \in (0,\infty)$, *with* $k \in L^1(0,1)$;
(iii) $G = G(t,u)$ *is locally Lipschitz continuous in* u *(uniformly for* $0 \leq t < \infty$), *is integrable for* $0 \leq t < \infty$ *and* $u \in \mathbb{R}$, *and is such that* $uG(t,u) > 0$ *holds for all* (t,u);
(iv) $\dfrac{f(T)}{f(t)} \leq \dfrac{k(T-s)}{k(t-s)}$ *whenever* $0 \leq s \leq T < t$.

Then the solution $u(t)$ *of the VHIE (6.4.3) exists and is continuous for all* $t \geq 0$, *and it satisfies* $0 \leq u(t) \leq f(t)$ *for* $0 \leq t < \infty$.

Proof The function $\bar{G}$ defined by

$$
\bar{G}(t,u) := \begin{cases} G(t,u) & \text{if } u \geq 0, \\ 0 & \text{if } u < 0, \end{cases}
$$

possesses the same continuity and integrability properties as $G(t,u)$. Suppose that $\bar{u}(t)$ is a solution of (6.4.3), with $G(s,u)$ replaced by $\bar{G}(s,u)$, that exists on a maximal interval $[0,\delta)$, and define the (possibly empty) set

$$
\bar{S} := \{t \in [0,\delta) : \bar{u}(t) < 0\}.
$$

If $\bar{S}$ is not empty, $\bar{S}$ is open and contains a maximal interval (t_0, t_1) with $t_0 > 0$. It follows that $\bar{u}(t_0) = 0$. Moreover, for $t_0 < t < t_1$ we have

$$
\begin{aligned}
0 > \bar{u}(t) &= f(t) - \int_0^{t_0} k(t - s)\bar{G}\big(s, \bar{u}(s)\big)\, ds \\
&= \frac{f(t)}{f(t_0)}\left(f(t_0) - \int_0^{t_0} k(t - s) f(t_0)\bar{G}\big(s, \bar{u}(s)\big)\big(f(t)\big)^{-1}\, ds \right).
\end{aligned}
$$

Assumption (iv) implies that that

$$
0 > f(t_0) - \int_0^{t_0} k(t_0 - s)\bar{G}\big(s, \bar{u}(s)\big)\, ds = \bar{u}(t_0) = 0,
$$

which contradicts the hypothesis that $\bar{S}$ is non-empty. In other words, we must have $\bar{u}(t) \geq 0$ for $t \in [0, \delta)$. Thus, by definition of $\bar{G}(t, u)$ and the fact that $\bar{u}(t) \geq 0$, we see that $\bar{u}(t)$ solves the original VIE (6.4.3) on $[0, \delta)$. The verification of the assertion that $\bar{u}(t) \leq f(t)$, as long as $\bar{u}(t)$ exists, and the continuation of $\bar{u}(t)$ to the entire interval $[0, \infty)$, is left to the reader. □

Remark 6.4.3 The paper by Friedman (1963, cf. p. 384 and p. 387) contains a number of variants of Theorem 6.4.2 that are based on e.g. the assumptions that f and k are continuously differentiable on $[0, \infty)$ and $(0, \infty)$, respectively. In this case, the assumption (iv) in the above theorem is replaced by

$$
\frac{k'(a)}{k(a)} < \frac{f'(b)}{f(b)} \quad \text{for all } a, b \text{ with } 0 < a \leq b < \infty.
$$

◇

Theorem 6.4.2 is concerned with the boundedness of (positive) solutions $u(t)$ of (6.4.3). The following theorem (Friedman, 1963, pp. 391–392) describes conditions under which the solution $u(t)$ of (6.4.3) (but now with $G = G(u)$) tends to zero as $t \to \infty$.

Theorem 6.4.4 *Assume that the given functions in*

$$
u(t) + \int_0^t k(t - s)G(u(s))\, ds = f(t), \quad t \geq 0, \tag{6.4.4}
$$

satisfy the following assumptions:

(i) k, G and f are continuous on $0 < z < \infty$, $\mathbb{R}$ and $0 \leq t < \infty$, respectively;

(ii) $k(z) \geq \kappa_0 > 0$ $(z > 0)$, $k \in L^1(0, 1)$ is monotone non-increasing, and $k(z) \leq \bar{k} < \infty$;

(iii) $uG(u) > 0$ if $u \neq 0$;

(iv) f is monotone non-decreasing, and $f \le \bar{f} < \infty$ for $t \in [0, \infty)$.

If $u(t)$ is a (positive) solution of (6.4.4), it has the property that $\lim_{t \to \infty} u(t) = 0$.

Proof If the assertion is false, there must exist a strictly increasing sequence $\{t_n\}$ in $\mathbb{R}^+$ with $t_n \to \infty$, as $n \to \infty$, and $t_{n+1} - t_n \ge 1$, so that $u(t_n) \ge u_0 > 0$ for some $u_0 > 0$. We will show that we can find a $\delta \in (0, 1)$ such that $u(t) > u_0/2$ for $t_n < t < t_n + \delta$. To verify this, let

$$J(\tau) := \int_0^\tau k(z)\,dz.$$

By assumption (ii) we have $J(\tau) > 0$ for $\tau > 0$. Hence, by (6.4.4) and for $t > t_n$ we obtain

$$
\begin{aligned}
\frac{u(t) - u(t_n)}{J(t - t_n)} =\ & \frac{f(t) - f(t_n)}{J(t - t_n)} \\
& - \frac{1}{J(t - t_n)} \int_0^{t_n} \big(k(t - s) - k(t_n - s)\big) G(u(s))\,ds \\
& - \frac{1}{J(t - t_n)} \int_{t_n}^{t} k(t - s) G(u(s))\,ds \\
=:\ & A_n(t) - B_n(t) - C_n(t),
\end{aligned}
$$

with $A_n(t) \ge 0$, $B_n(t) \ge 0$, $C_n(t) \ge 0$. Moreover, $C_n(t) \le \bar{G}_n$, where $\bar{G}_n$ is the least upper bound of $G(u(s))$ on $t_n \le s \le t$, and hence we have $\bar{G}_n \le \tilde{G}$, with $\tilde{G}$ denoting the least upper bound of $G(u)$ for $0 \le u \le \bar{f}$. It then follows that

$$u(t) \ge u(t_n) - \tilde{G} J(t - t_n) > u_0/2,$$

whenever $J(\delta) \le u_0/(2\tilde{G})$. If γ is the greatest lower bound of $G(u)$ on $u_0/2 \le u \le \bar{f}$, there holds

$$\int_0^t k(t - s) G(u(s))\,ds \ge \bar{k}\gamma \sum_{t_n < t - \delta} 1 \to \infty, \quad \text{as } t \to \infty.$$

Recalling the VIE (6.4.4) we obtain a contradiction to the initial hypothesis, namely,

$$\bar{f} \ge f(t) = u(t) + \int_0^t k(t - s) G(u(s))\,ds \to \infty, \quad \text{as } t \to \infty.$$

$\square$

6.5 Exercises and Research Problems

Exercise 6.5.1 Consider the VIE

$$u(t) = f(t) + \int_0^t k(t-s)u(s)\,ds, \quad t \geq 0.$$

(a) Let $k(z) = \lambda z^m$, with $\lambda < 0$ and $m \in \mathbb{N}$. Find the Laplace transform of the resolvent kernel $r(z)$ of $k(z)$. Is $r \in L^1(\mathbb{R}^+)$?

(b) Find the resolvent kernel $r(z)$ of the kernel $k(z) = e^{-az}\cos(bz)$, where $a \geq 0$, $b \geq 0$. Is $r \in L^1(\mathbb{R}^+)$?

(c) Let $0 < T < \infty$. Do the same for the kernel

$$k(z) = \begin{cases} T - z & \text{if } z \in [0, T], \\ 0 & \text{if } z > T. \end{cases}$$

Exercise 6.5.2 Prove the last statement in Theorem 6.2.1.

Exercise 6.5.3 Assume that $a \in L^1(\mathbb{R}^+)$ and $b \in L^p(\mathbb{R}^+)$ for some $p \in [1, \infty]$. Show that the convolution $(a*b)(z) = \int_0^z a(z-s)b(s)ds$ of a and b lies in $L^p(\mathbb{R}^+)$.

Exercise 6.5.4 Assume that $a \in L^1(\mathbb{R}^+)$ and $b \in BC_0(\mathbb{R}^+)$ (the space of continuous and bounded functions that have limit zero as $t \to \infty$). Show that their convolution $a * b$ is in $BC_0(\mathbb{R}^+)$. Then use this result to prove assertion (b) in Theorem 6.3.2.

Exercise 6.5.5 Assume that h is a non-negative function that is completely monotone on $\mathbb{R}^+$.

(i) Show that h is a convex function.

(ii) Show that its Laplace transform $H(\zeta)$ exists and satisfies $\lim_{\zeta \to \infty} H(\zeta) = 0$.

Exercise 6.5.6 Prove that the resolvent kernel r corresponding to the kernel $k \in L^1_{loc}(\mathbb{R}^+)$ in the convolution VIE (6.1.1) depends continuously on k with respect to the norm in $L^1_{loc}(\mathbb{R}^+)$ (cf. Theorem 6.2.1).

Exercise 6.5.7 Discuss the asymptotic behaviour of the solution of the VIE

$$u(t) = 1 + \lambda \int_0^t (t-s)^\beta u(s)\,ds, \quad t \geq 0,$$

when $\lambda < 0$ and $\beta > 1$.

(*Hint:* The solution of Exercise 1.7.9 and a look at the asymptotic properties of the Mittag-Leffler function $E_{1+\beta}(z)$ will prove helpful. See also Pollard, 1948 and Friedman, 1963, pp. 409–410.)

Exercise 6.5.8 Show that a convolution kernel $k(z)$ is completely monotone on $(0, \infty)$ if, and only if, it is of the form $\int_0^\infty e^{-sz} d\mu(s)$, where μ is a bounded, non-decreasing function.

Exercise 6.5.9 Prove the statement (b) in Corollary 6.2.13.

Exercise 6.5.10 Consider the VIE (6.3.1), but with $f(t) \equiv 1$ replaced by a more general function $f(t)$. For which f does the result of Theorem 6.3.3 on the complete monotonicity of the solution remain valid?

Exercise 6.5.11 Consider the VFIE

$$u(t) = u_0 + \int_0^{\theta(t)} K(t, s) u(s) \, ds, \ \ t \geq 0,$$

where f and K are continuous and non-negative, with continuous $\partial K / \partial t$. Show that if $u_0 > 0$, θ is non-negative, continuously differentiable and such that $\theta(t) \leq t$ on $\mathbb{R}^+$, the solution of the VFIE

$$u(t) = u_0 + \int_0^{\theta(t)} K(t, s) u(s) \, ds, \ \ t \geq 0,$$

is bounded on $\mathbb{R}^+$ if $\lim_{t \to \infty} \int_0^{\theta(t)} K(t, s) \, ds < \infty$. (See also Lipovan, 2006.)

Exercise 6.5.12 (*Research Problem*)
The paper by Brunner & Ou (2015) studies the asymptotic stability of solutions of linear VFIEs with proportional delay qt $(0 < q < 1)$. Extend their analysis to VFIEs with weakly singular kernels,

$$u(t) = f(t) + \int_{qt}^t (t - s)^{\alpha-1} k(t - s) u(s) \, ds, \ \ t \geq 0,$$

where $0 < q < 1$, $0 < \alpha < 1$ and $k(0) \neq 0$.

Exercise 6.5.13 Extend Theorem 6.4.1 and its proof to the Volterra–Hammerstein equation (6.4.1) when the kernel $K(t, s)$ is replaced by the weakly singular kernel $K_\alpha(t, s) := (t - s)^{\alpha-1} K(t, s)$ $(0 < \alpha < 1)$ or $K_0(t, s) := \log(t - s) K(t, s)$.

6.6 Notes

Section 6.1: *Introduction*
The seminal 1990 monograph by Gripenberg, Londen & Staffans contains a wide spectrum of results on the asymptotic properties of linear and non-linear VIEs; it is the authoritative source of such results, along with Miller (1971) and Nohel (1976), for their early development. Tsalyuk (1979) is a wide-ranging survey paper (see especially Chapter 2 on stability); its list of references contains more than 500 items.

Section 6.2: *Asymptotic Properties of Resolvent Kernels*
Extensions of the classical results in Paley & Wiener (1934) to VIEs of Hammerstein type were established in Shea & Wainger (1975; this paper also contains an illuminationg review of the context of the result by Paley–Wiener) and Jordan & Wheeler (1976). The asymptotic properties of resolvent kernels for linear VIEs with convolution kernels were studied in Grossman (1974), Levin (1977), Gripenberg (1979a on rapidly decaying resolvents r with $e^{\rho t} r(t)$ in $L^p(\mathbb{R}^+)$ for some $\rho > 0$ and $1 \leq p \leq \infty$), Gripenberg (1982 on convexity of $\log(k(z))$), Gripenberg (1982a on decay estimates), and Gripenberg (1982b, 1987); see also Gripenberg, Londen & Staffans (1990, Chapters 5 and 6). Analogous studies for non-convolution VIEs can be found in e.g. Gripenberg (1980a) and Staffans (1985); the latter paper discusses an application to VFIEs with constant delays. Gripenberg, Londen & Staffans (1990) contains, in Chapter 5, an excellent introduction to completey monotone functions. Studies of properties of resolvent kernels corresponding to non-convolution kernels can be found in Miller (1968b, 1971), the three papers by Vinokurov (1969a), and Staffans (1985).

Section 6.3: *Asymptotic Behaviour of Solutions*
Friedman (1963) and Levin (1963, 1963) contain a large range of results on the asymptotic behaviour of solutions to VIEs with convolution kernels. As already mentioned, Tsalyuk (1979) is an important review paper on the early development of the stability analysis of solutions to VIEs; it contains a comprehensive list of references of papers, especially by Soviet authors. Among other early papers on the asymptotic behaviour of solutions to VIEs are Tsalyuk (1968), Bownds & Cushing (1972 on VIEs with Pincherle–Goursat (finite-rank) kernels $K(t, s) = \sum_{n=1}^{p} A_n(x) B_n(t)$; their results are generalised in Bownds & Cushing (1973)), Bownds & Cushing (1975: systems of linear VIEs), Bownds & Cushing (1975b), Brauer (1975, 1976a), Cooke (1976: epidemics/population growth). See also Tsalyuk (1989) on the asymptotic behaviour of solutions to renewal equations (with proofs resorting to the Paley–Wiener theorem). Among the recent studies of the asymptotic behaviour of solutions to VIEs with convolution kernels and a special class of functions $f(t)$ is Lobanova & Tsalyuk (2015). The papers by Clément & Nohel (1979, 1981) deal with positive kernels and the preservation of positivity in solutions to such VIEs; this topic is also examined in Gripenberg, Londen & Staffans (Chapter 16).

Periodic solutions of VIEs with convolution kernels are studied in Friedmann (1965), Cooke & Kaplan (1976), Kang & Zhang (2003), Wong & Boey (2004 on the modelling of infectious diseases; cf. Section 9.2.7 in of the present book), and in Burton & Zhang (2011; see also for additional references).

Section 6.4: *VIEs of Hammerstein Form*
Non-linear Hammerstein-type perturbations of linear VIEs are studied in Miller, Nohel & Wong (1969) and Strauss (1970). An abstract version of the variation-of-constants formula (used in the analysis of non-linear perturbations of linear VIEs) in Theorem 6.4.1 is discussed in Miller (1971) (pp. 253–255).

The asymptotic behaviour of solutions to Hammerstein VIEs with convolution kernels is analysed in Friedman (1963; see also Ling, 1978), Miller (1968a, 1968b), Gripenberg (1979b), Clément & Nohel (1979, 1981: preservation of positivity). Chapter 15 in Gripenberg, Londen & Staffans (1990) contains a very general treatment of this problem. Nohel & Shea (1976) is an important paper on frequency domain methods.

Non-linear VIEs are studied in more general settings in Diekmann & van Gils (1984: VIEs on invariant manifolds), Castillo & Okrasiński (1993), Arias & Castillo (1999), Arias & Benítez (2003a, 2003b), Arias, Benítez & Bolós (2005, 2007) and Györi & Hartung (2010: asymptotically exponential solutions). The monograph by Miller & Sell (1970a) presents an authoritative study of semiflows associated with VIEs of Hammerstein type.

Volterra Functional Integral Equations:
An important early paper on the asymptotic behaviour of solutions to Hammerstein VIEs with constant delay,

$$u(t) = f(t) + \int_{\theta(t)}^{t} k(t-s)G(u(s))\,ds, \quad t \geq 0 \tag{6.6.1}$$

(where either $\theta(t) = t - \tau(t)$ ($\tau(t) \geq 0$) or $\theta(t) = qt$ ($0 < q < 1$)) is by Cooke (1976), who considered such mathematical models with constant delay $\tau > 0$; see also Cooke & Yorke (1973) and Cooke & Kaplan (1976). More recently a related linear version of (6.6.1),

$$u(t) = f(t) + \int_{0}^{t} k_0(t-s)u(s)\,ds + \int_{0}^{\theta(t)} k_1(t-s)u(s)\,ds, \quad t \geq 0,$$

with strictly increasing delay functions satisfying $\theta(0) = 0$ and $\theta(t) < t$ ($t > 0$), was studied by Brunner & Ou (2015) (see also Exercise 6.5.12).

While the analysis of the asymptotic behaviour of solutions of the state-dependent analogue of (6.6.1),

$$u(t) = \int_{\theta(u(t))}^{t} k(t-s)G(u(s))\,ds, \quad t > 0, \tag{6.6.2}$$

is essentially open, Bélair (1991, pp. 171–172) established a first result for a specific example of (6.6.2) ($k(t-s) \equiv 1$). The papers by Brunner & Maset (2009, 2010) introduced the notion of time transformation for analysing the existence and asymptotic properties of (analytical and numerical) solutions to delay differential equations with very general delays (including state-dependent delays): their techniques potentially provide the basis for the study of the asymptotic properties of solutions of (6.6.1) and state-dependent VFIEs (6.6.2).

7

Cordial Volterra Integral Equations

Summary

Volterra integral operators with certain weakly singular (integrable) kernels (so-called cordial Volterra integral operators) may be non-compact and hence possess uncountable spectra. We present an introduction to such Volterra operators and to the solvability of corresponding second-kind and first-kind cordial VIEs. This theory has also applications in the analysis of a class of VIEs of the third kind. The chapter concludes with a discussion of cordial VIEs with highly oscillatory kernels.

7.1 Cordial Volterra Integral Operators

The study of cordial VIEs is motivated by two particular Volterra integral operators with certain kernel singularities we encountered in Sections 1.5.3 and 2.1.5. The one corresponding to the weakly singular kernel $(t^2 - s^2)^{-1/2} K(t, s)$, with $K \in C(D)$, can be written as

$$(\mathcal{V}_\varphi u)(t) = \int_0^t t^{-1} \varphi(s/t) K(t, s) u(s) \, ds,$$

with its (integrable) *core* given by $\varphi(s/t) = \left(1 - (s/t)^2\right)^{-1/2}$. The other one arises in the analysis of third-kind VIEs of the form

$$a(t)u(t) = f(t) + \int_0^t (t - s)^{\alpha-1} K(t, s) u(s) \, ds, \quad t \in (0, T],$$

with $a(t) = t^\beta$ $(\beta > 0)$, $0 < \alpha \leq 1$ and $K \in C(D)$. For $\beta = \alpha$ the corresponding Volterra integral operator is

$$(\mathcal{V}_\varphi u)(t) = \int_0^t t^{-1} \varphi(s/t) K(t, s) u(s) \, ds.$$

Its (integrable) *core* is $\varphi(s/t) = \left(1 - (s/t)\right)^{\alpha-1}$. If $\alpha = 1$ (no kernel singularity) the core is constant: $\varphi(s/t) \equiv 1$.

The above integral operators are examples of so-called *cordial* Volterra integral operators. It will be seen that when $K(0, 0) \neq 0$ they are *non-compact* operators on $C(I)$ and possess uncountable spectra (cf. Appendix A.2.3).

7.1.1 Basic Properties of Cordial Volterra Integral Operators

Definition 7.1.1 A Volterra integral operator of the form

$$(\mathcal{V}_\varphi u)(t) := \int_0^t t^{-1} \varphi(s/t) K(t, s) u(s)\, ds, \quad t \in I, \tag{7.1.1}$$

is called a (linear) *cordial* Volterra integral operator with *core* φ if $\varphi \in L^1(0, 1)$ and if $K \in C^m(D)$ for some $m \geq 0$. $\diamond$

In Section 2.1.5 we had a brief look at a VIE described by the Volterra integral operator

$$
\begin{aligned}
(\mathcal{V}_\varphi u)(t) &:= \int_0^t \left(t^\gamma - s^\gamma\right)^{-1/\gamma} u(s)\, ds \\
&= \int_0^t t^{-1} \varphi(s/t) u(s)\, ds, \quad t \in I := [0, T], \tag{7.1.2}
\end{aligned}
$$

where $\varphi(z) := (1 - z^\gamma)^{-1/\gamma}$ and with kernel singularity corresponding to the value $\gamma = 2$. This operator (as an operator from $C(I)$ to $C(I)$) is linear and bounded, with

$$\|\mathcal{V}_\varphi\|_\infty := \sup_{f \neq 0} \frac{\|\mathcal{V}_\varphi f\|_\infty}{\|f\|_\infty} = \pi/2.$$

However, $\mathcal{V}_\varphi$ is not a compact operator since its spectrum is not countable, as the following result shows (Atkinson, 1976, pp. 30–31; see also Example 7.1.21 below).

Theorem 7.1.2 *The spectrum of the Volterra integral operator $\mathcal{V}_\varphi$ in (7.1.2) with $\gamma = 2$ is uncountable: for every real number $r \geq 0$ the function $\phi_r(t) = t^r$ is an eigenfunction of $\mathcal{V}_\varphi$, corresponding to the eigenvalue*

$$\hat{\varphi}(r) := \int_0^{\pi/2} \sin^r(\theta)\, d\theta = \tfrac{1}{2} B(\tfrac{1+r}{2}, \tfrac{1}{2}) \in [0, \pi/2]. \tag{7.1.3}$$

Here, $B(w, z) = \Gamma(w)\Gamma(z)/\Gamma(w + z)$ denotes the Euler beta function.

While the above statements can be verified by a straightforward calculation, they will also follow from the results stated in Theorem 7.1.10 below, since the above Volterra integral operator belongs to the class of so-called *cordial* Volterra integral operators. Such operators (as operators on $C^m(I)$ for some $m \geq 0$) are *not compact* and have *uncountable spectra* if $K(0, 0) \neq 0$.

Remark 7.1.3 In the subsequent analysis the cordial operator with constant kernel $K(t, s) = 1$,

$$(\mathcal{V}_\varphi^1 u)(t) := \int_0^t t^{-1} \varphi(s/t) u(s) \, ds = \int_0^1 \varphi(x) u(tx) \, dx, \quad t \in [0, T], \quad (7.1.4)$$

will play a key role. We will refer to $\mathcal{V}^1$ as the *basic* cordial Volterra integral operator if φ is in $L^1(0, 1)$. If the form of a cordial Volterra integral operator is clear from the context in which it occurs, we shall not distinguish between the notations $\mathcal{V}_\varphi$ and $\mathcal{V}_\varphi^1$. ◇

It will be assumed throughout this chapter that the cores φ of the cordial Volterra integral operators $\mathcal{V}_\varphi$ and $\mathcal{V}_\varphi^1$ are *real-valued* (and non-trivial).

One of the principal aims of this chapter is to analyse the existence and uniqueness, as well as the regularity properties, of the solutions of cordial VIEs of the second kind,

$$u(t) = f(t) + (\mathcal{V}_\varphi u)(t), \quad t \in I, \quad (7.1.5)$$

and the first kind,

$$(\mathcal{V}_\varphi u)(t) = g(t), \quad t \in I. \quad (7.1.6)$$

Example 7.1.4 The Volterra integral operator

$$(\mathcal{V}^1 u)(t) = \int_0^t t^{-1} u(s) \, ds = \int_0^1 u(tx) \, dx$$

is cordial with *constant* core $\varphi(z) = 1$. As we shall see in Section 7.2.3 (cf. equation (7.2.13)) this cordial Volterra operator arises in the analysis of certain VIEs of the third kind.

Example 7.1.5 Consider the Volterra integral operator

$$
\begin{aligned}
(\mathcal{V} u)(t) &= \int_0^t t^{-\nu} s^{-\beta} (t^\gamma - s^\gamma)^{\alpha - 1} u(s) \, ds, \\
&= \int_0^t t^{-1} t^{-\nu + 1 + \gamma(\alpha - 1)} s^{-\beta} (1 - (s/t)^\gamma)^{\alpha - 1} u(s) \, ds,
\end{aligned}
$$

with $\beta < 1$, $\gamma > 0$, $0 < \alpha \leq 1$, and $\nu + \beta + \gamma(1-\alpha) = 1$. (We observe that $\mathcal{V}$ contains, in addition to the diagonal singularity, a boundary singularity (cf. Section 2.1.4).) Under the above assumptions $\mathcal{V}$ is a cordial Volterra integral operator with the L^1 core $\varphi(z) = z^{-\beta}(1 - z^\gamma)^{\alpha-1}$. We have seen a particular case of such a Volterra integral operator before, namely

$$(\mathcal{V}u)(t) = \int_0^t \left(t^\gamma - s^\gamma\right)^{\alpha-1} u(s)\,ds = \int_0^t t^{-1} t^{1-\gamma(1-\alpha)} \left(1 - (s/t)^\gamma\right)^{\alpha-1} u(s)\,ds,$$

with $\gamma > 1$ and $0 < \alpha < 1$. If $1 - \gamma(1-\alpha) = 0$. i.e. $1 - \alpha = 1/\gamma$, then this is a basic cordial Volterra operator with the L^1 core

$$\varphi(z) = \left(1 - z^\gamma\right)^{-1/\gamma} \quad (\gamma > 1). \tag{7.1.7}$$

This operator is non-compact (see Theorem 7.1.3(a) below).

If $1 - \gamma(1-\alpha) > 0$, then $\mathcal{V}$ may be regarded as a general cordial Volterra integral operator (7.1.3) with continuous kernel $K(t, s) = t^{1-\gamma(1-\alpha)}$. Since it satisfies $K(0, 0) = 0$, this operator is compact (see Theorem 7.1.10(b)).

Finally, if $1 - \gamma(1-\alpha) < 0$, the Volterra integral operator $\mathcal{V}$ has a strongly (Cauchy-type) singular kernel $K(t, s) = t^{1-\gamma(1-\alpha)}$; thus it is not a cordial Volterra integral operator.

Example 7.1.6 Another cordial Volterra integral operator that is related to a VIE of the third kind with a weakly singular kernel is

$$(\mathcal{V}u)(t) = \frac{1}{t^{\alpha+\beta}} \int_0^t (t - s)^{\alpha-1} s^\beta u(s)\,ds \quad (\alpha > 0, \ \beta > -1)$$

(see equations (7.2.18) and (7.2.19) in Section 7.2.3). It is readily verified that its core is

$$\varphi(z) = (1 - z)^{\alpha-1} z^\beta. \tag{7.1.8}$$

Example 7.1.7 The L^1 core

$$\varphi(z) = (1 - z)^{-\alpha} \quad (0 < \alpha < 1)$$

is associated with the Volterra integral operator

$$(\mathcal{V}u)(t) = \int_0^t t^{\alpha-1}(t - s)^{-\alpha} u(s)\,ds$$

$$= \int_0^t t^{-1}\left(1 - s/t\right)^{-\alpha} u(s)\,ds \quad (0 < \alpha < 1).$$

This cordial operator is related to Volterra integral equations whose kernels possess boundary and diagonal singularities (Pedas & Vainikko, 2006; cf. Section 2.1.3).

Example 7.1.8 The Volterra integral operator

$$(\mathcal{V}u)(t) = \int_0^t t^{-\beta}s^{\beta-1}u(s)\,ds \quad (\beta > 0)$$

is also cordial; its L^1 core φ is given by

$$\varphi(z) = z^{\beta-1} \quad (\beta > 0).$$

Weakly singular Volterra integral equations corresponding to the above core were studied in Han (1994); see also Diogo & Lima (2008).

We will now study various mapping properties of the cordial Volterra integral operators $\mathcal{V}^1$ and $\mathcal{V}$. In particular, we shall see that $\mathcal{V}^1 : C(I) \to C(I)$ is a non-compact operator, while $\mathcal{V} : C(I) \to C(I)$ is non-compact for all kernels $K \in C(D)$ with $K(0,0) \neq 0$. The results in the following theorems are due to Vainikko (2009, 2010a).

Theorem 7.1.9 *Assume that the core φ and the kernel $K(t,s)$ of the cordial Volterra integral operator $\mathcal{V}$ defined in (7.1.1) satisfy $\varphi \in L^1(0,1)$ and $K \in C(D)$. Then:*

(a) $\mathcal{V}_\varphi$ is a bounded linear operator that maps $C(I)$ into itself, and we have

$$\|\mathcal{V}_\varphi\|_\infty \le \|\varphi\|_1 \|K\|_\infty,$$

where $\|K\|_\infty := \max\{|K(t,s)| : (t,s) \in D\}$.

(b) If $K \in C^m(D)$ for some $m \ge 1$, then $\mathcal{V}_\varphi$ is a bounded linear operator from $C^m(I)$ into itself.

Proof Assertion (a) follows from the definition of the operator norm $\mathcal{V}_\varphi$ on the Banach space $C(I)$: we have $\|\mathcal{V}_\varphi\|_\infty = \sup\limits_{\|u\|_\infty \le 1} \|\mathcal{V}_\varphi u\|_\infty$, where

$$(\mathcal{V}_\varphi u)(t) := \int_0^t t^{-1}\varphi(s/t)K(t,s)u(s)\,ds = \int_0^1 \varphi(z)K(t,tz)u(tz)\,dz, \quad t \in I.$$

The verification of (b) is left as an exercise. $\qquad\square$

The next theorem is concerned with the non-compactness and some additional basic properties of the basic cordial Volterra integral operator $\mathcal{V}^1_\varphi$.

Theorem 7.1.10 *Assume that the core φ and the kernel K of the Volterra integral operator (7.1.1) are such that $\varphi \in L^1(0,1)$ and $K \in C(D)$. Then the following statements are true:*

(a) *The basic cordial Volterra operator* $\mathcal{V}_\varphi^1$ *has the property that for* $u_\lambda(t) :=$
    t^λ $(t \in I, \; \lambda \in \mathbb{C}, \; Re(\lambda) \geq 0)$,

$$(\mathcal{V}^1 u_\lambda)(t) = \hat{\varphi}(\lambda) u_\lambda(t), \quad \text{where} \quad \hat{\varphi}(\lambda) := \int_0^1 \varphi(x) x^\lambda \, dx.$$

   This holds in particular for all real $\lambda = r \geq 0$.
(b) *The basic cordial Volterra integral operator* $\mathcal{V}^1$ *is a non-compact operator*
    on $C(I)$.
(c) *If* $K(0,0) \neq 0$, *the general cordial operator* $\mathcal{V}_\varphi : C(I) \to C(I)$ *is not*
    compact; it is a compact operator for all $K \in C(D)$ *with* $K(0,0) = 0$.
(d) *The operator* $\mathcal{V}_\varphi - K(0,0)\mathcal{V}_\varphi^1 : C(I) \to C(I)$ *is compact.*

Proof We have seen in Theorem 7.1.2 that the basic cordial Volterra integral operator $\mathcal{V}_\varphi^1$ with the core $\varphi(z) = (1 - z^2)^{-1/2}$ has an uncountable set of eigenvalues, and hence an uncountable spectrum. Thus, this special cordial operator cannot be a compact operator on $C(I)$ (cf. Theorem A.2.27). This is also true for the general cordial Volterra integral operator $\mathcal{V}_\varphi^1$ with non-trivial core $\varphi \in L^1(0,1)$ since

$$\int_0^t t^{-1} \varphi(s/t) s^r \, ds = \int_0^1 \varphi(x) x^r \, ds \cdot t^r =: \hat{\varphi}(r) t^r$$

for any $r \geq 0$. Moreover, it is readily seen that the above equation remains valid if we replace $r \geq 0$ by $\lambda \in \mathbb{C}$ with $\operatorname{Re}\lambda \geq 0$:

$$\int_0^t t^{-1} \varphi(s/t) s^\lambda \, ds = \hat{\varphi}(\lambda) t^\lambda.$$

Moreover, $\hat{\varphi}(\lambda)$ is analytic for $\operatorname{Re}\lambda > 0$, and continuous for $\operatorname{Re}\lambda = 0$, and it can be shown that

$$\lim_{\operatorname{Re}\lambda \geq 0, |\lambda| \to \infty} |\hat{\varphi}(\lambda)| = 0.$$

The verification of (c) and (d) is left as an exercise (cf. Exercise 7.5.3).   □

Example 7.1.11 In Section 2.2.4 we briefly discussed a weakly singular VIE whose underlying Volterra integral operator,

$$(\mathcal{V}_\varphi u)(t) = \lambda \int_0^t \left(t^2 - s^2\right)^{-1/2} s u(s) \, ds = \lambda \int_0^t t^{-1} (1 - (s/t)^2)^{-1/2} s \, u(s) \, ds$$

corresponds to the core $\varphi(z) = (1 - z^2)^{-1/2}$ and the kernel $K(t, s) = \lambda s$ ($\lambda \neq 0$). Since $K(0,0) = 0$ this operator is a *compact operator*.

Remark 7.1.12 Additional (more general) mapping properties of cordial Volterra integral operators $\mathcal{V}_\varphi$, including operators with cores $\varphi \in L^p(0, 1)$ $(p > 1)$, and results on their spectra can be found in Vainikko (2010a; cf. Theorem 2.1) and, especially, in Yang (2015). ◇

The next theorem (Vainikko, 2009, p. 1150) shows that the composition of two basic cordial Volterra integral operators is again a cordial integral operator.

Theorem 7.1.13 *Let $\mathcal{V}_\varphi^1$ and $\mathcal{V}_\psi^1$ be (basic) cordial Volterra integral operators with L^1 cores φ and ψ. Then the composition $\mathcal{V}_\varphi^1\mathcal{V}_\psi^1$ of these operators is the cordial Volterra integral operator $V\varphi \star \psi$ with core $\varphi \star \psi$ given by*

$$(\varphi \star \psi)(z) = \int_z^1 x^{-1}\varphi(x)\psi(z/x)\,dx, \quad z \in (0, 1). \tag{7.1.9}$$

It has the properties

$$\varphi \star \psi = \psi \star \varphi \quad and \quad \|\varphi \star \psi\|_1 \le \|\varphi\|_1 \|\psi\|_1. \tag{7.1.10}$$

Proof The composition of the cordial Volterra integral operators $\mathcal{V}_\varphi^1$ and $\mathcal{V}_\psi^1$ is given by

$$
\begin{aligned}
(\mathcal{V}_\varphi^1\mathcal{V}_\psi^1)u(t) &= \int_0^t t^{-1}\varphi(s/t)(\mathcal{V}_\psi^1 u)(s)\,ds \\
&= \int_0^t t^{-1}\varphi(s/t)\left(\int_0^s s^{-1}\psi(v/s)u(v)dv\right)ds \\
&= \int_0^t t^{-1}\left(\int_v^t s^{-1}\varphi(s/t)\psi(v/s)ds\right)u(v)\,dv \\
&=: \int_0^t t^{-1}\sigma(t, v)u(v)\,dv.
\end{aligned}
$$

(When changing the order of integration we have used the fact that the integrals converge absolutely for any $u \in C(I)$.) The kernel $\sigma(t, v)$ depends only on the ratio t/v: for any constant $c > 0$ we obtain

$$
\begin{aligned}
\sigma(ct, cv) &= \int_{cv}^{ct} s^{-1}\varphi(s/ct)\psi(cv/s)ds = \int_v^t w^{-1}\varphi(w/t)\psi(v/w)dw \\
&= \sigma(t, v) = \sigma(1, v/t) =: \chi(v/t).
\end{aligned}
$$

We thus have found that

$$(\mathcal{V}_\varphi^1\mathcal{V}_\psi^1)u(t) = \int_0^t t^{-1}\chi(s/t)u(s)\,ds =: (V_{\varphi\star\psi}u)(t),$$

with the core

$$\chi(z) = (\varphi \star \psi)(z) = \int_z^1 x^{-1}\varphi(x)\psi(z/x)\,dx.$$

An obvious change of the variable of integration shows that $(\psi \star \varphi)(z) = (\varphi \star \psi)(z)$, and hence we have $\mathcal{V}_\psi^1 \mathcal{V}_\varphi^1 = \mathcal{V}_\varphi^1 \mathcal{V}_\psi^1$. We note for later use (cf. Section 7.2.2) that if $z = s/t$ the above expression for the composite core χ becomes

$$\chi(s/t) = \int_s^t w^{-1}\varphi(w/t)\psi(s/w)\,dw, \quad 0 < s < t \leq T.$$

It remains to show that $\chi \in L^1(0, 1)$. Using the assumption that φ and ψ are L^1 cores we obtain

$$\int_0^1 \chi(z)\,dz = \int_0^1 (\varphi \star \psi)(z)\,dz = \int_0^1 \left(\int_z^1 x^{-1}\varphi(x)\psi(z/x)\,dx \right) dz$$

$$= \int_0^1 \left(\int_0^x \psi(z/x)\,dz \right) x^{-1}\varphi(x)\,dx$$

$$= \int_0^1 \left(x \int_0^1 \psi(w)\,dw \right) x^{-1}\varphi(x)\,dx$$

$$= \int_0^1 \left(\int_0^1 \psi(w)\,dw \right) \varphi(x)\,dx = \int_0^1 \varphi(x)\,dx \int_0^1 \psi(w)\,dw.$$

The inequality $|(\varphi \star \psi)(z)| \leq (|\varphi| \star |\psi|)(z)$ then leads to

$$|\chi(z)| \leq \int_0^1 |(\varphi \star \psi)(z)|\,dz \leq \int_0^1 (|\varphi| \star |\psi|)(z)\,dz$$

$$= \int_0^1 |\varphi(z)|\,dz \int_0^1 |\psi(z)|\,dz = \|\varphi\|_1 \|\psi\|_1 < \infty.$$

$$\square$$

Remark 7.1.14

(i) The Banach space $L^1(0, T)$ with norm $\| \cdot \|_1$ and product operation $\star$ defined in (7.1.9) forms a commutative Banach algebra (cf. Appendix A.2.3). This algebra is *not unital*: the existence of an identity element ψ with $\psi \star \varphi = \varphi$ for all $\varphi \in L^1(0, 1)$ would imply that for $\varphi(z) = 1$, $\int_z^1 x^{-1}\psi(x)dx = 1$, or (by differentiation) $z^{-1}\psi(z) = 0$ $(0 < z < 1)$. The result $\psi(z) = 0$ leads to a contradiction.

(ii) Theorem 7.1.13 is readily extended to the composition of general cordial Volterra integral operators with cores $\varphi(z)$, $\psi(z)$ and continuous kernels $K(t, s)$, $H(t, s)$. We leave the details as an exercise (cf. Exercise 7.5.1).◇

Example 7.1.15 The composition of two basic cordial Volterra integral operators with constant (real) cores $\varphi(z) = \lambda_1$ and $\psi(z) = \lambda_2$ is a cordial Volterra integral operator $\mathcal{V}_{\varphi \star \psi}$ with core

$$\chi(z) = (\varphi \star \psi)(z) = \lambda_1 \lambda_2 \int_z^1 x^{-1} \, dx = -\lambda_1 \lambda_2 \log(z), \quad z \in (0, 1).$$

Hence, $\chi(s/t) = -\lambda_1 \lambda_2 \log(s/t) \ (0 < s < t \le T)$.

Example 7.1.16 For $\mathcal{V}_\varphi^1$ and $\mathcal{V}_\psi^1$ with cores $\varphi(z) = z^{\beta-1}$ $(\beta > 0, \ \beta \ne 1)$ and $\psi(z) = \lambda_2$ we obtain the cordial Volterra integral operator $\mathcal{V}_{\varphi \star \psi}$ whose core is

$$\chi(z) = (\varphi \star \psi)(z) = \lambda_2 \int_z^1 x^{-1} x^{\beta-1} \, dx = \frac{\lambda_2}{\beta - 1} \left(1 - z^{\beta-1}\right), \quad z \in (0, 1).$$

For $z = s/t$ this becomes

$$\chi(s/t) = \frac{\lambda_2}{\beta - 1} \left(1 - (s/t)^{\beta-1}\right) \ (0 < s < t \le T).$$

7.1.2 The Spectrum of a Cordial Volterra Integral Operator

Consider the cordial VIEs

$$\mu u(t) = f(t) + (\mathcal{V}_\varphi^1 u)(t), \quad t \in I, \tag{7.1.11}$$

and

$$\mu u(t) = f(t) + (\mathcal{V}_\varphi u)(t), \quad t \in I, \tag{7.1.12}$$

where $\mathcal{V}_\varphi^1$ and $\mathcal{V}_\varphi$ denote respectively the cordial Volterra integral operators defined in (7.1.4) and (7.1.1), and where $f \in C(I)$ and (in the case (7.1.1)) $K \in C(D)$ are given (real-valued) functions. For which values of $\mu \ne 0$ do the cordial VIEs (7.1.11) and (7.1.12) possess unique solutions in $C(I)$ or, equivalently, when are the operators $\mu \mathcal{I} - \mathcal{V}_\varphi^1$ and $\mu \mathcal{I} - \mathcal{V}_\varphi$ (with $\mathcal{I}$ denoting the identity operator on $C(I)$) invertible (as bounded operators on $C(I)$)? In the following we denote the *spectrum* of the cordial Volterra integral operator $\mathcal{V}_\varphi$ by $\sigma(\mathcal{V}_\varphi)$; the set $\rho(\mathcal{V}_\varphi) := \mathbb{C} \setminus \sigma(\mathcal{V}_\varphi)$ is the *resolvent set* of $\mathcal{V}_\varphi$ (see also Appendix A.2.3).

The following theorem shows that the spectra of the basic cordial Volterra integral operator $\mathcal{V}_\varphi^1$ and the general Volterra operator $\mathcal{V}_\varphi$ with $K(0, 0) \ne 0$ are closely related.

Theorem 7.1.17 *Assume that $K(0,0) \neq 0$. Then the spectrum of V_φ is uncountable. It is related to the spectrum of the basic cordial Volterra operator V_φ^1 (corresponding to $K(t,s) \equiv 1$) by*

$$\sigma(V_\varphi) = K(0,0)\sigma(V_\varphi^1); \qquad (7.1.13)$$

that is,

$$\sigma(V_\varphi) = \{0\} \cup \{K(0,0)\hat{\varphi}(\lambda) : \lambda \in \mathbb{C}, \; Re\,\lambda \geq 0\}, \qquad (7.1.14)$$

where

$$\hat{\varphi}(\lambda) := \int_0^1 \varphi(x)x^\lambda \, dx \quad (Re\,\lambda \geq 0). \qquad (7.1.15)$$

In (7.1.14), $\{0\}$ belongs to the closure of the set $\{K(0,0)\hat{\varphi}(\lambda) : \lambda \in \mathbb{C}, \; Re\,\lambda \geq 0\}$, and we have

$$\lim_{Re(\lambda)\geq 0, \; |\lambda|\to\infty} |\hat{\varphi}(\lambda)| = 0. \qquad (7.1.16)$$

Proof Assume that $\varphi(z) \not\equiv 0$ and let, without loss of generality, $K(0,0) = 1$. One first proves that the hypothesis $\mu \in \rho(V_\varphi^1)$, $\mu \in \sigma(V_\varphi^1)$ leads to a contradiction. Then one uses the fact (Theorem 7.1.10(d)) that the cordial Volterra integral operator

$$(\tilde{V}_\varphi u)(t) := \int_0^t t^{-1}\varphi(s/t)\big(K(t,s) - K(0,0)\big)u(s)\,ds, \; t \in I,$$

is a compact operator on $C(I)$. (Note that $0 \in \sigma(V_\varphi)$ but $\mu = 0$ is not an eigenvalue of V_φ.) The technical details of the proof can be found in Vainikko (2010a, pp. 204–206). $\qquad \square$

Remark 7.1.18 More refined results on the spectra of cordial Volterra integral operators V_φ (in particular on properties of eigenvalues lying on the boundary of $\sigma(V_\varphi)$) can be found in Yang (2015). $\qquad \diamond$

The following corollary follows from (7.1.15) and is an obvious generalisation of Theorem 7.1.2.

Corollary 7.1.19 *Let $\varphi \in L^1(0,1)$. Then for every real number $r \geq 0$ the function $\phi_r(t) = t^r$ is an eigenfunction of V_φ^1 corresponding to the eigenvalue*

$$\hat{\varphi}_r := \int_0^1 \varphi(x)x^r \, dx.$$

We illustrate the result (7.1.14) of Theorem 7.1.17 by means of the four cordial Volterra integral operators introduced in Examples 7.1.4–7.1.8.

Example 7.1.20 If the basic cordial Volterra integral operator (7.1.4) has the *constant* core $\varphi(z) = 1$, then its spectrum is given by

$$\sigma(\mathcal{V}_\varphi^1) = \{0\} \cup \{(1 + \lambda)^{-1} : \operatorname{Re} \lambda \geq 0\}. \tag{7.1.17}$$

Example 7.1.21 The cordial Volterra integral operator $\mathcal{V}_\varphi^1$ with the core

$$\varphi(z) = (1 - z^\gamma)^{-1/\gamma} \quad (\gamma > 1)$$

has the spectrum

$$\sigma(\mathcal{V}_\varphi^1) = \{0\} \cup \left\{ \gamma^{-1} \Gamma(1 - \gamma^{-1}) \Gamma(\gamma^{-1}(\lambda + 1)) / \Gamma(\gamma^{-1}\lambda + 1) : \ \gamma > 1, \right.$$
$$\left. \operatorname{Re} \lambda \geq 0 \right\}. \tag{7.1.18}$$

For the value $\gamma = 2$ we obtain

$$\sigma(\mathcal{V}_\varphi^1) = \{0\} \cup \left\{ \frac{1}{2} \frac{\Gamma(\frac{1}{2})\Gamma(\frac{1+\lambda}{2})}{\Gamma(1 + \frac{\lambda}{2})} : \operatorname{Re} \lambda \geq 0 \right\},$$

where $\Gamma(\frac{1}{2}) = \pi^{1/2}$. This extends the result of Theorem 7.1.2 to complex λ. The case $\gamma = 3/2$ is related to Lighthill's (non-linear) Volterra integral operator (cf. Lighthill, 1950 and Curle, 1978, as well as Example 9.2.9 in Section 9.2.3). The spectrum of the corresponding operator $\mathcal{V}_\varphi$ is given by

$$\sigma(\mathcal{V}_\varphi^1) = \{0\} \cup \left\{ \frac{2}{3} \frac{\Gamma(\frac{2(1+\lambda)}{3})\Gamma(\frac{1}{3})}{\Gamma(1 + \frac{2\lambda}{3})} : \operatorname{Re} \lambda \geq 0 \right\}.$$

Example 7.1.22 We saw in Example 7.1.6 that the cordial Volterra operator with core

$$\varphi(z) = (1 - z)^{\alpha - 1} z^\beta \quad (\alpha > 0, \ \beta > -1)$$

arises in the analysis of certain third-kind VIEs (see Section 7.2.3 below). By (7.1.13) it possesses the spectrum

$$\sigma(\mathcal{V}_\varphi^1) = \{0\} \cup \left\{ \frac{\Gamma(\alpha)\Gamma(1 + \beta + \lambda)}{\Gamma(1 + \alpha + \beta + \lambda)} : \operatorname{Re} \lambda \geq 0 \right\}. \tag{7.1.19}$$

For $\beta = 0$ and core $\varphi(z) = (1 - z)^{\alpha - 1}$ the spectrum reduces to

$$\sigma(\mathcal{V}_\varphi^1) = \{0\} \cup \left\{ \frac{\Gamma(\alpha)\Gamma(1 + \lambda)}{\Gamma(1 + \alpha + \lambda)} : \operatorname{Re} \lambda \geq 0 \right\}. \tag{7.1.20}$$

The third-kind VIEs studied in Grandits (2008) are, for $a(t) = t$, related to cordial Volterra integral operators whose cores correspond respectively to the values $\alpha = 1$, $\beta = 0$ and $\alpha = 3/2$, $\beta = 0$ (see Section 7.2.3 for details).

Example 7.1.23 Consider the cordial Volterra operator $\mathcal{V}_\varphi^1$ with core

$$\varphi(z) = z^{\beta-1} \quad (\beta > 0)$$

(cf. Example 7.1.8). Its spectrum is

$$\sigma(\mathcal{V}_\varphi^1) = \{0\} \cup \{1/(\beta + \lambda) : \ \operatorname{Re}\lambda \geq 0\}. \tag{7.1.21}$$

For $\beta = 1$ we obtain the result in Example 7.1.20.

Remark 7.1.24 Second-kind VIEs described by particular cases of the cordial Volterra integral operator

$$(\hat{\mathcal{V}}_\varphi u)(t) := \int_0^t t^{-1}\varphi(s/t)K(t, s, s/t)u(s)\,ds, \ \ t \in I, \tag{7.1.22}$$

with kernel K that is at least continuous with respect to the first two variables $(t, s) \in D$ and with $\varphi \in L^1(0, 1)$, have been considered in e.g. Han (1993) and Diogo, McKee & Tang (1991). It can be shown that Vainikko's theory of cordial Volterra integral operators can be extended to encompass $\hat{\mathcal{V}}_\varphi$: this operator is bounded (as an operator on $C(I)$), it is compact when $K(0, 0, z) = 0$ for $z \in [0, 1]$, and non-compact when $K(0, 0, z) \neq 0$ for $z \in [0, 1]$. ◇

7.2 Linear Cordial Volterra Integral Equations

7.2.1 Cordial VIEs of the Second Kind

Consider the cordial VIE

$$\mu u(t) = f(t) + (\mathcal{V}_\varphi u)(t), \ \ t \in I = [0, T], \tag{7.2.1}$$

where $\mu \in \mathbb{R}$ ($\mu \neq 0$) and the kernel $K(t, s)$ of $\mathcal{V}_\varphi$ is such that $K \in C(D)$, with $K(0, 0) \neq 0$. We recall that $\rho(\mathcal{V}_\varphi) := \mathbb{C} \setminus \sigma(\mathcal{V}_\varphi)$ is the *resolvent set* of $\mathcal{V}_\varphi$ (that is, the set of values μ for which $(\mu\mathcal{I} - \mathcal{V}_\varphi)^{-1}$ exists as a linear bounded operator on the space $C(I)$). The proof of the following theorem is obvious.

Theorem 7.2.1 *If $\mu \notin \sigma(\mathcal{V}_\varphi)$ then the cordial VIE (7.2.1) with $K \in C(D)$ and $K(0, 0) \neq 0$ possesses a unique solution $u \in C(I)$ for any $f \in C(I)$.*

Remark 7.2.2 What can be said if μ lies on the *boundary* of the spectrum of $\mathcal{V}_\varphi$? This subtle question (whose answer is related to the Fredholm Alternative for $\mathcal{V}_\varphi$) is explored in detail in Yang, 2015, especially for the case $\varphi \in L^p(0, 1)$ $(p > 1)$ (cf. Sections 3 and 4; compare also Vainikko, 2009). $\diamond$

Example 7.2.3 The cordial VIE

$$u(t) = f(t) + \kappa \int_0^t t^{-1}(s/t)^{\beta-1} u(s)\,ds, \quad t \in I \ (\beta > 0),$$

has the core $\varphi(z) = z^{\beta-1}$ and the constant kernel $K(t, s) = \kappa \neq 0$. By (7.1.3) the spectrum of the underlying cordial Volterra integral operator is

$$\sigma(\mathcal{V}_\varphi) = \{0\} \cup \{\kappa/(\beta + \lambda) : \ \mathrm{Re}\,\lambda \geq 0\}$$

(cf. Example 7.1.23). We note in passing that the above cordial VIE can be written in the form

$$t^\beta u(t) = t^\beta f(t) + \kappa \int_0^t s^{\beta-1} u(s)\,ds, \quad t \in I.$$

This is a Volterra integral equation of the *third kind*; we shall explore the connection between certain VIEs of the third kind and second-kind cordial VIEs in Section 7.2.3.

Example 7.2.4 Let $\gamma > 1$ and $\alpha \in (0, 1)$ be such that $\gamma(1 - \alpha) = 1$. The VIE

$$
\begin{aligned}
u(t) &= f(t) + \int_0^t (t^\gamma - s^\gamma)^{\alpha-1} K(t, s) u(s)\,ds \\
&= f(t) + \int_0^t t^{-1}\big(1 - (s/t)^\gamma\big)^{-1/\gamma} K(t, s) u(s)\,ds, \quad t \in I,
\end{aligned}
$$

corresponds to the core $\varphi(z) = (1 - z^\gamma)^{-1/\gamma}$. According to Example 7.1.21 and (7.1.18) the spectrum of the underlying cordial Volterra integral operator is given by

$$
\begin{aligned}
\sigma(\mathcal{V}_\varphi) = \{0\} \cup \big\{ &K(0, 0)\gamma^{-1}\Gamma(1 - \gamma^{-1}) \\
&\times \Gamma\big(\gamma^{-1}(\lambda + 1)\big)/\Gamma(\gamma^{-1}\lambda + 1) : \ \mathrm{Re}\,\lambda \geq 0 \big\}.
\end{aligned}
$$

Suppose now that f and K in (7.2.1) are smooth functions: $f \in C^m(I)$ and $K \in C^m(D)$ for some $m \geq 1$. We have seen in Section 2.2.1 that for such (non-trivial) functions the solution of the weakly singular VIE

$$u(t) = f(t) + \int_0^t (t - s)^{\alpha-1} K(t, s) u(s)\,ds, \quad t \in I,$$

with $0 < \alpha < 1$ is continuous but not smooth on $[0, T]$: its derivative $u'(t)$ behaves like $t^{\alpha-1}$ as $t \to 0^+$. For the VIE (7.2.1) with operator $\mathcal{V}_\varphi$ corresponding to $\varphi \in L^1(0, 1)$, $K \in C^m(D)$, $K(0, 0) \neq 0$, and $f \in C^m(I)$, the situation is rather different: here, smooth data lead to smooth solutions. The regularity result in the following theorem (which we state for the case $K(t, s) \equiv 1$; that is, for the basic cordial operator $\mathcal{V}^1$) makes this precise (Vainikko, 2010a, p. 214); it also refines Theorem 7.2.1.

Theorem 7.2.5 *Assume that* $\varphi \in L^1(0, 1)$, $f \in C^m(I)$ *and* $K \in C^m(D)$ $(m \geq 1)$. *If* $\mu \in \rho(\mathcal{V}_\varphi)$ *then the unique solution of* $\mu u = f + \mathcal{V}_\varphi u$ *lies in* $C^m(I)$.

Proof This regularity result is a consequence of the mapping properties of the cordial Volterra integral operator satisfying the above assumptions: we have (as shown in Theorem 8.1.11(a)) that $\mathcal{V}_\varphi$ is a bounded linear operator from $C^m(I)$ into itself. $\square$

We conclude this section with the cordial analogue of the Gronwall lemma for VIEs with weakly singular kernels (cf. Lemma 1.3.13). It is concerned with the cordial integral inequalities

$$z(t) \leq f(t) + (\mathcal{V}_\varphi^1 z)(t) = f(t) + \lambda \int_0^t t^{-1}\varphi(s/t)z(s)\,ds \quad t \in I, \qquad (7.2.2)$$

and

$$z(t) \leq f(t) + (\mathcal{V}_\varphi z)(t) = f(t) + \int_0^t t^{-1}\varphi(s/t)K(t, s)z(s)\,ds, \quad t \in I,$$

and is due to Vainikko (2016). In order to state and prove it, the notation

$$C_+(I) := \{f \in C(I): \ f(t) \geq 0 \text{ for } t \in I\}$$

and

$$C_+^+(I) := \{f \in C_+(I): \ f \text{ is non-decreasing for } t \in I\}$$

will be used.

Lemma 7.2.6 *Assume that* $\lambda > 0$, $\varphi \in L^1(0, 1)$, $\varphi(x) \geq 0$ $(x \in (0, 1)$, *and* $\lambda\|\varphi\|_1 < 1$.

(a) *If* $z \in C_+(I)$ *satisfies the inequality (7.2.2) described by the basic cordial Volterra integral operator* $\mathcal{V}_\varphi^1$ *and if* $f \in C_+^+(I)$, *then*

$$z(t) \leq \left(1 - \lambda\|\varphi\|_1\right)^{-1} f(t), \quad t \in I.$$

(b) *Let $z \in C_+(I)$ satisfy the integral inequality corresponding to the general cordial operator $\mathcal{V}_\varphi$, with $K \in C(D)$ such that $0 \leq K(t,s) \leq Mt^\beta$ for some $\beta > 0$. Then for any $f \in C_+^+(I)$ there holds*

$$z(t) \leq \Phi_\varphi(\lambda) f(t), \quad t \in I,$$

where

$$\Phi_\varphi(\lambda) := 1 + \sum_{j=1}^\infty \hat{\varphi}(0)\hat{\varphi}(\beta) \cdots \hat{\varphi}((j-1)\beta) M^j \lambda^j t^{j\beta},$$

with

$$\hat{\varphi}(j\beta) := \int_0^1 \varphi(x) x^{j\beta} \, dx \to 0 \text{ as } j \to \infty.$$

Proof In order to prove (a), recall from Theorem 7.1.9 that $\mathcal{V}_\varphi \leq \|\varphi\|_1$. It follows from (7.2.2) by iteration that

$$z(t) \leq f(t) + \sum_{j=1}^n ((\lambda \mathcal{V}_\varphi^1)^j f)(t) + ((\lambda \mathcal{V}_\varphi^1)^{n+1} f)(t) \ (n \geq 1).$$

The assertion then follows from $\lambda \|\mathcal{V}_\varphi^1\| \leq \lambda \|\varphi\|_1$. The proof of (b) is left to the reader. $\qquad\square$

Remark 7.2.7 If $\lambda \|\varphi\}_1 \geq 1$, it is easy to verify that the statement (a) in Lemma 7.2.6 cannot be true for any $f \in C_+^+(I)$ (choose, for example, $f(t) = t^r$ for $r > 0$). $\qquad\diamond$

7.2.2 Cordial VIEs of the First Kind

We now analyse the existence of solutions of cordial VIEs of the first kind,

$$(\mathcal{V}_\varphi u)(t) = \int_0^t t^{-1} \varphi(s/t) H(t,s) u(s) \, ds = g(t), \quad t \in I, \tag{7.2.3}$$

and

$$(\mathcal{V}_\varphi^1 u)(t) = \int_0^t t^{-1} \varphi(s/t) u(s) \, ds = g(t), \quad t \in I. \tag{7.2.4}$$

We begin our discussion on the sovability of cordial VIEs of the first kind with a simple example.

Example 7.2.8 For any $g \in C^1[0, T]$ the cordial VIE with the constant core $\varphi(z) \equiv 1$,

$$\int_0^t t^{-1} u(s)\, ds = g(t), \quad t \in I$$

possesses the unique continuous solution

$$u(t) = \frac{d}{dt}\big(tg(t)\big) = g(t) + tg'(t), \quad t \in I.$$

More generally, if $H \in C(D)$, $\partial H/\partial t \in C(D)$, and $H(t,t) \neq 0$ for $t \in I$, the cordial VIE (7.2.4) has also a unique solution $u \in C(I)$, since it is equivalent to the VIE

$$u(t) = \big(tg(t)\big)'/H(t,t) - \int_0^t \big(H(t,t)\big)^{-1} \frac{\partial H(t,s)}{\partial t} u(s)\, ds, \quad t \in I.$$

Let us next consider a first-kind cordial VIE with a non-constant core we have encountered already in Section 1.5.3, namely

$$\int_0^t \frac{(t^\gamma - s^\gamma)^{\alpha-1}}{\Gamma(\alpha)} u(s)\, ds$$

$$= \frac{1}{\Gamma(\alpha)} \int_0^t t^{-1} \big(1 - (s/t)^\gamma\big)^{\alpha-1} t^{1-\gamma(1-\alpha)} u(s)\, ds = g(t) \qquad (7.2.5)$$

($t \in [0,T]$), with $\gamma > 1$ and $0 < \alpha < 1$. If $1 - \gamma(1-\alpha) \geq 0$, it corresponds to the cordial Volterra integral operator

$$(\mathcal{V}_\varphi u)(t) = \frac{1}{\Gamma(\alpha)} \int_0^t t^{-1} \big(1 - s/t^\gamma\big)^{\alpha-1} t^{1-\gamma(1-\alpha)} u(s)\, ds \qquad (7.2.6)$$

with core $\varphi(z) = (1 - z^\gamma)^{\alpha-1}$ and the (continuous) kernel $H(t,s) = (1/\Gamma(\alpha))t^{1-\gamma(1-\alpha)}$. (Note that for $1 - \gamma(1-\alpha) > 0$ this cordial integral operator is compact.) If $\gamma(1-\alpha) = 1$ the solution of (7.2.5) presented in Corollary 1.5.12 can be written in the form

$$u(t) = \frac{d}{dt}\left(\frac{\gamma}{\Gamma(1/\gamma)} \int_0^t \big(1 - (s/t)^\gamma\big)^{-(1-1/\gamma)} (s/t)^{\gamma-1} g(s)\, ds\right). \qquad (7.2.7)$$

The underlying solution operator in (7.2.7) is the weakly singular Volterra integral operator

$$(\mathcal{V}g)(t) := \int_0^t \big(1 - (s/t)^\gamma\big)^{-(1-1/\gamma)} (s/t)^{\gamma-1} g(s)\, ds. \qquad (7.2.8)$$

We recognise in (7.2.7) the resolvent representation, or *variation-of-constants formula*, for the solution of the cordial first-kind VIE (7.2.5) (with $\gamma(1-\alpha) = 1$); the kernel of the integral operator (7.2.8) is its *resolvent function*.

We summarise these observations in the following theorem.

Theorem 7.2.9 *Assume that $\gamma > 1$ and $0 < \alpha < 1$ in the first-kind cordial VIE (7.2.5) are such that $\gamma(1 - \alpha) = 1$. For any $g \in C^1(I)$ this VIE possesses a unique solution $u \in C(I)$, and this solution is given by the variation-of-constants formula*

$$u(t) = \frac{d}{dt}(\mathcal{V}g)(t), \quad t \in I,$$

where the Volterra integral operator $\mathcal{V}$ is the one defined by (7.2.8).

It is not difficult to see that such a variation-of-constants formula for the solution of a first-kind cordial VIE with a more general core does not always exist. Such cordial VIEs can be reduced to equivalent VIEs of the second kind, as shown in the following theorem. It is a particular case of a result by Vainikko (2012; cf. Theorem 4.2 with $m = 0$): it reveals that under certain assumptions on the core $\varphi(z)$ the first-kind cordial VIE (7.2.4) is equivalent to a second-kind VIE. We employ the notation

$$(D_\star g)(t) := \frac{d}{dt}(tg(t)), \quad \psi_b(z) := z\varphi'(z) + b\varphi(z) \ (b \in \mathbb{R}),$$

$$g_b(t) := (D_\star g)(t) - bg(t).$$

Recall that (cf. (7.1.15))

$$\hat{\varphi}(\lambda) := \int_0^1 \varphi(x)x^\lambda \, dx, \quad \lambda \in \mathbb{C}, \ \ \mathrm{Re}\,\lambda \geq 0.$$

Theorem 7.2.10 *Assume that the (real-valued) core φ is in $L^1(0, 1)$ and $g \in C^1(I)$. Then the following statements are true for $b < 1$:*

(a) The first-kind cordial VIE (7.2.4) and the second-kind cordial VIE

$$\varphi(1)u(t) = g_b(t) + (\mathcal{V}_{\psi_b}u)(t), \quad t \in I, \tag{7.2.9}$$

posess the same solution sets in $C(I)$.

(b) If $\varphi(1) \neq 0$, and $\hat{\varphi}(\lambda) \neq 0$ for $\mathrm{Re}\,\lambda \geq 1$ and $\lambda = 0$, the operator $\varphi(1)\mathcal{I} - \mathcal{V}_{\psi_b}$ is invertible, and

$$\left(D_\star - b\mathcal{I}\right)\left(\varphi(1)\mathcal{I} - \mathcal{V}_{\psi_b}\right)^{-1}$$

is a bounded linear operator from $C^1(I)$ to $C(I)$.

(c) If $\hat{\varphi}(0) > 0$ and $\psi_b(z) \geq 0 \ (0 < z < 1)$, the operator $\varphi(1)\mathcal{I} - \mathcal{V}_{\psi_b} \in \mathcal{L}(C(I))$ is invertible, and

$$\left\|\left(\varphi(1)\mathcal{I} - \mathcal{V}_{\psi_b}\right)^{-1}\right\|_{\mathcal{L}(C(I))} \leq \frac{1}{(1 - b)\hat{\varphi}(0)}.$$

The reader is referred to Vainikko (2012) for the proof: it is a consequence
of a more general result on the equivalence (in the sense of coinciding
solution sets) of the first-kind cordial VIE (7.2.4) and the second-kind cordial VIE (7.2.9) where the underlying technical framework is that of certain
weighted spaces consisting of functions $u \in C^m(I)$ for which the limits
$\lim_{t \to 0} t^{j-r} u^{(j)}(t)$ $(j = 0, \ldots, m)$ exist. Instead, we illustrate the meaning of
Theorem 7.2.10 by two examples.

Example 7.2.11 It is easy to verify directly that for $g \in C^1(I)$ the cordial VIE

$$\int_0^t t^{-1}(s/t)^{\beta-1} u(s)\, ds = g(t), \quad t \in [0, T],$$

with core $\varphi(z) = z^{\beta-1}$ $(\beta > 0)$ has the solution

$$u(t) = \frac{d}{dt}\left(t^\beta g(t)\right) = \beta g(t) + t g'(t), \quad t \in (0, T].$$

For $\beta = 1$ we obtain of course the result of Example 7.2.3. On the other hand,
if we resort to Theorem 7.2.10 and the definitions preceding it, we find that

$$\psi_b(z) = (b + \beta - 1)z^{\beta-1}, \quad \text{and}$$

$$g_b(t) = \frac{d}{dt}\left(t g(t)\right) - b g(t) = (1 - b)g(t) + t g'(t).$$

Hence, since $\varphi(1) = 1$, the class of equivalent cordial VIE of the second kind is

$$u(t) = (1 - b)g(t) + t g'(t) + \int_0^t t^{-1}(b + \beta - 1)(s/t)^{\beta-1} u(s)\, ds \ (b < 1).$$

If we choose $b = 1 - \beta$ (which satisfies the condition $b < 1$) the above
expression reduces to the solution expression found by the direct approach.
(See also Example 7.2.14 below.)

Example 7.2.12 Let $\varphi(z) = 2 - z$ (cf. Vainikko, 2012, p. 692). It is easy to
verify directly that the solution of the corresponding first-kind cordial VIE,

$$\int_0^t t^{-1}(2 - s/t)u(s)\, ds = g(t), \quad t \in [0, T],$$

coincides with the one of the equivalent *second-kind* cordial VIE

$$u(t) = 2g(t) + t g'(t) - 2\int_0^t t^{-1} u(s)\, ds, \quad t \in [0, T].$$

If we employ Theorem 7.2.10(a), we see that

$$\psi_b(z) = 2b - (b + 1)z \quad \text{and} \quad g_b(t) = (t g(t))' - b g(t) \ (b < 1).$$

The resulting class of equivalent second-kind cordial VIEs is thus given by

$$u(t) = g_b(t) + \int_0^t t^{-1}(b+1)\big(2b - (s/t)\big)u(s)\,ds \ \ (b < 1).$$

For the value $b = -1$ we are led to the second-kind VIE found earlier in this example.

Is there an analogous result for the general first-kind cordial VIE (7.2.3) with non-constant kernel $H \in C(D)$? We have seen in Section 1.5 (cf. Theorem 1.5.7) that the weakly singular first-kind VIE

$$\int_0^t (t-s)^{\alpha-1} H(t,s)u(s)\,ds = g(t), \ \ t \in I \ (0 < \alpha < 1),$$

can be transformed into a VIE of the second kind, provided H and g are continuously differentiable and satisfy $H(t,t) \neq 0$ $(t \in I)$, $g(0) = 0$. Does a similar result hold for the general cordial VIE of the first kind,

$$\int_0^t t^{-1}\varphi(s/t)H(t,s)u(s)\,ds = g(t), \ \ t \in I \ ? \tag{7.2.10}$$

Example 7.2.13 If the core in (7.2.10) is constant, e.g. $\varphi(z) \equiv 1$, and if $H(t,t) \neq 0$ for $t \in I$, then this first-kind cordial VIE is equivalent to the second-kind VIE

$$H(t,t)u(t) = \frac{d}{dt}\big(tg(t)\big) - \int_0^t \frac{\partial H(t,s)}{\partial t}u(s)\,ds, \ \ t \in I, \tag{7.2.11}$$

provided that the derivatives exist and are continuous. It now follows from the classical theory of second-kind VIEs (Section 1.2.1) that under the above assumptions this VIE possesses a unique solution $u \in C(I)$. For $H(t,s) \equiv 1$ we obtain the solution

$$u(t) = \frac{d}{dt}\big(tg(t)\big) = g(t) + tg'(t), \ \ t \in I.$$

Example 7.2.14 Consider the cordial first-kind VIE (7.2.10) with core $\varphi(z) = z^{\beta-1}$ $(\beta > 0,\ \beta \neq 1)$ and $H \in C^1(D)$.

(i) Assume first that $H(t,s) \equiv 1$. The VIE (7.2.10) is then equivalent to the first-kind VIE

$$\int_0^t s^{\beta-1}u(s)\,ds = t^\beta g(t), \ \ t \in I,$$

whose unique continuous solution is given by $u(t) = \beta g(t) + tg'(t)$, $t \in I$. This is the result we have already found in Example 7.2.11.

(ii) If $H(t, s)$ is a general (non-constant) C^1 kernel, it is easy to verify that the VIE is equivalent to the second-kind VIE

$$H(t, t)u(t) = \frac{d}{dt}\left(t^\beta g(t)\right) - \int_0^t t^{-1}(s/t)^{\beta-1} \frac{\partial H(t, s)}{\partial t} t\, u(s)\, ds,$$

(7.2.12)

provided that $g \in C^1(I)$. If $H(t, t) \neq 0$ for all $t \in I$, this is a second-kind cordial VIE with kernel $K(t, s) := -\dfrac{\partial H(t, s)}{\partial t} t$. Since $K(0, 0) = 0$, the underlying cordial Volterra integral operator is compact, and hence we arrive at the same conclusion as in Example 7.2.13.

The following theorem (a particular case of Theorem 5.2 in Vainikko, 2014, p. 1627) reveals that under certain conditions on the core φ and the kernel $H(t, s)$ the general first-kind cordial VIEs (7.2.10) possesses a unique solution $u \in C(I)$.

Theorem 7.2.15 *Assume that*

(i) $\displaystyle\int_0^1 |\varphi(x)|\, dx < \infty$ *and* $\displaystyle\int_0^1 x(1-x)\varphi'(x)\, dx < \infty;$

(ii) $\hat{\varphi}(0) := \displaystyle\int_0^1 \varphi(x)\, dx > 0;$

(iii) $\psi_b(z) := b\varphi(z) + z\varphi'(z) > 0$ *for* $z \in (0, 1)$ *and* $b < 1;$

(iv) $H \in C(D),\ \partial H/\partial t \in C(D),$ *and* $H(t, t) \neq 0$ *for* $t \in I.$

Then the cordial first-kind VIE (7.2.10) has a unique solution $u \in C(I)$ for any $g \in C^1(I)$.

Proof   The assertion in Theorem 7.2.15 is a consequence of the equivalence of the first-kind cordial VIE (7.2.10) and the second-kind cordial VIE (7.2.12), and Theorem 7.2.1.   □

Example 7.2.16   Consider the Abel-type VIE

$$\int_0^t \left(t^\gamma - s^\gamma\right)^{-1/\gamma} H(t, s)u(s)\, ds = g(t),\quad t \in I\ (\gamma > 1).$$

The L^1 core of the equivalent first-kind cordial VIE (cf. the discussion preceding Theorem 7.2.9) is $\varphi(z) = \left(1 - z^\gamma\right)^{-1/\gamma}$, with $\varphi'(z) = \left(1 - z^\gamma\right)^{-1/\gamma-1} z^{\gamma-1}$ ($z \in (0, 1)$). It follows that for $\gamma > 1$ the assumptions (i)–(iii) are satisfied; in particular, we have

$$\int_0^1 z(1-z)\varphi(z)\,dz = \int_0^1 (1-z)\big(1-z^\gamma\big)^{-1/\gamma-1}z^\gamma\,dz < \infty$$

and

$$\psi_b(z) = \big(1-z^\gamma\big)^{-1/\gamma-1}\big(b(1-z^\gamma)+z^\gamma\big) > 0 \ \ (z \in (0,1),\ 0 \le b < 1).$$

Thus the above first-kind VIE possess a unique solution $u \in C(I)$ whenever H and g fulfill the conditions stated in Example 7.2.14.

The extension of Theorem 7.2.15 to smoother H and g, leading to a result on the regularity of the solution of (7.2.10), is now clear: it follows again from the equivalence of (7.2.10) and (7.2.12), recalling the regularity result of Theorem 7.2.5. We note that the following theorem is a particular case of Theorem 5.2 in Vainikko (2014).

Theorem 7.2.17 *Assume that the given function in (7.2.10) have the following properties:*

(i) $H \in C^{m+1}(D)$ ($m \ge 1$), $H(t,t) \ne 0$ for $t \in I$, and $g \in C^{m+1}(I)$;
(ii) φ is subject to the assumptions (i)–(iii) in Theorem 7.2.15.

Then the first-kind cordial VIE (7.2.10) possesses a unique solution $u \in C^m(I)$.

7.2.3 VIEs of the Third Kind (II)

For the convenience of the reader we recall from Section 1.6 the definition of a VIE of the third kind.

Definition 7.2.18 The VIE

$$a(t)u(t) = f(t) + \int_0^t (t-s)^{\alpha-1}K(t,s)u(s)\,ds, \ \ t \in I \ (0 < \alpha \le 1),$$

(7.2.13)

is said to be a (linear) VIE of the *third kind* if the function a vanishes on a non-empty proper subset of I (for example, a finite number of points, or a compact subinterval). ◇

An important subclass such third-kind VIEs corresponds to functions a with $a(0) = 0$ and $a(t) \ne 0$ when $t > 0$. The basic representative is the equation

$$tu(t) = f(t) + \int_0^t (t-s)^{\alpha-1}K(t,s)u(s)\,ds, \ \ t \in [0,T] \ (0 < \alpha \le 1),$$

(7.2.14)

with $f \in C(I)$ and $K \in C(D)$. As we have already indicated at the end of Section 1.6, a natural framework for analysing the solvability of such third-kind VIEs in e.g. the space $C(I)$ is the one of *cordial* VIEs.

A necessary condition for the existence of a solution $u \in C(I)$ for (7.2.14) is obviously $f(0) = 0$. (This is of course not sufficient, as the example $f(t) = t^r$ with $0 < r < 1$ shows.)

We will first consider the VIE (7.2.14) with regular kernel $K(t, s)$ (i.e. with $\alpha = 1$). This VIE can be written in the cordial form

$$u(t) = f_1(t) + \int_0^t t^{-1} K(t, s) u(s) \, ds, \quad t \in (0, T], \tag{7.2.15}$$

where we have set $f_1(t) := t^{-1} f(t)$. As we have seen in Section 7.2, this is a second-kind cordial VIE with constant core $\varphi(z) = 1$; its unique solvability in $C(I)$ depends crucially on the value of the kernel $K(t, s)$ at the point $(t, s) = (0, 0)$.

If $K(0, 0) = 0$, the cordial Volterra integral operator in (7.2.8),

$$(\mathcal{V}_\varphi u)(t) = \int_0^t t^{-1} K(t, s) u(s) \, ds, \quad t \in I,$$

is a compact (and quasi-nilpotent) operator on $C(I)$. Since it has no eigenvalues, the classical Volterra theory (cf. Theorem 1.2.7) applies, and thus the third-kind VIE (7.2.15) possesses a unique solution $u \in C(I)$ for all f for which $f_1 \in C(I)$.

For K with $K(0, 0) \neq 0$ the Volterra integral operator $\mathcal{V}_\varphi$ is non-compact, as we have seen in Theorem 7.1.10(b). Its spectrum is given by

$$\sigma(\mathcal{V}_\varphi) = \{0\} \cup \{K(0, 0)(1 + \lambda)^{-1} : \text{Re}(\lambda \geq 0\}$$

(recall Example 7.1.15 and Theorem 7.1.13). Hence, assuming that $f_1 \in C(I)$, the VIE (7.2.15) will have a unique solution $u \in C(I)$ only if 1 does not belong to the spectrum of $\mathcal{V}_\varphi$ (recall Theorem 7.2.1 with $\mu = 1$).

These observations allow us to state the following theorem.

Theorem 7.2.19 *Assume that $K \in C(D)$ and $f \in C(I)$ is such that $f_1 \in C(I)$. Then the following statements are true:*

(a) *If K is such that $K(0, 0) = 0$, the third-kind VIE (7.2.14) possesses a unique solution $u \in C(I)$.*

(b) *If K has the property that $K(0, 0) \neq 0$, the third-kind VIE (7.2.14) has a unique solution $u \in C(I)$ if the value $\mu = 1$ (i.e. the coefficient of $u(t)$ on the left-hand side of (7.2.15)) is not in the spectrum $\sigma(\mathcal{V}_\varphi)$.*

The above analysis can be used to deal with the third-kind VIE (7.2.13) (with $\alpha = 1$) when $a(t)$ is of the form

$$a(t) = \sum_{j=1}^{n} a_j t^j = t \sum_{j=1}^{n} a_j t^{j-1}, \quad \text{with } a_1 \neq 0 \text{ and } a(t) \neq 0 \ (t \in (0, T]).$$

$$(7.2.16)$$

We note that the subsequent analysis is readily extended to functions $a \in C^d(I)$ for some $d \geq 1$, with $a(0) = 0$ and $a(t) \neq 0$ for $t \in (0, T]$. Setting $f_1(t) := t^{-1} f(t)$ and

$$\tilde{f}(t) := \left(\sum_{j=1}^{n} a_j t^{j-1} \right)^{-1} f_1(t), \quad \tilde{K}(t, s) := \left(\sum_{j=1}^{n} a_j t^{j-1} \right)^{-1} K(t, s),$$

$$(7.2.17)$$

we obtain the equivalent VIE

$$u(t) = \tilde{f}(t) + \int_0^t t^{-1} \tilde{K}(t, s) u(s) \, ds, \quad t \in (0, T],$$

where $\tilde{K}(0, 0) = a_1^{-1} K(0, 0)$. This yields the following generalisation of Theorem 7.2.17.

Theorem 7.2.20 *Assume that $a(t)$ satisfies (7.2.16), and let f and K be subject to the assumptions in Theorem 7.2.17. Then the assertions (a) and (b) in Theorem 7.2.17 remain valid for the third-kind VIE (7.2.13) with $\alpha = 1$.*

Proof If $\alpha = 1$ the two assertions are a direct consequence of Theorem 7.2.17, with $\tilde{f}$ and $\tilde{K}$ replacing f and K, since $\tilde{K}(0, 0) = a^{-1} K(0, 0)$. □

We will now look at the linear third-kind VIE (7.2.14) with *weakly singular kernel* and $a(t) = t^\beta$ with $\beta > 0$,

$$t^\beta u(t) = f(t) + \int_0^t (t - s)^{\alpha - 1} K(t, s) u(s) \, ds, \quad t \in [0, T], \qquad (7.2.18)$$

where $0 < \alpha < 1$ and $f \in C(I)$, $K \in C(D)$. The cordial form of the above VIE is

$$u(t) = f_\beta(t) + \int_0^t t^{-1} t^{\alpha - \beta} (1 - s/t)^{\alpha - 1} K(t, s) u(s) \, ds,$$

$$t \in (0, T] \ (0 < \alpha < 1), \qquad (7.2.19)$$

with $f_\beta(t) := t^{-\beta} f(t)$. The solvability of (7.2.18) in $C(I)$ is of course governed by the properties of the underlying Volterra integral operator

$$(\mathcal{V}^{\alpha,\beta} u)(t) := \int_0^t t^{-1} t^{\alpha-\beta} (1 - s/t)^{\alpha-1} K(t, s) u(s) \, ds, \qquad (7.2.20)$$

where $K \in C(D)$. The following results were established in Seyed Allaei, Yang & Brunner (2015).

Theorem 7.2.21 *Assume that* $0 < \alpha \leq 1$, $\beta > 0$, *and* $K \in C(D)$.

(a) *If* $0 < 1 - \alpha + \beta < 1$, *then* $\mathcal{V}^{\alpha,\beta}$ *is a bounded and compact operator from* $C(I)$ *to* $C(I)$, *and (7.2.19) possesses a unique solution* $u \in C(I)$ *whenever* $f_\beta \in C(I)$.

(b) *If* $1 - \alpha + \beta = 1$, $\mathcal{V}^{\alpha,\beta}$ *is a bounded operator on* $C(I)$. *It is compact (and quasi-nilpotent) if* $K(0,0) = 0$; *otherwise, it is a non-compact operator with the uncountable spectrum*

$$\sigma(\mathcal{V}^{\alpha,\beta}) = \{0\} \cup \left\{ K(0,0) \frac{\Gamma(\alpha)\Gamma(1+\lambda)}{\Gamma(1+\alpha+\lambda)} : \operatorname{Re} \lambda \geq 0 \right\}.$$

If $\mu = 1 \notin \sigma(\mathcal{V}^{\alpha,\beta})$ *the VIE (7.2.19) has a unique solution* $u \in C(I)$ *whenever* $f_\beta \in C(I)$.

(c) *If* $1 - \alpha + \beta > 1$ *and the kernel* K *has the form*

$$K(t, s) = s^{\beta-\alpha} H(t, s), \quad \text{with} \ \ H \in C(D), \qquad (7.2.21)$$

then $\mathcal{V}^{\alpha,\beta} : C(I) \to C(I)$ *is a bounded operator. It is compact if* $H(0,0) = 0$; *otherwise, it is non-compact with the uncountable spectrum given by*

$$\sigma(\mathcal{V}^{\alpha,\beta}) = \{0\} \cup \left\{ H(0,0) \frac{\Gamma(\alpha)\Gamma(1-\alpha+\beta+\lambda)}{\Gamma(1+\beta+\lambda)} : \operatorname{Re} \lambda \geq 0 \right\}.$$

If $\mu = 1 \notin \sigma(\mathcal{V}^{\alpha,\beta})$ *the VIE (7.2.19) possesses a unique solution* $u \in C(I)$ *for all* f *for which* f_β *is in* $C(I)$.

Proof The results (a), (b) and (c) follow immediately from the results in Sections 7.1.2 and 7.1.3 on cordial Volterra integral operators and the solvability of cordial VIEs.

(a) If $0 < 1 - \alpha + \beta < 1$, $\mathcal{V}^{\alpha,\beta}$ may be viewed as a cordial Volterra integral operator with kernel $\tilde{K}(t, s) := t^{\alpha-\beta} K(t, s)$, with $\alpha - \beta > 0$. Since $\tilde{K}(0,0) = 0$, the compactness of $\mathcal{V}_{\alpha,\beta}$ follows, and hence the existence of a unique solution $u \in C(I)$ is a consequence of the classical Volterra theory of Chapter 1.

(b) For $1 - \alpha + \beta = 1$ we have $\beta = \alpha$ and hence by (7.2.21), $\mathcal{V}^{\alpha,\beta}$ is a cordial Volterra integral operator with L^1 core $\varphi(z) = (1 - z)^{\alpha-1}$. Its spectrum is given by Example 7.1.11.

(c) If $1 - \alpha + \beta > 1$ and the kernel $K(t, s)$ is of the form (7.2.21), it is readily seen that now $\mathcal{V}^{\alpha,\beta}$ is a cordial Volterra integral operator with L^1 core $\varphi(z) = z^{\beta-\alpha}(1 - z)^{\alpha-1}$ and kernel $H(t, s)$. The spectrum of this cordial operator was presented in Example 7.1.11. □

Remark 7.2.22 If $\mu = 1$ lies in the spectrum of $\mathcal{V}^{\alpha,\beta}$, the cordial VIE (7.2.19) either has no solution or it has multiple solutions. The detailed answer requires a deeper understanding of the connection between the (complex) spectrum of this cordial Volterra integral operator and the eigenvalues that lie on its boundary. The interested reader will find a comprehensive analysis of this problem in the seminal paper of Yang (2015). ◇

We conclude this discussion of the solvability of third-kind VIEs by looking at an example that was studied by von Wolfersdorf (2011, following an earlier paper by Berg & von Wolfersdorf, 2005; see also Janno & von Wolfersdorf, 2005). It concerns the VIE (7.2.18) with $\beta = \alpha > 0$ and $K(t, s) = \kappa > 0$,

$$t^\alpha u(t) = f(t) + \kappa \int_0^t (t - s)^{\alpha-1} u(s)\, ds, \quad t \in [0, T], \tag{7.2.22}$$

and the question on the existence of non-trivial solutions to the *homogeneous* version of the above third-kind VIE.

Theorem 7.2.23 *The homogeneous VIE corresponding to (7.2.22) possesses the solution $u_0(t) = t^\rho$ $(t > 0)$, where $\rho > -1$ is the unique root of the equation*

$$\kappa B(\alpha, \rho + 1) = 1, \quad \text{with } \kappa > 0. \tag{7.2.23}$$

Moreover, the only real solution $u_0(t)$ that is continuous on $[0, T]$ is the constant function $u_0(t) = \gamma$ $(\gamma \in \mathbb{R})$.

Proof Substitution of $u_0(t)$ in the homogeneous integral equation corresponding to the VIE (7.2.22) shows that $u_0(t) = t^\rho$ is indeed a solution on $(0, T]$. To show that this solution is unique for given $\kappa > 0$ and $\alpha > 0$, we resort the following properties of the Euler beta function: for any $\alpha > 0$,

(i) $dB(\alpha, \rho + 1)/d\rho < 0 \ (\rho > -1)$;
(ii) $B(\alpha, \rho + 1) \to \infty$ as $\rho \to -1^+$; and
(iii) $B(\alpha, \rho + 1) \to 0$ as $\rho \to \infty$.

It follows from (i)–(iii) that for any given $\kappa > 0$ the equation (7.2.23) has a unique real root $\beta > -1$ (compare also Appendix 2 in von Wolfersdorf, 2011). □

We now employ the above results to analyse the solvability of the non-homogeneous VIE (7.2.22) and look for solutions of the form $u(t) = t^\beta w(t)$, where $\rho > -1$ is the root of (7.2.23) and $w \in C(I)$. It is readily verified that w must satisfy the VIE

$$w(t) := t^{-(\alpha+\rho)} f(t) + \kappa \int_0^t t^{-1} \underbrace{(1 - (s/t))^{\alpha-1}(s/t)^\rho}_{=:\varphi(s/t)} w(s)\, ds, \ \ t \in (0, T].$$

(7.2.24)

Since the core $\varphi(z) = (1 - z)^{\alpha-1} z^\rho$ is in $L^1(0, 1)$, this is a cordial VIE whose underlying Volterra operator is non-compact. It follows from Example 7.1.22 that its spectrum is given by

$$\sigma(\mathcal{V}_\varphi) = \{0\} \cup \left\{ \kappa \frac{\Gamma(\alpha)\Gamma(1 + \rho + \lambda)}{\Gamma(1 + \alpha + \rho + \lambda)} : \ \text{Re}\, \lambda \geq 0 \right\}.$$

7.3 Non-Linear Cordial VIEs

The non-linear version of the cordial VIE (7.1.5) is

$$u(t) = f(t) + \int_0^t t^{-1} \varphi(s/t) k(t, s, u(s))\, ds \tag{7.3.1}$$

$$= f(t) + \int_0^1 \varphi(x) k(t, tx, u(tx))\, dx, \ \ t \in I = [0, T],$$

where $\varphi \in L^1(0, 1)$. Its Hammerstein analogue is

$$u(t) = f(t) + \int_0^t t^{-1} \varphi(s/t) K(t, s) G(s, u(s))\, ds, \ \ t \in I. \tag{7.3.2}$$

Let $\mu \neq 0$ denote a (real or complex) number. We will study the existence and uniqueness of continuous solutions of the non-linear cordial VIE

$$\mu u(t) = f(t) + \int_0^t t^{-1} \varphi(s/t) k(t, s, u(s))\, ds, \ \ t \in I. \tag{7.3.3}$$

For kernel functions $k = k(t, s, u)$ that have a continuous derivative with respect to u the kernel

$$a(t, s) := \frac{\partial k(t, s, u)}{\partial u}\bigg|_{u=u(s)} \tag{7.3.4}$$

will play a key role in the subsequent analysis. Note that for the Hammerstein VIE (7.3.2) it has the form

$$a(t, s) = K(t, s)\frac{\partial G(s, u)}{\partial u}\bigg|_{u=u(s)}. \tag{7.3.5}$$

As we shall see, the analysis of the existence of a continuous solution for (7.3.1) is closely related to that of a corresponding linear cordial VIE with kernel $a(t, s)$ along a particular solution of (7.3.2).

Definition 7.3.1 If $a(t, s)$ exists, is continuous on D, and satisfies $a(0, 0) \neq 0$, then

$$(\mathcal{V}_{\varphi}u)(t) := \int_0^t t^{-1}\varphi(s/t)k(t, s, u(s))\,ds, \quad t \in I, \tag{7.3.6}$$

is called a non-linear cordial Volterra integral operator, and we say that the VIE (7.3.1) is a *non-linear cordial VIE*. ◇

Example 7.3.2 The non-linear Volterra integral operator

$$(\mathcal{V}_{\varphi}u)(t) := \int_0^t \left(t^{3/2}-s^{3/2}\right)^{-2/3}su^4(s)\,ds = \int_0^1 \underbrace{\left(1-x^{3/2}\right)^{-2/3}}_{=\varphi(x)} tx\,u^4(tx)\,dx,$$

arising in the analysis of the VIE

$$u(t) = 1 - \frac{3\sqrt{3}}{\pi}\int_0^t \left(t^{3/2} - s^{3/2}\right)^{-2/3}su^4(s)\,ds$$

that is equivalent to Lighthill's equation (cf. Example 9.2.9), has the linearised kernel

$$a(t, s) = \frac{\partial(su^4)}{\partial u}\bigg|_{u=u(s)} = 4su^3(s).$$

In this case we have $a(0, 0) = 0$ for all $u = u(s)$.

Assume that $\sigma(\mathcal{V}_{\varphi}^1)$ is the spectrum of the basic cordial Volterra integral operator $\mathcal{V}_{\varphi}^1$ with core φ, and denote by $\rho(\mathcal{V}_{\varphi}^1) := \mathbb{C} \setminus \sigma(\mathcal{V}_{\varphi}^1)$ its resolvent set. We observe that if $u^*(t)$ is a solution of (7.3.3), $w^* := u^*(0)$ must be a root of the non-linear algebraic equation

$$\mu w^* = f(0) + k(0, 0, w^*)\int_0^1 \varphi(z)\,dz. \tag{7.3.7}$$

The following result on the existence of a unique continuous solution for
(7.3.3) is due to Vainikko (2011, Theorem 7.1).

Theorem 7.3.3 *Assume that*

(i) $\varphi \in L^1(0, 1)$;
(ii) $k \in C(D \times \Omega)$, $\partial k(t, s, u)/\partial u \in C(D \times \Omega)$ *(where Ω is some open set in $\mathbb{R}$);*
(iii) the algebraic equation (7.3.7) has a root $w^* \in \Omega$;
(iv) $\mu \in a^*(0, 0)\rho(\mathcal{V}_\varphi)$, *with* $a^*(0, 0) = \dfrac{\partial k(0, 0, u)}{\partial u}\Big|_{u=w^*}$.

Then there exists a $T_1 \in (0, T]$ so that for each $f \in C(I)$ the non-linear cordial VIE (7.3.2) possesses a unique solution $u^* \in C[0, T_1]$ satisfying $u^*(0) = w^*$.

Proof The key ingredient is the Fréchet derivative of the non-linear Volterra integral operator

$$(\mathcal{V}_\varphi u)(t) := \int_0^t t^{-1}\varphi(s/t)k(t, s, u(s))\,ds$$

underlying the cordial VIE (7.3.2). As we mention in Example A.3.2 in the Appendix, it is given by

$$(\mathcal{V}_\varphi' u)(t) = \int_0^t t^{-1}\varphi(s/t)a(t, s)u(s)\,ds,$$

where $a(t, s)$ is defined in (7.3.4). One then shows that, under the assumption (iv), the operator $\mu \mathcal{I} - \mathcal{V}_\varphi'$ has a bounded inverse on $C(I)$ (see Vainikko, 2011, pp. 98, for details). Also, it is readily seen that the root w^* of the algebraic equation (7.3.7) is simple (consider the function

$$r(\xi) := \mu\xi - k(0, 0, \xi)\int_0^1 \varphi(z)\,dz - f(0) \;:$$

the assumption that $r'(w^*) = 0$ leads to a contradiction with the assumption (iv)). □

Remark 7.3.4 Sato (1953b) studied the non-linear third-kind VIE

$$tu(t) = f(t) + \int_0^t k(t, s, u(s))\,ds, \quad t \geq 0, \tag{7.3.8}$$

where f and k are assumed to be sufficiently regular (in fact: analytic) in a neighbourhood of $t = 0$. This VIE is closely related to the non-linear cordial

VIE (7.3.2) since, in analogy to the linear third-kind VIE (7.2.14), the non-linear third-kind VIE (7.3.8) can be rewritten as a non-linear cordial VIE,

$$u(t) = f_0(t) + \int_0^t t^{-1} k(t, s, u(s)) \, ds, \quad t \geq 0, \tag{7.3.9}$$

with *constant core* $\varphi(z) = 1$, provided f is such that the function $f_0(t) := t^{-1} f(t)$ is continuous at $t = 0$. We observe that Sato's condition

$$c_0 = a_1 + k(0, 0, c_0)$$

(Sato, 1953, p. 148), which assumes that (7.3.8) has a (local) solution of the form $u(t) = c_0 + c_1 t + c_2 t^2 + \cdots$, corresponds to the algebraic equation (7.3.7), since he assumes that

$$f(t) = \sum_{j=0}^{\infty} a_j t^j \quad (\text{with } a_0 = 0), \quad \text{and} \quad u(t) = \sum_{j=0}^{\infty} c_j t^j.$$

◇

In view of applications we also state the result of Theorem 7.3.3 for the Volterra–Hammerstein version of (7.3.2),

$$\mu u(t) = f(t) + \int_0^t t^{-1} \varphi(s/t) K(t, s) G(s, u(s)) \, ds, \quad t \in I. \tag{7.3.10}$$

It follows from (7.3.7) and (7.3.5) that if $u^*(t)$ is a solution of (7.3.10), $w^* := u^*(0)$ must solve the non-linear algebraic equation

$$\mu w^* = f(0) + K(0, 0) G(0, w^*) \int_0^1 \varphi(x) \, dx. \tag{7.3.11}$$

Theorem 7.3.3 yields the following existence result for (7.3.10); its proof is left to the reader.

Corollary 7.3.5 *Assume that*

(i) $\varphi \in L^1(0, 1)$, $f \in C(I)$, and $K \in C(D)$ with $K(0, 0) \neq 0$;
(ii) $\partial G / \partial u \in C(I \times \Omega)$;
(iii) the algebraic equation (7.3.11) possesses the root w^* in Ω;
(iv) $\mu \in a^*(0, 0) \rho(\mathcal{V}_\varphi^1)$, where $a^*(0, 0) = K(0, 0) \dfrac{\partial G(0, u)}{\partial u}\bigg|_{u = w^*}$.

Then there exists $T_1 \in (0, T]$ so that the non-linear cordial VIE (7.3.10) possesses a unique solution $u^* \in C[0, T_1]$ which satisfies $u^*(0) = w^*$.

In Theorem 7.2.5 we showed that if the functions f and K in a linear cordial VIE are smooth, the solution inherits this regularity. An analogous regularity results is true for the non-linear cordial VIE (7.3.2), albeit in general only on some subinterval $[0, T_1]$ of $[0, T]$. This is made precise in the next theorem (Vainikko, 2011).

Theorem 7.3.6 *Assume that the assumptions (i), (iii) and (iv) of Theorem 7.3.3 hold, and replace assumption (ii) by (ii') $k \in C^m (D \times \Omega)$ for some $m \geq 1$. Then there exists $T_1 \in [(0, T]$ so that for every $f \in C^m (I)$ the non-linear cordial VIE (7.3.2) possesses a unique solution $u^* \in C^m [0, T_1]$ satisfying $u^*(0) = w^*$. If the solution can be extended to the whole interval $I = [0, T]$, we have $u^* \in C^m (I)$.*

Proof By resorting to Theorem 7.2.5 (see also Theorem 8.1.9) on the mapping properties of linear cordial Volterra integral operators and the regularity of solutions of corresponding cordial VIEs, the assertion in Theorem 7.3.6 can be verified by applying those results to the linear cordial VIE with kernel $a(t, s)$ defined in (7.3.4), and by observing assumption (iv). The technical details can be found in Vainikko (2011, pp. 100–103). □

Remark 7.3.7 A first regularity result for non-linear cordial VIEs (7.3.2) was established implicitly in Sato (1953b) for the non-linear third-kind VIE (7.3.8), under the assumption that f and k are analytic functions for $0 \leq s \leq t < \rho_0$, $|u - f(t)| < \rho_1$. As mentioned in Remark 7.3.4 this equation is formally equivalent to the non-linear cordial VIE (7.3.2) with constant core $\varphi(z) \equiv 1$.

◇

7.4 Cordial VIEs with Highly Oscillatory Kernels

7.4.1 The Spectra of Highly Oscillatory Cordial Volterra Operators

According to (7.1.13) the spectrum of the general linear cordial Volterra integral operator

$$(\mathcal{V}_\varphi u)(t) = \int_0^t t^{-1} \varphi(s/t) K(t, s) u(s) \, ds, \quad t \in [0, T],$$

is given by

$$\sigma(\mathcal{V}_\varphi) = \{0\} \cup \{K(0, 0)\hat{\varphi}(\lambda) : \lambda \in \mathbb{C}, \ \text{Re}(\lambda) \geq 0\},$$

where

$$\hat{\varphi}(\lambda) = \int_0^1 \varphi(x)x^\lambda\,dx.$$

How is the spectrum $\sigma(\mathcal{V}_{\omega,\varphi})$ of the corresponding *highly oscillatory* general cordial Volterra operator

$$(\mathcal{V}_{\omega,\varphi}u)(t) = \int_0^t t^{-1}\varphi(s/t)K_\omega(t,s)u(s)\,ds \quad (\omega \gg 1), \tag{7.4.1}$$

with

$$K_\omega(t,s) := K(t,s)e^{i\omega g(t,s)}, \tag{7.4.2}$$

related to $\sigma(\mathcal{V}_\varphi)$?

Theorem 7.4.1 *Assume that the oscillator $g(t,s)$ defining the highly oscillatory kernel $K_\omega(t,s)$ of the general cordial Volterra integral operator $\mathcal{V}_{\omega,\varphi}$ of (7.4.1) is smooth, non-negative, and independent of ω, and let $K \in C(D)$ be independent of ω and satisfy $K(0,0) \neq 0$. Then the spectrum $\mathcal{V}_{\omega,\varphi}$ depends on ω if, and only if, $g(0,0) \neq 0$: we have*

$$\sigma(\mathcal{V}_{\omega,\varphi}) = K(0,0)e^{i\omega g(0,0)}\sigma(\mathcal{V}_\varphi). \tag{7.4.3}$$

The eigenvalues $\hat{\varphi}_\omega(\lambda)$ of $\mathcal{V}_{\omega,\varphi}$ satisfy

$$|\hat{\varphi}_\omega(\lambda)| = |K_\omega(0,0)||\hat{\varphi}(\lambda)|.$$

Proof Theorem 7.1.17 implies that the spectrum of $\mathcal{V}_{\omega,\varphi}$ is given by

$$\sigma(\mathcal{V}_{\omega,\varphi}) = K_\omega(0,0)\sigma(\mathcal{V}_\varphi),$$

where, by (7.4.2),

$$K_\omega(0,0) = K(0,0)e^{i\omega g(0,0)},$$

with $K(0,0) \neq 0$. Hence the second statement in Theorem 7.3.3 follows from Theorem 7.1.17. $\qquad\square$

As we have indicated in Section 4.1.3, the analysis of the asymptotic behaviour of the spectra of the highly oscillatory *Fredholm* integral operators

$$(\mathcal{F}_\omega u)(t) = \int_{-1}^1 K(x,y)e^{i\omega g(x,y)}u(y)\,dy \quad (\omega \gg 1)$$

with e.g. oscillators $g(x,y) = (x-y)^2$ or $g(x,y) = xy$ is quite difficult. This is completely different for highly oscillatory cordial *Volterra* integral operators (7.4.1) with the analogous oscillators, as the following corollary shows.

Corollary 7.4.2 *For any* $\omega > 0$ *the spectra of the highly oscillatory cordial Volterra integral operators* $\mathcal{V}_{\omega,\varphi}$ *in (7.4.1), with* $K(t,s) \equiv 1$ *and oscillators*

$$g(t,s) = (t-s)^2 \quad or \quad g(t,s) = ts,$$

are given by

$$\sigma(\mathcal{V}_{\omega,\varphi}) = \sigma(\mathcal{V}_\varphi) \ (= \sigma(\mathcal{V}_\varphi^1)).$$

Proof Since the above oscillators $g(t,s)$ all have the property $g(0,0) = 0$, the statement (7.4.3) of Theorem 7.4.1 yields the desired result. $\square$

Remark 7.4.3 As a simple example of an oscillator $g(t,s)$ that does not vanish at $(0,0)$ we consider, similarly to Iserles (2005), the 'irregular oscillator'

$$g(t,s) = \tfrac{1}{2}(a+t-s)^2 \ (a>0), \tag{7.4.4}$$

for which we have $g(0,0) = a^2/2$. By Theorem 7.4.1, the spectrum of $\mathcal{V}_{\omega,\varphi}$ with kernel $K(t,s) \equiv 1$ and the above oscillator is given by

$$\sigma(\mathcal{V}_{\omega,\varphi}) = e^{i\omega a^2/2}\sigma(\mathcal{V}_\varphi).$$

Thus, if $\mathcal{V}_\varphi = \mathcal{V}_{0,\varphi}$ is as in Example 7.1.8 with $\gamma = 2$ (and core $\varphi(z) = (1-z^2)^{-1/2}$), its *real* eigenvalues $\hat\varphi(r) = \tfrac{1}{2}B(\tfrac{1+r}{2},\tfrac{1}{2}) \in [0,\pi/2]$ $(r \geq 0)$ become complex when $\omega > 0$: they are

$$\hat\varphi_\omega(r) = \hat\varphi(r)e^{i\omega a^2/2} \ (r \geq 0),$$

with $\hat\varphi(0) = \pi/2$ and, by (7.1.14) of Theorem 7.1.17, $\hat\varphi(r) \to 0$ as $r \to \infty$. $\diamond$

The above analysis shows that the derivation of the asymptotic behaviour of the spectra of highly oscillatory cordial Volterra integral operators is in sharp contrast to the one of the behaviour of the spectra of analogous highly oscillatory Fredholm integral operators with oscillators like $g(x,y) = (x-y)^2$ (cf. Section 4.1.3).

7.4.2 Second-Kind Cordial VIEs with Highly Oscillatory Kernels

We now consider the cordial VIE

$$\mu u(t) = f(t) + (\mathcal{V}_{\omega,\varphi}u)(t), \ t \in I, \tag{7.4.5}$$

where the highly oscillatory cordial Volterra integral operator $\mathcal{V}_{\omega,\varphi}$ is defined in (7.4.1) and $\mu \neq 0$. We assume as always that $K(0,0) \neq 0$, and that f, K and the oscillator g do not depend on ω.

Theorem 7.4.4 *Assume that the given functions in the highly oscillatory cordial VIE (7.4.5) do not depend on ω and are subject to the conditions $f \in C(I)$, $K \in C(D)$, $K(0,0) \neq 0$, with g smooth and non-negative on D. If $\mu \notin \sigma(\mathcal{V}_\varphi)$ (that is, the non-oscillatory cordial VIE*

$$\mu u(t) = f(t) + (\mathcal{V}_\varphi u)(t), \quad t \in I,$$

has a unique solution $u \in C(I)$), the highly oscillatory cordial VIE (7.4.5) has a unique solution $u \in C(I)$ for all $\omega > 0$, regardless of whether $g(0,0) = 0$ or $g(0,0) \neq 0$.

Proof It follows from Theorem 7.4.1 that for oscillators $g(t,s)$ in (7.4.1) with $g(0,0) = 0$ there holds

$$\sigma(\mathcal{V}_{\omega,\varphi}) = K(0,0)\sigma(\mathcal{V}_\varphi^1) \quad \text{for all } \omega > 0.$$

Thus, the highly oscillatory cordial second-kind VIE (7.4.5) possesses a unique solution for any $\omega > 0$ whenever $\mu \notin \sigma(\mathcal{V}_\varphi)$.

If $g(0,0) \neq 0$, Theorem 7.4.1 implies that the eigenvalues $\hat{\varphi}_\omega(\lambda)$ of $\mathcal{V}_{\omega,\varphi}$ have the moduli

$$|\hat{\varphi}_\omega(\lambda)| = |\hat{\varphi}(\lambda)| \quad (\mathrm{Re}(\lambda) \geq 0)$$

for all $\omega > 0$, and hence the assertion follows. $\qquad\square$

Example 7.4.5 The cordial VIE

$$\mu u(t) = f(t) + \int_0^t (t^2 - s^2)^{-1/2} u(s)\,ds \quad (\mu \neq 0) \tag{7.4.6}$$

corresponds to the cordial Volterra integral operator $\mathcal{V}_\varphi^1$ with core $\varphi(z) = (1 - z^2)^{-1/2}$ (recall Example 7.1.8). The spectrum $\sigma(\mathcal{V}_\varphi^1)$ of this operator is

$$\sigma(\mathcal{V}_\varphi^1) = \{0\} \cup \{\hat{\varphi}(\lambda) = \tfrac{\sqrt{\pi}}{2}\Gamma(\tfrac{1}{2}(1+\lambda))/\Gamma(1 + \tfrac{1}{2}\lambda) : \mathrm{Re}(\lambda) \geq 0\}.$$

Since the spectrum of the corresponding highly oscillatory cordial Volterra operator $\mathcal{V}_{\omega,\varphi}$ is $\sigma(\mathcal{V}_{\omega,\varphi}) = \sigma(\mathcal{V}_\varphi)$ for all $\omega > 0$ (Theorem 7.4.1), the highly oscillatory cordial VIE

$$\mu u(t) = f(t) + \int_0^t (t^2 - s^2)^{-1/2} e^{i\omega g(t,s)} u(s)\,ds$$

possesses a unique solution $u \in C(I)$ for any $\omega > 0$, whenever $\mu \notin \sigma(\mathcal{V}_\varphi^1)$.

7.4.3 Cordial First-Kind VIEs with Highly Oscillatory Kernels

The highly oscillatory version of (7.2.3) is

$$(\mathcal{V}_{\omega,\varphi}u)(t) = f(t), \quad t \in I, \tag{7.4.7}$$

where $\mathcal{V}_{\omega,\varphi}$ is as in (7.4.1).

Example 7.4.6 Using the well-known inversion formula

$$u(t) = \frac{2}{\pi} \left(f(0) + t \int_0^t (t^2 - s^2)^{-1/2} f'(s)\,ds \right), \quad 0 < t \leq T \tag{7.4.8}$$

for the first-kind Abel-type VIE

$$\int_0^t (t^2 - s^2)^{-1/2} u(s)\,ds = f(t), \quad t \in I,$$

with $f \in C^1(I)$ (see Section 1.5.3), together with an obvious modification of the substitutions (4.1.10), we see that the solution of the highly oscillatory first-kind VIE

$$\int_0^t (t^2 - s^2)^{-1/2} e^{i\omega g(t,s)} u(s)\,ds = f(t), \quad t \in I, \tag{7.4.9}$$

with $f \in C^1(I)$, $f(0) = 0$, and separable oscillator $g(t,s) = g_0(t) - g_0(s)$ as in (4.1.5), has the form

$$u(t) = \frac{2t}{\pi} \int_0^t (t^2 - s^2)^{-1/2} e^{i\omega \left(g_0(t) - g_0(s) \right)} f'(s)\,ds \tag{7.4.10}$$

$$- \frac{2ti\omega}{\pi} \int_0^t (t^2 - s^2)^{-1/2} e^{i\omega \left(g_0(t) - g_0(s) \right)} f(s) g_0'(s)\,ds, \quad t \in I.$$

We observe that this solution structure is analogous to the one shown in Example 4.2.7: for $\omega > 0$ we have again an 'ω-perturbation' of the classical solution corresponding to $\omega = 0$.

7.5 Exercises and Research Problems

Exercise 7.5.1 Let $\mathcal{V}_\varphi$ and $\mathcal{V}_\psi$ be cordial Volterra integral operators of the form (7.1.1) with L^1 cores φ, ψ and continuous kernels $K(t,s)$, $H(t,s)$. Determine the core and the kernel of their composition $\mathcal{V}_\varphi \circ \mathcal{V}_\psi$.

Exercise 7.5.2 For which (real-valued) non-constant cores $\varphi(z)$ is 1 an eigenvalue of the basic cordial Volterra integral operator $\mathcal{V}_\varphi^1$?

Exercise 7.5.3 Assume that $\varphi \in L^1(0, 1)$ and $K \in C(D)$, with $K(0, 0) \neq 0$. Show that the Volterra integral operator $\mathcal{V}_\varphi - K(0, 0)\mathcal{V}_\varphi^1$ is a compact operator from $C(I)$ into $C(I)$.

Exercise 7.5.4 Let

$$(\mathcal{V}_{\alpha,\beta}u)(t) := \int_0^t (t - s)^{\alpha-1}s^{-\beta}K(t, s)u(s)\,ds, \quad t \in [0, 1],$$

with $0 < \alpha < 1$, $\beta < 1$ and $K \in C(D)$. For which values of α, β is $\mathcal{V}_{\alpha,\beta}$ compact (as an operator on $C[0, 1]$)? When is it a cordial Volterra integral operator?

Exercise 7.5.5 Let the non-negative integers m and m' satisfy $m > m'$ and consider the basic cordial Volterra integral operators $\mathcal{V}_\varphi^1 : C^m(I) \to C^m(I)$ and $\mathcal{V}_\varphi^1 : C^{m'}(I) \to C^{m'}(I)$. Show that their spectra are related by $\sigma_m(\mathcal{V}_\varphi^1) \subset \sigma_{m'}(\mathcal{V}_\varphi^1)$. What can be said if $m \to \infty$?

Exercise 7.5.6 Prove Theorem 7.2.5.

Exercise 7.5.7 Assume that the kernel $K(t, s)$ in the third-kind VIE

$$\mu t u(t) = f(t) + \int_0^t K(t, s)u(s)\,ds, \quad t \in [0, T],$$

is of the form $K(t, s) = (1 - s/t)^{1/2}\Gamma(t, s)$, where the Γ is smooth on D (see also Grandits, 2008). If f is smooth on I and satisfies $f(0) = 0$, when does the above VIE possess a unique solution $u \in C(I)$? Discuss its regularity when $f \in C^d(I)$ and $\Gamma \in C^d(D)$ for some $d \geq 2$.

Exercise 7.5.8 Consider the Volterra integral operator $\mathcal{V}_\varphi : C(I) \to C(I)$ defined by

$$(\mathcal{V}_\varphi u)(t) := \int_0^t K(t, s)p(t, s)u(s)\,ds, \quad t \in I,$$

with smooth K and

$$p(t, s) := \pi^{-1/2}\frac{1}{\sqrt{\log(t/s)}}\left(\frac{s}{t}\right)^\beta\left(\frac{1}{s}\right) \quad (\beta > 0)$$

(cf. Han, 1994). Show that $\mathcal{V}_\varphi$ is a cordial Volterra integral operator and find its core. Determine its spectrum if $K(0, 0) \neq 0$.

Exercise 7.5.9 Is the adjoint Volterra integral operator

$$(\mathcal{V}^*u)(t) := \int_t^T (s^2 - t^2)^{-1/2} u(s)\, ds, \quad t \in [0, 1],$$

a compact and quasi-nilpotent operator on $C[0, 1]$? If not, find its spectrum.

Exercise 7.5.10 Kilbas & Saigo (1995) considered the second-kind Abel–Volterra integral equation

$$u(t) = f(t) + \frac{b(t)}{\Gamma(\alpha)} \int_0^t (t - s)^{\alpha - 1} u(s)\, ds, \quad t \in (0, T] \ (0 < \alpha < 1),$$

where b and f have the asymptotic expansions

$$b(t) = \sum_{j=-1}^{\infty} \frac{b_j}{\Gamma(1 + j\alpha)} t^{j\alpha}, \quad f(t) = \sum_{j=-1}^{\infty} \frac{f_j}{\Gamma(1 + j\alpha)} t^{j\alpha}.$$

Analyse the solvability (in $C(I)$) of this VIE when $a_{-1} \neq 0$, $a_j = 0 \ (j \geq 0)$, and $f_{-1} = f_0 = 0$.

Exercise 7.5.11 Is the problem of solving the cordial VIE of the first kind,

$$\int_0^t t^{-1} \varphi(s/t) u(s)\, ds = g(t), \quad t \in I = [0, T],$$

with core $\varphi \in L^1(0, 1)$ and $g \in C^1(I)$, well posed?

Exercise 7.5.12 Let $\mathcal{V}_\varphi^1$ be a (basic) linear cordial Volterra integral operator with given core $\varphi \in L^1(0, 1)$. Show that there does not exist a basic cordial Volterra integral operator $\mathcal{V}_\psi$ with L^1 core ψ so that $\mathcal{V}_\chi := \mathcal{V}_\varphi^1 \mathcal{V}_\psi^1$ is a cordial integral operator with *constant* core $\chi(z)$.

Exercise 7.5.13 Assuming that $g \in C^{m+1}(I)$ for some $m \geq 0$ derive the inversion formula for the solution of the (adjoint) VIE of the first kind,

$$2 \int_t^T t^{-1} \big((s/t)^2 - 1\big)^{-1/2} s u(s)\, ds = g(t), \quad t \in [0, T].$$

Discuss the regularity of the solution, both for $g(T) = 0$ and $g(T) \neq 0$.

Exercise 7.5.14 Suppose that the given functions in the first-kind VIE

$$\int_0^t K(s/t) u(s)\, ds = g(t), \quad t \in I, \tag{7.5.1}$$

are such that $K \in C^1[0, 1]$, $K(1) \neq 0$, $g \in C^1(I)$, $g(0) = 0$. Show that (7.5.1) possesses a solution $u \in C(I)$ if, and only if, u solves the second-kind VIE

$$K(1)u(t) = g'(t) + \int_0^1 K'(s)su(ts)\,ds, \quad t \in I$$

(see also Fenyö & Stolle, 1984, volume 3, pp. 317–319).

Exercise 7.5.15 (*Research Problem*)
Assume that $g \in C^{m+1}(I)$ and $K \in C^{m+1}(D)$, with $K(t, t) \neq 0$ for $t \in I$. For which cores $\varphi \in L^1(0, 1)$ does the cordial first-kind VIE $(\mathcal{V}_\varphi u)(t) = g(t)$, $t \in I$, possess a unique solution $u \in C^m(I)$? (Compare also Vainikko, 2014, Theorem 5.1.)

Exercise 7.5.16 Assume that $H \in C(D)$, $\partial H/\partial H \in C(D)$, $H(t, t) \neq 0$ $(t \in I)$, and $g \in C^1(I)$. Does the first-kind VIE

$$\int_0^t t^{-1}(t - s)^{\alpha-1} H(t, s) e^{i\omega(t-s)} u(s)\,ds = g(t), \quad t \in I \ (0 < \alpha < 1, \ \omega \geq 0),$$

possesses a unique solution $u \in C(I)$?

Exercise 7.5.17 Analyse the solvability in $C(I)$ of the first-kind VIE

$$\int_0^t \left(\log(t/s) \right)^{-1/2} (s/t)^\beta s^{-1} u(s)\,ds = g(t), \quad t \in [0, T],$$

assuming that $\beta > 0$ and $g \in C^1(I)$ (see also Exercise 7.5.8).

7.6 Notes

Section 7.1: *Cordial Volterra Integral Operators*
The (mapping and spectral) theory of cordial Volterra integral operators (and the label 'cordial') are due to Vainikko (2009, 2010a). Section 6 in Vainikko (2009) contains a good geometrical description of the sets $\{\hat\varphi(\lambda) : \text{Re}(\lambda) \geq 0\}$, with $\hat\varphi(\lambda)$ defined in (7.1.15) certain cores $\varphi(z)$. For a discussion of cordial Volterra operators with $\varphi \in L^p(0, 1)$ $(p > 1)$ see Vainikko (2010a, pp. 202+) and Yang (2015).

Atkinson (1976; see also Atkinson, 1997, p. 20) discusses an Abel-type Volterra integral operator with kernel singularity $(t^2 - s^2)^{-1/2}$ that is equivalent to a non-compact cordial Volterra integral operator; this example inspired Vainikko's studies of general cordial integral operators. Integral operators that are equivalent to cordial Volterra integral operators were also studied by Han (1994; Chandler & Graham, 1988 analysed an analogous non-compact Fredholm integral operator). The paper by Graham & Sloan (1979) deals with non-compact Fredholm integral operators.

Sections 7.2: *Linear Cordial VIEs*
In the papers by Vainikko (2012, 2014) the theory of second-kind cordial VIEs (Vainikko, 2009, 2010a) is extended to first-kind cordial VIEs. The regularity of solutions to linear second-kind cordial VIEs is studied in Vainikko (2009, 2010a); see also Yang (2015). Regularity results for first-kind cordial VIEs are established in Vainikko (2012, 2014). See also Atkinson (1974) on the regularity of solutions of Abel-type first-kind VIEs of a somewhat more general form that (7.2.4) (compare also Section 2.2.3 in the present book). Cordial VIEs with cores $\varphi \in L^P(0, 1)$ ($p > 1$) are analysed in Vainikko (2010a) and Yang (2015).

The connection between a class of third-kind VIEs and cordial VIEs of the second kind is explored in Seyed Allaei, Yang & Brunner (2015). Von Wolfersdorf (2007) discusses certain non-compact Volterra integral operators (with kernel singularities different from the ones considered in the present chapter) that arise in the analysis of third-kind VIEs (see also Section 1.5.2).

Section 7.3: *Non-Linear Cordial VIEs*
Sato (1953b) studied the solvability of a class of non-linear VIEs of the third kind that are equivalent to non-linear cordial VIEs of the second kind. Vainikko (2011) contains a general theory of non-linear second-kind cordial VIEs; the paper includes regularity results for their solutions.

Section 7.4: *Cordial VIEs with Highly Oscillatory Kernels*
In the paper by Brunner (2014) a section is dedicated to the study of properties of solutions of cordial VIEs with highly oscillatory kernels.

Numerical Analysis of Cordial VIEs
The applicability spline collocation methods to (linear and non-linear) second-kind cordial VIEs by is described and analysed in Vainikko (2010b, 2011) and in Diogo & Vainikko (2013). The paper by Seyed Allaei, Yang & Brunner (2016) presents a convegence analysis of collocation methods for certain third-kind VIEs that are equivalent to cordial VIEs. The numerical analysis of cordial VIEs of the first kind remains to be studied.

8

Volterra Integral Operators on Banach Spaces

Summary

We discuss a selection of topics related to Volterra integral operators on Banach spaces like C^d ($d \geq 0$), the Hölder spaces C^β ($\beta \in (0, 1)$) and the Lebesgue spaces L^p ($p \geq 1$). Among these topics are the mapping properties, the existence of resolvent operators and the singular values of such integral operators. In addition, we review classical and recent results on the asymptotic behaviour of the norms of powers of such integral operators. Relevant definitions and theorems from Functional Analysis are stated in the Appendix.

8.1 Mapping Properties

8.1.1 Volterra Integral Operators on $C(I)$

We first consider the linear Volterra integral operator $\mathcal{V}_\alpha : C(I) \to C(I)$,

$$(\mathcal{V}_\alpha f)(t) = \int_0^t k_\alpha(t - s) K(t, s) f(s)\, ds, \quad t \in I = [0, T] \ (0 \leq \alpha \leq 1),$$

$$(8.1.1)$$

as an operator on the Banach space $C(I)$ (with norm $\|f\|_\infty = \max\{|f(t)| : t \in I\}$), assuming that $K \in C(D)$ and

$$k_\alpha(t - s) := \begin{cases} (t - s)^{\alpha - 1} & \text{if } 0 < \alpha \leq 1, \\ \log(t - s) & \text{if } \alpha = 0. \end{cases}$$

As in Chapter 1 we employ the short notation $\mathcal{V} := \mathcal{V}_1$ when $\alpha = 1$.

Theorem 8.1.1 *For any $\alpha \in [0, 1]$ and $K \in C(D)$ the linear Volterra integral operator $\mathcal{V}_\alpha$ is a compact (and hence a bounded) operator on $C(I)$.*

279

Proof As we have already seen in Sections 1.2.1 and 1.3.2, the boundedness of $\mathcal{V}_\alpha$ is a consequence of the facts that $\|K\|_\infty = \max\{|K(t,s)| : (t,s) \in D\} =: \bar{K} < \infty$ and $\int_0^t |k_\alpha(t-s)|ds < \infty$ (since the interval $I = [0,T]$ is bounded). This result can also be established by deriving an upper bound for the operator norm $\|\mathcal{V}_\alpha\| = \sup\{\|\mathcal{V}_\alpha f\|_\infty : \|f\|_\infty \leq 1\}$ (cf. Definition A.2.2(iii)). Under the above assumptions on its kernel operator $\mathcal{V}_\alpha$ is a compact operator on $C(I)$ (see also Section A.2.2). $\qquad\square$

As we have seen in Section 7.1.1 (cf. Theorem 7.1.10(c)) the compactness result of Theorem 8.1.1 is in general not true for the cordial Volterra integral operator

$$(\mathcal{V}_\varphi f)(t) := \int_0^t t^{-1}\varphi(s/t)K(t,s)f(s)\,ds, \quad t \in I, \qquad (8.1.2)$$

with core $\varphi \in L^1(0,1)$ and $K \in C(D)$. While $\mathcal{V}_\varphi : C(I) \to (C(I)$ is again a bounded operator it is, in contrast to $\mathcal{V}_\alpha$, a compact operator only if $K(0,0) = 0$. For any K with $K(0,0) \neq 0$ (for example, for non-zero constant kernels) it is not compact. As we have seen, a particular example is the Volterra integral operator with the weakly singular kernel $(t^2 - s^2)^{-1/2}$,

$$(\mathcal{V}_\varphi f)(t) = \int_0^t (t^2 - s^2)^{-1/2}f(s)\,ds = \int_0^t t^{-1}\varphi(s/t)f(s)\,ds, \quad t \in I,$$

with L^1 core $\varphi(z) = (1 - z^2)^{-1/2}$ and kernel $K(t,s) \equiv 1$ (cf. Section 1.5.3). For the sake of comparison with classical Volterra integral operators we summarise these properties of the cordial Volterra integral operator (8.1.2) in the following theorem.

Theorem 8.1.2 *Assume that $K \in C(D)$ and $\bar{K} := \max\{|K(t,s)| : (t,s) \in D\}$. Then the corresponding linear cordial Volterra integral operator $\mathcal{V}_\varphi : C(I) \to C(I)$ defined by (8.1.2) is bounded, with $\|\mathcal{V}_\varphi\|_\infty \leq \|\varphi\|_1 \bar{K}$. If $K(0,0) = 0$ the operator is compact; it is non-compact when $K(0,0) \neq 0$.*

8.1.2 Volterra Integral Operators on Hölder and L^p-spaces

The first detailed study of the compactness of linear Volterra integral operators $\mathcal{V}_\alpha$ with $\alpha = 1$ and $K \in C(D)$ or $K \in L^p(D)$ $(p > 1)$ appears to be the one by Stefan Banach in his book of 1932 (see pp. 98–99). If $p = 2$ this integral operator is a particular case of a *Hilbert–Schmidt integral operator* (studied by David Hilbert in a paper of 1904 and by Erhard Schmidt, who improved and simplified Hilbert's approach, in 1907. See also Stewart, 2011).

If $K \in L^2((0, T) \times (0, T))$ the (Hilbert–Schmidt) norm of the integral operator is given by

$$\|\mathcal{V}\|_2 = \left(\int_0^T \left(\int_0^T |K(t, s)|^2 \, ds \right)^2 \right)^{1/2}. \qquad (8.1.3)$$

Theorem 8.1.3 *Let $\mathcal{V}$ be the Volterra integral operator defined in 8.1.1), with $\alpha = 1$. Then:*

(a) *If $K \in C(D)$, $\mathcal{V}$ is a compact (and hence bounded) operator from $L^p(0, T)$ to $L^p(0, T)$ for any $p \in [1, \infty)$.*
(b) *If $K \in L^2(D)$, $\mathcal{V}$ is a compact (and hence bounded) operator from $L^2(0, T)$ to $L^2(0, T)$, with norm (8.1.3).*

The reader is invited to show that the statements in Theorem 8.1.2 are in general no longer true for $\mathcal{V}_\alpha$ with any $\alpha \in [0, 1)$.

Remark 8.1.4 Results on the mapping properties of linear Volterra integral operators $\mathcal{V}$ with general L^p kernels can be found in Gripenberg, Londen & Staffans (1990, cf. Section 9.7).   ◇

We now present a selection of results on the mapping properties of the Volterra integral operator

$$(\mathcal{V}_\alpha f)(t) := \int_0^t (t - s)^{\alpha-1} K(t, s) f(s) \, ds, \quad t \in I = [0, T],$$

where $0 < \alpha < 1$. These properties were studied extensively in the seminal paper by Hardy & Littlewood (1928). A good overview of the results is given in the book by Gorenflo & Vessella (1991, Chapter 4).

Suppose first that $\mathcal{V}_\alpha$ acts on a Hölder space $C^\beta(I)$ with $0 < \beta < 1$. The following result can be found in Gorenflo & Vessella (1991, pp. 70–72).

Theorem 8.1.5 *Assume that $f \in C^\beta(I)$ with $0 < \beta < 1 - \alpha$. If $K \in C(D)$ and $f(0) = 0$ then $\mathcal{V}_\alpha f$ is in $C^{\alpha+\beta}(I)$.*

Proof Without essential loss of generality, assume that $K(t, s) \equiv 1$. The core of the proof consists in deriving an upper bound for

$$|(\mathcal{V}_\alpha f)(t) - (\mathcal{V}_\alpha f)(t - h)| \quad (0 < h \le t \le T),$$

where the constant in the bound depends on α and β. This can be done by using the decomposition

$$\int_0^t (t-s)^{\alpha-1} f(s)\,ds = \int_0^h (t-s)^{\alpha-1} f(s)\,ds + \int_h^t (t-s)^{\alpha-1} f(s)\,ds,$$

together with the result

$$\int_\tau^t (t-\tau)^{\alpha-1} f(t)\,ds = \frac{1}{\alpha}(t-\tau)^\alpha f(t)$$

with $\tau = 0$ and $\tau = h$. We hence obtain (for $K(t,s) \equiv 1$ in (8.1.1))

$$
\begin{aligned}
&(\mathcal{V}_\alpha f)(t) - (\mathcal{V}_\alpha f)(t-h) \\
&= \int_0^t (t-s)^{\alpha-1} f(s)\,ds - \int_0^{t-h} (t-h-s)^{\alpha-1} f(s)\,ds \\
&= \int_0^t s^{\alpha-1} f(t-s)\,ds - \int_h^t (t-s)^{\alpha-1} f(s-h)\,ds \\
&= \frac{1}{\alpha}\big(t^\alpha - (t-h)^\alpha\big) f(t) - \int_0^h s^{\alpha-1}\big(f(t) - f(t-s)\big)\,ds \\
&\quad - \int_h^t \big(s^{\alpha-1} - (s-h)^{\alpha-1}\big)\big(f(t) - f(t-s)\big)\,ds.
\end{aligned}
$$

By establishing upper bounds for the absolute values of the three terms on the right-hand side of the above equation (using the concavity of the fractional powers) we are led to the mapping result in Theorem 8.1.5. (The additional technical details can be found in the paper cited above.) The step from the constant kernel to a general continuous kernel $K(t,s)$ is then straightforward.

We note that the above result is no longer true when $\alpha + \beta = 1$ (cf. Exercise 8.6.2). $\qquad\qquad\square$

We now consider the mapping properties of the Volterra integral operator $\mathcal{V}_\alpha$, with $0 < \alpha \leq 1$ and $K \in C(D)$, when it acts on $L^p(0,T)$ with $1 < p \leq \infty$. Recall that for $1 \leq p < q \leq \infty$, the inclusion relation $L^q(0,T) \subset L^p(0,T)$ holds (cf. Section A.1.3).

Theorem 8.1.6 *Assume that $f \in L^p(0,T)$ with $p > 1$. If $1 \leq p \leq 1/\alpha$ and $K \in C(D)$, then $\mathcal{V}_\alpha f \in L^q(0,T)$ for every $q \in [1, p/(1-\alpha p))$.*

Proof We leave the proof of Theorem 8.1.6 as an exercise (see also Hardy & Littlewood, 1928, p. 575 or Gorenflo & Vessella, 1991, pp. 80–81). $\qquad\square$

Remark 8.1.7 The image of $f \in L^p(0, T)$ under $\mathcal{V}_\alpha$ with $0 < \alpha < 1$ is not always in an L^q space. We shall see below that for $p > 1/\alpha$ $(0 < \alpha < 1)$ the image of $f \in L^p(0, T)$ lies in a Hölder space: $\mathcal{V}_\alpha f \in C^{\alpha - 1/p}(I)$. ◇

Theorem 8.1.8 *If* $f \in L^p(0, T)$ *with* $p > 1/\alpha$ $(0 < \alpha < 1)$ *then* $\mathcal{V}_\alpha f \in C^{\alpha - 1/p}(I)$.

Proof The above mapping property of $\mathcal{V}_\alpha$ can be established by using Young's inequality (with $1/p + 1/q = 1$) together with Hölder's inequality (cf. Appendix A.1.3): setting $\bar{K} := \max \{ |K(t, s)| : (t, s) \in D \}$, we find that

$$\|\mathcal{V}_\alpha f\|_\infty \le \frac{\bar{K} T^{\alpha - 1/p}}{\Gamma(\alpha)} \left(\frac{\alpha p - 1}{p - 1} \right)^{1/1/p - 1} \|f\|_p .$$

The technical details may be found in Gorenflo & Vessella (1991, pp. 67–68). See also Exercise 8.6.6 for the case $p = 1/\alpha$. □

Hardy & Littlewood (1928, p. 578) showed that the following is true when $\mathcal{V}_\alpha$ acts on the space $L^1(0, T)$. The proof is again based on Young's inequality.

Theorem 8.1.9 *For* $f \in L^1(0, T)$ *we have* $\mathcal{V}_\alpha f \in L^q(0, T)$, *with* $q = \dfrac{1}{1 - \alpha} - \varepsilon$ *for* $\varepsilon \in (0, \alpha/(1 - \alpha)]$.

Remark 8.1.10 We note (cf. Gorenflo & Vessella, 1991, Section 4.1; see also Exercise 8.4.1) that $\mathcal{V}_\alpha$ is not a continuous operator from $L^1(0, T)$ to $L^{1/(1-\alpha)}(0, T)$. Also, it is not a continuous operator from $L^{1/\alpha}(0, T)$ to $L^\infty(0, T)$. ◇

We conclude this section by describing some of the mapping properties of the linear cordial Volterra integral operator

$$(\mathcal{V}_\varphi f)(t) := \int_0^t t^{-1} K(t, s) f(s) \, ds, \quad t \in I, \tag{8.1.4}$$

with core $\varphi \in L^1(0, 1)$ and $K \in C(D)$. The following result can be found in Vainikko (2010a; see also Yang, (2015 for cordial Volterra integral operators with cores $\varphi \in L^p(0, 1)$ with $p > 1$).

Theorem 8.1.11 *The cordial Volterra integral operator (8.1.4) has the following properties:*

(a) *If $\varphi \in L^1(0,1)$, $K \in C^m(D)$ $(m \geq 0)$ then φ is a bounded operator from $C^m(I)$ into $C^m(I)$. For kernels (t,s) with $K(0,0) = 0$ the operator is compact and quasi-nilpotent.*

(b) *If $\varphi \in L^1(0,1)$ and $K \in C(D)$, with $K(0,0) = 0$, then $V_\varphi : L^\infty(0,T) \to C(I)$ is compact.*

Proof The statement (a) is a consequence of Theorems 7.1.9 and 7.1.10(c). The proof of (b), as well as further mapping properties (for instance, when $\varphi \in L^p(0,1)$ with $p > 1$ and V_φ acts on a Hölder space) can be found in Vainikko (2010a, Section 2). □

8.2 Quasi-Nilpotency

Definition 8.2.1 Let X be a Banach space. An operator $A \in \mathcal{L}(X)$ is called *nilpotent* if $A^m = 0$ for some finite positive integer m. It is said to be *quasi-nilpotent* if $\lim_{m\to\infty} \|A^m\|^{1/m} = 0$. ◇

The quasi-nilpotency of Volterra integral operators (8.1.1) with continuous or weakly singular kernels is a consequence of Theorems 1.2.3 (see also Remark 1.2.4), Theorem 1.3.5, and Theorem 1.3.10; compare also Theorem A.2.31).

Theorem 8.2.2 *Assume that the kernel $K(t,s)$ of the Volterra integral operator $V_\alpha : C(I) \to C()I)$ defined in (8.1.1) is continuous on D. Then V_α is quasi-nilpotent for all $\alpha \in [0,1]$, and hence its spectrum is $\sigma(V_\alpha) = \{0\}$.*

The quasi-nilpotency of the Volterra integral operator with delay argument $\theta(t)$,

$$(V_\theta f)(t) := \int_0^{\theta(t)} f(s)\,ds, \quad t \in [0,1], \tag{8.2.1}$$

where $\theta \in C[0,1]$ satisfies $\theta(t) \leq t$ for all $t \in [0,1]$, was studied by Lyubich (1984) and Whitley (1987; see also Domasov, 2007), both as an operator on $C[0,1]$ and on $L^p(0,1)$ $(1 \leq p < \infty)$. While Lyubich proved the sufficiency of the condition $\theta \leq t$ and conjectured its necessity for quasi-nilpotency, Whitley also proved the latter. We summarise their results in the following theorem but refer the reader to Whitley's paper for the proof.

Theorem 8.2.3 *(a) Let θ be a continuous function from* [0, 1] *to* [0, 1]. *A necessary and sufficient condition for* $\mathcal{V}_\theta$: *C*[0, 1] → *C*[0, 1] *to define a quasi-nilpotent operator is that* $\theta(t) \leq t$ *for all* $t \in [0, 1]$.

(b) If θ is a (Lebesgue) measurable map of [0, 1] *into* [0, 1], *a necessary and sufficient condition for* $\mathcal{V}_\theta$ *to define a quasi-nilpotent operator on* $L^p(0, 1)$ *for any p with* $1 \leq p < \infty$ *is that* $\theta(t) \leq t$ *for almost every* $t \in [0, 1]$.

The particular case corresponding to $\theta(t) = t^\beta$ $(0 < \beta < 1)$ was considered by Whitley (1987) and Zima (1992); see also Exercise 8.6.9.

8.3 Resolvent Kernels and Resolvent Operators

8.3.1 Resolvent Kernels for Second-Kind VIEs

In Section 6.1 we described various asymptotic properties of resolvent kernels $r(t - s)$ corresponding to second-kind VIEs with convolution kernels $k(t - s)$. We now turn to analogous results for the resolvent kernel $R(t, s)$ of the kernel $K(t, s)$ describing the (scalar) linear VIE

$$u(t) + \int_0^t K(t, s)u(s)\,ds = f(t), \quad t \geq 0. \tag{8.3.1}$$

The result plays a role in the analysis of its non-linearly perturbed counterpart,

$$u(t) + \int_0^t K(t, s)\big(u(s) + G(s, u(s))\big)\,ds, \quad t \geq 0. \tag{8.3.2}$$

The following definition, introduced (for matrix kernels) in Miller (1971, Chapter IV; see also Gripenberg, Londen & Staffans, 1990, Section 9.2) describes the kind of kernels we have in mind.

Definition 8.3.1 Assume that $1 < p < \infty$, and let q be such that $1/p + 1/q = 1$. A function $K = K(t, s)$ is said to be a Volterra kernel (on $D := \{(t, s) : 0 \leq s \leq t \leq T\}$) of *type* (L^p, T) if

(i) $K(t, s)$ is integrable in t and s, and $K(t, s) = 0$ when $s > t$;

(ii) for almost all $t \in [0, T]$, $K(t, \cdot) \in L^q(0, T)$;

(iii) for almost all $s \in [0, T]$, $K(\cdot, s) \in L^p(0, T)$;

(iv) $\int_0^T \left(\int_0^T |K(t, s)|^q \, ds \right)^{p/q} dt < \infty, \ \int_0^T \left(\int_0^T |K(t, s)|^p \, dt \right)^{q/p} ds < \infty.$

◇

Example 8.3.2

(a) If $K \in C(D)$, K is of type (L^p, T) for all $p > 1$.
(b) If $K \in L^2(D)$, K is of type (L^2, T).

Assume that $K(t, s)$ is a Volterra kernel of type (L^p, T), and consider the functions $R_n = R_n(t, s)$ defined by $R_1(t) := K(t, s)$ and

$$R_{n+1}(t, s) := \int_s^t K(t, v) R_n(v, s) \, dv, \quad (t, s) \in Di \, (n \geq 1).$$

It is easy to verify (by using Hölder's inequality; cf. Section A.1.3) that each $R_n(t, s)$ is also a Volterra kernel of type (L^p, T) and that $R(t, s) := \lim_{n \to \infty} R_n(t, s)$ possesses the same property. Moreover, as shown in the next theorem (Miller, 1971, pp. 197–201), $R(t, s)$ satisfies abstract resolvent equations that are the analogues of the resovent equations (1.2.9) and (1.2.10) for $K \in C(D)$.

Theorem 8.3.3 *Let* $1 < p < \infty$ *and assume that* $K = K(t, s)$ *is a Volterra kernel of type* (L^p, T). *Then the following statements are true:*

(a) *There exists a Volterra kernel* $R = R(t, s)$ *of type* (L^p, T) *which solves the resolvent equation*

$$R(t, s) = -K(t, s) + \int_s^t K(t, v) R(v, s) \, dv \text{ almost everywhere in } D.$$

(b) *This function* R, *the resolvent kernel of* K, *also solves the complementary resolvent equation*

$$R(t, s) = -K(t, s) + \int_s^t R(t, v) K(v, s) \, dv \text{ almost everywhere in } D.$$

Remark 8.3.4 If D is extended to $\{(t, s) : 0 \leq s \leq t < \infty\}$, the above theorem remains valid for Volterra kernels $K(t, s)$ of type L_{loc}^p (that is, if $K(t, s)$ is of type (L^p, T) for any $T < \infty$).   ◇

8.3.2 Resolvent Kernels for First-Kind VIEs

We have seen in Theorem 1.5.1 (cf. equation (1.5.2)) that the solution of the Abel integral equation

$$\int_0^t \frac{(t - s)^{\alpha - 1}}{\Gamma(\alpha)} u(s) \, ds = g(t), \quad t \in I \ (0 < \alpha < 1), \tag{8.3.3}$$

is given by the 'resolvent representation' (or variation-of-constants formula)

$$u(t) = \frac{d}{dt} \left(\int_0^t \frac{(t-s)^{-\alpha}}{\Gamma(1-\alpha)} g(s) \, ds \right). \tag{8.3.4}$$

In the following we will refer to the kernel $(t-s)^{-\alpha} / \Gamma(1-\alpha)$ in (8.1.6) as the first-kind *resolvent* (or resolvent function) of the first kind of the VIE (8.3.5) (cf. Gripenberg, 1980b and Gripenberg, Londen & Staffans, 1991, Section 5.5).

In (7.2.6) an analogous variation-of-constants formula was found for the cordial VIE of the first kind,

$$\frac{1}{\Gamma(\alpha)} \int_0^t t^{-1} \left(1 - (s/t)^\gamma\right)^{\alpha-1} t^{1-\gamma(1-\alpha)} u(s) \, ds = g(t), \quad t \in I, \tag{8.3.5}$$

where $\gamma > 1$, $0 < \alpha < 1$ and $\gamma(1 - \alpha) = 1$: it is

$$u(t) = \frac{d}{dt} \left(\frac{\gamma}{\Gamma(1/\gamma)} \int_0^t \left(1 - (s/t)^\gamma\right)^{-(1-1/\gamma)} (s/t)^{\gamma-1} g(s) \, ds \right), \quad t \in (0, T]. \tag{8.3.6}$$

In analogy to the terminology introduced above we will call

$$\left(\gamma/\Gamma(1/\gamma)\right)\left(1 - (s/t)^\gamma\right)^{-(1-1/\gamma)} (s/t)^{\gamma-1}$$

in (8.3.6) the first-kind resolvent function associated with the cordial VIE (8.3.5).

The solution representations (8.3.5) and (8.3.8) raise the following question: does there exist a (possibly generalised) function $R = R(t, s)$ so that the solution of a first-kind VIE with (continuous or weakly singular) kernel can be written in the form.

$$u(t) = \frac{d}{dt} \left(\int_0^t R(t, s) g(s) \, ds \right), \quad t \in I ?$$

In particular, can the solution of such a VIE with a convolution kernel,

$$\int_0^t k(t-s) u(s) \, ds = g(t), \quad t \in I, \tag{8.3.7}$$

be expressed by

$$u(t) = \frac{d}{dt} \left(\int_0^t r(t-s) g(s) \, ds \right), \quad t \in I, \tag{8.3.8}$$

where the resolvent function r solves the *first-kind resolvent equation*

$$\int_0^t k(t-s) r(s) \, ds = \int_0^t r(t-s) k(s) \, ds = 1, \quad t \in I ? \tag{8.3.9}$$

In order to understand why the kernel k and the resolvent r might be related by an equation of the form (8.3.9) we follow the argument presented in

Gripenberg, Londen & Staffans (1990, Section 5.5). For a given small constant $\varepsilon > 0$ the solution u_ε of the singularly perturbed second-kind VIE

$$\varepsilon u(t) + \int_0^t k(t-s)u(s)\,ds = f(t), \quad t \in I, \tag{8.3.10}$$

is formally given by

$$u_\varepsilon(t) = \varepsilon^{-1}\left(f(t) - \int_0^t \rho_\varepsilon(t-s)f(s)\,ds \right), \quad t \in I,$$

where $\rho_\varepsilon(t)$ denotes the resolvent kernel of $k_\varepsilon(t) := \varepsilon^{-1}k(t-s)$. By (1.2.24) and (1.2.25) it satisfies the resolvent equations

$$\varepsilon\rho_\varepsilon(t) + \int_0^t \rho_\varepsilon(t-s)k(s)\,ds = \varepsilon\rho_\varepsilon(t) + \int_0^t k(t-s)\rho_\varepsilon(s)\,ds = k(t), \quad t \in I.$$

It can be shown that the function r_ε defined by

$$r_\varepsilon(t) := \varepsilon^{-1}\left(1 - \int_0^t \rho_\varepsilon(s)\,ds \right),$$

solves the equations

$$\varepsilon r_\varepsilon(t) + \int_0^t r_\varepsilon(t-s)k(s)\,ds = \varepsilon r_\varepsilon(t) + \int_0^t k(t-s)r_\varepsilon(s)\,ds = 1, \quad t \in I,$$

and that the solution u_ε of (8.3.12) can formally be written as

$$u_\varepsilon(t) = \frac{d}{dt}\left(\int_0^t r_\varepsilon(t-s)f(s)\,ds \right), \quad t \in I.$$

If $\lim_{\varepsilon \to 0^+} r_\varepsilon(t)$ exists and is some (generalised) function (e.g. a measure), we obtain the solution representation (8.3.10) and the resolvent equations (8.3.11).

Definition 8.3.5 Assume that $k \in L^1(0, T)$ (or $k \in L^1_{loc}(\mathbb{R}^+)$). If k and $r \in L^1(0, T)$ (or $r \in L^1_{loc}(\mathbb{R}^+)$) satisfy (8.1.9), then r is called a resolvent function of the first kind of the kernel k.   ◇

8.4 Singular Values of Volterra Integral Operators

If the kernel $K_\alpha(t, s) := (t-s)^{\alpha-1}K(t, s)$ of the Volterra integral operator $\mathcal{V}_\alpha$ describing the first-kind VIE $\mathcal{V}_\alpha u = g$ is such that $0 < \alpha \le 1$ and $K \in C(D)$, then the VIE is an example of an operator equation $Ax = f$ with *compact* operator $A \in \mathcal{L}(X)$, where X is a Banach space or a *Hilbert* space. Problems of this kind are usually ill posed.

The following definition can be found, for example, in Kress (2014, Section 15.4).

Definition 8.4.1 Assume that X and Y are (infinite-dimensional) Hilbert spaces, and let $A : X \to Y$ be a linear compact operator, with A^* denoting the adjoint operator. Then the square roots μ_n of the eigenvalues of the self-adjoint compact operator A^*A are called the *singular values* of the operator A. The multiplicity of a singular value μ_n is given by the dimension of the kernel of the operator $(\mu_n^2 \mathcal{I} - A^*A)$. ◇

We illustrate the above definition by looking at the singular values of two Volterra integral operators.

Example 8.4.2 (Faber & Wing, 1988, p. 58)
The Volterra integral operator $\mathcal{V} : L^2(0, 1) \to L^2(0, 1)$,

$$(\mathcal{V}f)(t) = \int_0^t f(s)\,ds, \quad t \in [0, 1],$$

has as its adjoint the operator

$$(\mathcal{V}^* f(t) = \int_t^1 f(s)\,ds, \quad t \in [0, 1].$$

Therefore, the operator $\mathcal{V}^*\mathcal{V}$ is given by

$$(\mathcal{V}^*\mathcal{V}f)(t) = \int_t^1 \int_0^s f(v)\,dv\,ds, \quad t \in [0, 1].$$

It is readily verified that the eigenvalue problem for the (self-adjoint) operator $\mathcal{V}^*\mathcal{V}$,

$$(\mathcal{V}^*\mathcal{V}\phi)(t) = \mu^2 \phi(t), \quad t \in [0, 1] \ (\phi(t) \not\equiv 0),$$

is equivalent to the Sturm–Liouville problem

$$\mu^2 \phi''(t) + f(t) = 0, \quad \phi'(0) = \phi(1) = 0.$$

It has the eigenvalues $\mu_n = \dfrac{2}{(2n + 1)\pi}$ ($n \in \mathbb{N}$). As $n \to \infty$, these singular values of $\mathcal{V}$ behave like $\mu_n \sim n^{-1}$.

Example 8.4.3 (Faber & Wing, 1986b, pp. 748–753)
Let $\mathcal{V}_\alpha : L^2(0, 1) \to L^2(0, 1)$ ($\alpha > 0$) be given by

$$(\mathcal{V}_\alpha f)(t) = \int_0^t \frac{(t - s)^{\alpha-1}}{\Gamma(\alpha)} f(s)\,ds, \quad t \in [0, 1].$$

As shown in the above paper, a rather lenghty calculation of the eigenvalues of the (self-adjoint) operator $V_\alpha^* V_\alpha$ reveals that the singular values $\mu_n(\alpha)$ of V_α behave like $\mu_n(\alpha) \sim 1/n^\alpha$ as $n \to \infty$.

For $\alpha = 1$ we recover the result of Example 8.4.2. If $\alpha = 2$ the singular values decay faster, namely like $\mu_n(2) \sim n^{-2}$. Comparing these observations with Example 1.4.9 ($r = \alpha = 1$ and $r = \alpha = 2$, respectively), we see that in the former case the degree of ν-smoothing is $\nu = 1$, while in the latter case we have $\nu = 2$. In other words, if the singular values decay slowly (for example, when $\alpha = 1$) we have a *mildly ill-posed* problem, while a more rapid asymptotic decay of the singular values ($\alpha = 2$) identifies a more *severely ill-posed* problem. We will make this observation more precise following the proof of Theorem 8.4.5.

The following two theorems and their proofs may be found in e.g. Kato (1995, pp. 26–262 or in Kress (2014, pp. 311–312; see also Section A.2.3 for the basic definitions).

Theorem 8.4.4 *Assume that A is a linear compact operator from a Hilbert space X to a Hilbert space Y, and let $\{\mu_n\}$ be the sequence of the (non-zero) singular values of A (repeated according to their multiplicities). Then there exist orthonormal sequences $\{x_n\}$ in X and $\{y_n\}$ in Y so that the following statements are true:*

(a) $Ax_n = \mu_n y_n$ *and* $A^* y_n = \mu_n x_n$ ($n \in \mathbb{N}$).
(b) *Each* $x \in X$ *possesses the* singular value decomposition

$$x = \sum_{n=1}^\infty \langle x, x_n \rangle x_n + Px,$$

where P is the orthogonal projector of X onto the null space of A. Moreover,

$$Ax = \sum_{n=1}^\infty \mu_n \langle x, x_n \rangle y_n .$$

Proof Let $\{x_n\}$ be an orthonormal sequence of eigenelements of the self-adjoint operator $A^* A$ (observe that since A is a compact operator, A^* is also compact). It follows that $A^* Ax_n = \mu_n^2 x_n$ ($n \geq 1$). Define the orthonormal sequence $\{y_n\}$ by $y_n := \mu^{-1} Ax)_n$ ($n \geq 1$). A simple calculation then shows that $Ax_n = \mu_n y_n$ and $A^* y_n = \mu_n x_n$ ($n \geq 1$), thus proving assertion (a).

Since $A^* A$ is a compact self-adjoint operator on the Hilbert space X we may employ the well-known result that any element $x \in X$ has an expansion of the form

$$x = \sum_{n=1}^{\infty} \langle x, x_n \rangle x_n + Qx,$$

where Q is the orthogonal projection operator $X \rightarrow \ker(A^*A)$ (see end of Section A.2.1, or Kress, 2014, Section 15.3). If $y \in \ker(A^*A)$ it follows that

$$\langle Ay, Ay \rangle = \langle y, A^*Ay \rangle = 0,$$

and therefore we obtain $\ker(A^*A) = \ker A$. Assertion (b) follows by considering Ax and using (a). $\square$

The result of Theorem 8.4.4 allows us to express the solution of the operator equation $Ax = y$ in terms of the singular system (μ_n, x_n, y_n), as shown in Kress (2014, pp. 311–312).

Theorem 8.4.5 *Let the linear compact operator $A : X \rightarrow Y$ (where X and Y are infinite-dimensional Hilbert spaces) possess the singular system (μ_n, x_n, y_n) (with $\mu_n \neq 0$) described in Theorem 8.4.4. Then the equation $Ax = y$ has a solution in X if, and only if, y lies in $(\ker A^*)^\perp$, the orthogonal complement of the null space of A^*, and satisfies*

$$\sum_{n=1}^{\infty} \mu_n^{-2} |\langle y, y_n \rangle| < \infty.$$

Under these conditions the solution is given by

$$x = \sum_{n=1}^{\infty} \mu_n^{-1} \langle y, y_n \rangle x_n.$$ (8.4.1)

Proof   This is left as an exercise. $\square$

Theorem 8.4.5 provides the mathematical foundation for the above observation regarding the role of the rate of decay of the singular values of a compact linear operator in characterising the degree of ill-posedness of a 'first-kind' operator equation $Ax = y$ (recall Example 8.4.2). If x is the solution of the given equation and x_ε denotes the solution of the perturbed system $Ax = y + \varepsilon y_n =: y_\varepsilon$ (where y_n is an element of the orthonormal sequence in the singular system (μ_n, x_n, y_n)), it follows from (8.4.1) that $\|x_\varepsilon - x\| / \|y_\varepsilon - y\| = \mu_n^{-1}$. This shows that (since $\mu_n \rightarrow 0$, as $n \rightarrow \infty$; see Theorem A.2.28) a faster rate of decay of the singular values of A magnifies the influence of the error εy_n in y on the solution of $Ax = y$.

8.5 Norms of Powers of $\mathcal{V}$

In Sections 1.2.1 and 1.3.2 we encountered the Neumann series representations

$$u = f + \sum_{m=1}^{\infty} \mathcal{V}^m f \quad \text{and} \quad u = f + \sum_{m=1}^{\infty} \mathcal{V}_\alpha^m f \qquad (8.5.1)$$

for the solutions of the VIEs $u = f + \mathcal{V}u$ and $u = f + \mathcal{V}_\alpha u$ corresponding the Volterra integral operators

$$(\mathcal{V}u)(t) := \int_0^t K(t,s)u(s)\,ds \qquad (8.5.2)$$

and

$$(\mathcal{V}_\alpha u)(t) := \int_0^t (t-s)^{\alpha-1} K(t,s)u(s)\,ds \quad (0 < \alpha < 1) \qquad (8.5.3)$$

respectively, on $C(I)$.

In the proofs of the uniform convergence of the associated Neumann series we employed the norm bounds

$$\|\mathcal{V}^m\| \le \|\mathcal{V}\|^m \quad \text{and} \quad \|\mathcal{V}_\alpha^m\| \le \|\mathcal{V}_\alpha\|^m \quad (m \ge 2).$$

There arises the natural question regarding the asymptotic behaviour of the *norms of the powers* of $\mathcal{V}^m$ and $\mathcal{V}_\alpha^m$ of these Volterra operators, as $m \to \infty$.

Motivated by Problem 188 in Halmos (1982) – which asks to find the norm of $\mathcal{V}$ (cf. (8.5.2)) with constant kernel $K(t,s) \equiv 1$ and with the underlying function space now being $L^2(0,1)$ – the above question has received considerable attention since the mid-1990s. Before describing this development in some detail in the following sections, we note that the iterates of $\mathcal{V}$ and $\mathcal{V}_\alpha$ in (8.5.2) and (8.5.4) are given by

$$(\mathcal{V}^m f)(t) = \frac{1}{(m-1)!} \int_0^t (t-s)^{m-1} f(s)\,ds \quad (m \ge 1) \qquad (8.5.4)$$

when $\alpha = 1$, and

$$(\mathcal{V}_\alpha^m f)(t) = \frac{\Gamma^m(\alpha)}{\Gamma(m\alpha)} \int_0^t (t-s)^{m\alpha-1} f(s)\,ds \quad (m \ge 1) \qquad (8.5.5)$$

when $0 < \alpha < 1$.

8.5.1 The Basic Volterra Integral Operator $\mathcal{V}$

In the first edition (1967) of his celebrated Hilbert space problem book of 1982, Halmos showed that the (operator) norm of the basic Volterra integral operator $\mathcal{V} : \ L^2(0,1) \to L^2(0,1)$ given by

$$(\mathcal{V}f)(t) := \int_0^t f(s)\,ds, \quad t \in [0, 1], \tag{8.5.6}$$

is $\|\mathcal{V}\|_2 = \frac{2}{\pi}$ (cf. Halmos, 1982, pp. 300–301). Recall that if $A : L^2(0, 1) \to L^2(0, 1)$ is a linear operator, its (operator) norm is defined by

$$\|A\|_2 := \sup \{\|Af\|_2 : \|f\|_2 \le 1\},$$

where

$$\|f\|_2 := \left(\int_0^1 |f(t)|^2 \, dt \right)^{1/2}.$$

Lao & Whitley (1997) and Thorpe (1998) computed the values of $\|\mathcal{V}^m\|_2$ for $m = 2, \ldots, 10$ and $m = 2, \ldots, 20$, respectively. Moreover, it was conjectured in the former paper that

$$\lim_{m \to \infty} m! \, \|\mathcal{V}^m\|_2 = \frac{1}{2}. \tag{8.5.7}$$

The conjecture was proved by Kershaw (1999) and (using a different, simpler approach) by Little & Reade (1998; compare also Thorpe, 1998 where m in (8.5.7) is allowed to be any positive real number, with $\Gamma(m + 1)$ replacing $m!$). More general Volterra operators were then studied in Eveson (2003, 2005; see Section 8.5.2, below). Some of these results were considerably sharpened (especially the ones on lower bounds for $\|\mathcal{V}^m\|$) by Adell & Gallardo-Gutièrrez (2007) and by Böttcher & Dörfler (2009, 2010a, 2010b).

The following theorem contains the formal statement of the result (8.5.7).

Theorem 8.5.1 *Let $\mathcal{V} : L^2(0, 1) \to L^2(0, 1)$ be as in (8.5.6). Then*

$$\lim_{m \to \infty} m! \, \|\mathcal{V}^m\|_2 = \frac{1}{2}. \tag{8.5.8}$$

Remark 8.5.2 The above asymptotic result is also true for the *Hilbert–Schmidt* norms of $\mathcal{V}^m$,

$$\|\mathcal{V}^m\|_* := \left(\int_0^1 \!\! \int_0^1 \left(K_m(t, s) \right)^2 ds\, dt \right)^{1/2} \tag{8.5.9}$$

(Kershaw, 1999; see also Thorpe, 1998, Little & Reade, 1998 and Eveson, 2003). Here, the kernels $K_m(t, s)$ $(m \ge 1)$ are defined by

$$K_m(t, s) := \begin{cases} \dfrac{1}{(m-1)!}(t - s)^{m-1} & \text{if } s \le t \\[2ex] 0 & \text{if } s \ge t \end{cases}$$

(cf. (8.5.4)).   ◇

Proof We follow the approach employed in Little & Reade (1998).
(I) Using the adjoint operator $\mathcal{V}^*$ of the Volterra operator $\mathcal{V}$ defined in (8.5.6),

$$(\mathcal{V}^* f)(t) = \int_t^1 f(s)\, ds, \quad t \in [0, 1],$$

the composite operator $\mathcal{A} := \mathcal{V}\mathcal{V}^*$ is positive, compact and and self-adjoint
(cf. Appendix A.2.2). It is easy to verify that for $f \in C([0, 1])$,

$$\mathcal{A}f = g \quad \text{if and only if} \quad g'' = -f \quad \text{and} \quad g(0) = g'(1) = 0.$$

It thus follows that the eigenvalue problem

$$(\mathcal{A}\phi)(t) = \lambda\phi(t), \quad t \in [0, 1],$$

for

$$(\mathcal{A}f)(t) := (\mathcal{V}\mathcal{V}^* f)(t) = \int_0^t \left(\int_s^1 f(v)\, dv \right) ds$$

is equivalent to a Sturm–Liouville problem on $[0, 1]$, namely

$$-\phi''(t) = \lambda\phi(t), \quad \phi(0) = 0, \quad \phi'(1) = 0.$$

The eigenvalues and eigenfunctions of this classical problem are well known.
Hence we are led to the result stated in the following lemma.   □

Lemma 8.5.3 *The eigenvalues and eigenfunctions of $\mathcal{A}$ are given respectively by*

$$\lambda_n = \left((n + \tfrac{1}{2})^2 \pi^2 \right)^{-1} \quad \text{and} \quad \phi_n(t) = \sin((n + \tfrac{1}{2})\pi t) \quad (n = 0, 1, \dots).$$

Since the largest eigenvalue of the operator $\mathcal{A} = \mathcal{V}\mathcal{V}^*$ is given by $\lambda_0 = (\tfrac{2}{\pi})^2$,
it follows that

$$\|\mathcal{V}\|_2 = \|\mathcal{A}\|_2^{1/2} = \tfrac{2}{\pi}.$$

(II) In order to derive analogous results for $\|\mathcal{V}^m\|_2$ when $m \geq 2$ we have
to resort to a different analysis, based on the derivation of upper and lower
bounds for $\|\mathcal{V}^m\|_2$ that will coincide when $m \to \infty$.

The derivation of *upper bounds* is straightforward. The following result is
due to Little & Reade (1998).

Lemma 8.5.4 *As $m \to \infty$,*

$$\|\mathcal{V}^{m+1}\|_2 \leq \frac{1}{2(m+1)!} \left(1 + \mathcal{O}(m^{-1}) \right).$$

Proof It follows from (8.5.9) that

$$(\mathcal{V}^{m+1} f)(t) = \int_0^1 K_m(t, s) f(s) \, ds,$$

and hence

$$\|\mathcal{V}^{m+1}\|_2 \le \left(\int_0^1 \int_0^1 (K_m(t, s))^2 ds \, dt \right)^{1/2} = \frac{1}{m!} \left(\int_0^1 \int_0^t (t - s)^{2m} ds \, dt \right)^{1/2}$$

$$= \frac{1}{m!} \left(\frac{1}{(2m + 1)(2m + 2)} \right)^{1/2} = \frac{1}{2(m + 1)!} \left(\frac{m + 1}{m + \frac{1}{2}} \right)^{1/2}$$

$$= \frac{1}{2(m + 1)!} \left(1 + \mathcal{O}(\tfrac{1}{m}) \right), \quad \text{as } m \to \infty.$$

$\square$

The next lemma (cf. Little & Reade, 1998) presents a *lower bound* for $\|\mathcal{V}^{m+1}\|_2$ as $m \to \infty$.

Lemma 8.5.5 *For any* $\varepsilon \in (0, 1/2)$ *there exists an integer* $M = M(\varepsilon)$ *so that*

$$\|\mathcal{V}^m\|_2 \ge \frac{1}{(m + 1)!} \left(\frac{1}{2} - \varepsilon \right) \quad \text{for all } m > M.$$

Proof For the function $f_m(t) := (1 - t)^m / (2 - t)$ ($m \ge 1$, $t \in [0, 1]$) we obtain, using obvious substitutions,

$$(\mathcal{V}^{m+1} f_{m+1})(t) = \frac{1}{m!} \int_0^t \frac{(t - s)^m (1 - s)^{m+1}}{2 - s} \, ds$$

$$= \frac{t^{m+1}}{m \cdot m!} \int_0^m \left(1 - \frac{z}{m} \right)^m \left(1 - \frac{tz}{m} \right)^m \frac{m - tz}{2m - tz} \, dz.$$

We wish to find the limit, as $m \to \infty$, of $\frac{m \cdot m!}{t^{m+1}} (\mathcal{V}^{m+1} f_{m+1})(t)$. For $0 \le t \le 1$, $0 \le z \le m$, the three terms under the above integral sign possess the asymptotic behaviour

$$(1 - z/m)^m \to e^{-z}, \quad (1 - tz/m)^m \to e^{-tz}, \quad (m - tz)/(2m - tz) \to 1/2,$$

as $m \to \infty$, and in each case the convergence is monotonic with m. Thus, by the Lebesgue Monotone Convergence Theorem and Dini's Theorem for uniform convergence (Rudin, 1976),

$$\lim_{m \to \infty} \frac{m \cdot m!}{t^{m+1}} (\mathcal{V}^{m+1} f_{m+1})(t) = \frac{1}{2} \int_0^\infty e^{-(1+t)z} \, dz = \frac{1}{2(1 + t)}.$$

Let now $\mathcal{U} : L^2(0, 1) \to L^2(0, 1)$ be the unitary operator defined by $(\mathcal{U} f)(t) := f(1 - t)$. It follows from $(\mathcal{U} f_m)(t) = t^m / (1 + t)$ that

$$m \cdot m! \frac{(\mathcal{V}^{m+1} f_{m+1})(t)}{(\mathcal{U} f_{m+1})(t)} \to \frac{1}{2}, \quad \text{as } m \to \infty,$$

uniformly for $t \in [0, 1]$. In order to obtain the desired lower bound for $\|\mathcal{V}^{m+1}\|_2$, we first observe that the above analysis implies that, for given $\varepsilon \in (0, 1/2)$, there exists $M = M(\varepsilon)$ so that

$$m \cdot m! \, (\mathcal{V}^{m+1} f_{m+1})(t) > (\tfrac{1}{2} - \varepsilon) \| f_{m+1} \|_2,$$

for all $m > M$ and $t \in [0, 1]$. We also have, since $\mathcal{U}$ is unitary,

$$m \cdot m! \|\mathcal{V}^{m+1} f_{m+1}\|_2 > (\tfrac{1}{2} - \varepsilon) \|\mathcal{U} f_{m+1}\|_2 = (\tfrac{1}{2} - \varepsilon) \| f_{m+1} \|_2$$

$(m > M)$. Using the norm inequality $\|\mathcal{V}^{m+1} f_{m+1}\|_2 \leq \|\mathcal{V}^{m+1}\|_2 \| f_{m+1} \|_2$, we arrive at the estimate

$$\|\mathcal{V}^{m+1}\|_2 \geq \frac{1}{m \cdot m!} (\tfrac{1}{2} - \varepsilon) \geq \frac{1}{(m + 1)!} (\tfrac{1}{2} - \varepsilon).$$

The results of Lemmas 8.5.3 and 8.5.4 imply the assertion in Theorem 8.5.1. There now arise two obvious questions:

(i) How does the result of Theorem 8.5.1 change if the Volterra operator $\mathcal{V}$ defined in (8.5.6) is replaced by the more general Volterra integral operator

$$(\mathcal{V} f(t) = \int_0^t k(t - s) f(s) \, ds, \quad t \in [0, 1], \tag{8.5.10}$$

with (integrable) convolution kernel, for example, by the weakly singular operator

$$(\mathcal{V}_\alpha f)(t) := \int_0^t (t - s)^{\alpha-1} f(s) \, ds, \quad t \in [0, 1] \;\; (0 < \alpha < 1) ? \tag{8.5.11}$$

(ii) What are the analogues of the asymptotic results of Theorem 8.5.1 and the ones for more general convolution-type Volterra integral operators when the L^2-norm is replaced by the L^p-norm $(1 \leq p \leq \infty)$?

These questions are answered by Eveson (2003, 2005); we will discuss his results in the following section.

8.5.2 Volterra Integral Operators with Convolution Kernels

Eveson (2003, 2005) considered Volterra integral operators

$$(\mathcal{V}f)(t) := \int_0^t k(t-s)f(s)\,ds, \quad t \in [0,1], \qquad (8.5.12)$$

with convolution kernels of the form

$$k(t-s) = (t-s)^r h(t-s) \quad (r > -1), \qquad (8.5.13)$$

where the function $h = h(z)$ is in $L^1(0,1)$, is differentiable at $z = 0$, and satisfies $h(0) \neq 0$. The constant

$$C_p := \begin{cases} \dfrac{1}{p^{1/p}q^{1/q}} & \text{if } 1 < p < \infty, \\[2mm] 1 & \text{if } p = 1 \text{ or } p = \infty, \end{cases} \qquad (8.5.14)$$

corresponding to a given L^p-space will assume the role of the constant $1/2$ in (8.5.8) of Theorem 8.5.1. Here, q denotes the Hölder conjugate of p (corresponding to the dual space L^q of L^p), i.e. $1/p + 1/q = 1$.

Here is one of Eveson's main result (Corollary 3.4 in Eveson, 2005). The result for the case $p = 2$ was established in Eveson (2003).

Theorem 8.5.6 *Assume that the Volterra integral operator* $\mathcal{V}$ *defined in (8.3.12), with* $k(z) = z^r e^{\mu z}$ $(r > -1,\ \mu \in \mathbb{R})$, *acts on* $L^p(0,1)$ $(1 \leq p \leq \infty)$. *Then the* L^p *norms of the iterates* $\mathcal{V}^m$ *behave like*

$$\|\mathcal{V}^m\|_p \sim C_p e^{\mu} \frac{(\Gamma(r+1))^m}{\Gamma(m(r+1)+1)}, \quad \text{as } m \to \infty \text{ through } \mathbb{R}^+, \qquad (8.5.15)$$

with the constant C_p *defined in (8.5.14).*

Evenson's proof is based on the fact that the asymptotic behaviour of the norms $\|\mathcal{V}^m\|_p$ is the same as that of the norms of the iterates of the simpler Volterra operator (8.5.10) with convolution kernel $k(z) = h(0)z^r e^{(h'(0)/h(0))z}$. $\qquad \square$

8.6 Exercises and Research Problems

Exercise 8.6.1 Consider the Volterra integral operator

$$(\mathcal{V}_\alpha f)(t) := \int_0^t \frac{(t-s)^{\alpha-1}}{\Gamma(\alpha)} f(s)\,ds, \quad t \in [0,1]\ (0 < \alpha < 1).$$

Show that $\mathcal{V}_\alpha : L^1(0,1) \to L^{1/(1-\alpha)}(0,1)$ is not a continuous operator.

Exercise 8.6.2 Prove that the mapping result in Theorem 8.1.5 is not true when $\alpha + \beta = 1$.

Exercise 8.6.3 Consider the cordial Volterra integral operator

$$(\mathcal{V}_\varphi u)(t) = \int_0^t t^{-1} \varphi(s/t) K(t,s) u(s)\, ds, \quad t \in I,$$

with core $\varphi \in L^1(0,1)$ and $K \in C(D)$.

(a) Prove that the operator $\mathcal{V}_\varphi : C(I) \to C(I)$ is compact if $K(0,0) = 0$.
(b) Assume that $K(0,0) \neq 0$. Show that $\mathcal{V}_\varphi - K(0,0)\mathcal{V}_\varphi^1$ (where $\mathcal{V}_\varphi^1$ denotes the basic cordial Volterra operator defined in (7.1.4)) is a compact operator on $C(I)$ (cf. Theorem 7.1.10(d)).

Exercise 8.6.4 In his paper of 1897 (pp. 162–163), Volterra studied the integral equation with two variable limits of integration,

$$u(t) = f(t) + (\tilde{\mathcal{V}} u)(t), \quad t \in [-T, T]. \tag{8.6.1}$$

See also pp. 92–94 in his book of 1913, and Exercise 1.7.3 in Chapter 1. It corresponds to the integral operator

$$(\tilde{\mathcal{V}} f)(t) := \int_{-t}^t K(t,s) f(s)\, ds,$$

where $K \in C(\tilde{D})$ ($\tilde{D} := \{(t,s) : -t \leq s \leq t \ (t \in [0,T])\}$ (recall Example 1.2.4). Is this operator compact (as an operator on $C[-T, T]$) ?

Exercise 8.6.5 Consider the linear Volterra integral operator (8.1.1) with logarithmic singularity,

$$(\mathcal{V}_0 f)(t) := \int_0^t \log(t - s) K(t,s) f(s)\, ds, \quad t \in I,$$

where $K \in C(D)$. If $f \in L^p(0,T)$ ($p \geq 1$), in which space does its image $\mathcal{V}_0 f$ lie? Discuss in particular the case $p = 2$.

Exercise 8.6.6 (Hardy & Littlewood, 1928, p. 578)
Let $f \in L^p(0,T)$, and assume that $0 < \alpha < 1$ and $p = 1/\alpha$. Show that the statement '$(\mathcal{V}_\alpha f)(t)$ is bounded' is false.

Exercise 8.6.7 Assume that $K(t,s) \equiv 1$ in Exercise 8.6.4. Halmos (1982, cf. p. 100 and p. 301) calls the integral operator $\tilde{\mathcal{V}}$ with this kernel a *skew-symmetric* Volterra operator (as an operator on $C[-1, 1]$ or $L^2(-1, 1)$).

(a) Find $\|\tilde{\mathcal{V}}\|$.

(b) Show that $\tilde{\mathcal{V}}$ is *nilpotent* of *index* 2 (that is, $\tilde{\mathcal{V}}^2 = 0$). What is $\sigma(\tilde{\mathcal{V}})$?

(*Hint:* Show that the range of $\tilde{\mathcal{V}}$ is contained in the set $\{f \in L^2(-1, 1) : f$ is odd$\}$, and that f odd implies $\tilde{\mathcal{V}}f = 0$.)

Exercise 8.6.8 Assume that the kernel $K(t, s)$ is of type (L^p, T) $(1 < p < \infty)$, and define the functions $R_n(t, s)$ by $R_1(t, s) := K(t, s)$ and

$$R_{n+1}(t, s) := \int_s^t K(t, v)R_n(v, s)\, dv \quad (t, s) \in D \quad (n \geq 1).$$

Show that each $R_n(t, s)$ is of type (L^p, T). Then use this result to prove Theorem 8.3.3.

Exercise 8.6.9 Consider the integral operator $\mathcal{V}_\theta : C[0, 1] \to C[0, 1]$ given by

$$(\mathcal{V}_\theta f)(t) := \int_0^{\theta(t)} f(s)\, ds, \quad t \in [0, 1],$$

where $\theta \in C[0, 1]$. Show that a necessary condition for $\mathcal{V}_\theta$ to be *quasi-nilpotent* is that $\theta(t) \leq t$ for all $t \in [0, 1]$. In particular, discuss the specal case $\theta(t) = t^\beta$ $(0 < \beta < 1)$. (See also Domanov, 2007.)

Exercise 8.6.10 Is the auto-convolution operator $\mathcal{A} : L^2(0, 1) \to L^2(0, 1)$,

$$(\mathcal{A}u)(t) := \int_0^t u(t - s)u(s)\, ds, \quad t \in [0, 1],$$

with domain $\{u \in L^2(0, 1) : u(t) \geq 0 \text{ a.e. on } [0, 1]\}$, compact? (See also Fleischer & Hofmann, 1997.)

Exercise 8.6.11 A related class of non-linear VIEs of the *first kind* (to which we have briefly alluded in Section 3.4.3) is

$$(\mathcal{A}u)(t) := \int_0^t u(t - s)u(s)\, ds = g(t), \quad t \in I = [0, T].$$

Here, the *auto-convolution* operator $\mathcal{A} : U \subset X \to Y$ from a domain U in a Banach space X into a Banach space Y is not always compact (see, for example, Gorenflo & Hofmann, 1994 and Dai & Lamm, 2008 for more recent references). Among the results established in the former paper are:

(a) If $\mathcal{A} : U \subset C(I) \to C(I)$, and if U contains only monotone functions, then $\mathcal{A}$ is a compact integral operator.

(b) If $\mathcal{A} : U \subset L^2(0, T) \to L^2(0, T)$, and if U contains only non-negative, non-decreasing functions, then $\mathcal{A}$ is compact.

Prove these assertions.

Exercise 8.6.12 (*Research Problem*)
Let $\mathcal{V}_\varphi^1$ be the basic cordial integral operator defined by

$$(\mathcal{V}_\varphi^1 f)(t) := \int_0^t t^{-1} \varphi(s/t) f)s)\, ds, \quad t \in I,$$

with core $\varphi \in L^1(0, 1)$. Analyse the behaviour of $\|(\mathcal{V}_\varphi^1)^m\|$ as m tends to infinity; that is, establish the analogue of Theorem 8.5.1 for the norms of powers of $\mathcal{V}_\varphi^1$).

Exercise 8.6.13 Prove Theorem 8.4.5.

Exercise 8.6.14 Discuss the connection between the rate of decay of the singular values of a linear Volterra integral operator $\mathcal{V}_\alpha$ ($0 < \alpha \leq 1$) and its degree of ν-smoothing (cf. Definition 1.4.8).

8.7 Notes

Section 8.1: *Mapping Properties*
Banach (1932, p. 98 and p. 162) discusses Volterra integral operators that are compact ('totalement continue') as operators on $C(I)$ and $L^p(I)$ ($p > 1$). Mapping properties of Volterra integral operators (on Hölder and L^p spaces) with weakly singular kernels form part of the fundamental paper by Hardy & Littlewood (1928). These properties are also discussed in the books by Hille & Phillips (1957), Gripenberg, Londen & Staffans (1990, pp. 227–241) and Gorenflo & Vessella (1991; the latter also analyses the mapping properties of weakly singular Volterra integral operators on Sobolev spaces of fractional orders). The book by Jörgens (1970) contains, in Section 11, a very general discussion of integral operators on Lebesgue spaces L^p. In addition, see the book by Krasnosel'skiĭ et al. (1976), as well as the papers by Kwapisz (1991), Howard & Schep (1990, on norms of Volterra integral operators on L^p-spaces) and Fixman (2000). The case $p = 2$ is studied in Tricomi (1957, pp. 8–15), Miller (1971, e.g. pp. 193–201), and Corduneanu (1991, pp. 119–123). Bedivan & O'Regan (2000) analyse the mapping properties of Volterra integral operators on Fréchet spaces, while cordial Volterra integral operators on L^∞ and on Hölder spaces are discussed in Vainikko (2010a) and Vainikko (2014, Section 2).

Properties and applications of integral operators of *Volterra–Stieltjes* type are the subject of Banaś & Dronka (2000) and Banaś & Caballero Mena (2008).

The following represent a selection of papers and monographs that deal with various *advanced topics* regarding Volterra integral operators: Ringrose (1962 on triangular representation of integral operators); Osher (1967 on the similarity of Volterra integral operators); Gohberg & Kreĭn (1970: this important monograph contains a wide-ranging theory of abstract Volterra operators on Hilbert spaces); Klebanov & Sleeman (1996 on axiomatic theory of VIEs); Väth (1998a, 1998b, 1999: the 1999 monograph contains a study of abstract Volterra equations by means of topological and algebraic methods); Jung (1999 on duality theory, corresponding to the classical one for semigroups of operators); Corduneanu (2000: a survey of abstract Volterra equations); Shkarin (2006 on anti-supercyclic Volterra operators); Wu (2011 on Volterra operators and Carleson measures).

Section 8.2: *Quasi-Nilpotency*
Halmos (1982, pp. 98–99) gives a concise introduction to the *quasi-nilpotency* of Volterra integral operators. Analogous results for Volterra operators of the form $\int_0^{\theta(t)} u(s)\,ds$, with $\theta : [0,1] \to [0,1]$, were conjectured and established in Lyubich (1984), Whitley (1987), Zima (1992), Shkarin (2006a, 2006b) and Domanov (2007, 2008).

Quasi-nilpotent operators on Banach spaces play an important role in the invariant subspace problem; see, for example, Read (1997) and Pietsch (2007, pp. 2002–204).

Section 8.3: *Existence of Resolvent Kernels*
Good introductions to the abstract *resolvent theory* (including the uniqueness of the solution of the resolvent equations in $C(D)$ and more general spaces like $L^p(D)$) may be found in Miller (1971, Chapter IV) and Corduneanu (1991, Section 4.2). The most general treatments are those in Gripenberg, Londen & Staffans (1990; see in particular Sections 2.3–2.5, and Chapters 6 and 9) and in Prüss (2012, where the abstract resolvent theory is presented in Chapters I.1, I.2 and III.10).

Grimmer (1982), Grimmer & Pritchard (1983) and Staffans (1983) studied the existence of resolvent operators for abstract VIEs of the forms

$$u(t) = f(t) + \int_0^t A(t,s)u(s)\,ds, \quad t \in I = [0,T], \tag{8.7.1}$$

and

$$u(t) = f(t) + \int_0^t a(t-s)u(s)\,ds, \quad t \in I, \tag{8.7.2}$$

on a Banach space X. Here, $A(t,s)$ and $a(t-s)$ are closed operators with dense domains that do not depend on (t,s). Their analysis also explored the connection of resolvent operators with C_0 semigroups. For related papers see Grimmer & Prüss (1985) and Lizama (1993).

Plato (1997) derives resolvent estimates of Abel-type integral operators

$$(Au)(t) := \frac{\beta^{1-\alpha}}{\Gamma(\alpha)} \int_0^t \left(t^\beta - s^\beta\right)^{\alpha-1} s^{\beta-1} u(s)\,ds \ (0 < \alpha < 1, \ \beta > 0)$$

acting on L^p spaces.

The monograph by Engel & Nagel (2000) contains an illuminating and comprehensive treatment of various aspects of strongly continuous *semigroups* of linear operators on Banach spaces (Volterra integral operators associated with such semigroups are discussed in Section III.1). The 2006 book by the same authors serves as a concise introduction to the principal aspects of this topic. A fascinating connection between perturbations of strongly continuous semigroups and renewal equations of Stieltjes type is explored in Diekmann, Gyllenberg & Thieme (1993).

The solvability of weakly singular first-kind VFIEs with constant delay is studied in Kappel & Zhang (1986) and in Burns, Herdman & Turi (1990), using an L^2 semigroup setting (such VFIEs occur in connection with neutral Volterra functional differential equations). See also Staffans (1984a). Semigroups associated with non-linear VIEs are discussed in Corduneanu (1991, Section 5.4).

Of the papers and monographs on *abstract VIEs* we mention Osher (1967), Miller & Sell (1970a) (topolgical dynamics; see also Väth (1999)), Corduneanu (1991, Ch. 5), Vasin (1996), Klebanov & Sleeman (1996 on axiomatic theory of VIEs), and Väth (1998a, 1998b). Jung (1999) studied the abstract Volterra integral equation

$$u(t) = u_0 + A \int_0^t a(t-s)u(s)\,ds, \quad t \geq 0,$$

where A is a closed operator with dense domain $D(A)$ in a Banach space X; the papers extends the duality theory on semigroups of operators corresponding to $a \equiv 1$ (cf. Hille & Phillips, 1957). In Chapter V of Miller (1971) the behaviour of solutions of non-linearly perturbed Hammerstein VIE is analysed by using the setting of admissible operators in Fréchet spaces. See also Corduneanu (1991, Ch. 4) on admissibility techniques.

Szufla (1984) contains a study of non-linear VIEs in L^p spaces. Abstract *non-linear* VIEs are studied in Panin (2015: local solvability and blow-up for a Hammerstein VIE with operator-valued kernel) and in Emmrich & Vallet (2016: VIEs of the form $Au + B\mathcal{H}u = f$, where A and B are operators on real, reflexive and separable Banach spaces, and $\mathcal{H}$ is a Hammerstein–Volterra integral operator). The fundamental paper by Friedman & Shinbrot (1967) represents an important milestone in the theory of *VIEs in Banach spaces*: it is concerned with VIEs of the form

$$u(t) = f(t) - \int_0^t k(t-s)(A(s)u)(s)\,ds,$$

where u and f have values in a (complex) Banach space X, k is a scalar function, and A is an unbounded operator on X which generates an analytic semigroup (assuming that the resolvent $[\lambda\mathcal{I} - A(t)]^{-1}$ exists for all $t \in I$ and $\mathrm{Re}(\lambda) \leq 0$, and

$$\|[\lambda\mathcal{I} - A(t)]^{-1}\| \leq C(1 + |\lambda|)^{-1},$$

with C independent of t and λ). VIEs in Banach spaces are also analysed in Grimmer & Miller (1977), Cléments & Nohel (1979), Barbu (1979) ($0 < \alpha < 1$), Szufla (1984: existence of L^p-solutions of Volterra–Hammerstein VIEs), Grimmer & Prüss (1985), Kwapisz (1991a, 1991b, 1991c, 1993), Desch & Prüss (1993) and the 1993 monograph by Prüss. VIEs in abstract spaces are also treated in Corduneanu (1991, Ch. 5). In addition, see Fenyö & Stolle (1984, Volume 3, Section 13.1), Dudkiewicz (2006), and the monograph by Askhabov (2009, which deals with monotone operators in various Banach spaces, with applications to implicit VIEs with power linearity we discussed in Section 3.4.3).

Hönig (1975) is the classical monograph on Volterra–Stieltjes integral equations. This topic is also treated in Mingarelli (1983).

Problems 186–189 in Halmos (1982) address VIEs in Hilbert spaces (especially in L^2). See also Witte (1997).

The book by Precup (2002) discusses various fixed-point theorems (e.g. Schauder and Leray–Schauder) for non-linear integral operators on Banach spaces. In addition, see Pulyaev & Tsalyuk (1991) on various Banach spaces of admissible functions (including the space of continuous and bounded functions on $\mathbb{R}^+$,

$$BC := \{u \in C[0, \infty) : \sup_{t \geq 0} \|u(t)\| < \infty\}),$$

and Vessella (1985) on the stability and continuous dependence of solutions to Abel integral equations in L^p spaces. A related problem (on the regularisation of first-kind VIEs in L^p spaces) is studied in Brookes & Lamm (2011). The ill-posedness of operator equations $Ax = y$ with non-compact operator A (with application to first-kind VIEs) is the subject of the 2010 paper by Hofmann & Kindermann.

Section 8.4: *Singular Values of Volterra Integral Operators*
Singular values of integral operators form the content of the fundamental paper by Smithies (1938). (I am grateful to Prof Jia Zhongxia of Tsinghua University, Beijing, for pointing out this reference to me.) The papers by Hille & Tamarkin (1931), Faber et al. (1986), Faber & Wing (1986a, 1986b, 1988), and Tuan & Gorenflo (1994) are concerned with various properties of singular values for the Volterra integral operator $\mathcal{V}_\alpha$ ($0 < \alpha \leq 1$), while Hofmann (1994) discusses singular values of *non-linear* Volterra integral operators.

Section 8.5: *Norms of Powers of Volterra Integral Operators*
Problem 188 in Halmos (1982, p. 147) was arguably the motivation that led to the papers by Lao & Whitley (1997), Kershaw (1999), Thorpe (1998), Little & Reade (1998) and Eveson (2003, 2005) on the asymptotic behaviour of the norms $\|\mathcal{V}^n\|$ of the (integer) powers of a Volterra integral operator $\mathcal{V}$. The reader may also wish to consult the illuminating related papers by Böttcher & Dörfler (2009, 2010a, 2010b).

9

Applications of Volterra Integral Equations

Summary

The aim of this chapter is to be a guide (with extensive references) to a wide and representative spectrum of applications where VIEs and VFIEs arise either as mathematical models or as key tools for the analysis (existence, uniqueness, regularity) of solutions of such models (often time-dependent PDEs). When describing these applications, in the form of examples, we do not, owing to limitation of space, in general include a description of the derivation of the model equations; instead we provide the interested reader with appropriate references on their physical background.

9.1 VIEs of the First Kind

9.1.1 Integral Equations of Abel Type

Example 9.1.1 (Abel, 1823, 1826) The papers by Niels Henrik Abel (1801–1829) mark the origin of the theory and applications of (weakly singular) first-kind integral equations with variable upper limit of integration. In these papers Abel studied the following mechanical problem (the so-called *tautochrone problem*) which led to the weakly singular first-kind integral equation (1.5.1) (that now bears his name) with $\alpha = 1/2$. Suppose that a particle of mass m moves, under the influence of gravity, along some (unknown) curve Γ in the vertical (x, y)-plane from a point P_h at height $y = h > 0$ to a point P_0 at level $y = 0$, taking the (prescribed) time $\tau = \tau(h)$ to slide from P_h to P_0. Let Γ be described by the equation $x = \Phi(y)$, with $P \in \Gamma$ having the coordinates $(\Phi(y), y)$. It follows from the conservation of energy,

$$\frac{m}{2}v^2 + mgy = \text{const} = mgh$$

304

(where g denotes the gravitational constant) that the velocity of the particle satisfies

$$v = \frac{ds}{dt} = \left(2g(h - y)\right)^{1/2}, \quad 0 \le y \le h.$$

Hence the total time the particle needs to complete its journey on Γ is

$$\tau(h) = \int_{P_0}^{P_1} \frac{ds}{dv} = \int_0^h \left(\frac{1 + (\Phi'(y))^2}{2g(h - y)}\right)^{1/2} dy.$$

If we set

$$u(y) := \left(1 + (\Phi'(y))^2\right)^{1/2} \quad \text{and} \quad f(h) := \tau(h)(2g)^{1/2},$$

the problem of finding the curve Γ is reduced to finding the solution $u = u(y)$ of the first-kind 'Abel' integral equation

$$\int_0^h (h - y)^{-1/2} u(y)\, dy = f(h) \tag{9.1.1}$$

for given $h > 0$. It follows from the physical context that $f(0) = 0$.

First-kind VIEs with different kernel singularities occur as mathematical models in a variety of applications. A typical example is described in the next example.

Example 9.1.2 (Gorenflo & Vessella, 1991, Ch. 3) The integral equation

$$2 \int_x^R (r^2 - x^2)^{-1/2} r u(r)\, dr = g(x), \quad 0 \le x \le R < \infty, \tag{9.1.2}$$

models the gas discharge in a cylindrical tube (with circular cross section of radius $R > 0$) whose intesity $u(r)$ depends only on the distance from the axis of the cylinder. The aim is to determine $u(r)$ ($0 \le r \le R$) by measurements of $g(x)$ from outside the cylinder.

First-kind VIEs of the form (9.1.2) arise also as mathematical models in tomography (i.e. the reconstruction in non-destructive testing and diagnosis (CAT scanners) from 2D projections), seismology, and geometric probability; see Anderssen (1977), Andersson & de Hoog (1990) and Gorenflo & Vessella (1991, Chapters 2 and 3) for these and other applications.

9.1.2 General First-Kind VIEs

Example 9.1.3 (Peirce & Siebrits, 1996, 1997) The VIE

$$\int_0^t \cos\left(\omega(t - s)\right) u(s)\, ds = g(t), \quad t \ge 0,$$

describes the time evolution of one of the spatial modes in the Fourier transform of the solution of the wave equation

$$\frac{\partial w(t, r)}{\partial t^2} - c^2 \Delta w(t, r) = F(t, r), \quad r \geq 0,$$

that is related to the elastodynamic equations in the geosciences (for modelling geological prospecting and for assessing the effect of fault movement and fracture propagation on surface structures and mining excavation). Here, ω is a (possibly very large) positive parameter (see also Section 4.3.3). Detailed references on the geophysical background of these models can be found in the above papers by Peirce & Siebrits.

Example 9.1.4 (Davies & Duncan, 2004 and references) The integral equation

$$\int_\Gamma \frac{w(t - |\xi - x|, \xi)}{|\xi - x|} \, d\xi = a(t, x), \quad (t, x) \in (0, T) \times \Gamma,$$

where $\Gamma \subset \mathbb{R}^3$ is a two-dimensional surface, is the single-layer potential equation for transient accoustic scattering from the surface Γ. Knowledge of w permits the computation of the scattered field in $\mathbb{R}^3$. The use of Fourier transform techniques lead to the first-kind VIE for the transformed solution $\hat{u}(t - s, \omega)$,

$$2\pi \int_0^t J_0\big(\omega(t - s)\big)\hat{u}(t - s, \omega) \, ds = \hat{a}(t, \omega) \ (t \in (0, T), \ \omega \in \mathbb{R}^2),$$

where $J_0(z) = \pi^{-1} \int_0^\pi \cos\big(z \sin(\theta)\big)d\theta$ is the Bessel function of order zero.

Example 9.1.5 (de Hoog & Anderssen, 2008, 2011 and references) First-kind VIEs with convolution kernels,

$$\int_0^t k(t - s)u(s) \, ds = g(t), \quad t \geq 0 \ (g(0) = 0),$$

occur in the analysis of causal processes in e.g. viscoelasticity (stress-strain response), exponential forgetting, and hydrological applications (see also Loy & Anderssen, 2014).

Example 9.1.6 (Lamm, 2000 and its references) If an unknown heat source $u = u(t)$ is applied at the left end $x = 0$ of an insulated semi-infinite bar, one wants to determine the values $u(t)$ from measurements $g = g(t)$ $(0 \leq t \leq T)$ of the temperature at the position $x = 1$ of the bar. This problem corresponds to solving the *inverse heat equation* (also called the sideways heat equation).

As indicated in Example 1.4.10, $u(t)$ is the solution of the severely ill-posed first-kind VIE

$$(2\pi^{1/2})^{-1} \int_0^t (t-s)^{-3/2} e^{-1/(4(t-s))} u(s)\,ds = g(t), \quad t \in [0, T].$$

Example 9.1.7 (Apartsyn, 2014 and its references) First-kind VFIEs of the form

$$\sum_{i=1}^r \int_{a_{i-1}(t)}^{a_i(t)} K_i(t,s) u(s)\,ds = g(t), \quad t \in [0, T],$$

with delays satisfying $0 \le a_0(t) < \cdots < a_r(t) = t$ $(a_i(0) = 0)$ represent age-structured models of e.g. electric power systems (a particular case of vintage capital models with endogenous swapping of obsolete vintages; cf. Ivanov et al., 2004 and its references). Details and specific examples may also be found in the book by Hritonenko & Yatsenko (1996) and in the paper by Apartsyn & Sidler (2013).

9.2 VIEs of the Second Kind

9.2.1 The Renewal Equation

Example 9.2.1 The renewal equation is a linear second-kind VIE with (non-negative) convolution kernel,

$$u(t) = f(t) + \int_0^t k(t-s) u(s)\,ds = f(t) + \int_0^t k(s) u(t-s)\,ds, \quad t \ge 0. \quad (9.2.1)$$

It arises as a mathematical model in the theory of industrial replacement and in population theory (cf. Lotka, 1939, 1939a; also for an extensive list of early references), Lotka (1942, 1945) and the 1941 book by Feller.

In the theory of *industrial replacement* (see, for example, Hritonenko & Yatsenko, 1996 and Ivanov et al., 2004), if $k(s)$ denotes the density of the probability (at the moment of instalment) that an individual will drop out at age s, then

$$f(t) = \int_0^t k(t-\tau) \eta(\tau)\,d\tau \quad (t \ge 0)$$

represents the rate of dropping out (at time t) of individuals belonging to the parent population. Here, $\eta(\tau)$ is the age distribution of the population at the moment $t = 0$, and $u(t)$ stands for the rate of dropping out (at time t) of individuals of the total population. Such an individual belongs either to the parent population, or it joined the population by the process of replacement at some moment $t - \tau$ $(0 < \tau < t)$. We then have $\int_0^\infty k(s)ds = 1$.

9.2.2 Population Growth Models

Example 9.2.2 (Brauer, 1975) The Hammerstein-type VIE

$$u(t) = f(t) + \int_0^t P(t - s)G(u(s))\,ds, \quad t \geq 0, \qquad (9.2.2)$$

arises in the mathematical modelling of the growth of a population or the spread of a disease in a population. In the former case $u(t)$ denotes the size of the population at time t; $G(u)$ is the number of members added to the population in unit time when the population size is u; $P(t)$ is the probability that a member of the population survives to age t; and $f(t)$ denotes the members of the population that were already present at time $t = 0$ and are still alive at age t. We note that $G(u)/u$ is the growth rate of the population size in unit time per unit population size. A typical function G is given by the logistic function $G(u) = au - bu^2$ $(a, b > 0)$ with $0 \leq u \leq a/b$.

In the context of modelling the spread of a disease, the solution $u(t)$ of (9.2.2) represents the number of members of the population that are afflicted by the disease. It is assumed that the recovery from the disease conferring negligible immunity (that is, the growth rate depends on the number of members with the disease), and that that the disease has a negligible incubation period; that is, there is no delay term in the integral in (9.2.2) (in contrast to the VIE in Example 9.2.4 below).

Example 9.2.3 (Brauer, 1976a) If there is harvesting in a population whose growth is modelled by (9.2.2), the resulting growth is now described by the VIE

$$u(t) = f(t) + \int_0^t P(t - s)G(u(s))\,ds - EH(t), \quad t \geq 0. \qquad (9.2.3)$$

Here, $EH(t)$ represents the number of members of the population harvested up to time t, with $E > 0$ denoting the constant rate of harvesting.

Example 9.2.4 (Cooke & Yorke, 1973, Cooke, 1976, Smith, 1977) A basic model of single-species population growth in which there is immigration into the population at any prescribed rate, and with any prescribed age distribution and given life span τ, is a non-linear (Hammerstein) VFIE with (not necessarily constant) delay $\tau > 0$ or with variable delay $\tau(t) > 0$,

$$u(t) = f(t) + \int_{t-\tau(t)}^{t} P(t-s)G(u(s))\,ds, \quad t \geq 0. \qquad (9.2.4)$$

Here, the convolution kernel $P(s)$ represents the proportion (or probability) of the population surviving to age at least s. Related VFIE models were studied by Gripenberg (1983), Torrejón (1990) and Cañada & Zertiti (1994).

Example 9.2.5 (Bélair, 1991) The main assumption underlying the mathematical model (9.2.4), namely that the delay $\tau(t) > 0$ depends only on t is often too restrictive. A more realistic model will have to take into account the biological fact that the life span τ may also depend on the size $u(t)$ of the population at the time $t > 0$. Such a population growth model was studied by Bélair (1991): it is described by a VFIE with *state-dependent delay*, namely

$$u(t) = \int_{t-\tau(u(t))}^{t} P(t-s)G(u(s))\,ds, \quad t \geq 0, \qquad (9.2.5)$$

with $P(t) \equiv 1$. Here, the delay τ in the delay function $\theta(t) := t - \tau(u(t))$ represents the life span of the population which now depends on the size $u(t)$ of the population at time t (reflecting e.g. crowding effects). It is assumed that the number of births is a function of the population size only (that is, the birth rate is density-dependent but not age-dependent).

Remark 9.2.6 For the kernel $P(t-s) \equiv 1$ it is tempting to 'simplify' the delay VIE (9.2.5), by differentiating it with respect to t, to obtain the state-dependent (but 'local') delay differential equation

$$u'(t) = \frac{G(u(t)) - G(u(t - \tau(u(t))))}{1 - \tau'(u(t))G(u(t - \tau(u(t))))}. \qquad (9.2.6)$$

While *any* constant function $u(t) = u_c$ solves the above DDE, this is *not* true in the original VFIE (9.2.5) with $P(t) \equiv 1$: it is easily verified that $u(t) = u_c$ is a solution of the VIE if, and only if, $u_c = G(u_c)\tau(u_c)$. This observation implicitly contains a warning to numerical analysts, namely that the use of the the DDE (9.2.6) as the basis for the ('indirect') numerical solution of the state-dependent delay VIE (9.2.5) may lead to approximations for $u(t)$ that do not correctly reflect the dynamics of the original VFIE (9.2.5).   ◇

9.2.3 Heat Transfer, Diffusion Models and Shock Wave Problems

Example 9.2.7 (Mann & Wolf, 1951, Roberts & Mann, 1951) An important class of non-linear (Hammerstein-type) VIEs arises in the mathematical modelling of heat transfer between a gas at constant unit temperature and a semi-infinite solid, when Newton's law of cooling ceases to be linear but is replaced by the weaker condition that the net rate of heat exchange from the gas to the solid is a function $G(w(t, 0))$ of the surface temperature $w(t, 0)$. The temperature distribution $w = w(t, x)$ within the solid is then completely determined by the surface (interface) temperature $u(t) := w(t, 0)$, and hence the original initial-boundary-value problem for $w(t, x)$ can be reduced to a non-linear VIE for the surface temperature,

$$u(t) = \pi^{-1/2} \int_0^t (t - s)^{-1/2} G(u(s))\, ds, \quad t \geq 0. \tag{9.2.7}$$

Here, the function $G(u)$ comes from the (Neumann) boundary condition at $x = 0$,

$$w_x(t, 0) = -G(w(t, 0)), \quad t > 0.$$

(This reduction of an initial-boundar-value problem for a parabolic PDE to a VIE is reminiscent of what we encountered in Section 3.3 on quenching.)

A representative example comes from the mathematical modelling of the temperature $w(x, t)$ of a non-linearly radiating semi-infinite solid $x \geq 0$. Under the assumptions that heat is radiated away from the end $x = 0$ at a rate proportional to $w^n(x, t)$ ($n > 1$) and that an external source supplies heat at a rate proportional to a given function $f(t) > 0$, the mathematical model is given by the initial-boundary-value problem

$$w_t = w_{xx}, \quad x > 0, \ t > 0,$$
$$w_x(t, 0) = \gamma w^n(t, 0) - f(t), \quad t > 0,$$
$$w(0, x) = 0, \quad x \geq 0,$$
$$\lim_{x \to \infty} w(t, x) = 0, \quad t \geq 0$$

(see, for example, Keller & Olmstead, 1960, Handelsman & Olmstead, 1972, Keller & Olmstead, 1972, Olmstead & Handelsman, 1976, Olmstead, 1977) Gorenflo, 1977, as well as Levinson, 1960 on the theory of superfluidity).

In order to determine the temperature $w(t, x)$ at any point (t, x) ($x > 0$) it is sufficient to know $u(t) := w(t, 0)$ ($t \geq 0$). In this case the VIE (9.2.7) for the surface temperature $u(t)$ becomes

$$u(t) = \pi^{-1/2} \int_0^t (t - s)^{-1/2} \big(f(s) - \gamma u^n(s) \big)\, ds, \quad t \geq 0,$$

where the constant $\gamma > 0$ represents the ratio of the radiative properties to the conductive properties of the solid material.

Example 9.2.8 (Olmstead & Handelsman, 1972, Angell & Olmstead, 1987) If in Example 9.2.7 the parameter γ (which governs the rate $\gamma w^n(t, 0)$ at which heat is being radiated away from the end $x = 0$ of the solid) is large ($\gamma \gg 1$) then, setting $\varepsilon^n := \gamma^{-1}$, the VIE (9.2.9) becomes a singularly perturbed VIE of the type

$$\varepsilon u(t) = \pi^{-1/2} \int_0^t (t - s)^{-1/2} \big(h(s) - u^n(s) \big) \, ds, \quad t \geq 0.$$

It follows that when $\varepsilon > 0$ and h is continuous, the solution satisfies $u(0) = 0$. However, for $\varepsilon = 0$ and $h(0) > 0$ we obtain $u(t) = \big(h(t) \big)^{1/n}$. This implies that the solution of the above VIE exhibits a boundary layer at $t = 0^+$. Related references are the PhD thesis by Angell (1985) and the survey article by Kauthen (1997, which contains an extensive list of references).

Example 9.2.9 (Lighthill, 1950, p. 371, Noble, 1964, pp. 308–309, Curle, 1978) A related Hammerstein VIE is the celebrated Lighthill equation we have mentioned before (cf. Section 7.3, Example 7.3.2),

$$u^4(t) = -\frac{1}{2t^{1/2}} \int_0^t \big(t^{3/2} - s^{3/2} \big)^{-1/3} u'(s) \, ds, \quad u(0) = 1; \quad u(0) = 1.$$

It describes the heat transfer across a laminar boundary, for example, in the steady flight of a projectile or aircraft, from a deliberately heated wall (e.g. to prevent icing). Here, $u(t)$ is a non-dimensional measure of the temperature distribution on the surface of the projectile or aircraft. By resorting to a generalisation of Abel's inversion formula (recall Theorem 1.5.11 with $h(t) = t^{3/2}$ and $\alpha = 1/3$), the above VIE can be written in the form

$$u(t) = 1 - \frac{3\sqrt{3}}{2\pi} \int_0^t \big(t^{3/2} - s^{3/2} \big)^{-2/3} s u^4(s) \, ds, \quad t \geq 0.$$

As we have seen in Example 7.3.2, the underlying integral operator is a (compact) cordial Volterra–Hammerstein integral operator.

Example 9.2.10 (Keller, 1981, Okrasiński, 1991b, pp. 1014–1015) In the mathematical modelling of shock waves in a gas-filled shock wave tube one is interested in finding the axial component of the particle behind the shock wave (see the references in Keller, 1981). In the analysis of the resulting mathematical model equations one encounters an implicit VIE of the form

$$u^2(t) = C_p \int_0^t (t-s)^{-1/2} u(s)\, ds, \ t \geq 0$$

(Keller, 1981, p. 173), where the constant C_p depends on certain physical constants (including a boundary-layer parameter and the Prandtl number). This VIE is a particular case of the implicit VIE we studied in Section 3.4.3 (cf. Remark 3.4.15). Thus, setting $z(t) := u^2(t)$ we obtain the homogeneous non-linear VIE

$$z(t) = C_p \int_0^t (t-s)^{-1/2} \big(z(s)\big)^{1/2}\, ds, \ t \geq 0.$$

This equation belongs to the class of Hammerstein VIEs whose solvability (that is, the existence of non-trivial solutions) is described in Theorem 3.1.19 and Example 3.1.21.

9.2.4 Blow-Up and Quenching Phenomena

We first present some examples of applications leading to VIEs whose solutions may *blow up* in finite time.

Example 9.2.11 (Roberts, Lasseigne & Olmstead, 1993) Thermal ignition in solid combustible materials can be modelled mathematically by non linear VIEs. A representative example is the formation of shear bands in steel: subjection to very high strain rates is accompanied by a dramatic rise in the temperature. A basic mathematical model describing this phenomenon is the Hammerstein-type VIE

$$u(t) = \gamma \pi^{-1/2} \int_0^t (t-s)^{-1/2} (1+s)^q \big(u(s)+1\big)^p\, ds, \ t \geq 0 \ (p > 1)$$

(recall Example 3.2.7), where the non-negative material parameters γ, q and p are related to the constitutive law for plastic straining. Details on the derivation of this VIE model can be found in the list of references of the paper cited above. The blow-up of the solution $u(t)$ has been analysed in Section 3.2.1 (Example 3.2.7).

Example 9.2.12 (Olmstead, Kirk & Roberts, 2010) The following VIE occurs as a mathematical model for blow-up in a subdiffusive medium with a localised high-energy source. In this model the classical diffusion operator is replaced by a heat operator containing a temporal fractional derivative, complemented by an advection term. The source term grows non-linearly as the temperature increases (cf. the list of references in the above paper). We will consider the spatially one-dimensional case. Setting

$$(D_t^{1-\alpha} w)(t, x) := \frac{1}{\Gamma(\alpha)} \frac{\partial}{\partial t} \int_0^t (t-s)^{\alpha-1} w(s, x) \, ds \quad (0 < \alpha < 1),$$

the resulting non-local fractional PDE has the form

$$w_t(t, x) - w_x(t, x) = \frac{\partial^2}{\partial x^2} \left(D_t^{1-\alpha} w(t, x) \right) + \lambda L_0(x) G(w(t, 0)), \ t > 0, \ x \in \mathbb{R},$$

$$(9.2.8)$$

with $\lambda > 0$. The spatially localised source term is described by the localising function (in a small region Ω about $x = 0$)

$$L_0(x) := \begin{cases} 1, & \text{if } x \in \Omega, \\ 0, & \text{otherwise.} \end{cases}$$

The equation (9.2.8) is complemented by the boundary and initial conditions

$$z(t, x) \to 0 \ \text{ as } |x| \to \infty, \quad \text{and} \quad z(0, x) = 0, \ x \in \mathbb{R}.$$

Under appropriate assumptions on the regularity of the given functions the solution of this initial-boundary-value problem is completely determined by its value $u(t) := w(t, 0)$ at $x = 0$. The VIE for $u(t)$ can be written as

$$u(t) = \int_0^t k(t-s) G(u(s)) \, ds, \ t \geq 0,$$

where the non-linearity G is the one in the localised source term. The convolution kernel $k(t-s)$ depends on the Green's function $H(t, 0; x, 0)$ associated with the above initial-boundary-value problem (cf. Olmstead, Kirk & Roberts, 2010, pp. 1658–59).

Example 9.2.13 (Olmstead & Roberts, 1996) The mathematical modelling of certain reaction processes governed by a combination of localised and non-local heat sources leads to systems of two non-linear VIEs whose solution may or may not blow-up in finite time (see, for example, Chadam & Yin, 1993). As a typical example we will consider the temperature distribution $w = w(t, x)$ in a strip of length $L > 0$, depending on a non-linear source which combines a localised contribution $\delta(x - a) G(w(t, x))$ (for some fixed $a \in (0, L)$) and a global contribution $[W(t)]^r$. Here, $r > 1$ and

$$W(t) := \|w(t, \cdot)\|_1 = L^{-1} \int_0^t |w(t, x)| \, dx.$$

The corresponding mathematical model consists of the semilinear diffusion equation

$$w_t(t, x) - w_{xx}(t, x) = \delta(x - a) G(z(t, x)) [W(t)]^r, \ x \in (0, L), \ t > 0,$$

the homogeneous Neumann or Dirichlet boundary conditions

$$w_x(t, 0) = w_x(t, L) = 0 \quad \text{(Neumann)} \quad \text{or}$$
$$w(t, 0) = w(t, L) = 0 \quad \text{(Dirichlet)}, \quad t > 0,$$

and an appropriate initial condition. If $H = H_N(t, x; s, \xi)$ and $H = H_D(t, x; s, \xi)$ stand respectively for the Green's functions associated with the Neumann or Dirichlet boundary conditions, then $u(t) := w(t, a)$ and $W(t)$ can be shown to be the solution of a *system of VIEs*, namely

$$u(t) = f(t, a) + \int_0^t H(t, a; s, a)G(u(s))[W(s)]^r \, ds,$$

$$W(t) = F(t, a) + \int_0^t \left(L^{-1} \int_0^t H(t, x; s, a)dx \right) G(u(s)) \, ds,$$

with

$$f(t, x) := \int_0^t H(t, x; 0, \xi)v_0(\xi) \, d\xi \quad \text{and} \quad F(t) := L^{-1} \int_0^t f(t, x) \, dx.$$

The finite-time blow-up of the solution of this system depends on the type of boundary condition: for Neumann conditions the solution always blow up in finite time, while this is not necessarily true for Dirichlet conditions (Olmstead & Roberts, 1996).

Example 9.2.14 (Roberts & Olmstead, 1999) In Example 3.3.5 we described a VIE quenching problem that has its origin in an initial-boundary-value problem for the linear diffusion equation with (non-linear) Neumann boundary conditions. The original problem consists in analysing the behaviour of the classical solution $v = v(t, x)$ of

$$\begin{aligned} w_t &= w_{xx}, \quad t > 0, \quad x \in (0, 1), \\ w(0, x) &= w_0(x) \in [0, 1), \quad x \in (0, 1), \\ w_x(t, 1) &= \lambda G(w(t, 1)), \quad t > 0, \\ w_x(t, 0) &= 0, \quad t > 0, \end{aligned} \tag{9.2.9}$$

where $\lambda \in \mathbb{R}$ denotes a non-zero parameter (see also Levine, 1983, Deng & Xu, 1999 and Yang & Brunner, 2014, Section 4.3). At the right boundary $x = 1$, the solution $w(t, 1) =: u(t)$ satisfies the VIE

$$u(t) = f(t) + \lambda \int_0^t k(t - s)G(u(s)) \, ds, \quad t \geq 0, \tag{9.2.10}$$

where G is the non-linear function in the right boundary condition (9.2.9). The function f and the convolution kernel k depend on the *Green's function* $\Gamma_N(t, x; \tau, \xi)$ of the given Neumann problem: as we indicated in Example 3.3.4 we have

$$f(t) = \int_0^t \Gamma_N(t, 1; 0, \xi) w_0(\xi) \, d\xi,$$

$$k(z) = \Gamma_N(z, 1; 0, 1) = 1 + 2 \sum_{n=1}^{\infty} e^{-n^2 \pi^2 z}. \qquad (9.2.11)$$

Since

$$K(t) = 1 + \frac{2}{\pi^2} \sum_{n=1}^{\infty} \frac{1}{n^2} \left(1 - \exp(-n^2 \pi^2 t) \right) \to \infty \quad \text{as} \ t \to \infty,$$

it follows from Theorem 3.3.7(b) that the solution $u(t) = v(t, 1)$ of (9.2.10) always quenches in finite time.

9.2.5 American Option Pricing

Example 9.2.15 (Jiang, 2005) The mathematical modelling of American option pricing leads to a free boundary-value problem where the free boundary (the *optimal exercise boundary*) is the solution of a non-linear VIE. Paraphrasing Jiang (2005, p. 113), an American option gives its holder the right to exercise the option at any time, and this may lead to a better profiteering opportunity compared to European option offers, since the price of an American option cannot be less than that of an equivalent European option. Whether or not the holder can actually benefit from this right at the cost of a higher premium will depend on whether she/he can take advantage of the early exercise term and exercise the option at an optimal time. The (unknown) free boundary in the mathematical model is the curve that divides the underlying domain $\{0 \le S < \infty, 0 \le t \le T\}$ into two parts: the *continuation region* and the *stopping region*.

The American option price can be decomposed into the sum of the European option price and a term that represents the extra premium required by the early exercising of the option. The European option price is determined by the Black–Scholes formula, while the extra premium depends on the location of the optimal exercise boundary $S(t)$. It can be shown (see Jiang, 2005, pp. 132–134 for details) that $S(t)$ is the solution of the rather complex (non-standard) non-linear VIE

$$
S(t) = K + S(t) e^{-q(T-t)} N\left(-\frac{-\ln \frac{S(t)}{K} + \beta_2(T-t)}{\sigma \sqrt{T-t}} \right)
$$
$$
- K e^{-r(T-t)} N\left(-\frac{-\ln \frac{S(t)}{K} + \beta_1(t-t)}{\sigma \sqrt{T-t}} \right)
$$

$$-Kr \int_t^T e^{-r(s-t)} \left[1 - N \left(\frac{\ln \frac{S(t)}{S(s)} + \beta_1(s-t)}{\sigma \sqrt{s-t}} \right) \right] ds$$

$$+ qS(t) \int_t^T e^{-q(s-t)} \left[1 - N \left(\frac{\ln \frac{S(t)}{S(s)} + \beta_2(s-t)}{\sigma \sqrt{s-t}} \right) \right] ds,$$

with the option strike price K, the time to maturity (i.e. the option's expiration date) T, the dividend rate q, the risk-free interest rate r, the volatility σ, and

$$\beta_1 := r - q - \frac{\sigma^2}{2}, \quad \beta_2 := r - q + \frac{\sigma^2}{2}.$$

The VIE depends depends on the fundamental solution of the Black–Scholes equation via the function

$$N(z) := (2\pi)^{-1/2} \int_{-\infty}^z e^{-\alpha^2/2} \, d\alpha.$$

Using the variable transformation $t = T - \tau$ and setting $u(\tau) := \ln S(T - \tau)$ we obtain the equivalent *implicit* VIE

$$e^{u(\tau)} = K + e^{u(\tau)} e^{-q\tau} N \left(-\frac{u(\tau) - \ln K + \beta_2 \tau}{\sigma \sqrt{\tau}} \right)$$

$$- K e^{-r\tau} N \left(-\frac{u(\tau) - \ln K + \beta_1 \tau}{\sigma \sqrt{\tau}} \right)$$

$$- Kr \int_0^\tau e^{-r(\tau - \eta)} N \left(-\frac{u(\tau) - u(\eta) + \beta_1(\tau - \eta)}{\sigma \sqrt{\tau - \eta}} \right) d\eta$$

$$+ q e^{u(\tau)} \int_0^\tau e^{-q(\tau - \eta)} N \left(-\frac{u(\tau) - u(\eta) + \beta_2(\tau - \eta)}{\sigma \sqrt{\tau - \eta}} \right) d\eta.$$

9.2.6 Optimal Control Problems

In this section we will adopt the standard notation used in optimal control: $x(t)$ will denote the solution, while $u(t)$ will denote the control function.

Example 9.2.16 (Gripenberg, 1983b) This problem arises as an investment model in which $x(t)$ denotes the flow of available resources (determined by previous investments and exterior inputs) and $G(x(t))$ stands for the investements. The aim is to minimise the total inputs $\int_0^\infty u(t)dt$ so that self-sustained growth is achieved (that is, the returns on the investements are sufficient for consumption and reinvestments). The resulting optimal control problem is the following: find a non-negative function $u = u(t)$ on $[0, \infty)$ so that $\int_0^\infty u(t)dt$ is mimimised under the condition that

$$\lim_{t \to \infty} \inf x(t) \ge \inf \left\{ \omega \in [0, \infty) : G(\omega) \int_0^\infty a(z)\, dz > \omega \right\}$$

when

$$x(t) = u(t) + \int_0^t k(t-s) G(x(s))\, ds, \quad t \in [0, \infty).$$

Here, the kernel $k(z)$ is assumed to be integrable, non-negative and non-increasing on $[0, \infty)$, and G is non-negative and concave on $[\omega_0, \infty)$, with $G(\omega) = 0$ on $[0, \omega_0]$ for some $\omega_0 \ge 0$.

A related control problem is studied in Gripenberg (1987): it asks to maximise the integral $\int_0^T F(t, x(t), u(t)) dt$, where $x(t)$ satisfies the VIE

$$x(t) = \int_0^t k(t, s, x(s), u(s))\, ds, \quad t \in [0, T],$$

subject to a constraint of the form $Q(t, x(t), u(t)) \ge 0$ $(t \in [0, T])$, where x and u are in $\mathbb{R}^d$. The notes at the end of the chapter contain additional references optimal control problems involving VIEs.

9.2.7 A Brief Review of Further Applications

- *Periodic Solutions of Infectious Disease Models:*
  The paper by Wong & Boey (2004) investigates the existence of a non-trivial non-negative periodic solution of a VFIE with constant delay $\tau > 0$,

  $$u(t) = \int_{t-\tau}^t G(s, u(s))\, ds, \quad t \ge 0,$$

  that models the spread of infectious disease. Related VFIEs are studied in e.g. Cooke (1976) and Nussbaum (1977).
- *Identification of Memory Kernels:* The mathematical modelling of heat conduction or viscoelastic phenomena in materials with memory leads to initial-boundary-value problems for parabolic or hyperbolic Volterra integro-differential equations where the (convolution) memory kernel is not known. The analysis of the existence of a solution to this inverse problem is based on a pair of (non-linear) first-kind and second-kind VIEs. Details can be found in the papers by von Wolfersdorf (1994), Unger & von Wolfersdorf (1995), Janno & Wolfersdorf (1995, 1996, 1997a, 1997b, 1998) (the paper 1997b contains extensive references, also on work by Grasselli, Lorenzi and Kabanikhin dealing with inverse problems for wave equations). The 1999 monograph by Kabanikhin & Lorenzi contains, in Chapter 1, an

overview of a broad class of inverse problems for hyperbolic equations that reduce to Volterra operator equations of the first or second kind.

- *Linear Viscoelasticity:* The VIE

$$u(t) = f(t) + \int_0^t \frac{k(t-s)}{1-p(t)} u(s(ds, \quad t \geq 0,$$

with $0 \leq p(t) < 1$, $k(z) > 0$ and $\int_0^\infty k(z)dz < 1$, occurs in the description of the modes in a linear viscoelastic rod, placed horizontally with fixed end points and subject to quasi-static bending (Reynolds, Appleby & Györi, 2007).

- *Theory of Water Percolation:* A mathematical model of water percolation from a cylindrical reservoir into the surrounding unsaturated region is given by the implicit VIE

$$u^2(x) = f(x) + \int_0^x k(t-s)u(s)\,ds \quad x \geq 0,$$

where $u(x)$ describes the height of the water table, with the functions k and f depending on certain physical parameters (Goncerzewicz et al., 1978, Okrasiński, 1980, 1989). An analogous implicit VIE,

$$Bu^{1/2}(t) = 1 - \int_0^t (t-s)^{-1/2}u(s)\,ds \quad (B > 0),$$

is discussed in Consiglio (1940; see also Miller, 1971, p. 72); it represents an early mathematical model for the study of turbulence. A similar implicit VIE arises in the mathematical modelling of the propagation of simple non-linear waves in gas-filled tubes with friction (Keller, 1981).

- *Simplified Formulation of the Fokker–Planck–Kolmogorov Equation:* The VIE with 'space map' g,

$$u(t,x) = f(t,x) + \int_0^t k(t-s)u(s, g(t,s,x))\,ds,$$

represents an alternative formulation of the Fokker–Planck–Kolmogorov equation for e.g. piecewise deterministic processes (cf. Annunziato & Messina, 2010 and Annunziato, Brunner & Messina, 2012, and their references). It is a model for distribution probability functions of a class of piecewise deterministic processes (i.e. a deterministic motion resulting the action of a discrete semi-Markov process). The kernel $k(t-s)$ represents the memory of the process, and the 'space map' $g(t, s, x)$ reflects a deterministic transport depending on the spatial variable x.

• *Asymptotic Membrane Theory of Hyperbolic Shells:* The VFIE

$$u(x) = f(x) + \frac{\gamma}{\theta(x)} \int_0^{\theta(x)} u(s)\, ds, \quad x \in [0, 1],$$

where, for given $a > 1$ and $0 < \gamma < 1$,

$$\theta(x) := \begin{cases} ax & \text{if } 0 \leq x \leq a^{-1} \\ 1 & \text{if } a^{-1} \leq x \leq 1, \end{cases}$$

is a simple model for certain problems in elasticity theory. A similar functional integral equation arises in the asymptotic membrane theory of hyperbolic shells where it is the key to understanding the regularity properties of the asymptotic replacement field at the corners of the shell (Pilla, 1996, Pilla & Pitkäranta, 1996).

• *Transient Behaviour of Semiconductor Devices:* First-kind and (systems of) second-kind VIEs occur as mathematical models of switching of semiconductor PN-diodes from forward to reverse bias (Schmeiser, 1993, Schmeiser, Unterreiter & Weiss, 1993). Similar VIEs also arise in connection with the bipolar drift diffusion equation which can, under certain conditions, be reduced to a coupled system of non-linear second-kind VIEs; cf. Unterreiter (1996).

• *Auto-Convolution VIEs:* An interesting application of auto-convolution VIEs of the *second kind* is described in von Wolfersdorf (2008), namely its use for the computation of special functions like the generalised Mittag-Leffler function. See also von Wolfersdorf & Janno (1995) and von Wolfersdorf (2000: this paper contains a large list of references). Auto-convolution VIEs of the *first kind*,

$$\int_0^t u(t - s)u(s)\, ds = g(t), \quad t \in I,$$

arise in stochastics, probability theory, and spectroscopy; they are severely ill posed (Hoffman, 1994). See also Gorenflo & Hofmann (1994) and von Wolfersdorf (2000 and its references), as well as Fleischer & Hofmann (1996, 1997) and Dai & Lamm (2008).

• *Non-Ageing Linear Isothermal Quasi-Static Isotropic Compressible Solid Viscoelasticity:* Mathematical models are partial VIEs of the second kind. Shaw & Whiteman (1997) contains an illuminating survey; see also Shaw, Warby & Whiteman (1997) and Ilhan (2012) for an abstract setting of such VIEs.

9.3 VIEs of the Third Kind

Example 9.3.1 (von Wolfersdorf & Hofmann, 2008) In the theory of elasticity the following inverse problem is encountered: find the convolution kernel $k(t - s)$ of a linear second-kind VIE when the quotient $q(z) := r(z)/k(z)$ of its resolvent kernel $r(t - s)$ and its kernel is known. As shown in the paper mentioned above, this leads to an auto-convolution VIE of the third kind for k,

$$a(z)k(z) - \int_0^z q(s)k(z - s)k(s)\,ds, \quad z \in I,$$

with $a(z) := 1 - q(z)$. Such VIEs were studied in Berg & von Wolfersdorf (2005).

In certain singular optimal control problems over an infinite interval, as encountered in mathematical economics (see Grandits, 2008 for details and relevant references), the value function is the solution of a free boundary-value problem. It turns out that the free boundary, subject to so-called smooth fitting conditions, is the solution of a non-linear VIE involving the Green's function of the heat operator. If this VIE is linearised (using the Fréchet derivative of the non-linear integral operator), there results a third-kind VIE of the form described in Example 1.6.3.

9.4 Systems of Integral-Algebraic VIEs

Example 9.4.1 (Cannon, 1984, Jumarhon et al., 1996) Diffusion of a chemical in a bulk liquid with a chemical reaction on e.g. some electrode; also: in antibody/antigen technology and the development of a medical product (Jones, Jumarhon, McKee & Scott, 1996). A representative mathematical model is given by an initial-boundary-value problem for the linear diffusion equation:

$$u_t(t, x) = u_{xx}(t, x), \quad x \in (0, 1), \quad t > 0; \tag{9.4.1}$$

$$u(0, x) = u_0(x), \quad x \in (0, 1);$$

$$u_x(t, 0) = G_0(t, u(t, 0));$$

$$u_x(t, 1) = G_1(t, u(t, 1)), \quad t > 0. \tag{9.4.2}$$

Let $F(t, x) := (4\pi t)^{-1/2}e^{-x^2/4t}$ denote the fundamental solution of the (one-dimensional) linear heat equation. In order to obtain an expression for the solution of the above initial-boundary-value problem we require the functions

$$\theta(t, x) := \sum_{n=-\infty}^{\infty} F(t, x + 2n) \quad (t > 0)$$

and

$$w(t, x) := \int_0^1 \big(\theta(t, x - \xi) + \theta(t, x + \xi)\big) u_0(\xi) \, d\xi \, .$$

The solution then can be written in the form

$$u(t, x) = w(t, x) - 2 \int_0^t \theta(t - s, x) G_0(s, \phi_0(s)) \, ds$$

$$+ 2 \int_0^t \theta(t - s, x - 1) G_1(s, \phi_1(s)) \, ds \qquad (9.4.3)$$

if, and only if, ϕ_0 and ϕ_1 are (piecewise) continuous functions that solve the system of VIEs

$$\phi_0(t) = w(t, 0) - 2 \int_0^t \theta(t - s, 0) G_0(s, \phi_0(s)) \, ds$$

$$+ 2 \int_0^t \theta(t - s, -1) G_1(s, \phi_1(s)) \, ds, \qquad (9.4.4)$$

$$\phi_1(t) = w(t, 1) - 2 \int_0^t \theta(t - s, 1) G_0(s, \phi_0(s)) \, ds$$

$$+ 2 \int_0^t \theta(t - s, 0) G_1(s, \phi_1(s)) \, ds \qquad (9.4.5)$$

(cf. Cannon, 1984, Sections 6.5–7.2, and pp. 72–75). If $G_0(s, \cdot)$ and $G_1(s, \cdot)$ are Lipschitz continuous, the solution u is unique.

We observe that (9.4.4), (9.4.5) represent a system of VIEs of the second kind. If the second boundary condition (9.4.2) is replaced by the Dirichlet condition $u(t, 1) = h(t)$, it can be shown that the solution representation (9.4.3) becomes

$$u(t, x) = w(t, x) - 2 \int_0^t \theta(t-s, x) G_0(s, \phi_0(s)) ds + 2 \int_0^t \theta(t-s, x-1) \phi_1(s) \, ds,$$

where ϕ_0 and ϕ_1 now are the solution of a system consisting of a second-kind VIE and a first-kind VIE (that is, a *system of IAEs*),

$$\phi_0(t) = w(t, 0) - 2 \int_0^t \theta(t - s, 0) G_0(s, \phi_0(s)) ds$$

$$+ 2 \int_0^t \theta(t - s, -1) \phi_1(s) ds, \qquad (9.4.6)$$

$$h(t) = w(t, 1) - 2 \int_0^t \theta(t - s, 1) G_0(s, \phi_0(s)) ds$$

$$+ 2 \int_0^t \theta(t - s, 0) \phi_1(s) ds \qquad (9.4.7)$$

(cf. Cannon, 1984, p. 75). If $G_0(s, \cdot)$ is Lipschitz continuous, we have $\phi_0(t) = u(t, 0)$ and $\phi_1(t) = u_x(t, 1)$ $(t \geq 0)$ (Jumarhon et al., 1996).

9.5 Notes

Section 9.1: *VIEs of the First Kind*

An early paper on applications of first-kind Abel integral equations is Rothe (1931; cf. pp. 381–387). The review paper by Anderssen (1977) contains extensive lists of applications modelled by first-kind Abel-type VIEs with kernel singularities $(t^2 - s^2)^{-1/2}$ $(s < t)$ or $(s^2 - t^2)^{-1/2}$ $(s > t)$. A large list of references can also be found in Anderssen & de Hoog (1990) and in Gorenflo & Vessella (1991). Analogous VIEs of the second kind arise in stereology; cf. Anderssen & de Hoog (1990), p. 379.

A mathematical model of the Hertz contact problem (identation of an elastic half-space by an axisymmetric punch under a monotonically applied normal force, assuming Coulomb friction in the region of contact; cf. Gauthier, Knight & McKee, 2007 and Linz & Noble, 1971) is given by a system of two first-kind Abel-type integral equations with kernel singularities $(s^2 - x^2)^{-1/2}$ $(s > x)$.

Applications of more general VIEs of the first kind are described in the books by Bukhgeim (1983, 1999), Hritonenko & Yatsenko (1996) and Apartsyn (2003). A selection of papers on this subject are von Wolfersdorf (1995 on the Darboux problem), Peirce & Siebrits (1996, 1997), Apartsyn (2013), Apartsyn & Sidler (2013 on VFIE arising in age-structured models of electric power systems; see also Apartsyn, 2014 and its references). Other first-kind VIEs arising in applications can be found in Davis & Duncan (2004, related to a retarded potential integral equation), de Hoog & Anderssen (2006 on variable lower limit of integration αt $(0 < \alpha < 1)$), Anderssen, Davies & de Hoog (2008, 2011 on linear viscoelasticity) and de Hoog & Anderssen (2010, 2012a on the effect of kernel perurbations on g and u, with applications in e.g. hydrology and viscoelastic processes).

Another source of (systems of) first-kind VIEs is the *identification of memory kernels* in Volterra-type integro-differential equations describing heat conduction or viscoelastic behaviour in materials with memory; see, for example, the papers by von Wolfersdorf (1994), Unger & von Wolfersdorf (1995), Janno & Wolfersdorf (1996, 1997a, 1997b, 1998, 2001, 2002), and their references. Of interest also is the book by Kabanikhin & Lorenzi (1999: Ch. 1 describes the reduction of inverse problems for hyperbolic PDEs to (operator) VIEs). Systems of first-kind VIEs with convolution kernels also arise in system identification problems that require the construction of symmetric kernels $h_i(z_1, \ldots, z_i)$ for the VIE

$$\sum_{i=1}^{n} \int_0^t \cdots \int_0^t h_i(t - s_1, \ldots, t - s_i)u(s_1)\ldots u(s_i)\, ds = g(t)$$

with given input $u(t)$ and output $g(t)$ (Brenner & Xu, 2002).

Section 9.2: *VIEs of the Second Kind*

There exists extensive literature on the *renewal equation*, going back to Lotka (1939, 1939a) and Feller (1941); see, for example, Lotka (1939/1998, 1942, 1945), Feller (1941, 1971), Brauer (1976b), Miller (1975a), Tsalyuk (1989), Gripenberg, Londen & Staffans (1990: Section 15.7 deals with non-linear renewal equations). The paper by Ney (1977) is related to stochastic growth models.

Second-kind VIEs arise as *population growth models* and models for the spreading of *epidemics*, as shown in the papers by Brauer (1975, 1976a), Thieme (1979), Diekmann (1979), Thieme (1977), Diekmann (1978), Busenberg & Cooke (1980), Hethcote & Tudor (1980), Gripenberg (1981b, 1983a), Metz & Diekmann (1986), and Hethcote, Lewis & van den Driessche (1989). Classical references are the books by Waltham (1974) and Brauer & Castillo-Chávez (2001).

In addition (also on VFIEs with constant or variable delays) see Cooke & Yorke (1973), Cooke (1976), Cooke & Kaplan (1976), Smith (1977), Torrejón (1990), Cañada & Zertiti (1994), Hethcote & van den Driessche (1995, 2000), Brauer & van den Driessche (2003), Bélair (1990 on state-dependent VFIE), and Gourley, Liu & Lou (2016 on where VIEs are used as a tool to analyse the boundedness of solutions of a model describing intra-specific and insect larval development).

Of the numerous papers discusssing *heat transfer and diffusion models* we mention the ones by Lighthill (1950; see also Noble, 1964 and Curle, 1978), Mann & Wolf (1951), Roberts & Mann (1951: extension of the analysis in Mann & Wolf, 1951), Levinson (1960 on superfluidity), Keller & Olmstead (1971), Handelsman & Olmstead (1972), Olmstead & Handelsman (1976), Olmstead (1977), Gorenflo (1987) and Berrone (1995). A non-local diffusion problem leading to a two-dimensional Volterra integral equation of the second kind with two weak kernel singularities is described in McKee & Cuminato (2015).

On *blow-up phenomena* see the review papers by Levine (1990), Brunner & Bandle (1998) and Deng & Levine (2000) on finite-time blow-up in semilinear parabolic PDEs), and on blow-up in non-linear VIEs, Olmstead (1983, 1997, 2000), Roberts, Lasseigne & Olmstead (1983, 1997, 2000), Olmstead & Roberts (1994, 1996, 2008), Olmstead, Roberts & Deng (1995), Roberts & Olmstead (1996), Roberts (1997, 2007), Mydlarczyk (1999: an extension of the analysis in Roberts, Lasseigne & Olmstead, 1993), Roberts (1998, 2007: review papers), Olmstead, Kirk & Roberts (2010), Kirk, Olmstead & Roberts (2013 on systems of VIEs), Brunner & Yang (2013 on general theory and references), Yang & Brunner (2013a on blow-up in VFIEs) and Yang & Brunner (2013b on numerical analysis of blow-up).

Quenching in PDEs is described by Levine (1983, 1992: these papers contain extensive lists of references). For VIE models with quenching of solutions see Olmstead & Roberts (1993), Deng & Roberts (1997), Roberts & Olmstead (1999), Kirk & Roberts (2002, 2003) and Roberts (2007: review paper). The paper by Yang & Brunner (2014) presents the general theory of quenching and contains numerous references on applications.

VIEs arising in the modelling of *American option pricing* are derived in the books by Jiang (2005) and Kwok (2008: the latter also contains an overview of numerical methods). I am grateful to Professor Zhang Kai (Jilin University) for pointing out these references to me.

Optimal control problems involving VIEs are described in Nelson & Young (1968, minimising natural fuel requirement for nuclear reactor power systems), in the three papers by Vinokurov (1969; see also the comments by Neustadt & Warga, 1970), in Bakke (1975, 1976), Angell (1976, 1983), Medhin (1986), Carlson (1987, 1990), Gripenberg (1983b, 1989), Belbas (1999, 2008), Jung (1999), Belbas & Schmidt (2005) and Apartsyn (2008).

Section 9.4: *Systems of Integral-Algebraic VIEs*
The mathematical modelling of e.g. the time evolution of the temperature at the surface of a conductiong solid with high thermal loss leads to a (non-linear) *singularly perturbed* VIE. Problems of this kind were studied by Olmstead (1972), Olmstead & Handelsman (1972), Angell (1985: PhD thesis), Angell & Olmstead (1987), Skinner (1975, 1995: population growth model) and (2000: heat conduction with linear radiation condition). The review paper by Kauthen (1997) contains a comprehensive list of references on such appliactions. In addition, see Bijura (2002a, 2002b, 2002c, 2006, 2010) and Shubin (2006).

Stochastic VIEs
As we mentioned in the Preface, limitations of space prevent a detailed discussion of *stochastic VIEs* (in view of recent developments, a thorough introduction to this increasingly important topic would require a separate monograph). The reader wishing to acquire insight into such problem may wish to consult Tudor (1986), Shaikhet (1995), Chapter 11 in O'Regan & Meehan (1998), Appleby & Reynolds (2003), Karczewska (2007), Karczewska & Lizama (2009), Appleby & Riedle (2010), Øksendal & Zhang (2010), Zhang (2010: this paper contains an extensive list of references), Desch & Londen (2011, on L^p theory), Desch & Londen (2013, on regularity of solutions) and Cao & Zhang (2015 and its references on applications). See also Lewin (1994) on stochastic singularly perturbed VIEs. The paper by Cao & Zhang (2015) describes and analyses a collocation method for the numerical solution of stochastic VIEs.

Appendix
A Review of Banach Space Tools

Summary

The Appendix is intended as a concise guide to basic concepts and results from Functional Analysis that are used throughout this book, especially in Chapter 2 (regularity of solutions), Chapter 5 (integral-algebraic equations), Chapter 7 (cordial Volterra integral operators and integral equations), and Chapter 8 (Volterra integral operators on Banach spaces).

A.1 Banach Spaces in the Theory of VIEs

Before presenting a review of various definitions and key theorems from the theory of linear operators on Banach spaces we briefly look at some particular Banach spaces relevant to the theory of Volterra integral equations. The reader interested in seeing a more detailed treatment, and the proofs of the various theorems, may wish to consult one of the monographs mentioned in Section A.4 Notes, below.

We remind the reader that a (real or complex) linear normed space X is a *Banach space* if it is complete with respect to its norm; that is, if every Cauchy sequence $\{x_n\}$ of elements $x_n \in X$ converges to an alement $x \in X$. A special class of Banach spaces is given by the *Hilbert spaces* $X = H$. Here, the norm $\| \cdot \|$ is induced by an inner product $\langle \cdot, \cdot \rangle$ in H: we have $\|x\| = (\langle x, x \rangle)^{1/2}$, and H is complete with respect to this norm.

A.1.1 The Spaces $C^d(I)$

For an interval $I = [0, T]$ and a given integer $d \geq 0$, the space

$$C^d(I) := \{f : \ f^{(j)} \in C(I) \ (j = 0, 1, \ldots, d)\}$$

denotes the functions possessing continuous derivatives of order d on I. If $d = 0$ we write $C(I)$ instead of $C^0(I)$: it denotes the space of functions that are continuous on I. Under the norm

$$\|f\|_\infty := \max\{|f(t)| : t \in I\},$$

$C(I)$ is a Banach space. If $d \geq 1$ the space $C^d(I)$ becomes a Banach space under the norm

$$\|f\|_{d,\infty} := \sum_{j=0}^{d} \|f^{(j)}\|_\infty . \tag{A.1.1}$$

A.1.2 The Hölder Spaces $C^{d,\beta}(I)$

We have seen in Section 1.3 that the solutions of linear first- and second-kind VIEs with weakly singular kernels but otherwise smooth data are smooth on the left-open interval $(0, T]$ but have unbounded first derivatives at $t = 0$. These solutions belong to certain *Hölder spaces*. The general Hölder space $C^{d,\beta}(I)$ ($d \in \mathbb{N}$, $0 < \beta \leq 1$) consists of functions that have continuous derivatives up to order d on I, with the derivative of order d being Hölder-continuous of order β. This is made precise in the following definition (using the notation of Kufner et al., 1977).

Definition A.1.1 Let $0 < \beta < 1$.

(i) A function f defined on $I = [0, T]$ is said to be *Hölder-continuous* at $t = t_0 \in I$ with exponent β if its (local) *Hölder coefficient* at t_0,

$$|f|_{\beta,t_0} := \sup_{t \in I, t \neq t_0} \frac{|f(t) - f(t_0)|}{|t - t_0|^\beta},$$

is finite. It is (uniformly) Hölder-continuous in I if the *Hölder constant*

$$|f|_\beta := \sup_{t,s \in I, t \neq s} \frac{|f(t) - f(s)|}{|t - s|^\beta}$$

is finite. The space $C^\beta(I)$ of Hölder-continuous functions on I is called a *Hölder space*.

(ii) If $d \in \mathbb{N}$ ($d \geq 1$), we define

$$C^{d,\beta}(I) := \{f \in C^d(I) : f^{(d)} \in C^\beta(I)\}. \tag{A.1.2}$$

◇

Example A.1.2 The standard example of a function that is in $C^{d,\beta}(I)$ (for any $T > 0$) is $f(t) := t^{d+\beta}$ ($d \in \mathbb{N}$, $0 < \beta < 1$). If $d = 0$, it is uniformly Hölder-continuous on $I = [0, T]$.

The proof of the following theorem can be found, for example, in Kufner et al. (1977, pp. 22–26) or in Gilbarg & Trudinger (2001, pp. 52–53).

Theorem A.1.3 *For* $0 < \beta < 1$, $|f|_\beta$ *defines a* semi-norm *on* $C^\beta(I)$, *and* $\|f\|_{C^\beta} := \|f\|_\infty + |f|_\beta$ *defines a* norm *on* $C^\beta(I)$. *If* $d \geq 1$ *and* $\|f\|_{C^{d,\beta}} :=$
$$\sum_{j=0}^{d} \|f^{(j)}\|_\infty$$
then $\|f\|_{C^{d,\beta}} := \|f\|_{C^d} + |f^d|_\beta$ *defines a norm on* $C^{d,\beta}(I)$.
Endowed with these norms, the Hölder spaces $C^\beta(I)$ *and* $C^{d,\beta}(I)$ ($d \geq 1$) *are* Banach spaces.

We conclude this section by stating some additional properties of Hölder spaces. Their proofs may be found in e.g. Kufner (1977, pp. 22–27).

Theorem A.1.4 *For Hölder spaces the following inclusion relation holds: if* $d + \beta < d' + \beta'$ *then*
$$C^{d',\beta'}(I) \subset C^{d,\beta}(I).$$

In particular, if $f \in C^\beta(I)$ *then* $f \in C^{\beta'}(I)$ *whenever* $\beta' < \beta$. *Moreover, the product of two Hölder-continuous functions is again Hölder-continuous; that is, for* $0 < \alpha, \beta < 1$ *we have*
$$f \in C^\alpha(I), \ g \in C^\beta(I) \implies fg \in C^\gamma(I), \ \text{with} \ \gamma := \min\{\alpha, \beta\}.$$

A.1.3 The Lebesgue Spaces $L^p(0, T)$

It is assumed that the reader is familiar with the basic theory of Lebesgue measures and Lebesgue integrals (see, for example, Ciarlet, 2013, pp. 25–34 for a thorough introduction). In the following, Ω denotes a (bounded) open subset of $\mathbb{R}^n$. In e.g. Chapters 2, 7 and 8 the role of Ω is taken by $\Omega = (0, T)$. Thus, we will state the subsequent definitions and theorems in terms of the open interval $(0, T)$.

We remind the reader that two functions f and g that are Lebesgue measurable on Ω are said to be *equivalent* if they coincide on Ω except possibly on some subset of Ω of (Lebesgue) measure zero. We then write $f = g$ a.e. ('almost everywhere') on Ω.

Definition A.1.5 For $f : (0, T) \to \mathbb{R}$ (or $f : (0, T) \to \mathbb{C}$) we set

$$\|f\|_p := \begin{cases} \left(\displaystyle\int_0^T |f(t)|^p \, dt \right)^{1/p} & \text{if } 1 \leq p < \infty \\[2ex] \text{sup ess} \{|f(t)|\} & \text{if } p = \infty, \end{cases} \tag{A.1.3}$$

where the essential supremum of f is given by

$$\text{sup ess} \{|f(t)|\} := \inf\{c \geq 0 : |f(t)| \leq c \text{ a.e. on } (0, T)\}.$$

For $1 \leq p \leq \infty$ the Lebesgue space $L^p(0, T)$ is defined by

$$L^p(0, T) := \{f : \|f\|_p < \infty\},$$

and (A.1.3) defines a norm for $L^p(0, T)$. A function $f \in L^1(0, T)$ will be called (Lebesgue) *integrable* on $(0, T)$. If $p > 1$ then $f \in L^p(0, T)$ means that $|f|^p$ is (Lebesgue) integrable on $(0, T)$.   ◇

The proofs of the following properties of L^p spaces defined above can be found, for example, in Gorenflo & Vessella (1991) or in Ciarlet (2013).

(i) The spaces $L^p(0, T)$ ($1 \leq p \leq \infty$) are complete under the norms (A.1.3).

(ii) The following *inclusion relation* is true: if $1 \leq p < q \leq \infty$, then $L^q(0, T) \subset L^p(), T)$.

(iii) There holds the *Hölder's inequality*: if f and g are L^1 functions on I, and if $p \geq 1$ and q are such that $1/p + 1/q = 1$ holds, then

$$\|fg\|_1 = \int_0^T |f(t)g(t)| \, dt \leq \left(\int_0^T |f(t)|^p \, dt \right)^{1/p}$$
$$\times \left(\int_0^T |g(t)|^q \, dt \right)^{1/q} = \|f\|_p \|g\|_q .$$

For $p = 2$ we obtain the *Cauchy–Schwarz inequality*,

$$|\langle f, g \rangle| \leq \|f\|_2 \|g\|_2 .$$

(iv) Another important inequality is *Young's inequality* (see e.g. Gilbarg & Trudinger, 2001, p. 145 or Gorenflo & Vessella, 1991, p. 65), which we state in the following form (more abstract versions may be found in Kufner et al., 1977, pp. 64–65 and p. 135). Assume that $f \in L^p(\mathbb{R})$ and $g \in L^q(\mathbb{R})$, where $1 \leq p \leq \infty$ and $1 \leq q \leq \infty$ are such that $1/p + 1/q \geq 1$. If we denote by $f * g$ the convolution (on $\mathbb{R}$),

$$(f * g)(t) := \int_{\mathbb{R}} f(t - s)g(s) \, ds, \quad t \in \mathbb{R},$$

then $\|f * g\|_r \leq \|f\|_p \|g\|_q$, with $1/p + 1/q = 1 + 1/r$.

Theorem A.1.6 *Under the norms (A.1.3) the function spaces* $L^p(0, T)$ $(1 \leq p \leq \infty)$ *are* Banach spaces. *If* $p = 2$ *then* $L^2(0, T)$ *is a Hilbert space, corresponding to the inner product*

$$\langle f, g \rangle := \int_0^T f(t)\overline{g(t)} \, dt \quad \textit{and the induced norm} \quad \|f\|_2 := |\langle f, f \rangle|^{1/2}.$$

Here, $\overline{g(t)}$ denotes the complex conjugate value of $g(t)$.

A.1.4 The Sobolev Spaces $W^{d,p}(\Omega)$

These function spaces may be viewed as analogues of the Hölder spaces in Section A.1.2, in the sense that the conditions of continuous differentiability and Hölder continuity for the latter are now replaced by weak (or: distributional) differentiability and L^p integrability (see, for example, Gilbarg & Trudinger, Ch. 7 or Ciarlet, 2013, pp. 312–319 and 326–331). As in the previous section we will again state our definitions and theorems in terms of $\Omega = (0, T)$.

Definition A.1.7 For fixed $p \in [1, \infty]$ and $d \in \mathbb{N}$ the Sobolev space $W^{d,p}(0, T)$ is the space of all L^p functions possessing (weak) derivatives up to (and including) order d on $(0, T)$. ◇

Theorem A.1.8 *The expression*

$$\|f\|_{W^{d,p}} := \left(\sum_{j=0}^d \int_0^T |f^{(j)}(t)|^p \, dt \right)^{1/p} = \left(\sum_{j=0}^d \|f^{(j)}\|_p^p \right)^{1/p}$$

defines a norm in the Sobolev space $W^{d,p}(0, T)$. Under this norm $W^{d,p}(0, T)$ becomes a Banach space. In analogy to the L^p spaces, Sobolev spaces also satisfy the inclusion relation $W^{d,q}(0, T) \subset W^{d,p}(0, T)$ whenever $1 \leq p < q \leq \infty$.

Remark A.1.9 If $p = 2$, the notation $H^d(0, T) := W^{d,2}(0, T)$ is usually employed. These spaces are Hilbert spaces under the scalar product

$$\langle f, g \rangle := \int_0^T \sum_{j=0}^d f^{(j)}(t)\overline{g^{(j)}(t)} \, dt, \quad \text{with induced norm}$$

$$\|f\|_2 := |\langle f, f \rangle|^{1/2}.$$

For $d = 0$ we have $W^{0,2}(0, T) = H^0(0, T) = L^2(0, T)$. ◇

A.2 Linear Operators on Banach Spaces

A.2.1 Bounded Operators

Definition A.2.1 Let X and Y be normed spaces, and assume that A is a linear mapping (linear operator) from X to Y. Then:

(i) A is *injective* (or: one-to-one) if for $y \in Y$ there exists at most one $x \in X$ so that $y = Ax$.

(ii) A is said to be *surjective* (or: on to) if for $y \in Y$ there is at least one $x \in X$ such hat $y = Ax$.

(iii) A is called bijective if it is both injective and surjective. $\diamond$

In the following, X and Y will denote Banach spaces. Unless stated otherwise, we assume that the domain of A is the space X.

Definition A.2.2 Let $A : X \to Y$ be a linear operator.

(i) A is said to be *continuous* if for all sequences $\{x_n\}$ in X with $\|x - x_n\| \to 0$ we have $\|Ax - Ax_n\| \to 0$, as $n \to \infty$.

(ii) A is a *bounded* operator if there exists a constant C such that $\|Ax\| \le C\|x\|$ for all $x \in X$. We will denote the set of all bounded linear operators from X to Y by $\mathcal{L}(X, Y)$. If $X = Y$ we write $\mathcal{L}(X) := \mathcal{L}(X, X)$.

(iii) The *norm* of the operator A is given by

$$\|A\| := \sup \{\|Ax\| : x \in X, \ \|x\| \le 1\}.$$

(iv) A is a *closed operator* if its *graph*, $\operatorname{graph} A := \{(x, Ax) : x \in X\}$, is closed in $X \times Y$ with respect to coordinate-wise convergence; that is, A is closed if, and only if,

$$\lim_{n \to \infty} x_n = x \in X \quad \text{and} \quad \lim_{n \to \infty} Ax_n = y \quad \text{imply} \quad y = Ax.$$

(v) The set $\ker A := \{x \in X : Ax = 0\}$ is called the *null space* (or the *kernel*) of A, while

$$\operatorname{Im} A := \{y \in Y : \text{there exists } x \in X \text{ so that } Ax = y\}$$

is called its *image* (or *range*). $\diamond$

Theorem A.2.3 *A linear operator $A : X \to Y$ is bounded if, and only if, it is continuous. Also, if X is a normed space and Y is a Banach space, $\mathcal{L}(X, Y)$ is a Banach space. In particular, if $Y = X$ and X is a Banach space, $\mathcal{L}(X)$ is a Banach space.*

Theorem A.2.4 *A closed linear operator* $A : X \to Y$ *is continuous; that is,* $A \in \mathcal{L}(X, Y)$.

A key result in Banach space operator theory is the *Open Mapping Theorem* (Banach, 1932; see, for example, Ciarlet, 2013, pp. 255–257).

Theorem A.2.5 *Let X and Y be Banach spaces, and assume that the mapping* $A \in \mathcal{L}(X, Y)$ *is surjective. If M is an open subset of X, the image of M under A is an open subset of Y.*

The question on the solvability of the operator equation $Ax = y$, where $A \in \mathcal{L}(X, Y)$ and $y \in Y$, leads to the notion of the *inverse* of the operator $A : Y \to X$, and its continuity. What can be said about the existence and the properties of the inverse A^{-1} of an operator $A \in \mathcal{L}(X, Y)$? The following result is a consequence of the Open Mapping Theorem (cf. Ciarlet, 2013, pp. 255–257).

Corollary A.2.6 *Assume that $A \in \mathcal{L}(X, Y)$ is bijective. Then $A^{-1} \in \mathcal{L}(Y, X)$.*

Another key result in Banach space operator theory is the *Closed Graph Theorem* (Banach, 1932, pp. 41–42; Ciarlet, 2013, p. 259). It is, for example, used in proving that a projection operator (see Definition A.2.9 below) is a bounded operator.

The following theorem is a consequence of the Closed Graph Theorem (see, for example, Schechter, 2002, pp. 61–63).

Theorem A.2.7 *Assume that $A \in \mathcal{L}(X, Y)$ is such that* $\operatorname{Im} A = Y$ *and* $\ker A = \{0\}$. *Then $A^{-1} \in \mathcal{L}(Y, X)$.*

We have seen in Section 1.2 (Theorem 1.2.3) that if $X = C[0, T]$, with norm $\|\cdot\|_\infty$, and if $A : X \to X$ is a Volterra integral operator $\mathcal{V}$ with continuous or weakly singular kernel, the VIE $(\mathcal{I} - \mathcal{V})u = f$ has a unique solution, namely $u = (\mathcal{I} - \mathcal{V})^{-1}f \in X$ (where $\mathcal{I}$ denotes the identity operator on X) for every $f \in X$. The abstract basis for the invertibility of $\mathcal{I} - \mathcal{V}$ is contained in the following theorem.

Theorem A.2.8 *Let X be a Banach space and assume that $A \in \mathcal{L}(X)$ is such that* $\|A\| < 1$. *Then $\mathcal{I} - A \in \mathcal{L}(X)$ is bijective, and its inverse $(\mathcal{I} - A)^{-1}$ is in $\mathcal{L}(X)$. It is given by the Neumann series*

$$(\mathcal{I} - A)^{-1} = \sum_{n=0}^{\infty} A^n, \quad with \quad \|(\mathcal{I} - A)^{-1}\| \leq (1 - \|A\|)^{-1}.$$

A particular class of linear bounded operators, the *projection operators*, will play an important role in the decoupling theory of integral-algebraic equations (see Section 5.2).

Definition A.2.9 A linear operator $P : X \to M \subset X$ is called a *projection operator* in X if P is *idempotent* (that is, if $P^2 = P$). If $x \in X$, the element Px is called the projection of x on M.   ◇

Some of the basic properties of a projection operator P are described in the next theorem.

Theorem A.2.10 *Let X be a Banach space and $P : X \to M \subset X$ a projection operator.*

(a) If $x \in M$, then $Px = x$.
(b) The operator $Q := \mathcal{I} - P$ is a projection operator, and we have

$$\ker P = \operatorname{Im} Q, \quad \operatorname{Im} P = \ker Q.$$

(c) P is a bounded operator (i.e. $P \in \mathcal{L}(X)$), with $\|P\| = 1$.

If H is a *Hilbert space* we can define certain special projection operators. An *orthogonal projection* P is a projection operator whose range and whose null space are orthogonal subspaces of H; that is, if $\langle x, Py \rangle = \langle Px, Py \rangle = \langle x, Py \rangle$ for all $x, y \in H$. P is an orthogonal projection if, and only of, P is a self-adjoint operator.

A.2.2 Compact Operators

Definition A.2.11 Assume that X and Y are Banach spaces. A linear operator A from X to Y is said to be *compact* (or: *completely continuous*) if the image $\{Ax_n\}$ of any bounded sequence $\{x_n\}$ in X contains a convergent subsequence. We denote the set of compact linear operators from X to Y by $\mathcal{K}(X, Y)$.   ◇

Example A.2.12 The Fredholm integral operator: $\mathcal{F} : C(I) \to C(I)$ (with $I := [0, T]$),

$$(\mathcal{F}f)(t) := \int_0^T K(t, s) f(s) \, ds, \ t \in I,$$

where K continuous on $I \times I$, is compact. The same is true for $\mathcal{F} : L^2(0, T) \to L^2(0, T)$ with $K \in L^2((0, T) \times (0, T))$.

Example A.2.13 The Volterra integral operator: $\mathcal{V} : C(I) \to C(I)$,

$$(\mathcal{V}f)(t) = \int_0^t K(t, s) f(s) \, ds, \ \ t \in I,$$

with $K \in C(D)$ ($D := \{(t, s) : \ 0 \le s \le t \le T\}$), is a compact operator. If $\mathcal{V}$ has a kernel $K \in L^2(D)$, then $\mathcal{V}$ is a compact operator on $L^2(0, T)$.

Example A.2.14 The Fredholm integral operator $\mathcal{F}_\alpha : \ C(I) \to C(I)$ given by

$$(\mathcal{F}f)(t) := \int_0^T |t - s|^{\alpha-1} K(t, s) f(s) \, ds, \ \ t \in I,$$

with $K \in C(I \times I)$, is compact for any $\alpha \in (0, 1)$. The same is true if the singularity $|t - s|^{\alpha-1}$ is replaced by $\log |t - s|$.

Example A.2.15 The Volterra integral operator $\mathcal{V}_\alpha : \ C(I) \to C(I)$ with $0 < \alpha < 1$,

$$(\mathcal{V}_\alpha f)(t) = \int_0^t (t - s)^{\alpha-1} K(t, s) f(s) \, ds, \ \ t \in I,$$

is compact whenever $K \in C(D)$. With this assumption it is also compact if the kernel singularity is replaced by $\log |t - s|$.

In the following theorem we collect a number of basic properties of compact operators.

Theorem A.2.16 *Let X and Y be Banach spaces. Then the following statements are true:*

(a) *If $A : \ X \to Y$ is a compact linear operator, then A is bounded; that is, $\mathcal{K}(X, Y) \subset \mathcal{L}(X, Y)$.*

(b) *A projection P is compact if, and only if, the range* Im P *has finite dimension.*

(c) *An operator $A \in \mathcal{L}(X, Y)$ is compact if at least one of X and Y has finite dimension.*

(d) *Let $\mathcal{I}$ be the identity operator in the Banach space X. Then $\mathcal{I}$ is compact if, and only if, X is finite dimensional.*

(e) *Assume that the linear operator $A : \ X \to Y$ has finite rank (that is,* Im A *has finite dimension). Then A is compact if, and only if, it is bounded.*

(f) *Let X, Y and Z be Banach spaces, and assume that $A \in \mathcal{L}(X, Y)$ and $B \in \mathcal{L}(Y, Z)$. If either A or B is compact, then BA is compact. In particular, if $X = Y = Z$, both AB and BA are compact operators.*

The well-known *Fredholm Alternative* for linear operators in finite-dimensional vector spaces (e.g. in $\mathbb{R}^n$) has an analogue for compact operators in infinite-dimensional Banach spaces X (Gohberg & Goldberg, 1981, pp. 243–245).

Theorem A.2.17 *Let $A \in \mathcal{L}(X)$ be a compact operator. Then $\mathcal{I} - A$ has a closed range, and the dimension of its kernel is finite. In particular, the equation $(\mathcal{I} - A)x = y$ possesses a unique solution x for every $y \in X$ if, and only if, the equation $(\mathcal{I} - A)x = 0$ has only the trivial solution $x = 0$.*

We conclude this section by an application of some of the foregoing results, namely to the characterisation of the *ill-posed* (or *improperly posed*) problem we studied in Section 1.4. There, we already alluded to Theorem A.2.19, which is a consequence of the following theorem (cf. Kress, 2014, p. 28).

Theorem A.2.18 *Let $A \in \mathcal{L}(X, Y)$ be a compact operator from a Banach space X into a Banach space Y. Then A cannot have a bounded inverse (i.e. $A^{-1} \in \mathcal{L}(Y, X)$) unless X is finite dimensional.*

We recall the above-mentioned theorem for the convenience of the reader (Kress, 2014, p. 300).

Theorem A.2.19 *Let X and Y be Banach spaces, and assume that $A \in \mathcal{L}(X, Y)$ is a compact operator. Then the problem $Au = f$ is ill posed if $\dim X = \infty$.*

We conclude this section by reminding the reader of the definition of the adjoint of a bounded linear operator $A : X \to Y$, where X and Y are *Hilbert spaces*. Such operators play a role in Section 8.4 (singular values of Volterra integral operators) and Section 8.5 (asymptotic behaviour of norms of powers of a Volterra integral operator.

Definition A.2.20 Let X and Y be (real or complex) Hilbert spaces with inner products $\langle \cdot, \cdot \rangle_X$ and $\langle \cdot, \cdot \rangle_Y$, respectively.

(i) If $A \in \mathcal{L}(X, Y)$, the operator $A^* \in \mathcal{L}(Y, X)$ is said to be the *adjoint* of A if it satisfies

$$\langle Ax, y \rangle_Y = \langle x, A^* y \rangle_X \quad \text{for all} \ x \in X \ \text{and} \ y \in Y.$$

(ii) $A \in \mathcal{L}(X)$ is called *self-adjoint* (or *Hermitian*) if $A^* = A$; that is, if

$$\langle Ax, y \rangle = \langle x, Ay \rangle \quad \text{for all} \quad x \in X.$$

If X is a real Hilbert space, such an operator is often called *symmetric*. ◇

A.2.3 The Spectrum of Bounded Linear Operators

The space $\mathcal{L}(X)$ of linear bounded operators on a Banach space X is a prominent example of a *Banach algebra*, with the composition of two operators defining the operation of multiplication. Thus, the abstract setting of Banach algebras can be exploited to obtain information on the nature of the spectrum and the resolvent set of an operator $A \in \mathcal{L}(X)$.

We briefly remind the reader of the basic definitions and properties of a general *algebra* (compare also the notes in Section A.2 for a selection of references). An algebra $\mathcal{A}$ is a (real or complex) linear space for which an additional operation, namely multiplication of two elements of $\mathcal{A}$, is defined. For all elements a, b, $c \in \mathcal{A}$ and all $\lambda \in \mathbb{C}$ it has the properties of associativity $((ab)c = a(bc))$, distributivity $((a + b)c = ac + bc,\ a(b + c) = ab + ac)$, and $(\lambda a)b = a(\lambda b) = \lambda(ab)$.

An element $e \in \mathcal{A}$ is a *unital* element if $ae = ea = a$ for all $a \in \mathcal{A}$; it is unique if it exists. An algebra $\mathcal{A}$ with a unital element is called a *unital algebra*. In such an algebra an element a is said to be *invertible* if there exists $b \in \mathcal{A}$ so that $ab = ba = e$; this element b is unique, and we write $b = a^{-1}$.

The important special case of a *Banach algebra* (where the underlying linear space is a complete normed space) is used as a mathematical tool in Chapter 7 when analysing the spectrum of a (non-compact) cordial Volterra integral operator.

Definition A.2.21

(i) An algebra $\mathcal{A}$ is called a *normed algebra* if the underlying space is a normed linear space, and multiplication is continuous: $\|ab\| \le \|a\|\|b\|$ for all a, $b \in \mathcal{A}$. If $\mathcal{A}$ is a unital algebra, we require that $\|e\| = 1$.

(ii) If the normed space is complete, i.e. a Banach space, $\mathcal{A}$ is said to be a *Banach algebra*. ◇

Example A.2.22 If X is a Banach space, the space $\mathcal{L}(X)$ of bounded linear operators on X, with operator norm $\|A\| := \sup\limits_{x \neq 0} \dfrac{\|Ax\|}{\|x\|}$ (and identity operator $\mathcal{I}$ with $\|\mathcal{I}\| = 1$) is an important example of a unital (non-commutative) Banach algebra.

There exist Banach algebras that are *not unital*. We mention two such examples that arise in the spectral analysis of (non-compact) cordial Volterra integral operators (see Section 7.1.2).

Example A.2.23 If $\mathcal{V}_\varphi^1 : C(I) \to C(I)$ denotes the basic cordial Volterra integral operator

$$(\mathcal{V}_\varphi^1 f)(t) := \int_0^t t^{-1}\varphi(s/t)f(s)\,ds, \quad t \in I := [0, T],$$

with core $\varphi \in L^1(0, 1)$ (see (7.1.4)), let

$$\mathcal{A} := \{\mu\mathcal{I} - \mathcal{V}_\varphi^1 : \mu \in \mathbb{C}, \ \varphi \in L^1(0, 1)\},$$

where the norm is given by $\|\mu\mathcal{I} - \mathcal{V}_\varphi^1\|_{\mathcal{A}} := |\mu| + \|\varphi\|_1$. Because of the commutativity of basic cordial Volterra integral operators (Theorem 7.1.13), $\mathcal{A}$ is a commutative Banach algebra. However, it is not unital (see also Example A.2.24).

Example A.2.24 Let $\varphi \in L^1(0, 1)$ and define

$$\Phi := \{\mu - \varphi : \mu \in \mathbb{C}, \ \varphi \in L^1(0, 1)\},$$

with multiplication given by the star operation ($\star$) introduced in (7.1.9) and with the norm $\|\mu - \varphi\|_\Phi := |\mu| + \|\varphi\|_1$. While Φ is a commutative Banach algebra, it is not unital (as we have shown in Remark 7.1.14(i)). However, this non-unital B-algebra can be extended into a unital Banach algebra (see Vainikko, 2009, p. 1152).

In a given unital Banach algebra $\mathcal{A}$ one is often interested in the set of $\mu \in \mathbb{C}$ for which, for a given $a \in \mathcal{A}$, the element $\mu e - a$ is invertible. The next definition describes the relevant concepts.

Definition A.2.25

(i) The *resolvent set* $\rho(a)$ of an element in a unital Banach algebra $\mathcal{A}$ is the (open) set of all $\mu \in \mathbb{C}$ for which $\mu e - a$ is invertible, and the *resolvent* (or: the resolvent function) is defined by $r_a(\mu) := (\mu e - a)^{-1}$ ($\mu \in \rho(a)$).
(ii) The complement $\sigma(a) := \mathbb{C} \setminus \rho(a)$ is called the *spectrum* of a. The *spectral radius* $s_\sigma(a)$ of a is defined by $s_\sigma(a) := \max\{|a| : a \in \sigma(a)\}$.
(iii) The *point spectrum* (the set of *eigenvalues*) of a is the set of $\mu \in \mathbb{C}$ for which $\mu e - a$ is not one-to-one. ◇

We summarise some of the basic properties of $\rho(a)$ and $\sigma(a)$ in the following theorem (Douglas, 1998, Chapter 2; Megginson, 1998, Section 3.3).

Theorem A.2.26 *Assume that $\mathcal{A}$ is a unital Banach algebra (with respect to a complex Banach space X), and let $a \in \mathcal{A}$. Then:*

(a) $\rho(a)$ is an open set, and $\sigma(a)$ is compact.
(b) $\mu \in \sigma(a)$ implies that $|\mu| \leq \|a\|$. Also, if $\mu \in \mathbb{C}$ is such that $|\mu| > \|a\|$ then $\mu \in \rho(a)$.
(c) $s_\sigma(a) = \lim_{n \to \infty} \|a^n\|^{1/n}$.

As indicated in Example A.2.22 the Banach algebra that is relevant in the context of Chapter 8 is $\mathcal{L}(X)$, the space of linear bounded operators on a Banach space X where the norm of $A \in \mathcal{L}(X)$ is the operator norm. That example reveals that this is a *unital Banach algebra*: the composition AB of $A, B \in \mathcal{L}(X)$ corresponds to multiplication, with the identity operator $\mathcal{I}$ being its unital element. Thus, according to the above definition the resolvent set of $A \in \mathcal{L}(X)$ is given by

$$\rho(A) := \{\mu \in \mathbb{C} : (\mu\mathcal{I} - A)^{-1} \in \mathcal{L}(X)\},$$

and the *spectrum* of A is the set

$$\sigma(A) := \{\mu \in \mathbb{C} : \mu \notin \rho(A)\}.$$

A number $\mu \in \mathbb{C}$ for which $\ker(\mu\mathcal{I} - A) \neq 0$ is an *eigenvalue* of A, and the set of all eigenvalues of A is the *point spectrum* of the operator A. An *eigenelement* of A is non-zero element $x \in X$ for which $Ax = \mu x$ holds.

The following theorem (cf. Kato, 1995, p. 185 or Kress, 2014, pp. 40–41) describes some of the key properties of the spectrum of a compact operator $A \in \mathcal{L}(X)$ acting on an infinite-dimensional normed space X, in particular on an infinite-dimensional Banach space. A good discussion of various aspects of the spectrum of a linear operator can be found in Chapter 9 of Halmos (1982).

Theorem A.2.27 *Let X be an infinite-dimensional normed space. If $A \in \mathcal{L}(X)$ is compact, then $0 \in \sigma(A)$ and the set $\sigma(A)$ is countable, with no accumulation point other than 0.*

Much more can be said if the linear operator is A is compact and self-adjoint on a *Hilbert space X*. We recall that a Hilbert space is a Banach space equipped with an inner product $\langle \cdot, \cdot \rangle$, and the induced norm of an element $x \in X$ is given by $\|x\| = |\langle x, x \rangle|^{1/2}$. If X and Y are two Hilbert spaces and $A \in \mathcal{L}(X, Y)$, then $A^* \in \mathcal{L}(Y, X)$ is called the *adjoint* of A if $\langle Ax, y \rangle = \langle x, A^*y \rangle$ holds for all $x \in X$, $y \in Y$. A linear operator A on a (real) Hilbert space is said to be *self-adjoint* if $\langle Ax, y \rangle = \langle x, Ay \rangle$ for all $x, y \in X$.

A remarkable property of a self-adjoint operator on a Hilbert space is that it is continuous (and hence bounded). This result is known as the Hellinger–Toeplitz Theorem (cf. Ciarlet, 2013, p. 160).

The results contained in the next theorem (the Spectral Theorem for compact self-adjoint operators) play an important role in the analysis of the *singular values* of Volterra integral operators and the ill-posedness of first-kind operator equations (cf. Section 8.1.7).

Theorem A.2.28 *Assume that*

(i) *X is an infinite-dimensional inner product space (e.g. a Hilbert space), and*

(ii) *the linear operator $A : X \to X$ is compact and self-adjoint, with* dim (Im A) $= \infty$.

Then the following statements hold:

(a) *There exists an infinite sequence $\{\lambda_n\}_{n \geq 1}$ of non-zero eigenvalues of A so that*

$$|\lambda_1| = \|A\|, \quad |\lambda_1| \geq |\lambda_2| \geq \cdots \geq |\lambda_n| \geq \cdots, \quad \text{and} \quad \lim_{n \to \infty} \lambda_n = 0.$$

(b) *There exist corresponding eigenelements $\{x_n\}_{n \geq 1}$ with $x_n \neq 0$ satisfying*

$$Ax_n = \lambda_n x_n \ (n \geq 1) \quad \text{and} \quad \langle x_i, x_j \rangle = \delta_{ij} \ (i, j \geq 1).$$

(c) *For any $x \in X$ we have*

$$Ax = \sum_{n=1}^{\infty} \lambda_n \langle x, x_n \rangle x_n .$$

A detailed proof can be found in Ciarlet (2013, pp. 222–227).

A.2.4 Quasi-Nilpotent Operators

The following definition can be found in e.g. Halmos (1982, cf. pp. 52, 98–99, 293–300) or Kato (1995, p. 153).

Definition A.2.29 Let X be a Banach space and assume that $A \in \mathcal{L}(X)$.

(i) The *spectral radius* of A is defined by

$$r(A) := \lim_{n \to \infty} \|A^n\|^{1/n}.$$

(ii) A is said to be *quasi-nilpotent* if $r(A) = 0$. $\diamond$

Remark A.2.30 The spectral radius of A is independent of the norm used in its definition. $\diamond$

The spectrum of a quasi-nilpotent operator is rather special, as the following theorem reveals (cf. Halmos, 1982, p. 52 and p. 96: Corollary in Problem 179).

Theorem A.2.31 *A is quasi-nilpotent if, and only if, $\sigma(A) = \{0\}$. Moreover, a compact operator whose point spectrum is empty is quasi-nilpotent.*

The classical example of a quasi-nilpotent operator on $C(I)$ is the Volterra integral operator

$$(\mathcal{V}f)(t) := \int_0^t (t-s)^{\alpha-1} K(t,s) f(s) \, ds, \quad t \in I,$$

with $K \in C(D)$ and $0 < \alpha \le 1$. On the other hand, not every Volterra integral operator is quasi-nilpotent. A basic example is the cordial Volterra integral operator $\mathcal{V}_\varphi : C(I) \to C(I)$ given by

$$(\mathcal{V}_\varphi f)(t) := \int_0^t t^{-1} \varphi(s/t) f(s) \, ds, \quad t \in I,$$

with the core $\varphi(z) = (1 - z^2)^{-1/2}$. Since its spectrum is the compact interval $[0, \pi/2]$ (see Theorem 7.1.2) it cannot be a compact operator.

A.3 Non-Linear Operators on Banach Spaces

A.3.1 The Fréchet Derivative

Assume that X and Y are Banach spaces and $\Omega \subset X$ is some open set. The Fréchet derivative of a mapping from X to Y is defined as follows (see, for example, Ciarlet, 2013, Section 7.1).

Definition A.3.1 A mapping $F : X \to Y$ is said to be *differentiable* at a point $a \in X$ if there there exists $A \in \mathcal{L}(X, Y)$ such that

$$F(a + h) = F(a) + Ah + \|h\|_X \, \delta(h) \quad \text{with} \quad \lim_{h \to 0} \delta(h) = 0 \in Y,$$

for all $a + h \in \Omega$. The linear mapping A is called the *Fréchet derivative* of F at $a \in \Omega$: $A = F'(a)$.

We say that f is *Fréchet differentiable* in Ω if its Fréchet derivative exists at all points $a \in \Omega$.   ◇

Example A.3.2 Consider the non-linear cordial Volterra integral operator on $C(I)$,

$$(\mathcal{V}_\varphi u)(t) = \int_0^t t^{-1} \varphi(s/t) k(t, s, u(s)) \, ds, \quad t \in I,$$

where $\varphi \in L^1(0, 1)$ and $k \in C(D \times \mathbb{R})$. If $k = k(t, s, u)$ has a continuous partial derivative $\partial k / \partial u$, the Fréchet derivative of $F = \mathcal{V}_\varphi$ is

$$F'(u) = (\mathcal{V}'_\varphi u)(t) = \int_0^t t^{-1} \varphi(s/t) a(t, s) u(s) \, ds,$$

with

$$a(t, s) := \left. \frac{\partial k(t, s, u)}{\partial u} \right|_{u=u(s)}.$$

This result plays a key role in Section 7.3 when analysing the existence of a local solution of the non-linear cordial VIE $\mu u = f + \mathcal{V}_\varphi u$.

A.3.2 The Implicit Function Theorem

An important result in non-linear functional analysis is the celebrated implicit function theorem (see, for example, Dieudonné, 1960, Section 10.2 or Ciarlet, 2013, Section 7.13). It is used in Section 3.5 to establish the existence of local continuous solutions of non-linear VIEs of the first kind.

In the following, X, Y and Z are Banach spaces, U is an open subset of $X \times Y$, and F is a continuously differentiable mapping from U to Z.

Theorem A.3.3 *Assume that $(x_0, y_0) \in U$ and that F possesses the following properties:*

(i) $F(x_0, y_0) = 0$;

(ii) $\partial F / \partial y : U \to \mathcal{L}(Y, Z)$ *is a continuous linear mapping such that* $\dfrac{\partial F}{\partial y}(x, y)$ *exists at all $(x, y) \in U$;*

(iii) $\dfrac{\partial F}{\partial y}(x_0, y_0) \in \mathcal{L}(Y, Z)$ *is a bijection, with* $\left(\dfrac{\partial F}{\partial y}(x_0, y_0)\right)^{-1} \in \mathcal{L}(Z, Y)$.

Then the following statements are true:

(a) There exists an open neighbourhood V of x_0 in X, a neighbourhood W of y_0 in Y, and an implicit function $f : V \to W$ that is continuous and is such that

$$V \times W \subset U \quad and \quad \{(x, y) \in V \times W :$$
$$F(x, y) = 0\} = \{(x, y) \in V \times W : y = f(x)\}.$$

(b) The mapping f is differentiable at x_0, and there holds

$$f'(x_0) = -\left(\frac{\partial F}{\partial y}(x_0, y_0)\right)^{-1}\frac{\partial F}{\partial x}(x_0, y_0) \in \mathcal{L}(X, Y).$$

A.3.3 The Fixed-Point Theorems of Banach and Schauder

Definition A.3.4

(i) Let (X, d) be a metric space. The mapping $F : X \to X$ is a *contraction map* if there exists a constant $\gamma \in [0, 1)$ so that

$$d\big(F(x), F(y)\big) \leq \gamma d(x, y) \quad \text{for all} \quad x, y \in X.$$

(ii) If X and Y are normed spaces and $S \subset X$, the mapping $F : S \to Y$ is said to be *compact* if F is continuous and if the closure of the image $F(B)$ of any bounded subset of S is a compact subset of Y. (If dim $X < \infty$ any continuous mapping F is compact.) ◇

Banach's celebrated fixed point theorem dates from 1922 (see also Pietsch, 2007, p. 29, for its history and the formulation given by Cacciopoli in 1930). Its proof (for metric spaces) can be found, for example, in Ciarlet (2013), p. 153. Here, we state it for Banach spaces (where $d(x, y) = \|x - y\|$) (cf. Dieudonné (1960), pp. 260–261).

Theorem A.3.5 *Let S be a closed subset of a Banach space X, and assume that the continuous map $F : S \to S$ is a contraction map. Then F possesses a unique fixed point in S.*

Schauder's fixed point theorem of 1930 is an extension of Brouwer's fixed point theorem to infinite-dimensional normed spaces. We will state it for Banach spaces. Its version for normed spaces and compact, convex subsets can be found in Ciarlet (2013, Section 9.12).

Theorem A.3.6 *Assume that* S *is a closed and convex subset of a Banach space* X. *If* $F : S \to S$ *is a continuous mapping such that the closure* $\overline{F(S)}$ *is compact, then* F *possesses at least one fixed point.*

A related fixed-point theorem is the one by Schauder and Leray (cf. Ciarlet, 2013, p. 736).

Theorem A.3.7 *Let* X *be a Banach space and assume that the compact mapping* $F : X \times [0, 1] \to X$ *has the following properties:*

(i) $F(x, 0) = 0$ *for all* $x \in X$;
(ii) there exists $\rho > 0$ *such that the set* $\{x \in X : F(x, s) = x$ *for some* $s \in [0, 1]\}$ *is contained in the open ball* $B(0, r)$.

Then $F(\cdot, 1) : X \to X$ *has at least one fixed point in the closed ball* $\bar{B}(0, r)$.

A.4 Notes

Section A.1: *Banach Spaces in the Theory of VIEs*
On Hölder space theory see, for example, Kufner et al. (1977, pp. 22–27), Kress (2014, pp. 103–105), Cherrier & Milani (2013, pp. 7–8) and Bressan (2013, Ch. 3); see also Gilbarg & Trudinger (1983, Section 4.1).

Concise introductions to L^p spaces can be found in Gohberg & Goldberg (1981, Appendix 2), Kato (1995) or Megginson (1998); for a comprehensive treatment (including Sobolev spaces) see Adams & Fournier (2003, Chapter 2), Davies (2007), Brezis (2011, Chapter 4), Cherrier & Milani (2012) and, especially, Ciarlet (2013, pp. 25–33 and 61–76).

Section A.2: *Linear Operators on Banach Spaces*
The monumental monograph by Pietsch (2007) is the seminal treatise of the history and the current state of the art of the theory of Banach spaces and linear operators on such spaces. The reader may also wish to consult the books by Gohberg & Kreĭn (1970), Megginson (1998, where basic properties of compact operators are studied on pp. 319–338), and Schechter (2002, which contains, in Appendix B, a summary of the 'major theorems' of Functional Analysis). An illuminating discussion of compact Fredholm integral operators can be found in Graham & Sloan (1979).

Good introductions the theory of *Banach algebras* are given in Rudin (1991, Chapters 10 and 11), Kato (1995), Douglas (1998), Megginson (1998, pp. 305–319); see also Hagen, Roch & Silbermann (2001) for a wider perspective, including applications in Numerical Analysis.

Books on *semigroups of operators* (relevant to the analogous theory for VIEs) are Hille & Phillips (1957) and Engel & Nagel (2000), as well as the 'short course' by Engel & Nagel (2006); see also Renardy & Rogers (2003, Ch. 11) for a concise general introduction. VIEs are studied in the context of semigroups in Chapter 9 of Gripenberg, Londen & Staffans (1990).

Section A.3: *Non-Linear Operators on Banach Spaces*
An introduction to the theory of Fréchet (and Gâteau) derivatives can be found in e.g. Pietsch (2007, pp. 178–180) and in Ciarlet (2013, Section 7.1). Compact non-linear mappings are discussed in Ciarlet (2013, p. 736); see also Berger (1977, pp. 88–92). A concise summary of various fixed-point theorems can be found in Ciarlet (2013, pp. 735–737); see also Gohberg & Goldberg (1980, pp. 255–263).

References

Abel, N.H. (1823), Solution de quelques problèmes à l'aide d'intégrales définies, *Magazin Naturvidensk.* **1**, 55–68.

Abel, N.H. (1826), Auflösung einer mechanischen Aufgabe, *J. Reine Angew. Math.* **1**, 153–157.

Adams, R.A. and J.F. Fournier (2003), *Sobolev Spaces* (Second Edition) (Amsterdam: Elsevier/Academic Press).

Adell, J.A. and E.A. Gallardo-Gutièrrez (2007), The norm of the Riemann–Liouville operator on $L^p[0, 1]$: a probabilistic approach, *Bull. London Math. Soc.* **39**, 565–574.

Agarwal, R.P. and D. O'Regan (eds.) (2000), *Integral and Integrodifferential Equations. Theory, Methods and Applications*, Ser. Math. Anal. Appl. **2** (Amsterdam: Gordon and Breach).

Agarwal, R.P., D. O'Regan and P.J.V. Wong (2013), *Constant-Sign Solutions of Systems of Integral Equations* (New York: Springer).

Agyingi, E.O. and C.T.H. Baker (2013), Derivation of variation of parameters formulas for non-linear Volterra equations, using a method of embedding, *J. Integral Equations Appl.* **25**, 159–191.

Anderssen, R.S. (1977), Application and numerical solution of Abel-type integral equations, *MRC Tech. Summary Report* **1787**, Math. Research Center, University of Wisconsin, Madison.

Anderssen, R.S., A.R. Davies and F.R. de Hoog (2007), On the interconversion integral equation for relaxation and creep, *ANZIAM J.* **48**, C346–C363.

Anderssen, R.S., A.R. Davies and F.R. de Hoog (2008), On the Volterra integral equation relating creep and relaxation, *Inverse Problems* **24**, 035009, 13 pp.

Anderssen, R.S., A.R. Davies and F.R. de Hoog (2011), The effect of kernel perturbations when solving the interconversion convolution equation of linear viscoelasticity, *Appl. Math. Lett.* **24**, 71–75.

Anderssen, R.S. and F.R. de Hoog (1990), Abel integral equations, in *Numerical Solution of Integral Equations* (M.A. Golberg, ed.), pp. 373–410 (New York: Plenum Press).

Anderssen, R.S. and F.R. de Hoog (2006), Regularization of first kind integral equations with applications to Couette viscometry, *J. Integral Equations Appl.* **18**, 249–265.

Andreoli, G. (1914), Sulle equazioni integrali, *Rend. Circ. Mat. Palermo* **37**, 76–112. [$u(t) = f(t) + \int_0^{qt} K(t, s)u(s)ds$, $0 < q < 1$, and related delay VIEs.]

Anello, G. (2006), An existence theorem for an implicit integral equation with discontinuous right-hand side, *J. Inequal. Appl.*, Art. ID 71396, 1–8.

Ang, D.D. and R. Gorenflo (1991), A nonlinear Abel integral equation, in *Optimal Control of Partial Differential Equations (Irsee 1990)* (K.-H. Hoffmann and W. Krabs, eds.), pp. 26–37, Lecture Notes in Control and Inform. Sci. **149** (Berlin–New York: Springer-Verlag). [Extension of results in Branca (1978), Brunner & van der Houwen (1986), and Gorenflo & Vessella (1991); see also Gorenflo & Pfeiffer (1991).]

Angell, J.S. (1985), *Asymptotic Analysis of Singularly Perturbed Integral Equations*, PhD thesis (Evanston, IL: Northwestern University).

Angell, J.S. and W.E. Olmstead (1987), Singularly perturbed Volterra integral equations, *SIAM J. Appl. Math.* **47**, 1–14.

Annunziato, M. and E. Messina (2010), Numerical treatment of a Volterra integral equation with space maps, *Appl. Numer. Math.* **60**, 809–815. [VIE of the form $u(x,t) = f(x,t) + \int_0^t k(t-s)u(g(x,t,s),s)\,ds$.]

Annunziato, M., H. Brunner and E. Messina (2012), Asymptotic stability of solutions to Volterra-renewal integral equations with space maps, *J. Math. Anal. Appl.* **395**, 766–775.

Anselone, P.M. (ed.) (1964), *Nonlinear Integral Equations (Madison 1963)* (Madison: University of Wisconsin Press).

Apartsyn, A.S. (2003), *Nonclassical Linear Volterra Equations of the First Kind* (Utrecht: VSP). [Theory of equations with variable upper and lower limit of integration.]

Apartsyn, A.S. (2004), Polylinear Volterra equations of the first kind, *Autom. Remote Control* **65**, 263–269.

Apartsyn, A.S. (2008), Multilinear Volterra equations of the first kind and some control problems, *Autom. Remote Control* **69**, 545–558.

Apartsyn, A.S. (2014), On some classes of linear Volterra integral equations, *Abstr. Appl. Anal.* **2014**, Art. ID 532409, 6 pp.

Apartsyn, A.S. and I.V. Sidler (2013), Using nonclassical Volterra equations of the first kind to model developing systems, *Autom. Remote Control* **74**, 899–910.

Appell, J. and P.P. Zabrejko (1990), *Nonlinear Superposition Operators* (Cambridge University Press).

Appleby, J.A.D. and D.W. Reynolds (2003), Non-exponential stability of scalar stochastic Volterra equations, *Statist. Probab. Lett.* **62**, 335–343.

Appleby, J.A.D. and M. Riedle (2010), Stochastic Volterra equations in weighted spaces, *J. Integral Equations Appl.* **22**, 1–17.

Arias, M.R. (2000), Existence and uniqueness of solutions for nonlinear Volterra equations, *Math. Proc. Cambridge Philos. Soc.* **129**, 361–370.

Arias, M.R. and R. Benítez (2003a), Properties of solutions for nonlinear Volterra integral equations, *Discrete Contin. Dyn. Syst.* suppl., 42–47.

Arias, M.R. and R. Benítez (2003b), Aspects of the behaviour of solutions of nonlinear Abel equations, *Nonlinear Anal.* **54**, 1241–1249.

Arias, M.R., R. Benítez and V.J. Bolós (2005), Nonconvolution nonlinear integral Volterra equations with monotone operators, *Comput. Math. Appl.* **50**, 1405–1414.

Arias, M.R., R. Benítez and V.J. Bolós (2007), Attraction properties of unbounded solutions for a nonlinear Abel integral equation, *J. Integral Equations Appl.* **19**, 439–452.

Arias, M.R. and J.M.F. Castillo (1999), Attracting solutions of nonlinear Volterra integral equations, *J. Integral Equations Appl.* **11**, 299–309. [Properties of non-trivial solutions of $u(x) = \int_0^x k(x-s)g(u(s))ds, \ g(0) = 0.$]

Artstein, Z. (1975), Continuous dependence of solutions of Volterra integral equations, *SIAM J. Math. Anal.* **6**, 446–456. [See also Gyllenberg (1981).]

Asanov, A. (1998), *Regularization, Uniqueness and Existence of Solutions of Volterra Equations of the First Kind* (Zeist: VSP). [This monograph contains a comprehensive bibliography on Russian contributions.]

Askhabov, S.N. (1991), Integral equations of convolution type with power nonlinearity, *Colloq. Math.* **62**, 49–65.

Askhabov, S.N. (2009), *Nonlinear Equations of Convolution Type* (in Russian) (Moscow: FIZMATLIT).

Askhabov, S.N. and M.A. Betilgiriev (1993), A priori estimates for the solutions of a nonlinear integral equation of convolution type and their applications, *Math. Notes* **54**, 1087–1092. [$u^\alpha(t) = f(t) + \int_0^t k(t-s)u(s)ds, \ \alpha > 1.$]

Askhabov, S.N. and N.K. Karapetyants (1990), Integral equations of convolution type with power nonlinearity and systems of such equations, *Soviet Math. Dokl.* **41**, 323–327.

Atkinson, K.E. (1974), An existence theorem for Abel integral equations, *SIAM J. Math. Anal.* **5**, 729–736. [Regularity of solutions to $\int_0^t (t^p - s^p)^{-\alpha} y(s)ds = t^\beta g(t), \ \beta > -1, \ 0 < \alpha < 1, \ p \geq 1; \ g(0) \neq 0.$]

Atkinson, K.E. (1976), *A Survey of Numerical Methods for the Solution of Fredholm Integral Equations of the Second Kind* (Philadelphia, PA: Society for Industrial and Applied Mathematics).

Atkinson, K.E. (1997), *The Numerical Solution of Integral Equations of the Second Kind* (Cambridge University Press).

Bai, F. (2011), *Collocation Methods for Weakly Singular Volterra Integral Equations with Vanishing Delays*, MSc thesis (St John's: Memorial University of Newfoundland).

Baillon, J.-B. and Ph. Clément (1981), Ergodic theorems for nonlinear Volterra equations in Hilbert space, *Nonlinear Anal.* **5**, 789–801.

Baker, C.T.H. (1977), *The Numerical Treatment of Integral Equations* (Oxford: Clarendon Press).

Baker, C.T.H. (2000), A perspective on the numerical treatment of Volterra integral equations, *J. Comput. Appl. Math.* **125**, 201–215.

Baker, C.T.H., E.O. Agyingi, E.I. Parmuzin, F.A. Rihan and Y. Song (2006), Sense from sensitivity and variation of parameters, *Appl. Numer. Math.* **56**, 397–412.

Bakke, V.L. (1974), A maximum principle for an optimal control problem with integral constraints, *J. Optimization Theory Appl.* **13**, 32–55.

Bakke, V.L. (1976), Boundary arcs for integral equations, *J. Optim. Theory Appl.* **19**, 425–433.

Banach, S. (1932), *Théorie des Opérations Linéaires* (New York: Chelsea Publishing Co.; 1993 reprint of original edition: Sceaux, Éditions Jacques Gabay). [See Ch. X, pp. 145–164, for Volterra integral operators, also in L^p spaces. An English translation, *Theory of Linear Operations* (Amsterdam: North-Holland), appeared in 1987.]

Banaś, J. and J. Caballero Mena (2005), Some properties of nonlinear Volterra–Stieltjes integral operators, *Comput. Math. Appl.* **49**, 1565–1573.

Banaś, J. and J. Dronka (2000), Integral operators of Volterra-Stieltjes type, their properties and applications, *Math. Comput. Modelling* **32**, 1321–1331.

Banaś, J. and T. Zając (2011), A new approach to the theory of functional integral equations of fractional order, *J. Math. Anal. Appl.* **375**, 375–387. [Conversion of nonlinear VIE with weakly singular kernel into nonlinear Volterra–Stieltjes VIE.]

Bandle, C. and H. Brunner (1998), Blow-up in diffusion equations: A survey, *J. Comput. Appl. Math.* **97**, 3–22.

Barbu, V. (1979), Existence for nonlinear Volterra equations in Hilbert spaces, *SIAM J. Math. Anal.* **10**, 552–569.

Bart, G.R. and R.L. Warnock (1973), Linear integral equations of the third kind, *SIAM J. Math. Anal.* **4**, 609–622.

Bateman, H. (1910), Report on the history and present state of the theory of integral equations, *Report to the British Association for the Advancement of Science* (Sheffield, 1910), pp. 345–424.

Becker, L.C. (2011), Resolvents and solutions of weakly singular linear Volterra integral equations, *Nonlinear Anal.* **74**, 1892–1912.

Becker, L.C. (2012), Resolvents for weakly singular kernels and fractional differential equations, *Nonlinear Anal.* **75**, 4839–4861.

Becker, L.C. (2013), Resolvents and solutions of singular Volterra integral equations with separable kernels, *Appl. Math. Comp.* **219**, 11265–11277.

Bedivan, D.M. and D. O'Regan (2000), Fixed point sets for abstract Volterra operators on Fréchet spaces, *Appl. Anal.* **76**, 131–152.

Beesack, P.R. (1969), Comparison theorems and integral inequalities for Volterra integral equations, *Proc. Amer. Math. Soc.* **20**, 61–66.

Beesack, P.R. (1985), Systems of multidimensional Volterra integral equations and inequalities, *Nonlinear Anal.* **9**, 1451–1486.

Beesack, P.R. (1987), On some variation of parameter methods for integrodifferential, integral and quasilinear partial integrodifferential equations, *Appl. Math. Comput.* **22**, 189–215.

Bélair, J. (1991), Population models with state-dependent delays, in *Mathematical Population Dynamics* (O. Arino, D.E. Axelrod and M. Kimmel, eds.), pp. 165–176 (New York: Marcel Dekker).

Belbas, S.A. (1999), Iterative schemes for optimal control of Volterra integral equations, *Nonlinear Anal.* **37**, 57–79.

Belbas, S.A. (2008), Optimal control of Volterra integral equations in two independent variables, *Appl. Math. Comput.*, **202**, 647–665.

Belbas, S.A. and W.H. Schmidt (2005), Optimal control of Volterra equations with impulses, *Appl. Math. Comput.* **166**, 696–723.

Bellen, A. and N. Guglielmi (2009), Solving neutral delay differential equations with state-dependent delay, *J. Comput. Appl. Math.* **229**, 350–362.

Bellen, A., S. Maset, M. Zennaro and N. Guglielmi (2009), Recent trends in the numerical solution of retarded functional differential equations, *Acta Numer.* **18**, 1–110.

Bellen, A. and M. Zennaro (2003), *Numerical Methods for Delay Differential Equations* (Oxford University Press).

Bellman, R. and K.L. Cooke (1963), *Differential-Difference Equations* (New York: Academic Press). [Chapters 7 and 8 deal with the theory of (systems of) renewal equations; compare also *Math. Reviews*, 26, #5259.]

Berg, L. and L. v. Wolfersdorf (2005), On a class of generalized autoconvolution equations of the third kind, *Z. Anal. Anwendungen* **24**, 217–250.

Berger, M.A. and V.J. Mizel (1980), Volterra equations with Itô integrals. I,II, *J. Integral Equations* **2**, 187–245; 319–337.

Berger, M.S. (1977), *Nonlinearity and Functional Analysis* (New York: Academic Press).

Bernfeld, S.R. and M.E. Lord (1978), A nonlinear variation of constants method for integro-differential and integral equations, *Appl. Math. Comp.* **4**, 1–14. [A correction of some of these results can be found in Beesack (1987). See also Agyingi & Baker (2012).]

Berrone, L.R. (1995), Local positivity of the solution to Volterra integral equations and heat conduction in materials that may undergo changes of phase, *Math. Notae* **38**, 79–93.

Bijura, A.M. (2002a), Singularly perturbed Volterra integral equations with weakly singular kernels, *Int. J. Math. Math. Sci.* **30**, 129–143.

Bijura, A.M. (2002b), Singularly perturbed systems of Volterra equations, *J. Appl. Anal.* **8**, 221–244.

Bijura, A.M. (2002c), Rigorous results on the asymptotic solutions of singularly perturbed nonlinear Volterra integral equations, *J. Integral Equations Appl.* **14**, 119–149. [See also the related survey paper (Kauthen, 1997).]

Bijura, A.M. (2004), Error bound analysis and singularly perturbed Abel-Volterra equations, *J. Appl. Math.* **2004**, 479–494.

Bijura, A.M. (2006), Initial-layer theory and model equations of Volterra type, *IMA J. Numer. Anal.* **71**, 315–331.

Bijura, A.M. (2012), Systems of singularly perturbed fractional integral equations, *J. Integral Equations Appl.* **24**, 195–211.

Birkhoff, G. (ed.) (1973), *A Source Book in Classical Analysis* (Cambridge, MA: Harvard University Press). [Ch. 13: translations of the papers by Abel (1826), Volterra (1896), and Fredholm (1903).]

Blom, J.G. and H. Brunner (1987), The numerical solution of nonlinear Volterra integral equations of the second kind by collocation and iterated collocation methods, *SIAM J. Sci. Comput.* **8**, 806–830.

Blom, J.G. and H. Brunner (1991), Algorithm 689: Discretized collocation and iterated collocation for nonlinear Volterra integral equations, *ACM Trans. Math. Software* **17**, 183–190.

Bôcher, M. (1909), *An Introduction to the Study of Integral Equations*, Cambridge Tracts in Mathematics and Mathematical Physics, No. 10 (Cambridge, MA: Cambridge University Press; 1971 reprint of second edition (1914): New York: Haffner Publishing Co.). [This is the first monograph on integral equations.]

du Bois-Reymond, P. (1888), Bemerkungen über $\delta z = \partial^2 x / \partial x^2 + \partial^2 z / \partial y^2 = 0$, *J. Reine Angew. Math.* **102**, 204–229.

Böttcher, A., H. Brunner, A. Iserles and S.P. Nørsett (2010), On the singular values and eigenvalues of the Fox-Li and related operators, *New York J. Math.* **16**, 539–561.

Böttcher, A. and P. Dörfler (2009), On the best constants in inequalities of the Markov and Wirtinger types for polynomials on the half-line, *Linear Algebra Appl.* **430**, 1057–1069.

Böttcher, A. and P. Dörfler (2010a), Weighted Markov-type inequalities, norms of Volterra operators, and zeros of Bessel functions, *Math. Nachr.* **283**, 40–57.

Böttcher, A. and P. Dörfler (2010b), On the best constants in Markov-type inequalities involving Laguerre norms with different weights, *Monatsh. Math.* **161**, 357–367.

Bownds, J.M. and J.M. Cushing (1972), Some stability criteria for linear systems of Volterra integral equations, *Funkcial. Ekvac.* **13**, 101–117.

Bownds, J.M. and J.M. Cushing (1973), A representation formula for linear Volterra integral equations, *Bull. Amer. Math. Soc.* **79**, 532–536.

Bownds, J.M. and J.M. Cushing (1975a), Some stability theorems for systems of Volterra integral equations, *Applicable Anal.* **5**, 65–77.

Bownds, J.M. and J.M. Cushing (1975b), On preserving stability of Volterra integral equations under a general class of perturbations, *Math. Systems Theory* **9**, 117–131.

Bownds, J.M., J.M. Cushing and R. Schutte (1976), Existence, uniqueness, and extendibility of solutions to Volterra integral systems with multiple variable delays, *Funkcial. Ekvac.* **19**, 101–111.

Boyarintsev, Yu.E. and V.F. Chistyakov (1998), *Algebro-Differential Systems* (in Russian) (Novosibirsk, "Nauka", Sibirskoe Predpriyatie RAN). [Review of research of Irkutsk group at Russian Academy of Sciences; in particular: reformulation of DAE systems as VIEs, and results on non-equivalence. See also *MR 2002b:34005* for a description of the contents, and compare Bulatov (2002), Chistyakov (2006).]

Brakhage, H., K. Nickel and P. Rieder (1965), Auflösung der Abelschen Integralgleichung 2. Art, *Z. Angew. Math. Phys.* **16**, 295–298.

Branca, H.W. (1978), The nonlinear Volterra equation of Abel's kind and its numerical treatment, *Computing* **20**, 307–321. [See also Janikowski (1962), Gorenflo & Pfeiffer (1991), Ang & Gorenflo (1991), Deimling (1995) for generalization of existence results from $\alpha = 1/2$ to arbitray $\alpha \in (0, 1)$.]

Brauer, F. (1972), A nonlinear variation of constant formula to Volterra equations, *Math. Systems Theory* **6**, 226–234. [Compare also Beesack (1987).]

Brauer, F. (1975), On a nonlinear integral equation for population growth problems, *SIAM J. Math. Anal.* **6**, 312–317.

Brauer, F. (1976a), Constant rate harvesting of populations governed by Volterra integral equations, *J. Math. Anal. Appl.* **56**, 18–27.

Brauer, F. (1976b), Perturbations of the nonlinear renewal equation, *Adv. in Math.* **22**, 32–51.

Brauer, F. and C. Castillo-Chávez (2001), *Mathematical Models in Population Biology and Epidemiology* (New York, Springer-Verlag).

Brauer, F. and P. van den Driessche (2003), Some directions for mathematical epidemiology, in *Dynamical Systems and Their Applications in Biology (Cape Breton, 2001)* (S. Ruan, G.S.K. Wolkowicz and J. Wu, eds.), pp. 95–112, *Fields Institute Communications*, Vol. 36 (Providence, RI, American Mathematical Society). [Contains an extensive bibliography.]

Brenner, M. and Y.S. Xu (2002), A factorization method for identification of Volterra systems, *J. Comput. Appl. Math.* **144**, 105–117.

Brezis, H. (2011), *Functional Analysis, Sobolev Spaces and Partial Differential Equations* (New York: Springer).

Brezis, H. and F.E. Browder (1975), Existence theorems for nonlinear integral equations of Hammerstein type, *Bull. Amer. Math. Soc.* **81**, 73–78.

Brooks, C.D. and P.K. Lamm (2011), A generalized approach to local regularization of linear Volterra problems in L^p spaces, *Inverse Problems* **27**, 055010, 26pp.

Brunner, H. (1983), Nonpolynomial spline collocation for Volterra equations with weakly singular kernels, *SIAM J. Numer. Anal.* **20**, 1106–1119.

Brunner, H. (1987), Collocation methods for one-dimensional Fredholm and Volterra integral equations, in *The State of the Art in Numerical Analysis (Birmingham 1986)* (A. Iserles and M.J.D. Powell, eds.), 563–600 (New York: Oxford University Press).

Brunner, H. (1991), On implicitly linear and iterated collocation methods for Hammerstein integral equations, *J. Integral Equations Appl.* **3**, 475–488.

Brunner, H. (1997), 1896–1996: One hundred years of Volterra integral equations of the first kind, *Appl. Numer. Math.* **24**, 83–93.

Brunner, H. (2004), *Collocation Methods for Volterra Integral and Related Functional Equations* (Cambridge University Press).

Brunner, H. (2004a), The numerical analysis of functional integral and integro-differential equations of Volterra type, *Acta Numer.* **13**, 55–145.

Brunner, H. (2014), On Volterra integral operators with highly oscillatory kernels, *Discrete Contin. Dyn. Syst.* **34** (2014), 903–914.

Brunner, H., P.J. Davies and D.B. Duncan (2012), Global convergence and local super-convergence of first-kind Volterra integral equation approximations, *IMA J. Numer. Anal.* **32**, 1117–1146.

Brunner, H. and P.J. van der Houwen (1986), *The Numerical Solution of Volterra Equations*, CWI Monographs, Vol. 3 (Amsterdam: North-Holland).

Brunner, H., A. Iserles and S.P. Nørsett (2010), The spectral problem for a class of highly oscillatory Fredholm integral operators, *IMA J. Numer. Anal.* **30**, 108–130.

Brunner, H., A. Iserles and S.P. Nørsett (2011), The computation of the spectra of highly oscillatory Fredholm integral operators, *J. Integral Equations Appl.* **23**, 467–519.

Brunner, H. and H. Liang (2010), Stability of collocation methods for delay differential equations with vanishing delays, *BIT Numer. Math.* **50**, 693–711.

Brunner, H., Y.Y. Ma and Y.S. Xu (2015), The oscillation of solutions of Volterra integral and integro-differential equations with highly oscillatory kernels, *J. Integral Equations Appl.* **27**, 455–487.

Brunner, H. and S. Maset (2009), Time transformations for delay differential equations, *Discrete Contin. Dyn. Syst.* **25**, 751–775.

Brunner, H. and S. Maset (2010), Time transformations for state-dependent delay differential equations, *Commun. Pure Appl. Anal.* **9** (2010), 23–45.

Brunner, H. and C.H. Ou (2015), On the asymptotic stability of Volterra functional equations with vanishing delays, *Commun. Pure Appl. Anal.* **14**, 397–406.

Brunner, H., A. Pedas and G. Vainikko (1999), The piecewise polynomial collocation method for nonlinear weakly singular Volterra equations, *Math. Comp.* **68**, 1079–1095. [Regularity results for VIEs with algebraic or logarithmic kernel singularities.]

Brunner, H. and Z.W. Yang (2013), Blow-up behavior of Hammerstein-type Volterra integral equations, *J. Integral Equations Appl.* **24**, 487–512. [See also Yang & Brunner (2013a) on analogous results for delay VIEs.]

Buckwar, E. (1997), *Iterative Approximation of the Positive Solution of a Class of Nonlinear Volterra-type Integral Equations*, PhD Thesis (Berlin: Freie Universität / Logos Verlag). [Analysis of VIEs of the form $y^{\beta}(t) = \int_0^t (t - s)^{-\alpha} k(t - s) y(s) ds,\ \beta > 1,\ 0 \leq \alpha < 1$.]

Buckwar, E. (2000), On a nonlinear Volterra integral equation, in: Corduneanu & Sandberg (2000, pp. 157–167), pp. 157–162.

Buckwar, E. (2005), Existence and uniqueness of solutions of Abel integral equations with power-law non-linearities, *Nonlinear Anal.* **63**, 88–96.

Bukhgeim, A.L. (1983), *Volterra Equations and Inverse Problems* (in Russian) (Novosibirsk, "Nauka" Sibirsk. Otdel.). [Consult also the detailed review 86b:35193 in *Math. Reviews.*]

Bukhgeim, A.L. (1999), *Volterra Equations and Inverse Problems* (Zeist, VSP). [Compare also Bukhgeim (1983) and Asanov (1998).]

Bulatov, M.V. (2002), Regularization of degenerate systems of Volterra integral equations, *Comput. Math. Math. Phys.* **42**, 315–320.

Bulatov, M.V. and P.M. Lima (2011), Two-dimensional integral-algebraic systems: analysis and computational methods, *J. Comput. Appl. Math.* **236**, 132–140.

Burns, J.A., T.L. Herdman and J. Turi (1990), Neutral functional integro-differential equations with weakly singular kernels, *J. Math. Anal. Appl.* **145**, 371–401. [Contains an analysis of first-kind VFIEs with weakly singular kernels; see also Kappel & Zhang (1986).]

Burton, T.A. (1983), *Volterra Integral and Differential Equations* (New York: Academic Press; 2nd edition, 2005: Amsterdam: Elsevier). [See also for numerous applications.]

Burton, T.A. and D.P. Dwiggins (2011), Resolvents of integral equations with continuous kernels, *Nonlinear Stud.* **18**, 293–305.

Burton, T.A. and B. Zhang (2011), Periodic solutions of singular integral equations, *Nonlinear Dyn. Syst. Theory* **11**, 113–123.

Busenberg, S. and K.L. Cooke (1980), The effect of integral conditions in certain equations modelling epidemics and population growth, *J. Math. Biol.* **10**, 13–32.

Bushell, P.J. (1976), On a class of Volterra and Fredholm non-linear integral equations, *Math. Proc. Cambridge Philos. Soc.* **79**, 329–335.

Bushell, P.J. and W. Okrasiński (1989), Uniqueness of solutions for a class of nonlinear Volterra integral equations with convolution kernel, *Math. Proc. Cambridge Philos. Soc.* **106**, 547–552.

Bushell, P.J. and W. Okrasiński (1990), Nonlinear Volterra integral equations with convolution kernel, *J. London Math. Soc. (2)* **41**, 503–510.

Bushell, P.J. and W. Okrasiński (1992), Nonlinear Volterra integral equations and the Apéry identities, *Bull. London Math. Soc.* **24**, 478–484. [Existence of nontrivial solutions of $y^\beta(t) = \int_0^t k(t-s)g(y(s))ds$ ($\beta \geq 1$, $g(0) = 0$). See also Buckwar (1997).]

Bushell, P.J. and W. Okrasiński (1996), On the maximal interval of existence for solutions to some nonlinear Volterra integral equations with convolution kernel, *Bull. London Math. Soc.* **28**, 59–65.

Butzer, P.L. (1958), Die Anwendung des Operatorenkalküls von Jan Mikusiński auf lineare Integralgleichungen vom Faltungstyp, *Arch. Ration. Mech. Anal.* **2**, 114–128. [See p. 125 on quadratic VIEs.]

Cahlon, B. (1992), Numerical solution for functional equations with state-dependent delay, *Appl. Numer. Math.* **9**, 291–305.

Cahlon, B. (1993), Oscillatory solutions of Volterra integral equations with state-dependent delay, *Dynam. Systems Appl.* **2**, 461–469.

Cahlon, B. and D. Schmidt (1997), Stability criteria for certain delay integral equations of Volterra type, *J. Comput. Appl. Math.* **84**, 161–188.

Calabrò, F. and G. Capobianco (2009), Blowing up behavior for a class of nonlinear Volterra integral equations connected with parabolic PDEs, *J. Comput. Appl. Math.* **228**, 580–588.

Campbell, S.L. and C.W. Gear (1995), The index of general nonlinear DAEs, *Numer. Math.* **72**, 173–196.

Cañada, A. and A. Zertiti (1994), Methods of upper and lower solutions for nonlinear delay integral equations modelling epidemics and population growth, *Math. Models Methods Appl. Sci.* **4**, 107–119. [Existence of positive periodic solutions. See also Cooke & Kaplan (1976).]

Cannon, J.R. (1984), *The One-Dimensional Heat Equation* (Reading, MA: Addison-Wesley Publishing Company).

Cao, Y., T. Herdman and Y. Xu (2003), A hybrid collocation method for Volterra integral equations with weakly singular kernels, *SIAM J. Numer. Anal.* **41**, 364–381.

Cao, Y.Z. and R. Zhang (2015), A stochastic co.llocation method for stochastic Volterra equations of the second kind, *J. Integral Equations Appl.* **27**, 1–25.

Carleman, T. (1922), Über die Abelsche Integralgleichung mit konstanten Integrationsgrenzen, *Math. Z.* **15**, 111–120.

Carlson, D.A. (1987), An elementary proof of the maximum principle for optimal control problems governed by a Volterra integral equation, *J. Optim. Theory Appl.* **54**, 43–61.

Carlson, D.A. (1990), Infinite-horizon optimal controls for problems governed by a Volterra integral equation with state-and-control-dependent discount factor, *J. Optim. Theor. Appl.* **66**, 311–336.

Carpinteri, A. and F. Mainardi (1997), *Fractals and Fractional Calculus in Continuum Mechanics*, Lecture Notes, International Centre for Mechanical Sciences, Udine 1996 (Wien-New York: Springer-Verlag). [See in particular the contribution by Gorenflo (1996).]

Castillo, J.M.F. and W. Okrasiński (1993), Boyd index and nonlinear Volterra equations, *Nonlinear Anal.* **20**, 721–732.

Castillo, J.M.F. and W. Okrasiński (1994), A new proof of existence of solutions for a class of nonlinear Volterra equations, *J. Integral Equations Appl.* **6**, 191–196. [Compare Gripenber (1990) for the original result.]

Cerha, J. (1972), A note on Volterra integral equations with degenerate kernel, *Comment. Math. Univ. Carolinae* **13**, 659–672.

Cerha, J. (1976), On some linear Volterra delay equations, *Časopis Pěst. Mat.* **101**, 111–123. [L^p-solutions and resolvent equations.]

Chadam, J.M. and H.-M. Yin (1993), A diffusion equation with localized chemical reactions, *Prof. Edinburgh Math. Soc.* **37**, 101–118.

Chakrabarti, A. (2008), Solution of the generalized Abel integral equations, *J. Integral Equations Appl.* **20**, 1–11. [See von Wolfersdorf (1965) for a more comprehensive analysis of such equations.]

Chambers, Ll.G. (1990), Some properties of the functional equation $\phi(x) = f(x) + \int_0^{\lambda x} g(x, y, \phi(y)) dy$, *Internat. J. Math. Math. Sci.* **14**, 27–44. [Representation of solutions: analogue of "Neumann series" for $0 < \lambda < 1$; application to scalar and multidimensional pantograph equations.]

Chandler, G.A. and I.G. Graham (1988), Product integration-collocation methods for noncompact integral operator equations, *Math. Comp.* **50**, 125–138.

Chang, C.C. and T.S. Lundgren (1959/1960), Airfoil in a sonic shear flow jet: a mixed boundary value problem for the generalized Tricomi equation, *Quart. Appl. Math.* **17**, 375–392. [See also von Wolfersdorf (1965), Gakhov (1966).]

Chen, G. and R. Grimmer (1980), Semigroups and integral equations, *J. Integral Equations Appl.* **2**, 133–154.

Chen, G. and R. Grimmer (1982), Integral equations as evolution equations, *J. Differential Equations* **45**, 53–74.

Cherrier, P. and A. Milani (2012), *Linear and Quasi-linear Evolution Equations in Hilbert Spaces* (Providence, RI: American Mathematical Society). [Ch. 1: introduction to Hölder, Lebesgue and Sobolev spaces.]

Chistyakov, V.F. (2006), On some properties of systems of Volterra integral equations of the fourth kind with a convolution-type kernel, *Math. Notes* **80**, 109–113.

Ciarlet, P.G. (2013), *Linear and Nonlinear Functional Analysis with Applications* (Philadelphia, Society for Industrial and Applied Mathematics (SIAM)).

Clément, Ph. (1980), On abstract Volterra equations with kernels having a positive resolvent, *Israel J. Math.* **36**, 193–200.

Clément, Ph. and J.A. Nohel (1979), Abstract linear and nonlinear Volterra equations preserving positivity, *SIAM J. Math. Anal.* **10**, 365–388.

Clément, Ph. and J.A. Nohel (1981), Asymptotic behavior of solutions of nonlinear Volterra equations with completely positive kernels, *SIAM J. Math. Anal.* **12**, 514–535.

Cochran, J.A. (1972), *Analysis of Linear Integral Equations* (New York: McGraw-Hill).

Coen, S. (2008), The life of Vito Volterra as depicted in some relatively recent biographies, *Mat. Soc. Cult. Riv. Unione Mat. Ital. (I)* **1**, 443–476, 590.

Cochran, W.G., J.-S. Lee and J. Potthoff (1995), Stochastic Volterra equations with singular kernels, *Stochastic Process. Appl.* **56**, 337–349.

Condon, M., A. Deaño, A. Iserles and K. Kropielnicka (2012), Efficient computation of delay differential equations with highly oscillatory terms, *Math. Model. Numer. Anal.* **46**, 1407–1420.

Consiglio, A. (1940), Risoluzione di una equazione integrale non lineare presentatasi in un problema di turbolenza, *Atti Accad. Gioenia Catania (6)* **4**, no. XX, 13 p.

Cooke, K.L. (1976), An epidemic equation with immigration, *Math. Biosci.* **29**, 135–158. [A generalisation of this mathematical model to one with state-dependent delay may be found in Bélair (1991).]

Cooke, K.L. and J.L. Kaplan (1976), A periodicity threshold theorem for epidemics and population growth, *Math. Biosci.* **31**, 87–104. [Compare also Cañada & Zertiti (1994).]

Cooke, K.L. and J.A. Yorke (1973), Some equations modelling growth processes and gonorrhea epidemics, *Math. Biosci.* **16**, 75–101.

Coppel, W.A. (1965), *Stability and Asymptotic Behavior of Differential Equations* (Boston: D.C. Heath & Co.).

Corduneanu, C. (1963), Sur une équation intégrale de la théorie du rélage automatique, *C.R. Acad. Sci. Paris* **256**, 3564–3567.

Corduneanu, C. (1965), Problèmes globaux dans la théorie des équations intégrales de Volterra, *Ann. Mat. Pura Appl. (4)* **67**, 349–363.

Corduneanu, C. (1991), *Integral Equations and Applications* (Cambridge, Cambridge University Press). [See also for an illuminating introduction to the theory of abstract Volterra equations and extensive references.]

Corduneanu, C. (2000), Abstract Volterra equations: a survey, *Math. Comput. Modelling* **32**, 1503–1528.

Corduneanu, C. and I.W. Sandberg (eds.) (2000), *Volterra Equations and Applications*, Stability Control, Theory, Methods Appl. **10** (Amsterdam, Gordon and Breach). [Proceedings of the Volterra Centennial Symposium held at the University of Texas at Arlington, May 1996.]

Cotton, É. (1910), Équations différentielles et équations intégrales, *Bull. Soc. Math. France* **38**, 144–154.

Curle, S.N. (1978), Solution of an integral equation of Lighthill, *Proc. Roy. Soc. London Ser. A* **364**, 435–441. [Lighthill's original paper of 1950.]

Dai, Z.W. and P.K. Lamm (2008), Local regularization for the nonlinear inverse auto-convolution problem, *SIAM J. Numer. Anal.* **46**, 832–868. [Contains an extensive list of references on first-kind auto-convolution VIEs.]

Davies, E.B. (2007), *Linear Operators and their Spectra* (Cambridge University Press).

Davies, P.J. and D.B. Duncan (2004), Stability and convergence of collocation schemes for retarded potential integral equations, *SIAM J. Numer. Anal.* **42**, 1167–1188.

Davis, H.T. (1924), Fractional operations as applied to a class of Volterra integral equations, *Amer. J. Math.* **46**, 95–109. [See also Evans (1910).]

Davis, H.T. (1926), *The Present State of Integral Equations*, Indiana University Studies **XIII**, No. 70 (Bloomington). [Contains an extensive bibliography (25 pages) on the development of the theory of VIEs.]

Davis, H.T. (1927), *A Survey of Methods for the Inversion of Integrals of Volterra Type*, Indiana University Studies **XIV**, Nos. 76–77 (Bloomington). [Discussion of Volterra's *Nota III* and the connection with Fuchsian differential equations: pp. 10–23.]

Davis, H.T. (1930), *The Theory of Volterra Integral Equations of the Second Kind*, Indiana University Studies **XVII**, Nos. 88–90 (Bloomington).

Deaño, A., D. Huybrechs and A. Iserles (2017), *Computing Highly Oscillatory Integrals* (Philadelphia: Society of Industrial and Applied Mathematics).

Deimling, K. (1979), Fixed points of condensing maps, in Londen & Staffans (1979), pp. 67–82.

Deimling, K. (1995), Nonlinear Volterra integral equations of the first kind, *Nonlinear Anal.* **25**, 951–957. [Extension of Volterra's "smoothing transformation" in his *Nota II* to nonlinear first-kind VIEs with weakly singular kernels. Compare also Dixon et al. (1986) for a similar existence and uniqueness result.]

Deng, K. and H.A. Levine (2000), The role of critical exponents in blow-up theorems: the sequel, *J. Math. Anal. Appl.* **243**, 85–126.

Deng, K. and C.A. Roberts (1997), Quenching for a diffusion equation with concentrated singularity, *Differential Integral Equations* **10**, 369–379.

Deng, K. and M.X. Xu (1999), Quenching for a diffusive equation with a concentrated singularity, *Z. Angew. Math. Phys.* **50**, 574–584.

Denisov, A.M. and S.V. Korovin (1992), On Volterra's integral equation of the first kind, *Moscow Univ. Comput. Math. Cybernet.* **3**, 19–24. [Generalisation of Volterra's 1897 paper. See also Lalesco (1911) and Denisov & Korovin (1995).]

Denisov, A.M. and A. Lorenzi (1995), On a special Volterra integral equation of the first kind, *Boll. Un. Mat. Ital. B (7)* **9**, 443–457.

Denisov, A.M. and A. Lorenzi (1997), Existence results and regularization techniques for severely ill-posed integrofunctional equations, *Boll. Un. Mat. Ital. B (7)* **11**, 713–732. [Delay VIEs $z(t) = k(t)z(\alpha(t)) + \int_{\beta(t)}^{t} b(t, s)z(s)ds$, $0 < \alpha(t) \le qt$ $(0 < q < 1)$; $0 \le \beta(t) < t$.]

Desch, G. and S.-O. Londen (2013), Maximal regularity for stochastic integral equations, *J. Appl. Anal.* **19**, 125–140.

Desch, W. and S.-O. Londen (2011), An L_p-theory for stochastic integral equations, *J. Evol. Equ.* **11**, 287–317.

Desch, W. and J. Prüss (1993), Counterexamples for abstract linear Volterra equations, *J. Integral Equations Appl.* **5**, 29–45.

Diekmann, O. (1978), Thresholds and travelling waves for the geographical spread of infection, *J. Math. Bio.* **6**, 109–130. [See also Thieme & Zhao (1993) for models based on nonlinear VFIEs.]

Diekmann, O. (1979), Integral equations and population dynamics, in *Colloquium Numerical Treatment of Integral Equations* (H.J.J. te Riele, ed.), pp. 115–149, MC Syllabus **41** (Amsterdam: Mathematisch Centrum). [This expository paper is a rich source on applications of VIEs and VIDEs; extensive list of references.]

Diekmann, O. and S.A. van Gils (1981), A variation-of-constants formula for nonlinear Volterra integral equations of convolution type, in *Nonlinear Differential Equations: Invariance, Stability, and Bifurcation (Trento, 1980)* (P. de Mottoni and L. Salvadori, eds.), pp. 133–143 (New York: Academic Press).

Diekmann, O. and S.A. van Gils (1984), Invariant manifolds for Volterra integral equations of convolution type, *J. Differential Equations* **54**, 139–180.

Diethelm, K. (2010), *The Analysis of Fractional Differential Equations: An application-oriented exposition using differential operators of Caputo type*, Lecture Notes in Math. **2004** (Heidelberg-New York: Springer-Verlag).

Diethelm, K. and N.J. Ford (2012), Volterra integral equations and fractional calculus: do neighboring solutions intersect?, *J. Integral Equations Appl.* **24**, 25–37.

Dieudonné, J. (1960), *Foundations of Modern Analysis* (New York: Academic Press).

Dieudonné, J. (1981), *History of Functional Analysis* (Amsterdam: North-Holland).

Dinghas, A. (1958), Zur Existenz von Fixpunkten bei Abbildungen vom Abel-Liouvilleschen Typus, *Math. Z.* **70**, 174–189.

Diogo, T. and P. Lima (2008), Superconvergence of collocation methods for a class of weakly singular Volterra equations, *J. Comput. Appl. Math.* **218**, 307–316.

Diogo, T., S. McKee and T. Tang (1991), A Hermite-type collocation method for the solution of an integral equation with a certain weakly singular kernel, *IMA J. Numer. Anal.* **11**, 595–605.

Diogo, T. and G. Vainikko (2013), Applicability of spline collocation to cordial Volterra equations, *Math. Model. Anal.* **18**, 1–21.

Dixon, J. and S. McKee (1984), Repeated integral inequalities, *IMA J. Numer. Anal.* **4**, 99–107.

Dixon, J., S. Mckee and R. Jeltsch (1986), Convergence analysis of discretization methods for nonlinear first kind Volterra integral equations, *Numer. Math.* **49**, 67–80. [Contains result on global existence of solutions for nonlinear VIEs of the first kind.]

Doetsch, G. (1973), *Handbuch der Laplace-Transformation. III: Anwendungen der Laplace-Transform* (Basel-Stuttgart: Birkhäuser Verlag). [Corrected reprint of 1956 edition.]

Dolph, C.L. and G.J. Minty (1964), On nonlinear integral equations of the Hammerstein type, in Anselone (1964, pp. 99–154. [Extensive bibliography, including historical papers.]

Domanov, I. (2007), On the spectrum and eigenfunctions of the operator $(Vf)(x) = \int_0^{x^\alpha} f(t)dt$, in *Perspectives in Operator Theory (Warsaw 2004)* (W. Arendt et al., eds.), 137–142, Banach Center Publ. **75** (Warsaw: Polish Academy of Sciences, Institute of Mathematics).

Domanov, I. (2008), On the spectrum of the operator which is a composition of integration and substitution, *Studia Math.* **185**, 49–65.

Domínguez, V., I.G. Graham and T. Kim (2013), Filon-Clenshaw-Curtis rules for highly oscillatory integrals with algebraic singularities and stationary points, *SIAM J. Numer. Anal.* **51**, 1542–1566.

Douglas, R.G. (1998), *Banach Algebra Techniques in Operator Theory* (New York: Springer-Verlag).

Driver, R.D. (1963), Existence theory for a delay-differential system, *Contributions to Differential Equations* **1**, 317–336.

Dutkiewicz, A. (2006), On the existence of L^p-solutions of Volterra integral equations with weakly singular kernels, *Demonstratio Math.* **39**, 837–844.

Eggermont, P.P.B. (1988), On monotone Abel-Volterra integral equations on the half line, *Numer. Math.* **52**, 65–79.

El'sgol'ts, L.E. and S.B. Norkin (1973), *Introduction to the Theory and Application of Differential Equations with Deviating Arguments* (New York–London: Academic Press).

Emmrich, E. and G. Vallet (2016), On a nonlinear abstract Volterra equation, *J. Integral Equations Appl.* **28**, 75–89.

Engel, K.-J. and R. Nagel (2000), *One-Parameter Semigroups for Linear Evolution Equations* (New York: Springer-Verlag). [See pp. 435–452 on semigroups for Volterra integral equations. Compare also Corduneanu (1991), Prüss (2012) and, especially, Gripenber, Londen & Staffans (1990) Ch. 8, as well the concise "short course" contained in the following book.]

Engel, K.-J. and R. Nagel (2006), *A Short Course on Operator Semigroups* (Unitext: New York: Springer-Verlag).

Engquist, B., A. Fokas, E. Hairer and A. Iserles (eds.) (2009), *Highly Oscillatory Problems*, London Math. Soc. Lecture Note Ser. **366** (Cambridge University Press).

Erdélyi, A. (1955), *Higher Transcendental Functions*, Vol. III (New York, McGraw-Hill). [Discusses the Mittag-Leffler function on pp. 206–211; see also Hille & Tamarkin (1930), Gorenflo (1987), Gorenflo & Vessella (1991), Kiryakova (2000).]

Evans, G.C. (1909), The integral equation of the second kind, of Volterra, with singular kernel, *Bull. Amer. Math. Soc.* **16**, 130–136.

Evans, G.C. (1910), Volterra's integral equation of the second kind, with discontinuous kernel, *Trans. American Math. Soc.* **11**, 393–413. [Based on Evans's 1910 Harvard University PhD thesis. Evans's papers represent the first contributions to the theory of singular VIEs of the second kind. See also a sequel in the same journal, **12** (1911), 429–472.]

Eveson, S.P. (2003), Norms of iterates of Volterra operators on L^2, *J. Operator Theory* **50**, 369–386.

Eveson, S.P. (2005), Asymptotic behaviour of iterates of Volterra operators on $L^p(0, 1)$, *Integral Equations Operator Theory* **53**, 331–341. [See also Shkarin (2006a), Addell & Gallardo-Gutièrrez (2007).]

Faber, V., T.A. Manteuffel, A.B. White, Jr. and G.M. Wing (1986), Asymptotic behavior of singular values and singular functions of certain convolution operators, *Comput. Math Appl. Ser. A* **12**, 733–747.

Faber, V. and G.M. Wing (1986a), Asymptotic behavior of singular values of convolution operators, *Rocky Mount. J. Math.* **16**, 567–574.

Faber, V. and G.M. Wing (1986b), Singular values of fractional integral operators: a unification of theorems of Hille, Tamarkin and Chang, *J. Math. Anal. Appl.* **120**, 745–760.

Faber, V. and G.M. Wing (1988), Effective bounds for the singular values of integral operators, *J. Integral Equations Applications* **1**, 55–64.

Feldstein, A. and K.W. Neves (1984), High order methods for state-dependent delay differential equations, *SIAM J. Numer. Anal.* **21**, 844–863.

Feldstein, A., K.W. Neves and S. Thompson (2006), Sharpness results for state-dependent delay differential equations, *Commun. Pure Appl. Anal.* **25**, 472–487.

Feller, W. (1941), On the integral equation of renewal theory, *Ann. Math. Statist.* **12**, 243–267. [Compare also Lotka (1939) for an extensive list of earlier references.]

Feller, W. (1971), *An Introduction to Probability Theory and Its Applications*, Vol. II (New York: John Wiley & Sons).

Fényes, T. (1967), A note on the solution of integral equations of convolution type of the third kind by application of the operational calculus of Mikusiński, *Studia Sci. Math. Hungar.* **2**, 81–89.

Fényes, T. (1977), On the operational solution of a convolution type integral equation of the third kind, *Studia Sci. Math. Hungar.* **12**, 65–75.

Fenyö, S. and H.W. Stolle (1982), *Theory und Praxis der linearen Integralgleichungen* (Berlin, VEB Deutscher Verlag der Wissenschaften; Basel-Boston: Birkhäuser Verlag). [Band 1 (1982): Theory of linear operators; Band 2 (1983): Theory of linear second-kind integral equations; Band 3 (1984): Linear first-kind equations and integral equations of special type; Band 4 (1984): Numerical methods and applications; this volume also contains a very extensive bibliography of some 2000 items.]

Fixman, U. (2000), On the numerical range of a Volterra operator in L_p, *Integral Equations Operator Theory* **37**, 9–19.

Fleischer, G., R. Gorenflo and B. Hofmann (1999), On the autoconvolution equation and total variation constraints, *Z. Angew. Math. Mech.* **79**, 149–159.

Fleischer, G. and B. Hofmann (1996), On inversion rates for the autoconvolution equation, *Inverse Problems* **12**, 419–435.

Fleischer, G. and B. Hofmann (1997), The local degree of ill-posedness and the autoconvolution equation, *Nonlinear Anal.* **30**, 3323–3332.

Fredholm, I. (1903), Sur une classe d'équations fonctionnelles, *Acta Math.* **27**, 365–390.

Friedman, A. (1963), On integral equations of Volterra type, *J. Analyse Math.* **11**, 381–413. [Positivity and asymptotic properties of solutions to linear and nonlinear convolution equations, including equations with weakly singular kernels. See also Ling (1978).]

Friedman, A. (1965), Periodic behavior of solutions of Volterra integral equations, *J. Analyse Math.* **15**, 287–303.

Friedman, A. and M. Shinbrot (1967), Volterra integral equations in Banach spaces, *Trans. Amer. Math. Soc.* **126**, 131–179.

Gabbasov, N.S. (2011), New versions of the collocation method for integral equations of the third kind with singularities in the kernel, *Differ. Equ.* **47**, 1357–1364.

Gabbasov, N.S. and R.R. Zamaliev (2010), New versions of spline methods for integral equations of the third kind with singularities in the kernel, *Differ. Equ.* **46**, 1330–1338.

Gakhov, F.D. (1966), *Boundary Value Problems* (Oxford: Pergamon Press). [1990 reprint: New York: Dover Publications.]

Galajikian, H. (1913), On certain non-linear integral equations, *Bull. Amer. Math. Soc.* **19**, 342–346. [Part of the author's master's thesis, Cornell University, 1912.]

Galajikian, H. (1914), Non-linear integral equations of Volterra type, *Ann. of Math. (2)* **16**, 172–192.

Gallardo-Gutiérrez, E.A. and A. Montes-Rodríguez (2004), The Volterra operator is not supercyclic, *Integral Equations Operator Theory* **50**, 211–216.

Gauthier, A., P.A. Knight and S. McKee (2007), The Hertz contact problem, coupled Volterra integral equations and a linear complementarity problem, *J. Comput. Appl. Math.* **206**, 322–340.

Gear, C.W. (1990), Differential algebraic equations, indices, and integral algebraic equations, *SIAM J. Numer. Anal.* **27**, 1527–1534. [Compare also Campbell & Gear (1995).]

Ghermanesco, M. (1959), Équations intégrales aux deux limites variables, *C.R. Acad. Sci. Paris* **248**, 1104–1105; **249**, 1606–1607. [Limits of integration in Volterra operator are $-t$ and t.]

Ghermanesco, M. (1961), Équations intégrales aux deux limites variables, *Ann. Mat. Pura Appl. (4)* **54**, 33–56.

Ghoreishi, F., M. Hadizadeh & S. Pishbin (2012), On the convergence analysis of the spline collocation method for system of integral-algebraic equations of index-2, *Int. J. Comput. Methods* **9**,1250048, 22pp.

Gilbarg, D. and N.S Trudinger (2001), *Elliptic Partial Differential Equations of Second Order* (Reprint of the 1988 edition) (Berlin-Heidelberg: Springer).

Gohberg, I. and S. Goldberg (1980), *Basic Operator Theory* (Basel-Boston: Birkhäuser Verlag).

Gohberg, I.C. and M.G. Kreĭn (1970), *Theory and Applications of Volterra Operators in Hilbert Space*, Translations of Mathematical Monographs **24** (Providence, RI, American Mathematical Society).

Gołaszewska, A. and J. Turo (2010), On nonlinear Volterra integral equations with state-dependent delays in several variables, *Z. Anal. Anwendungen* **29**, 91–106.

Gollwitzer, H.E. and R.A. Hager (1970), The nonexistence of maximum solutions of Volterra integral equations, *Proc. Amer. Math. Soc.* **26**, 301–304. [See also Nohel's review in *Math. Review* of this paper, and its relation to Nohel (1962).]

Goncerzewicz, J., H. Marcinkowska, W. Okrasiński and K. Tabisz (1978), On the percolation of water from a cylindrical reservoir into surrounding soil, *Zastos. Mat.* **16**, 249–261.

Goodstein, J.R. (2007), *The Volterra Chronicles. The Life and Times of an Extraordinary Mathematician 1860–1940* (Providence, RI, American Mathematical Society).

Gorenflo, R. (1987), Newtonssche Aufheizung, Abelsche Integralgleichungen zweiter Art und Mittag-Leffler-Funktionen, *Z. Naturforsch.* **42a**, 1141–1146.

Gorenflo, R. (1996), *Abel Integral Equations with Special Emphasis on Applications*, Lecture Notes in Math. Sciences, Graduate School of Math. Sciences **13**, University of Tokyo.

Gorenflo, R. and B. Hofmann (1994), On autoconvolution and regularization, *Inverse Problems* **10**, 353–373.

Gorenflo, R. and A.A. Kilbas (1995), Asymptotic solution of a nonlinear Abel-Volterra integral equation of second kind, *J. Fract. Calculus* **8**, 103–117. [Asymptotic expansion of solution near $t = 0$, $t = \infty$ of $y(t) = c \int_0^t (t - s)^{\alpha-1}(f(s) - y^m(s))ds$, $\alpha > 0$, $m > 1$.]

Gorenflo, R., A.A. Kilbas and S.V. Rogosin (1998), On the generalized Mittag-Leffler type functions, *Integral Transform. Spec. Funct* **7**, 215–224.

Gorenflo, R., A.A. Kilbas and S.B. Yakubovich (1994), On Abel type integral equations of the first kind, *J. Fract. Calc.* **5**, 59–68.

Gorenflo, R. and F. Mainardi (1997), Fractional calculus: integral and differential equations of fractional order, in Carpinteri & Mainardi (1997), pp. 223–276. [The article contains also a section on the Mittag-Leffler function.]

Gorenflo, R. and A. Pfeiffer (1991), On analysis and discretization of nonlinear Abel integral equations of the first kind, *Acta Math. Vietnam.* **16**, 211–262.

Gorenflo, R. and S. Vessella (1991), *Abel Integral Equations: Analysis and Applications*, Lectures Notes in Math. **1461** (Berlin-Heidelberg: Springer-Verlag).

Gourley, S., R.S. Liu and Y.J. Lou (2016), Intra-specific competition and insect larval development: a model with time-dependent delay, *Proc. Roy. Soc. Edinburgh Sect. A*, to appear.

Goursat, É. (1903), Sur un problème d'inversion résolu par Abel, *Acta Math.* **27**, 129–133.

Graham, I.G. (1982), Singularity expansions for the solutions of second kind Fredholm integral equations with weakly singular convolution kernels, *J. Integral Equations* **4**, 1–30.

Graham, I.G. and I.H. Sloan (1979), On the compactness of certain integral operators, *J. Math. Anal. Appl.* **68**, 580–594.

Grandits, P. (2008), A regularity theorem for a Volterra integral equation of the third kind, *J. Integral Equations Appl.* **20**, 507–526.

Griepentrog, E. and R. März (1986), *Differential-Algebraic Equations and Their Numerical Treatment*, Teubner Texts in Mathematics **88** (Leipzig: B.G. Teubner Verlagsgesellschaft).

Grimmer, R.C. (1982), Resolvent operators for integral equations in a Banach space, *Trans. Amer. Math. Soc.* **273**, 333–349.

Grimmer, R.C. and R.K. Miller (1977), Existence, uniqueness, and continuity for integral equations in a Banach space, *J. Math. Anal. Appl.* **57**, 429–447.

Grimmer, R.C. and A.J. Pritchard (1983), Analytic resolvent operators for integral equations in Banach space, *J. Differential Equations* **50**, 234–259.

Grimmer, R.C. and J. Prüss (1985), On linear Volterra equations in Banach spaces, *Comput. Math. Appl.* **11**, 189–205. [Resolvent operators, well-posedness, Hille-Yosida type theorem.]

Gripenberg, G. (1978), On positive, nonincreasing resolvents of Volterra equations, *J. Differential Equation* **30**, 380–390.

Gripenberg, G. (1979), An abstract nonlinear Volterra equation, *Israel J. Math.* **34**, 198–212.

Gripenberg, G. (1979a), On rapidly decaying resolvents of Volterra equations, *J. Integral Equations* **1**, 241–249.

Gripenberg, G. (1979b), On the boundedness of solutions of Volterra equations, *Indiana Univ. Math. J.* **28**, 279–290.

Gripenberg, G. (1980a), On the resolvents of nonconvolution Volterra kernels, *Funkcial. Ekvac.* **23**, 83–95.

Gripenberg, G. (1980b), On Volterra equations of the first kind, *Integral Equations Operator Theory* **4**, 473–488. [Existence of resolvents for first-kind VIEs.]

Gripenberg, G. (1981a), Unique solutions of some Volterra integral equations, *Math. Scand.* **48**, 59–67.

Gripenberg, G. (1981b), On some epidemic models, *Quart. Appl. Math.* **39**, 317–327. [VIEs of the form $x(t) = k[p(t) - \int_0^t A(t-s)x(s)ds][f(t) + \int_0^t a(t-s)x(s)ds]$.]

Gripenberg, G. (1982), Volterra integral operators and logarithmic convexity, *Math. Scand.* **50**, 209–220.

Gripenberg, G. (1982a), Decay estimates for resolvents of Volterra equations, *J. Math. Anal. Appl.* **85**, 473–487.

Gripenberg, G. (1982b), Asymptotic estimates for resolvents of Volterra equations, *J. Differential Equations* **46**, 230–243.

Gripenberg, G. (1983a), An estimate for the solution of a Volterra equation describing an epidemic, *Nonlinear Anal.* **7**, 161–165.

Gripenberg, G. (1983b), The construction of the solution of an optimal control problem described by Volterra integral equations, *SIAM J. Control Optim.* **21**, 582–597.

Gripenberg, G. (1987), Asymptotic behaviour of resolvents of abstract Volterra equations, *J. Math. Anal. Appl.* **122**, 427–438.

Gripenberg, G. (1989), Optimal control and integral equations, in *Differential Equations (Xanthi, 1987)* (C.M. Dafermos, G. Ladas and G. Papanicolaou, eds.), pp. 283–290, Lecture Notes in Pure and Appl. Math. **118** (New York: Marcel Dekker).

Gripenberg, G. (1990), On the uniqueness of solutions of Volterra equations, *J. Integral Equations Appl.* **2**, 421–430. [$x(t) = \int_0^t k(t-s)g(x(s))ds$ with $g(0) = 0$: see also Castillo & Okrasiński (1994) for another proof, and Bushell & Okrasiński (1996) with its references.]

Gripenberg, G., S.-O. Londen and O. Staffans (1990), *Volterra Integral and Functional Equations* (Cambridge University Press).

Gronwall, T.H. (1919), Note on the derivatives with respect to a parameter of the solutions of a system of differential equations, *Ann. of Math. (2)* **20**, 292–296.

Grossman, S.I. (1974), Integrability of resolvents of certain Volterra integral equations, *J. Math. Anal. Appl.* **48**, 785–793.

Grossman, S.I. (1979), Some notes on the resolvents of Volterra integral equations, in *Volterra Equations (Otaniemi, 1978)* (S.-O. Londen and O.J. Staffans, eds.), 88–91, Lecture Notes in Math. **737** (Berlin-Heidelberg: Springer-Verlag).

Guan, Q.G., R. Zhang and Y.K. Zou (2012), Analysis of collocation solutions for nonstandard Volterra integral equations, *IMA J. Numer. Anal.* **32**, 1755–1785.

Guerraggio, A. (2006), The "modern" Vito Volterra (in Italian), in *Matematica, Cultura e Società 2005* (I. Gabbani, ed.), 87–108, Centro di Ricerca Matematica Ennio De Giorgi (CRM) Ser. 2 (Pisa: Edizioni della Normale).

Guerraggio, A. and G. Paoloni (2012), *Vito Volterra* (translation of the 2008 original) (Heidelberg: Springer).

Guglielmi, N. and E. Hairer (2008), Computing the breaking points in implicit delay differential equations, *Adv. Comput. Math.* **29**, 229–247.

Gyllenberg, M. (1981), A note on continuous dependence of solutions of Volterra integral equations, *Proc. Amer. Math. Soc.* **81**, 546–548.

Györi, I. and F. Hartung (2010), Asymptotically exponential solutions in nonlinear integral and differential equations, *J. Differential Equations* **249**, 1322–1352.

Hadizadeh, M., F. Ghoreishi and S. Pishbin (2011), Jacobi spectral solution for integral-algebraic equations of index-2, *Appl. Numer. Math.* **61**, 131–148.

Hagen, R., S. Roch and B. Silbermann (2001), *C*-Algebras and Numerical Analysis* (New York: Marcel Dekker).

Hairer, E. and Ch. Lubich (1984), On the stability of Volterra-Runge-Kutta methods, *SIAM J. Numer. Anal.* **21**, 123–135.

Hairer, E. and G. Wanner (2010), *Solving Ordinary Differential Equations II* (Second revised edition) (Berlin: Springer-Verlag).

Halanay, A. and J.A. Yorke (1971), Some new results and problems in the theory of functional-differential equations, *SIAM Review* **13**, 55–80.

Hale, J.K. and S.M. Verduyn Lunel (1993), *Introduction to Functional Differential Equations* (New York: Springer-Verlag).

Halmos, P.R. (1982), *A Hilbert Space Problem Book* (Second edition) (New York: Springer-Verlag).

Hammerstein, A. (1930), Nichtlineare Integralgleichungen nebst Anwendungen, *Acta Math.* (Ivar Fredholm in memoriam) **54**, 117–176.

Han, W. (1994), Existence, uniqueness and smoothness results for second-kind Volterra equations with weakly singular kernels, *J. Integral Equations Appl.* **6**, 365–384.

Handelsman, R.A. and W.E. Olmstead (1972), Asymptotic solution to a class of nonlinear Volterra integral equations, *SIAM J. Appl. Math.* **22**, 373–384.

Hansen, P.C. (1998), *Rank-Deficient and Discrete Ill-Posed Problems* (Philadelphia, SIAM).

Hardy, G.H. and J.E. Littlewood (1928), Some properties of fractional integrals, *Math. Z.* **27**, 565–606.

Hartung, F., T. Krisztin, H.-O. Walther and J.H. Wu (2006), Functional differential equations with state-dependent delays: theory and applications, in *Handbook of Differential Equations*, Vol. III (A. Cañada, P. Drábek and A. Fonda, eds.), pp. 435–545 (Amsterdam: Elsevier/North Holland). [Contains extensive list of references.]

Haubold, H.J., A.M. Mathai and R.K. Saxena (2011), Mittag-Leffler functions and their applications, *J. Appl. Math.* **2011**, Art. ID 298628, 51 pp.

Hellinger, E. and O. Toeplitz (1927), *Integralgleichungen und Gleichungen mit unendlichvielen Unbekannten*, Encyklopädie der Wissenschaften **11.3**, 1335–1661 (1953 reprint: New York: Chelsea Publ. Co.). [Still the best source of information on the early history and development of the theory of integral equations.]

Henrici, P. (1977), *Applied and Computational Complex Analysis*, Vol. II (New York: Wiley-Interscience).

Herdman, T.L. (1976), Existence and continuation properties of solutions of a nonlinear Volterra integral equation, in *Dynamical Systems, Vol. II* (L. Cesari, J.H. Hale and J.P. LaSalle, eds.), pp. 307–310 (New York: Academic Press).

Herdman, T.L. (1977), Behavior of maximally defined solutions of a nonlinear Volterra equation, *Proc. Amer. Math. Soc.* **67**, 297–302. [Blow-up of solutions to system $x(t) = f(t) + \int_0^t g(t, s, x(s))ds$: answer to a question in Miller (1971), p.145.]

Hethcote, H.W. and P. van den Driessche (1995), An SIS epidemic model with variable population size and a delay, *J. Math. Biol.* **34**, 177–194.

Hethcote, H.W. and P. van den Driessche (2000), Two SIS epidemiologic models with delays, *J. Math. Biol.* **40**, 3–26.

Hethcote, H.W., M.A. Lewis and P. van den Driessche (1989), An epidemiological model with a delay and a nonlinear incidence rate, *J. Math. Biol.* **27**, 49–64.

Hethcote, H.W. and D.W. Tudor (1980), Integral equation models for endemic infectious diseases, *J. Math. Biol.* **9**, 37–47.

Hilbert, D. (1912), *Grundzüge einer allgemeinen Theorie der linearen Integralgleichungen* (Lepizig: B.G. Teubner; 1953 reprint: New York: Chelsea Publ. Co.). [Contains his six "Mitteilungen" of 1904–1910 on the theory of Fredholm integral equations; compare also the next reference, and see Stewart (2011) for an English translation of the 1904 paper.]

Hilbert, D. and E. Schmidt (1989), *Integralgleichungen und Gleichungen mit unendlich vielen Unbekannten* (A. Pietsch, ed.) (Leipzig: Teubner).

Hille, E. and R.S. Phillips (1957), *Functional Analysis and Semigroups*, American Mathematical Society Colloquium Publication, Vol. 31 (Providence, RI: American Mathematical Society).

Hille, E. and J.D. Tamarkin (1930), On the theory of linear integral equations, *Ann. Math.* **31**, 479–528. [First use of Mittag-Leffler function for representation of solution of weakly singular VIE.]

Hofmann, B. (1994), On the degree of ill-posedness for nonlinear problems, *J. Inverse Ill-Posed Probl.* **2**, 61–76.

Hofmann, B. and S. Kindermann (2010), On the degree of ill-posedness for linear problems with non-compact operators, *Methods Appl. Anal.* **17**, 445–461.

Holmgren, E. (1900), Sur un théorème de M. V. Volterra sur l'inversion des intégrales définies, *Atti R. Accad. Sci. Torino* **35**, 570–580.

Hönig, C.S. (1975), *Volterra Stieltjes-Integral Equations* (Amsterdam: North-Holland).

de Hoog, F.R. and R.S. Anderssen (2006), Regularization of first kind integral equations with application to Couette viscometry, *J. Integral Equations Appl.* **18**, 249–265.

de Hoog, F.R. and R.S. Anderssen (2010), Kernel perturbations for Volterra convolution integral equations, *J. Integral Equations Appl.* **22**, 427–441.

de Hoog, F.R. and R.S. Anderssen (2012a), Kernel perturbations for convolution first kind Volterra integral equations, *J. Math-for-Ind.* **4A**, 1–4.

de Hoog, F.R. and R.S. Anderssen (2012b), Kernel perturbations for a class of second-kind convolution Volterra equations with nonnegative kernel, *Appl. Math. Lett.* **25**, 1222–1225.

Hoppensteadt, F.C. (1983), An algorithm for approximate solutions to weakly filtered synchronous control systems and nonlinear renewal processes, *SIAM J. Appl. Math.* **43**, 834–843.

Howard, R. and A.R. Schep (1990), Norms of positive operators on L^p-spaces, *Proc. Amer. Math. Soc.* **109** (1990), 135–146.

Hritonenko, N. and Yu. Yatsenko (1996), *Modeling and Optimization of the Lifetime of Technologies* (Dordrecht: Kluwer Academic Publishers). [Discusses numerous models involving delay VIEs; extensive list of references on related VIE models.]

Huybrechs, D. and S. Olver (2009), Highly oscillatory quadrature, in Huybrechs et al. (2009), pp. 25–50.

Ilhan, O.A. (2012), Solvability of some integral equations in Banach space and their applications to the theory of viscoelasticity, *Abstr. Appl. Anal.* **2012**, 13 pp.

Imanaliev, M.I. and A. Asanov (2007), Regularization and uniqueness of solutions of systems of nonlinear Volterra integral equations of the third kind, *Dokl. Math.* **76**, 490–493.

Iserles, A. (1993), On the generalized pantograph functional differential equation, *Europ. J. Appl. Math.* **4**, 1–38. [Illuminating introduction to the theory of delay differential equations with proportional delay qt $(0 < q < 1)$; references on applications.]

Iserles, A. (2005a), On the numerical quadrature of highly-oscillatory integrals. Irregular oscillators, *IMA J. Numer. Anal.* **25**, 25–44.

Iserles, A. (2005b), On the numerical analysis of rapid oscillation, in *Group Theory and Numerical Analysis (Montréal 2003)* (P. Winternitz et al., eds.), pp. 149–163, CRM Proc. Lecture Notes **39** (Providence, RI: American Mathematical Society). [Survey of mathematical foundations of discretising highly oscillatory ODEs and highly oscillatory integrals.]

Iserles, A. and S.P. Nørsett (2004), On quadrature methods for highly oscillatory integrals and their implementation, *BIT Numer. Math.* **44**, 755–772.

Ito, K. and J. Turi (1991), Numerical methods for a class of singular integro-differential equations based on semigroup approximation, *SIAM J. Numer. Anal.* **28**, 1698–1722.

Ivanov, D.V., I.V. Karaulova, E.V. Markova, V.V. Trufanov and O.V. Khamisov (2004), Control of power grid development: numerical solutions, *Autom. Remote Control* **65**, 472–482.

Janikowski, J. (1962), Équation intégrale non linéaire d'Abel, *Bull. Soc. Sci. Lett. Łódź* **13**, No. 11, 8 pp. [Existence of solutions to weakly singular first-kind VIEs; see also Branca (1978), Dixon et al. (1984), and Deimling (1995).]

Janno, J. (1997), On a regularization method for the autoconvolution equation, *Z. Angew. Math. Mech.* **77**, 393–394.

Janno, J. and L. v. Wolfersdorf (1995), Regularization of a class of nonlinear Volterra equations of a convolution type, *J. Inverse Ill-Posed Probl.* **3**, 249–257.

Janno, J. and L. v. Wolfersdorf (1996), Inverse problems for identification of memory kernels in heat flow, *J. Inverse Ill-Posed Probl.* **4**, 39–66.

Janno, J. and L. v. Wolfersdorf (1997a), Identification of weakly singular memory kernels in heat conduction, *Z. Angew. Math. Mech.* **77**, 243–257.

Janno, J. and L. v. Wolfersdorf (1997b), Inverse problems for identification of memory kernels in viscoelasticity, *Math. Methods Appl. Sci.* **20**, 291–314.

Janno, J. and L. v. Wolfersdorf (1998), Identification of weakly singular memory kernels in viscoelasticity, *Z. Angew. Math. Mech.* **78**, 391–403.

Janno, J. and L. v. Wolfersdorf (2001), Identification of a special class of memory kernels in one-dimensional heat flow, *J. Inverse Ill-Posed Probl.* **9**, 389–411.

Janno, J. and L. v. Wolfersdorf (2005), A general class of autoconvolution equations of the third kind, *Z. Anal. Anwendungen* **24**, 523–543.

Jiang, L.S. (2005), *Mathematical Modeling and Methods of Option Pricing* (Singapore: World Scientific Publishing Co.).

Jones, S., B. Jumarhon, S. McKee and J.A. Scott (1996), A mathematical model of a biosensor, *J. Engrg. Math.* **30**, 321–337.

Jordan, G.S. and R.L. Wheeler (1976), A generalization of the Wiener-Lévy theorem applicable to some Volterra equations, *Proc. Amer. Math. Soc.* **57**, 109–114. [Generalization of Shea and Wainger's 1975 result.]

Jörgens, K. (1970), *Lineare Integraloperatoren* (Stuttgart: B.G. Teubner). [English translation: *Linear Integral Operators* (1982), Boston, Pitman.]

Jumarhon, B., W. Lamb, S. McKee and T. Tang (1996), A Volterra type method for solving a class of nonlinear initial-boundary value problems, *Numer. Methods Partial Differential Equations* **12**, 265–281.

Jung, M. (1999), Duality theory for solutions to Volterra integral equations, *J. Math. Anal. Appl.* **230**, 112–134.

Kabanikhin, S.I. and A. Lorenzi (1999), *Identification Problems of Wave Phenomena: Theory and Numerics* (Utrecht: VSP).

Kadem, A., M. Kirane, C.M. Kirk and W.E. Olmstead (2014), Blowing-up solutions to systems of fractional differential and integral equations with exponential nonlinearities, *IMA J. Appl. Math.* **79**, 1077–1088.

Kamont, Z. and M. Kwapisz (1981), On nonlinear Volterra functional integral equations in several variables, *Ann. Polon. Math.* **40**, 1–29.

Kang, S.G. and G. Zhang (2003), Periodic solutions of a class of integral equations, *Topol. Methods Nonlinear Anal.* **22**, 245–252.

Kappel, F. and K.P. Zhang (1986), On neutral functional differential equations with nonatomic difference operator, *J. Math. Anal. Appl.* **113**, 311–343.

Karakostas, G. (1981), Convergence of the bounded solutions of a certain implicit Volterra integral equation, *Funkcial. Ekvac.* **24**, 351–361.

Karakostas, G., I.P. Stavroulakis and Y.M. Wu (1993), Oscillation of Volterra integral equations with delay, *Tohoku Math. J.* **45**, 583–605.

Karapetyants, N.K., A.A. Kilbas, M. Saigo and S.G. Samko (2000), Upper and lower bounds for solutions of nonlinear Volterra convolution integral equations with power nonlinearity, *J. Integral Equations Appl.* **12**, 421–448.

Karczewska, A. (2007), *Convolution Type Stochastic Volterra Equations*, Lecture Notes in Nonlinear Analysis, **10** (Toruń: Juliusz Schauder Center for Nonlinear Studies).

Karczewska, A. (2009), Regularity of solutions to stochastic Volterra equations of convolution type, *Integral Transforms Spec. Funct.* **20**, 171–176.

Karczewska, A. and C. Lizama (2009), Strong solutions to stochastic Volterra equations, *J. Math. Anal. Appl.* **349**, 301–310.

Kato, T. (1995), *Perturbation Theory for Linear Operators* (reprint of the 1980 edition) (Berlin: Springer-Verlag).

Kauthen, J.-P. (1993), ε-expansions of solutions of singularly perturbed weakly singular Volterra integral equations of the second kind, private communication.

Kauthen, J.-P. (1997), A survey of singularly perturbed Volterra equations, *Appl. Numer. Math.* **24**, 95–114.

Kauthen, J.-P. (2000), The numerical solution of integral-algebraic equations of index 1 by polynomial spline collocation methods, *Math. Comp.* **70**, 1503–1514.

Kawarada, H. (1975), On solutions of initial boundary value problem for $u_t = u_{xx} + 1/(1 - u)$, *Publ. Res. Inst. Math. Sci.* **10**, 729–736.

Keller, J.B. and W.E. Olmstead (1971), Temperature of a nonlinearly radiating semi-infinite solid, *Quart. Appl. Math.* **29**, 559–566.

Keller, J.J. (1981), Propagation of simple non-linear waves in gas filled tubes with friction, *J. Appl. Math. Phys. (ZAMP)* **32**, 170–181.

Kershaw, D. (1999), Operator norms of powers of the Volterra operator, *J. Integral Equations Appl.* **11**, 351–362.

Khasi, M., F. Ghoreishi and M. Hadizadeh (2014), Numerical analysis of a high-order method for state-dependent integral equations, *Numer. Algor.* **66**, 177–201.

Kilbas, A.A. and M. Saigo (1995), On solution of integral equation of Abel-Volterra type, *Differential Integral Equations* **8**, 993–1011. [Asymptotics of solutions of $y(t) = (a(t)/\Gamma(\alpha)) \int_0^t (t-s)^{\alpha-1} y(s) ds + g(t)$ $(0 < \alpha < 1)$, with $a(t) = (\Gamma(\alpha))^{-1} \sum_{j=-1}^{\infty} a_j t^{j\alpha}$.]

Kilbas, A.A. and M. Saigo (1999), On solution of nonlinear Abel-Volterra integral equation, *J. Math. Anal. Appl.* **229**, 41–60. [$y^r(t) = a(t) \int_0^t (t-s)^{\alpha-1} y(s) ds + g(t)$, $\alpha > 0$, $r \neq 0, -1, -2, \ldots$.]

Kirk, C.M., W.E. Olmstead and C.A. Roberts (2013), A system of nonlinear Volterra equations with blow-up solutions, *J. Integral Equations Appl.* **25**, 377–393.

Kirk, C.M. and C.A. Roberts (2002), A quenching problem for the heat equation, *J. Integral Equations Appl.* **14**, 53–72.

Kirk, C.M. and C.A. Roberts (2003), A review of quenching results in the context of nonlinear Volterra equations, *Dynam. Contin. Discrete Impuls. Syst. Ser. A Math. Anal.* **10**, 343–356.

Kirsch, A. (1996), *An Introduction to the Theory of Inverse Problems* (New York: Springer-Verlag).

Kiryakova, V.S. (1994), *Generalized Fractional Calculus and Applications, Pitman Research Notes in Math.* **301** (Harlow: Longman).

Kiryakova, V.S. (2000), Multiple (multiindex) Mittag-Leffler functions and relations to generalized fractional calculus, *J. Comput. Appl. Math.* **118**, 241–259.

Klebanov, B.S. and B.D. Sleeman (1996), An axiomatic theory of Volterra integral equations, *Differential Integral Equations* **9**, 397–408. [See also Väth (1998a,1998b,1999).]

Kolk, M. and A. Pedas (2013), Numerical solution of Volterra integral equations with singularities, *Front. Math. China* **8**, 239–259.

Kosel, U. and L. v. Wolfersdorf (1986), Nichtlineare Integralgleichungen, *Seminar Analysis*, 93–128, Karl-Weierstrass-Institut für Mathematik (Berlin: Akademie der Wissenschaften der DDR).

Kowalewski, G. (1930), *Integralgleichungen* (Berlin: W. de Gruyter & Co.). [Goursat problem: see pp. 83–90. Also, Chapter 1 contains many classical results often not found in more recent books.]

Krasnosel'skiĭ, M.A. and P.P. Zabreĭko (1984), *Geometric Methods of Nonlinear Analysis* (Berlin-Heidelberg-New York: Springer-Verlag).

Krasnosel'skiĭ, M.A., P.P. Zabreĭko, E.I. Pustyl'nik and P.E. Sobolevskiĭ (1976), *Integral Operators in Spaces of Summable Functions* (Leiden: Nordhoff International Publishing).

Krasnov, M., A. Kissélev and G. Makarenko (1977), *Équations Intégrales: Problèmes et Exercises* (Moscow: Éditions Mir). [Contains numerous examples and solved problems of Volterra integral equations.]

Kress, R. (2014), *Linear Integral Equations* (Third edition) (New York: Springer-Verlag).

Kufner, A., O. John and S. Fučík (1977), *Function Spaces* (Leyden: Noordhoff International Publishing). [See also the revised and extended edition (2013) by L. Pick, A. Kufner, O. John & S. Fučík, *Function Spaces. Vol 1* (Berlin, de Gruyter).]

Kunkel, P. and V. Mehrmann (2006), *Differential-Algebraic Equations: Analysis and Numerical Solution* (Zürich: European Mathematical Society).

Kwapisz, J. (1991a), Weighted norms and Volterra integral equations in L^p spaces, *J. Appl. Math. Stoch. Anal.* **4**, 161–164.

Kwapisz, M. (1991b), Bielecki's method, existence and uniqueness results for Volterra integral equations in L^p space, *J. Math. Anal. Appl.* **154**, 403–416.

Kwapisz, M. (1991c), Remarks on the existence and uniqueness of solutions of Volterra functional equations in L^p spaces, *J. Integral Equations Appl.* **3**, 383–397.

Kwapisz, J. (1993), A note on multidimensional Volterra integral equations in L^p spaces, *Houston J. Math.* **19**, 275–280.

Kwapisz, M. and J. Turo (1983), Existence and uniqueness of solution for some integral-functional equation, *Comment. Math. Prace Mat.* **23**, 259–267.

Kwok, Y.K. (2008), *Mathematical Models of Financial Derivatives* (Second Edition) (Berlin: Springer).

Lakshmikantham, V. (ed.) (1987), *Nonlinear Analysis and Applications*, Lecture Notes in Pure and Appl. Math. **109** (New York: Marcel Dekker). [Contains numerous papers on the analysis and application of Volterra integral and functional equations.]

Lalesco, T. (1908), Sur l'équation de Volterra, Thèse (Paris: Gauthier-Villars); *J. Math. Pures Appl. (6)* **4**, 123–202. [Study of nonlinear VIEs: pp. 165–168.]

Lalesco, T. (1911), Sur une équation intégrale du type Volterra, *C.R. Acad. Sci. Paris* **52**, 579–580. [$\int_{\alpha(x)}^{x} N(x,s)\phi(s)ds = F(x)$, with $\alpha(x)$ tangent to $y = x$; see also Denisov & Korovin (1992) and Denisov & Lorenzi (1995) for generalisations.]

Lalesco, T. (1912), *Introduction à la Théorie des Équations Intégrales* (Paris: Hermann & Fils). [Contains chronological bibliography, 1826–1911.]

Lamb, W. (1985), A spectral approach to an integral equation, *Glasgow Math. J.* **26**, 83–89. [Inversion formulas for first-kind VIEs with kernels of the form $(t/s)^n (\log(t/s))^{\alpha-1}$.]

Lamm, P.K. (2000), A survey of regularization methods for first-kind Volterra equations, in *Surveys on Solution Methods for Inverse Problems* (D. Colton, H. Engl et al., eds.), pp. 53–82 (Vienna: Springer–Verlag). [See also Ring & Prix (2000).]

Lamour, R., R. März and C. Tischendorf (2013), *Differential-Algebraic Equations: A Projector Based Analysis*, Differential-Algebraic Equations Forum (Heidelberg: Springer).

Landau, H. (1977), The notion of approximate eigenvalues applied to integral equations of laser theory, *Quart. Appl. Math.* **35**, 165–172.

Lange, C.G. and D.R. Smith (1988), Singular perturbation analysis of integral equations, *Stud. Appl. Math.* **79**, 1–63; Part II: **90**, 1–74.

Lao, N. and R. Whitley (1997), Norms of powers of the Volterra operator, *Integral Equations Operator Theory* **27**, 419–425.

Lauricella, G. (1908), Sulle equazioni integrali, *Ann. Mat. Pura Appl. (3)* **15**, 21–45. [Survey paper on early development of theory of integral equations.]

Le Roux, J. (1895), Sur les intégrales des équations linéaires aux dérivées partielles du second ordre à deux variables indépendantes (Thèse), *Ann. Sci. Ecole Normale Supér. (3)* **12**, 227–316. [Convergence of Picard iteration for second-kind VIEs: uniform bounds for Neumann series by *geometric* series.]

Levin, D. (1997), Analysis of a collocation method for integrating rapidly oscillating functions, *J. Comput. Appl. Math.* **78**, 131–138.

Levin, J.J. (1963), The asymptotic behavior of the solution of a Volterra integral equations, *Proc. Amer. Math. Soc.* **14**, 534–541.

Levin, J.J. (1965), The qualitative behavior of a nonlinear Volterra equation, *Proc. Amer. Math. Soc.* **16**, 711–718.

Levin, J.J. (1977), Resolvents and bounds for linear and nonlinear Volterra equations, *Trans. Amer. Math. Soc.* **228**, 207–222. [Consult also for references on related results for nonlinear VIEs.]

Levin, J.J. (1980), Nonlinearly perturbed Volterra equations, *Tôhoku Math. J. (2)* **32**, 317–335.

Levine, H.A. (1983), The quenching of solutions of linear parabolic and hyperbolic equations with nonlinear boundary conditions, *SIAM J. Math. Anal.* **14**, 1139–1153.

Levine, H.A. (1990), The role of critical exponents in blowup theorems, *SIAM Rev.* **32**, 262–288.

Levine, H.A. (1992), Advances in quenching, in *Nonlinear Diffusion Equations and Their Equilibrium States*, **3** (N.G. Lloyd et al., eds.), 319–346, Progress in Nonlinear Differential Equations (Boston: Birkhäuser).

Levinson, N. (1960), A nonlinear Volterra equation arising in the theory of superfluidity, *J. Math. Anal. Appl.* **1**, 1–11. [Compare also Mann & Wolf (1951) for a related simpler model.]

Lewin, M. (1994), On solutions of singularly perturbed stochastic Volterra equations, *Libertas Math.* **14**, 51–63.

Liang, H. and H. Brunner (2012), Discrete superconvergence of collocation solutions for first-kind Volterra integral equations, *J. Integral Equations Appl.* **24**, 359–391.

Liang, H. and H. Brunner (2013), Integral-algebraic equations: theory of collocation methods I, *SIAM J. Numer. Anal.* **51**, 2238–2259.

Liang, H. and H. Brunner (2016), Integral-algebraic equations: theory of collocation methods II, *SIAM J. Numer. Anal.* **54**, 2640–2663.

Lifshits, M.A. and W. Linde (2002), Approximation and entropy numbers of Volterra operators with application to Brownian motion, *Mem. Amer. Math. Soc.* **157**, No. 745.

Lighthill, M.J. (1950), Contributions to the theory of heat transfer through a laminar boundary layer, *Proc. Roy. Soc. London Ser. A* **202**, 359–377. [Compare also Curle (1978).]

Lin, Q. (1963), Comparison theorems for difference-differential equations, *Sci. Sinica* **12**, 449. [Vector VIEs with constant delay; stability results by comparison with scalar VIE.]

Ling, R. (1978), Integral equations of Volterra type, *J. Math. Anal. Appl.* **64**, 381–397. [Number of zeros, boundedness and monotonicity of solutions; see also Friedman (1963) for earlier results.]

Ling, R. (1982), Solutions of singular integral equations, *Int. J. Math. Math. Sci.* **5**, 123–131.

Linz, P. (1985), *Analytical and Numerical Methods for Volterra Equations* (Philadelphia: SIAM).

Linz, P. and B. Noble (1971), A numerical method for treating identation problems, *J. Engrg. Math.* **5**, 227–231.

Lipovan, O. (2006), Integral inequalities for retarded Volterra equations, *J. Math. Anal. Appl.* **322**, 349–358.

Lipovan, O. (2009), On the asymptotic behavior of solutions to some nonlinear integral equations of convolution type, *Dyn. Contin. Discrete Impuls. Syst. Ser. A Math. Anal.* **16**, 147–154. [Implicit VIE: $u^\beta(t) = f(t) + \int_0^t k(t-s)u(s)ds$ with $\beta > 1$.]

Little, G. and J.B. Reade (1998), Estimates for the norm of the nth indefinite integral, *Bull. London Math. Soc.* **30**, 539–542.

Lizama, C. (1993), On Volterra equations associated with a linear operator, *Proc. Amer. Math. Soc.* **118**, 1159–1166.

Lizama, C. and M.P. Velasco (2015), Abstract Volterra equations with state-dependent delay, *J. Integral Equations Appl.* **27**, 219–231.

Lobanova, M.S. and Z.B. Tsalyuk (2015), Asymptotics of the solution of a Volterra integral equation with difference kernel, *Math. Notes* **97**, 396–401.

Londen, S.-O. (1984), Asymptotic estimates for resolvents of some integral equations, in *Infinite-Dimensional Systems (Retzhof 1983)* (F. Kappel and W. Schappacher, eds.), pp. 139–146, Lecture Notes in Math. **1076** (New York: Springer-Verlag).

Londen, S.-O. and O.J. Staffans (eds.) (1979), *Volterra Equations (Otaniemi, 1978)*, Lecture Notes in Math. **737** (Berlin-Heidelberg-New York: Springer-Verlag).

Lotka, A.J. (1939), A contribution to the theory of self-renewing aggregates, with special reference to industrial replacement, *Ann. Math. Statist.* **10**, 1–25.

Lotka, A.J. (1939a), *Analytical Theory of Biological Populations* (New York: Plenum Press). [Translation of the 1939 French edition; review of renewal equation and mathematical demography of single populations.]

Lotka, A.J. (1942), The progeny of an entire population, *Ann. Math. Statist.* **13**, 115–126.

Lotka, A.J. (1945), in *Population Analysis as a Chapter in the Mathematical Theory of Evolution* (Essays on Growth and Form Presented to D'Arcy Wentworth Thompson) (W.E. Le Gros Clark and P.B. Medawar, eds.), 355–385 (Oxford: Clarendon Press). [Review of integral equation of renewal theory.]

Lowengrub, M. and J. Walton (1979), Systems of generalized Abel equations, *SIAM J. Math. Anal.* **4**, 794–807.

Loy, R.J. and R.S. Anderssen (2014), Interconversion relationships for completely monotone functions, *SIAM J. Math. Anal.* **46**, 2008–2032.

Lubich, Ch. (1983a), Runge-Kutta theory for Volterra and Abel integral equations of the second kind, *Math. Comp.* **41**, 87–102. [Contains also proofs of Paley-Wiener theorems for VIEs and VIDEs.]

Lubich, Ch. (1986), A stability analysis of convolution quadrature for Abel-Volterra integral equations, *IMA J. Numer. Anal.* **6**, 87–101.

Lubich, Ch. (1987), Fractional linear multistep methods for Abel-Volterra integral equations of the first kind, *IMA J. Numer. Anal.* **7**, 97–106.

Lyubich, Yu.I. (1984), Composition of integration and substitution, in *Linear and Complex Analysis Problem Book* (V.P. Havin et al., eds.), 249–250, Lecture Notes in Math. **1043** (New York: Springer-Verlag).

Magnickiĭ, N.A. (1979), Linear Volterra integral equations of the first and third kinds (in Russian), *Zh. Vychisl. Mat. i Mat. Fiz.* **19**, 970–988, 1083.

Mainardi, F. (2010), *Fractional Calculus and Waves in Linear Viscoelasticity* (London: Imperial College Press).

Mainardi, F. and R. Gorenflo (2000), On Mittag-Leffler-type functions in fractional evolution processes, *J. Comput. Appl. Math.* **118**, 283–299. [Survey and extensive list of references, including historical ones.]

Małolepszy, T. and W. Okrasiński (2007), Conditions for blow-up of solutions of some nonlinear Volterra integral equations, *J. Comput. Appl. Math.* **205**, 744–750.

Małolepszy, T. and W. Okrasiński (2008), Blow-up conditions for nonlinear Volterra integral equations with power nonlinearity, *Appl. Math. Lett.* **21**, 307–312.

Małolepszy, T. and W. Okrasiński (2010), Blow-up time for solutions to some nonlinear Volterra integral equations, *J. Math. Anal. Appl.* **366**, 372–384.

Mandal, N., A. Chakrabarti and B.N. Mandal (1996), Solution of a system of generalized Abel integral equations using fractional calculus, *Appl. Math. Lett.* **9**, 1–4.

Mann, W.R. and F. Wolf (1951), Heat transfer between solids and gases under nonlinear boundary conditions, *Quart. Appl. Math.* **9**, 163–184.

März, R. (1992), Numerical methods for differential algebraic equations, *Acta Numer.* **1**, 141–198.

März, R. (2002a), The index of linear differential-algebraic equations with properly stated leading terms, *Results Math.* **42**, 308–338.

März, R. (2002b), Differential algebraic equations anew, *Appl. Numer. Math.* **42**, 315–335.

März, R. (2004a), Solvability of linear differential algebraic equations with properly stated leading terms, *Results Math.* **45**, 88–105.

März, R. (2004b), Fine decouplings of regular differential algebraic equations, *Results Math.* **46**, 57–72.

McKee, S. (1982), Generalised discrete Gronwall lemmas, *Z. Angew. Math. Mech.*, **62**, 429–434.

McKee, S. and J.A. Cuminato (2015), Nonlocal diffusion, a Mittag-Leffler function and a two-dimensional Volterra integral equation, *J. Math. Anal. Appl.* **423**, 243–252.

Mann, W.R. and F. Wolf (1951), Heat transfer between solids and gases under nonlinear boundary conditions, *Quart. Appl. Math.*, **9**, 163–184.

Meehan, M. and D. O'Regan (1999), A comparison technique for integral equations, *Irish Math. Soc. Bull.* **42**, 54–71. [Extension of comparison results of Friedman Friedman (1963) and Miller (1968b).]

Megginson, R.E. (1998), *An Introduction to Banach Space Theory* (New York: Springer-Verlag).

Medhin, N.G. (1986), Optimal processes governed by integral equations, *J. Math. Anal. Appl.* **120**, 1–12.

Meis, T. (1978), Eine spezielle Integralgleichung erster Art, in *Numerical Treatment of Differential Equations (Oberwolfach 1976)* (R. Bulirsch, R.D. Grigorieff and J. Schröder, eds.), pp. 107–120, Lecture Notes in Math. **631** (Berlin-Heidelberg:

Springer-Verlag). [$\int_0^t k(t, s, y(s), y(a(t, s)))ds$, $0 \leq a(t, s) \leq \max\{qt, t - \tau\}$ $(q \in (0, 1))$.]

Metz, J.A.J. and O. Diekmann (eds.) (1986), *The Dynamics of Physiologically Structured Populations*, Lecture Notes in Biomath. **68** (Berlin-Heidelberg: Springer-Verlag). [See in particular Ch. IV on age dependence where VIEs are introduced as a "natural modelling tool".]

Mikhlin, S.G. and S. Prössdorf (1986), *Singular Integral Operators* (Berlin-Heidelberg-New York: Springer-Verlag).

Miller, R.K. (1966), Asymptotic behavior of solutions of nonlinear Volterra equations, *Bull. Amer. Math. Soc.* **72**, 153–156.

Miller, R.K. (1968a), On the linearization of Volterra integral equations, *J. Math. Anal. Appl.* **23**, 198–208.

Miller, R.K. (1968b), On Volterra integral equations with nonnegative integrable resolvents, *J. Math. Anal. Appl.* **22**, 319–340.

Miller, R.K. (1971), *Nonlinear Volterra Integral Equations* (Menlo Park, CA: Benjamin).

Miller, R.K. (1975a), A system of renewal equations, *SIAM J. Appl. Math.* **29**, 20–34.

Miller, R.K. (1975b), Volterra integral equations in a Banach space, *Funkcial. Ekvac.* **18**, 163–194.

Miller, R.K. (2000), Volterra integral equations at Wisconsin, in Corduneanu & Sandberg (2000), pp. 15–26.

Miller, R.K. and A. Feldstein (1971), Smoothness of solutions of Volterra integral equations with weakly singular kernels, *SIAM J. Math. Anal.* **2**, 242–258.

Miller, R.K., J.A. Nohel and J.S.W. Wong (1969), Perturbations of Volterra integral equations, *J. Math. Anal. Appl.* **25**, 676–691.

Miller, R.K. and G.R. Sell (1968), Existence, uniqueness and continuity of solutions of integral equations, *Ann. Mat. Pura Appl. (IV)* **80**, 135–152.

Miller, R.K. and G.R. Sell (1970), Existence, uniqueness and continuity of solutions of integral equations. An addendum, *Ann. Mat. Pura Appl. (IV)* **87**, 281–286.

Miller, R.K. and G.R. Sell (1970a), *Volterra Integral Equations and Topological Dynamics*, Memoirs Amer. Math. Soc. **102** (Providence, RI: American Mathematical Society).

Miller, R.K. and A. Unterreiter (1992), Switching behavior of PN-diodes: Volterra integral equation models, *J. Integral Equations Appl.* **4**, 257–272. [The paper complements Schmeiser et al. (1993) and Unterreiter (1996).]

Mingarelli, A. (1983), *Volterra–Stieltjes Integral Equations and Generalized Ordinary Differential Equations*, Lecture Notes in Math. **989** (Berlin-New York: Springer-Verlag).

Mittag-Leffler, G.M. (1903), Sur la nouvelle fonction $E_\alpha(x)$, *C.R. Acad. Sci. Paris Sér. II* **137**, 554–558. [Compare also Winan (1905), Hille & Tamarkin (1930), and Pollard (1948).]

Monna, A.F. (1973), *Functional Analysis in Historical Perspective* (Utrecht: Oosthoek Publishing Company).

Mureşan, V. (1999), On a class of Volterra integral equations with deviating argument, *Studia Univ. Babeş-Bolyai Math.* **XLIV**, 47–54. [Existence and dependence on data for $x(t) = x(0) + \int_0^t f(s, x(\lambda s))ds$, $0 < \lambda < 1$.]

Mydlarczyk, W. (1990), Galerkin methods for nonlinear Volterra type equations, *Zastos. Mat.* **20**, 625–638. [Regularity results for solutions of $x(t) - \int_0^t m(t, s)k(t - s)g(x(s), s)ds = f(t)$ in Hölder and Nikol'skii spaces.]

Mydlarczyk, W. (1991), The existence of nontrivial solutions of Volterra integral equations, *Math. Scand.* **68**, 83–88.

Mydlarczyk, W. (1992), The existence of nontrivial solutions for a class of Volterra equations with smooth kernels, *Math. Scand.* **71**, 261–266.

Mydlarczyk, W. (1994), A condition for finite blow-up time for a Volterra integral equation, *J. Math. Anal. Appl.* **181**, 248–253. [A related early blow-up result may be found in Miller's 1971 book; see also Herdman (1977).]

Mydlarczyk, W. (1996), The existence of solutions to a Volterra integral equation, *Ann. Polon. Math.* **64**, 175–182. [Compare also [147].]

Mydlarczyk, W. (1999), The blow-up solution of integral equations, *Colloq. Math.* **79**, 147–156. [Survey and generalisation of results in Mydlarczyk (1994) and Olmstead et al. (1995).]

Mydlarczyk, W. (2001), A nonlinear Abel integral equation on the whole line, *Nonlinear Anal.* **45**, 273–279.

Mydlarczyk, W. and W. Okrasińskii (2001), Positive solutions to a nonlinear Abel-type integral equation on the whole line, *Comput. Math. Appl.* **41**, 835–842. [Sequel to Mydlarczyk (2001).]

Mydlarczyk, W. and W. Okrasiński (2003), Nonlinear Volterra integral equations with convolution kernels, *Bull. London Math. Soc.* **35**, 484–490.
$[u(x) = \int_0^x k(x - s)[u(s)]^p ds + F(x)$, with $p > 0$, $F(0) = 0$, $k \geq 0$.]

Mydlarczyk, W., W. Okrasiński and C.A. Roberts (2005), Blow-up solutions to a system of nonlinear Volterra equations, *J. Math. Anal. Appl.* **301**, 208–218.

Nelson, P., Jr and G. Young (1968), Minimizing the natural fuel requirement for nuclear reactor power systems: a nonstandard optimal control problem, *J. Optmization Theory Appl.* **2**, 138–154.

Neustadt, L.W. and J. Warga (1970), Comments on the paper "Optimal control of processes described by integral equations.I" by V.R. Vinokurov, *SIAM J. Control* **8**, 572.

Neves, K.W. and A. Feldstein (1976), Characterization of jump discontinuities for state-dependent delay differential equations, *J. Math. Anal. Appl.* **56**, 689–707.

Ney, P. (1977), The asymptotic behavior of a Volterra-renewal equation, *Trans. Amer. Math. Soc.* **228**, 147–155.

Niedziela, M. and W. Okrasiński (2006), A note on Volterra integral equations with power nonlinearity, *J. Integral Equations Appl.* **18**, 509–519.

Niemytzki, W. (1934), Théorie d'existence des solutions de quelques équations intégrales non-linéaires, *Mat. Sb.* **41**, 438–452.

Noble, B. (1964), The numerical solution of nonlinear integral equations and related topics, in Anselone (1964), pp. 215–318. ["Inversion" of Lighthill's equation: p. 308.]

Nohel, J.A. (1962), Some problems in nonlinear Volterra integral equations, *Bull. Amer. Math. Soc.* **68**, 323–329.

Nohel, J.A. (1964), Problems in qualitative behavior of solutions of nonlinear Volterra equations, in Anselone (1964), pp. 191–214.

Nohel, J.A. (1976), Review of "C. Corduneanu, *Integral Equations and Stability of Feedback Systems* (Academic Press, New York, 1973)", *SIAM Rev.* **12**, 520–526.

[Describes also the state of the art and open problems in the qualitative behaviour of solutions of Volterra equations.]

Nohel, J.A. and D.F. Shea (1976), Frequency domain methods for Volterra equations, *Adv. in Math.* **22**, 278–304. [Global existence (etc.) of solutions for $x'(t) = -\int_0^t a(t-\xi)g(x(\xi))d\xi + f(t)$.]

Norbury, J. and A.M. Stuart (1987), Volterra integral equations and a new Gronwall inequality. I. The linear case; II. The nonlinear case, *Proc. Roy. Soc. Edinburgh Sect. A* **106**, 361–373; 375–384.

Nussbaum, R.D. (1977), Periodic solutions of some nonlinear integral equations, in *Dynamical Systems (Univ. Florida, Gainsville, 1976)* (A.R. Bednarek and L. Cesari, eds.), 221–249 (New York: Academic Press).

Nussbaum, R.D. (1980), A quadratic integral equation, *Ann. Scuola Norm. Sup. Pisa Cl. Sci. (4)* **7**, 375–480.

Nussbaum, R.D. (1981), A quadratic integral equation. II, *Indiana Univ. Math. J.* **30**, 871–906.

Nussbaum, R.D. and N. Baxter (1981), A nonlinear integral equation, *Nonlinear Anal.* **5**, 1285–1307. [$u(x) = 1 + \lambda \int_x^1 (u(y))^\alpha (u(y-x))^\alpha dy, \ \alpha > 0.$]

Okrasiński, W. (1979), On the existence and uniqueness of nonnegative solutions of a certain nonlinear convolution equation, *Ann. Polon. Math.* **36**, 61–72. [$u^2(t) = \int_0^t k(t-s)u(s)ds.$]

Okrasiński, W. (1980), On a non-linear convolution equation occurring in the theory of water percolation, *Ann. Polon. Math.* **37**, 223–229.

Okrasiński, W. (1984), Non-negative solutions of some nonlinear integral equations, *Ann. Polon. Math.* **44**, 209–218. [Special implicit VIEs: $W(u(t)) = \int_0^t k(t-s)u(s)ds$; see also Buckwar (1997,2000).]

Okrasiński, W. (1986), On a nonlinear Volterra equation, *Math. Methods Appl. Sci.* **8**, 345–350.

Okrasiński, W. (1989), Nonlinear Volterra equations and physical applications, *Extracta Math.* **4**, 51–80.

Okrasiński, W. (1991a), Nontrivial solutions for a class of nonlinear Volterra equations with convolution kernel, *J. Integral Equations Appl.* **3**, 399–409.

Okrasiński, W. (1991b), Nontrivial solutions to nonlinear Volterra integral equations, *SIAM J. Math. Anal.* **22**, 1007–1015.

Okrasiński, W. (2000), Uniqueness of problems for some classes of nonlinear Volterra equations, in Agarwal & O'Regan (2000), pp. 259–267.

Øksendal, B. and T.S. Zhang (2010), Optimal control with partial information for stochastic Volterra equations, *Int. J. Stoch. Anal.* **2010**, Art. ID 329185, 25 pp.

Olmstead, W.E. (1972), Singular perturbation analysis of a certain Volterra integral equation, *Z. Angew. Math. Phys.* **23**, 889–900.

Olmstead, W.E. (1977), A nonlinear integral equation associated with gas absorption in a liquid, *Z. Angew. Math. Phys.* **28**, 513–523.

Olmstead, W.E. (1983), Ignition of a combustible half space, *SIAM J. Appl. Math.* **43**, 1–15.

Olmstead, W.E. (1997), Critical speed for the avoidance of blow-up in a reactive-diffusive medium, *Z. Angew. Math. Phys.* **48**, 701–710.

Olmstead, W.E. (2000), Blow-up solutions of Volterra equations, in Corduneanu & Sandberg (2000), pp. 385–389.

Olmstead, W.E. and R.A. Handelsman (1972), Singular perturbation analysis of a certain Volterra integral equations, *Z. Angew. Math. Phys.* **23**, 889–900.

Olmstead, W.E. and R.A. Handelsman (1976a), Asymptotic solution to a class of nonlinear Volterra integral equations. II, *SIAM J. Appl. Math.* **30**, 180–189. [See Handelsman & Olmstead (1972) for Part I.]

Olmstead, W.E. and R.A. Handelsman (1976b), Diffusion in a semi-infinite region with nonlinear surface dissipation, *SIAM Review* **18**, 275–291.

Olmstead, W.E., C.M. Kirk and C.A. Roberts (2010), Blow-up in a subdiffusive medium with advection, *Discrete Contin. Dynam. Syst.* **28**, 1655–1667.

Olmstead, W.E. and C.A. Roberts (1994), Explosion in a diffusive strip due to a concentrated nonlinear source, *Methods Appl. Anal.* **1**, 434–445.

Olmstead, W.E. and C.A. Roberts (1996), Explosion in a diffusive strip due to a source with local and nonlocal features, *Meth. Appl. Analysis* **3**, 345–357.

Olmstead, W.E. and C.A. Roberts (2001), Critical speed for quenching, *Dyn. Contin. Discrete Impuls. Syst. Ser. A Math. Anal.* **8**, 77–88.

Olmstead, W.E. and C.A. Roberts (2008), Thermal blow-up in a subdiffusive medium, *SIAM J. Appl. Math.* **69**, 514–523.

Olmstead, W.E., C.A. Roberts and K. Deng (1995), Coupled Volterra equations with blow-up solutions, *J. Integral Equations Appl.* **7**, 499–516. [See also the survey papers by Roberts (1998,2007).]

O'Malley, R.E. Jr (1991), *Singular Perturbation Methods for Ordinary Differential Equations* (New York: Springer-Verlag).

O'Regan, D. and M.M. Meehan (1998), *Existence Theory for Nonlinear Integral and Integrodifferential Equations* (Dordrecht: Kluwer Academic Publishers).

Osher, S.J. (1967), Two papers on similarity of certain Volterra integral operators, *Memoirs Amer. Math. Soc.* **73**, 47 pp.

Pachpatte, B.G. (1998), *Inequalities for Differential and Integral Equations* (San Diego, CA: Academic Press).

Padmavally, K. (1958), On a non-linear integral equation, *J. Math. Mech.* **7**, 533–555. [Generalisation of results in Mann & Wolf (1951) and Roberts & Mann (1951).]

Panin, A.A. (2015), On local solvability and blow-up of solutions of an abstract nonlinear Volterra integral equation, *Math. Notes* **97**, 892–908.

Paley, R.E.A.C. and N. Wiener (1934), *Fourier Transforms in the Complex Domain, Colloquium Publications* **XIX** (Providence, RI: American Mathematical Society).

Paveri-Fontana, S.L. and R. Rigacci (1979), A singularly perturbed weakly-singular integro-differential problem from analytical chemistry, in *Numerical Analysis of Singular Perturbation Problems (Nijmegen, 1978)* (P.W. Hemker and J.J.H. Miller, eds.), 475–484 (London-New York: Academic Press).

Pedas, A. (ed.) (1999), *Differential and Integral Equations: Theory and Numerical Analysis* (dedicated to Professor Gennadi Vainikko on his 60th birthday) (Tartu: Estonian Mathematical Society).

Pedas, A. and G. Vainikko (2006), Integral equations with diagonal and boundary singularities of the kernel, *Z. Anal. Anwendungen* **25**, 487–516.

Peirce, A. and E. Siebrits (1996), Stability analysis of model problems for elastodynamic boundary element discretization, *Numer. Methods Partial Differential Equations* **12**, 585–613. [Model problems: first-kind VIEs with convolution kernels in one and two dimensions.]

Peirce, A. and E. Siebrits (1997), Stability analysis and design of time-stepping schemes for general elastodynamic boundary element models, *Internat. J. Numer. Methods Engrg.* **40**, 319–342.

Pereverzev, S.V. and S. Prössdorf (1997), A discretization of Volterra integral equations of the third kind with weakly singular kernels, *J. Inverse Ill-Posed Probl.* **5**, 565–577. $[p(t)z(t) + \int_0^t h(t, \tau)(t - \tau)^{\alpha-1}z(\tau)d\tau = f(t), \ 0 < \alpha < 1,$ with $p(t)$ vanishing on some subset of $[0, 1]$.]

Pereverzev, S.V., E. Schock and S.G. Solodky (1999), On the efficient discretization of integral equations of the third kind, *J. Integral Equations Appl.* **11**, 501–513.

Peskir, G. (2002), On integral equations arising in the first-passage problem for Brownian motion, *J. Integral Equations Appl.* **14**, 397–423.

Picard, É. (1890), Mémoire sur la théorie des équations aux dérivées partielles et la méthode des approximations successives, *J. Math. Pures Appl. (4)* **6**, 145–210. [See pp. 197–200 for the "Picard iteration" process.]

Picard, É. (1907), Sur une équation fonctionnelle se présentant dans la théorie de certaines équations aux dérivées partielles, *C.R. Acad. Sci. Paris* **144**, 1009–1012. $[f(x) - P(x)f(\beta x) + \int_0^x \psi(x, y)f(y)dy = \varphi(x), \ \beta < 1.]$

Picard, É. (1911, Sur les équations intégrales de troisième espèce, *Ann. Sci. École Norm. Sup. (3)* **28**, 459–472.

Picone, M. (1960), Sull' equazione integrale non lineare di Volterra, *Ann. Mat. Pura Appl. (4)* **49**, 1–10.

Pietsch, A. (2007), *History of Banach Spaces and Linear Operators* (Boston: Birkhäuser).

Piila, J. (1996), Characterization of the membrane theory of a clamped shell. The hyperbolic case, *Math. Methods Appl. Sci.* **6**, 169–194.

Piila, J. and J. Pitkäranta (1996), On the integral equation $f(x) - (c/L(x)) \int_{L(x)}^x f(y)dy = g(x)$, where $L(x) = \min\{ax, 1\}, \ a > 1, J. Integral Equations Appl.* **8**, 363–378.

Pimbley, G.H., Jr (1967), Positive solutions of a quadratic integral equation, *Arch. Rational Mech. Anal.* **24**, 107–127.

Pishbin, S. (2015a), Optimal convergence results of piecewise polynomial collocation solutions for integral-algebraic equations of index-3, *J. Comput. Appl. Math.* **279**, 209–224.

Pishbin, S. (2015b), On the numerical solution of integral equations of the fourth kind with higher index: differentiability and tractability index-3, *J. Math. Model.* **2**, 156–169.

Pishbin, S., F. Ghoreishi and M. Hadizadeh (2013), The semi-explicit Volterra integral algebraic equations with weakly singular kernels: the numerical treatment, *J. Comput. Appl. Math.* **245**, 121–132.

Plato, R. (1997), Resolvent estimates for Abel integral operators and the regularization of associated first kind integral equations, *J. Integral Equations Appl.* **9**, 253–278. [Kernels of the form $(t^\beta - s^\beta)^{-\alpha}s^{-\beta} \ 0 < \alpha < 1, \ \beta > 0$.]

Pollard, H. (1948), The completely monotonic character of the Mittag-Leffler function $E_\alpha(-x)$, *Bull. Amer. Math. Soc.* **54**, 1115–1116.

Polyanin, A.D. and A.V. Manzhirov (2008), *Handbook of Integral Equations* (Second edition) (Boca Raton, FL, Chapman & Hall). [Excellent source of examples of first-kind and second-kind VIEs.]

Porath, G. (1974), Lineare Volterra Integralgleichungen zweiter Art mit Kernen vom allgemeinen Typ, *Beiträge Numer. Math.* **2**, 147–162.

Precup, R. (2002), *Methods in Nonlinear Integral Equations* (Dordrecht: Kluwer Academic Publishers).

Prössdorf, S. and B. Silbermann (1991), *Numerical Analysis for Integral and Related Operator Equations* (Basel-Boston: Birkhäuser Verlag).

Prüss, J. (2012), *Evolutionary Integral Equations and Applications* (Basel-Boston: Birkhäuser Verlag). [See also for discussion of semigroup approaches for evolutionary integral equations and applications; extensive bibliography.]

Pukhnacheva, T.P. (1990), A functional equation with contracting argument, *Siberian Math. J.* **31**, 365–367. [Existence/regularity result for
$$u(x) - a(x)u(\omega(x)) = \int_0^x K(x,\xi)u(x-\xi)d\xi + f(x), \ |\omega(x)| \le x/k \ (k > 1).]$$

Pulyaev, V.F. and Z.B. Tsalyuk (1991), On the asymptotic behavior of solutions of Volterra integral equations in Banach spaces, *Soviet Math. (Iz. VUZ)* **35**, 48–55.

Rabier, P.J. and W.C. Rheinboldt (2002), Theoretical and numerical analysis of differential-algebraic equations, in *Handbook of Numerical Analysis, Vol. VIII* (P.G. Ciarlet and J.L. Lions, eds.), pp. 183–540 (Amsterdam: North-Holland).

Ramalho, R. (1976), Existence and uniqueness theorems for a nonlinear integral equation, *Math. Ann.* **221**, 35–44. [$u(x) = 1 + \rho \int_x^1 u(y-x)u(y)dy, \ \rho \in \mathbb{R}$. See also Nussbaum (1980, 1981), Nussbaum & Baxter (1981).]

Read, C.J. (1997), Quasinilpotent operators and the invariant subspace problem, *J. London Math. Soc. (2)* **56**, 595–606.

Reinermann, J. and V. Stallbohm (1971), Eine Anwendung des Edelsteinschen Fixpunktsatzes auf Integralgleichungen vom Abel-Liouvilleschen Typ, *Arch. Math. (Basel)* **22**, 642–647. [Compare also Dinghas (1958).]

Reuter, G.E.H. (1956), Über eine Volterrasche Integralgleichung mit totalmonotonem Kern, *Arch. Math. (Basel)* **7**, 59–66.

Reynolds, D. (1984), On linear weakly singular Volterra integral equations of the second kind, *J. Math. Anal. Appl.* **103**, 230–262. [Extension of results by Evans of 1910 and 1911, including an analysis of third-kind VIEs.]

Reynolds, D.W., J.A.D Appleby and I. Györi (2007), On exact rates of growth and decay of solutions of a linear Volterra equation in linear viscoelasticity, *Note Mat.* **27**, 215–228.

Riaza, R. and R. März (2008), A simpler construction of the matrix chain defining the tractability index of linear DAEs, *Appl. Math. Lett.* **21**, 326–331.

Richter, G. (1976), On the weakly singular Fredholm integral equations with displacement kernels, *J. Math. Anal. Appl.* **55**, 32–42.

Ring, W. and J. Prix (2000), Sequential predictor-corrector regularization methods and their limitations, *Inverse Problems* **16**, 619–633. [See the related paper by Lamm (2000).]

Ringrose, J.R. (1962), On the triangular representation of integral operators, *Proc. London Math. Soc. (3)* **12**, 385–399.

Roberts, C.A. (1997), Characterizing the blow-up solutions for nonlinear Volterra integral equations, *Nonlinear Anal.* **30**, 923–933.

Roberts, C.A. (1998), Analysis of explosion for nonlinear Volterra equations, *J. Comput. Appl. Math.* **97**, 153–166. [Survey paper with extensive list of references.]

Roberts, C.A. (2000), A method to determine growth rates of nonlinear Volterra equations, in Corduneanu & Sandberg (2000), pp. 427–431.

Roberts, C.A. (2007), Recent results on blow-up and quenching for nonlinear Volterra equations, *J. Comput. Appl. Math.* **205**, 736–743.

Roberts, C.A., D.G. Lasseigne and W.E. Olmstead (1993), Volterra equations which model explosion in a diffusive medium, *J. Integral Equations Appl.* **5**, 531–546. [Analysis of blow-up solutions in weakly singular VIEs.]

Roberts, C.A. and W.E. Olmstead (1996), Growth rates for blow-up solutions of nonlinear Volterra equations, *Quart. Appl. Math.* **54**, 153–159.

Roberts, C.A. and W.E. Olmstead (1999), Local and non-local boundary quenching, *Math. Methods Appl. Sci.* **22**, 1465–1484.

Roberts, J.H. and W.R. Mann (1951), On a certain nonlinear integral equation of the Volterra type, *Pacific J. Math.* **1**, 431–445.

Roos, H.-G., M. Stynes and L. Tobiska (2008), *Robust Numerical Methods for Singularly Perturbed Differential Equations* (Second edition) (Berlin: Springer-Verlag).

Rothe, R. (1931), Zur Abelschen Integralgleichung, *Math. Z.* **33**, 375–387.

Rudin, W. (1976), *Principles of Mathematical Analysis* (Third edition) (New York: McGraw-Hill).

Rudin, W. (1991), *Functional Analysis* (Second edition) (New York: McGraw-Hill).

Saigo, M. and A.A. Kilbas (1998), On Mittag-Leffler type function and application, *Integral Transform. Spec. Funct.* **7**, 97–112.

Sakaljuk, K.D. (1960), Abel's generalized integral equation, *Soviet Math. Dokl.* **1**, 332–335.

Sakaljuk, K.D. (1965), The generalized Abel integral equation with inner coefficients (Russian), *Kišinev. Gos. Univ. Učen. Zap.* **82**, 60–68. [Compare the review in *Math. Rev.* **37**, #3294.]

Samko, S.G. (1968), A generalized Abel equation and fractional integration operators, *Differ. Uravn.* **4**, 298–314.

Samko, S.G., A.A. Kilbas and O.I. Marichev (1993), *Fractional Integrals and Derivatives* (Yverdon, Gordon and Breach). [See Section 4.3; also Kilbas & Saigo (1999) for related results on solution representations for Abel-type IEs.]

Satco, B. (2009), Volterra integral equations governed by highly oscillatory functions on time scales, *An. Ştiinţ. Univ. Ovidus Constanţa Ser. Mat.* **17**, 233–240.

Sato, T. (1951), Détermination unique de solution de l'équation intégrale de Volterra, *Proc. Japan Acad.* **27**, 276–278.

Sato, T. (1953a), Sur l'équation non linéaire de Volterra, *Compositio Math.* **11**, 271–290.

Sato, T. (1953b), Sur l'équation intégrale $xu(x) = f(x) + \int_0^x K(x, t, u(t))dt$, *J. Math. Soc. Japan* **5**, 145–153.

Sato, T. and A. Iwasaki (1955), Sur l'équation intégrale de Volterra, *Proc. Japan Acad.* **31**, 395–398.

Schechter, M. (2002), *Principles of Functional Analysis* (Second edition) (Providence, RI: American Mathematical Society).

Schiff, J.L. (1999), *The Laplace Transform* (New York: Springer-Verlag).

Schmeidler, W. (1950), *Integralgleichungen mit Anwendungen in Physik und Technik* (Leipzig: Akad. Verlagsgesellschaft Geest & Portig).

Schmeiser, C., A. Unterreiter and R. Weiss (1993), The switching behavior of a one-dimensional *pn*-diodes in low injection, *Math. Models Methods Appl. Sci.* **3**, 125–144. [Mathematical models involving singularly perturbed VIEs with weakly singular kernels; see also Miller & Unterreiter (1992) and Unterreiter (1996) for related Volterra models.]

Schneider, C. (1979), Regularity of the solution to a class of weakly singular Fredholm integral equations of the second kind, *Integral Equations Operator Theory* **2**, 62–68.

Schock, E. (1985), Integral equations of the third kind, *Studia Math.* **8**, 1–11.

Seyed Allaei, S. (2015), *The Numerical Solutions of Volterra Integral Equations of the Second and Third Kind*, PhD Thesis, University of Lisbon, Portugal.

Seyed Allaei, S., Z.W. Yang and H. Brunner (2015), Existence, uniqueness and regularity of solutions for a class of third kind Volterra integral equations, *J. Integral Equations Appl.* **27**, 325–342.

Seyed Allaei, S., Z.W. Yang and H. Brunner (2016), Numerical analysis of collocation methods for third-kind Volterra integral equations, *IMA J. Numer. Anal.* (to appear).

Shaikhet, L.E. (1995), On the stability of solutions of stochastic Volterra equations, *Autom. Remote Control* **56**, 1129–1137.

Shaw, S., M.K. Warby and J.R. Whiteman (1997), Error estimates with sharp constants for a fading memory Volterra problem in linear solid viscoelasticity, *SIAM J. Numer. Anal.* **34**, 1237–1254.

Shaw, S. and J.R. Whiteman (1996), Discontinuous Galerkin method with a-posteriori $L_p(0, t_i)$ error estimate for second-kind Volterra problems, *Numer. Math.* **74**, 361–383.

Shaw, S. and J.R. Whiteman (1997), Applications and numerical analysis of partial Volterra equations: a brief survey, *Comput. Methods Appl. Mech. Engrg.* **150**, 397–409. [Extensive bibliography on applications.]

Shea, D.F. and S. Wainger (1975), Variants of the Wiener-Levy theorem, with applications to stability problems for some Volterra integral equations, *Amer. J. Math.* **97**, 312–343.

Shiri, B. (2014), Numerical solution of higher index nonlinear integral algebraic equations of Hessenberg type using discontinuous collocation methods, *Math. Model. Anal.* **19**, 99–117.

Shiri, B., S. Shahmorad & G. Hojjati (2013), Convergence analysis of piecewise continuous collocation methods for higher index integral algebraic equations, *Int. J. Appl. Math. Comput. Sci.* **23**, 341–355.

Shkarin, S. (2006a), On similarity of quasinilpotent operators, *J. Funct. Anal.* **241**, 528–556.

Shkarin, S. (2006b), Antisupercyclic operators and orbits of the Volterra operator, *J. London Math. Soc.* (2) **73**, 506–528.

Shubin, C. (2006), Singularly perturbed integral equations, *J. Math. Anal. Appl.* **313**, 234–250.

Skinner, L.A. (1995), Asymptotic solution to a class of singularly perturbed Volterra integral equations, *Methods Appl. Anal.* **2**, 212–221.

Skinnner, L.A. (2000), A class of singularly perturbed singular Volterra integral equations, *Asymptot. Anal.* **22**, 113–127.

Skinner, L.A. (2011), *Singular Perturbation Theory* (New York: Springer).

Sloss, B.G. (2002), A generalization of a Volterra integral equation, *Appl. Anal.* **81**, 1005–1018.

Sloss, B.G. and W.F. Blyth (1994), Corrington's Walsh function method applied to a nonlinear integral equation, *J. Integral Equations Appl.* **6**, 239–255.
[VIEs of the form $u(x) = \sum_{i=1}^{N} b_i \{a_i(x) + \int_0^x k_i(x,t)u(t)dt\}^i$.]

Smarzewski, R. and H. Malinowski (1978), Numerical solution of a class of Abel integral equations, *IMA J. Appl. Math.* **22**, 159–170.

Smith, D.R. (1985), *Singular-Perturbation Theory* (Cambridge University Press).

Smith, H.L. (1977), On periodic solutions of a delay integral equation modelling epidemics, *J. Math. Biol.* **4**, 69–80. [Compare also Cooke & Kaplan (1976) and Cañada & Zetrtiti (1994).]

Smithies, F. (1938), The eigen-values and singular values of integral equations, *Proc. London Math. Soc. (2)* **43**, 255–279.

Sneddon, I.H. (1972), *The Use of Integral Transforms* (New York: McGraw-Hill).

Srivastava, H.M. and R.G. Buschman (1977), *Convolution Integral Equations* (New York: Wiley Eastern Ltd/Wiley & Sons).

Staffans, O.J. (1984a), Semigroups generated by a convolution equation, in *Infinite Dimensional Systems (Retzhof 1983)* (W. Kappel and F. Schappacher, eds.), pp. 209–226, Lecture Notes in Math. **1046** (Berlin: Springer-Verlag).

Staffans, O.J. (1984b), A note on a Volterra equation with several nonlinearities, *J. Integral Equations* **7**, 249–252.

Staffans, O.J. (1985), On a nonconvolution Volterra resolvent, *J. Math. Anal. Appl.* **108**, 15–30.

Stewart, G.W. (2011), Three fundamental papers on integral equations: Fredholm, Hilbert, Schmidt, www.cs.umd.edu/~stewart/FHS.pdf.

Szufla, S. (1984), On the existence of L^p-solutions of Volterra integral equations in Banach spaces, *Funkcial. Ekvac.* **27**, 157–172.

Tamarkin, J.D. (1930), On integrable solutions of Abel's integral equation, *Ann. of Math. (2)* **31**, 219–229.

Thieme, H.R. (1977), A model for the spatial spread of an epidemic, *J. Math. Biol.* **4**, 337–351.

Thieme, H.R. (1979), Asymptotic estimates of the solutions of nonlinear integral equations and asymptotic speeds for the spread of populations, *J. Reine Angew. Math.* **306**, 94–121. [Nonlinear Volterra-Fredholm integral equation; see also Diekmann (1978), Thieme & Zhao (2003), and Zhao (2003).]

Thieme, H.R. and X.-Q. Zhao (2003), Asymptotic speeds of spread and traveling waves for integral equations and delayed reaction-diffusion models, *J. Differential Equations* **195**, 430–470.

Thorpe, B. (1998), The norm of powers of the indefinite integral operator on (0, 1), *Bull. London Math. Soc.* **30**, 543–548.

Tonelli, L. (1928a), Su un problema di Abel, *Math. Ann.* **99**, 183–199.

Tonelli, L. (1928b), Sulle equazioni funzionali del tipo di Volterra, *Bull. Calcutta Math. Soc.* **20**, 31–48.

Torrejón, R. (1990), A note on a nonlinear integral equation from the theory of epidemics, *Nonlinear Anal.* **14**, 483–488.

Torrejón, R. (1993), Positive almost periodic solutions of a state-dependent nonlinear integral equation, *Nonlinear Anal.* **20**, 1383–1416.

Trefethen, Ll.N. and M. Embree (2005), *Spectra and Pseudospectra. The Behaviour of Nonnormal Matrices and Operators* (Princeton, NJ: Princeton University Press).

Tricomi, F.G. (1957), *Integral Equations* (New York: Interscience Publishers; New York: Dover Publications, 1985). [VIEs in L^2: pp. 10–15.]

Tsalyuk, Z.B. (1968), Stability of Volterra equations, *Differential Equations* **4**, 1015–1021.

Tsalyuk, Z.B. (1970), Asymptotic properties of solutions of the regeneration equation, *Differential Equations* **6**, 1112–1114.

Tsalyuk, Z.B. (1979), Volterra integral equations, *J. Soviet Math.* **12**, 715–758. [Survey paper: contains references to some 500 papers on Volterra equations reviewed in *Referativnyi Zhurnal "Matematika"* between 1966 and 1976.]

Tsalyuk, Z.B. (1989), Asymptotic estimates for the solutions of the renewal equation, *Differential Equations* **25**, 239–243. [See also the related paper by Pulyaev and Tsalyuk (1991).]

Tuan, V.K. and R. Gorenflo (1994), Asymptotics of singular values of fractional integral operators, *Inverse Problems* **10**, 949–955.

Tudor, C. (1986), On Volterra stochastic equations, *Boll. Un. Mat. Ital. A (6)* **5**, 335–344.

Turo, J. (1995), Nonlinear stochastic functional integral equations in the plane, *J. Appl. Math. Stochastic Anal.* **8**, 371–399.

Tychonoff, A. (1938), Sur les équations fonctionnelles de Volterra et leurs applications à certains problémes de la physique mathématique, *Bull. Univ. d'État de Moscou Sér. Internat. Sér. A Math. Méchan.* **1**, 1–25. [Compare pp. 22–23 for the iterated kernels corresponding to weakly singular kernels with $0 < \alpha < 1$.]

Unger, F. and L. v. Wolfersdorf (1995), Inverse Probleme zur Identifikation von Memory-Kernen, *Freiberger Forschungsberichte Mathematik* **C458** (Freiberg: Technische Universität Bergakademie).

Unterreiter, A. (1996), Volterra integral equation models for semiconductor devices, *Math. Models Appl. Sci.* **19**, 425–450. [See pp. 448–449 for open numerical problems for VIEs arising in this model; compare also Schmeiser et al. (1993).]

Ursell, F. (1969), Integral equations with rapidly oscillating kernel, *J. London Math. Soc.* **44**, 449–459.

Vainikko, G. (1993), *Multidimensional Weakly Singular Integral Equations*, Lecture Notes in Math. **1549** (Berlin: Springer-Verlag).

Vainikko, G. (2009), Cordial Volterra integral equations 1, *Numer. Funct. Anal. Optim.* **30**, 1145–1172.

Vainikko, G. (2010a), Cordial Volterra integral equations 2, *Numer. Funct. Anal. Optim.* **31**, 191–219.

Vainikko, G. (2010b), Spline collocation for cordial Volterra integral equations, *Numer. Funct. Anal. Optim.* **31**, 313–338.

Vainikko, G. (2011), Spline collocation-interpolation method for linear and nonlinear cordial Volterra integral equations, *Numer. Funct. Anal. Optim.* **32**, 83–109. [Contains also regularity results for solutions of nonlinear cordial VIEs.]

Vainikko, G. (2012), First kind cordial Volterra integral equations 1, *Numer. Funct. Anal. Optim.* **33**, 680–704.

Vainikko, G. (2014), First kind cordial Volterra integral equations 2, *Numer. Funct. Anal. Optim.* **35**, 1607–1637.

Vainikko, G. (2016), Private communication (26 February 2016).

Vasin, V.V. (1996), Monotone iterative processes for nonlinear operator equations and their applications to Volterra equations, *J. Inv. Ill-Posed Problems* **4**, 331–340. [Newton-type iterative processes for nonlinear VIEs of the first kind.]

Väth, M. (1998a), Abstract Volterra equations of the second kind, *J. Integral Equations Appl.* **10**, 319–362.

Väth, M. (1998b), Linear and nonlinear abstract Volterra equations, *Funct. Differ. Equ.* **5**, 499–512.

Väth, M. (1999), *Volterra and Integral Equations of Vector Functions* (New York: Marcel Dekker). [Study of abstract Volterra equations via topological and algebraic methods.]

Vessella, S. (1985), Stability results for Abel equations, *J. Integral Equations* **9**, 125–134.

Vinokurov, V.R. (1969a), Certain questions in the theory of the stability of systems of Volterra integral equations I,II,III (in Russian), *Izv. Vyssh. Uchebn. Zaved. Matematika* **1969** (no. 6(85)), 24–34; **1969** (no. 7(86)), 28–38; **1971** (no. 4 (107)), 20–31.

Vinokurov, V.R. (1969b), Optimal control of processes described by integral equations I,II,III, *SIAM J. Control* **7**, 324–336, 337–45, 346–355. [See also the comments by Neustadt & Warga (1970) on paper I.]

Vivanti, G. (1929), *Elemente der Theorie der Linearen Integralgleichungen* (Hannover: Helwingsche Verlagsbuchhandlung). [Features extensive annotated bibliography, including a list of dissertations.]

Vogel, Th. (1965), *Théorie des Systèmes Évolutifs, Traité de Physique Théorique et de Physique Mathématique*, **XXII** (Paris: Gauthier-Villars). [See also MR 32, #8546.]

Volterra, V. (1896a), Sulla inversione degli integrali definiti, *Atti R. Accad. Sci. Torino* **31**, 311–323 (Nota I); 400–408 (Nota II).

Volterra, V. (1896b), Sulla inversione degli integrali definiti, *Atti R. Accad. Sci. Torino* **31**, 557–567 (Nota III); 693–708 (Nota IV).
[These four fundamental papers can also be found in Volterra (1954), *Opere* **II**, pp. 216–262.]

Volterra, V. (1896c), Sulla inversione degli integrali multipli, *Rend. R. Accad. Lincei (5)* **5**, 289–300. [Systems of VIEs. See also Volterra (1954), *Opere* **II**, pp. 263–275.]

Volterra, V. (1897), Sopra alcune questioni di inversione di integrali definite, *Ann. Mat. Pura Appl. (2)* **25**, 139–178. [Existence of solutions to first-kind VIE with delay qt, $0 < q < 1$.]

Volterra, V. (1913), *Leçons sur les Équations Intégrales et les Équations Integro-Différentielles* (Paris: Gauthier-Villars; 2008 reprint: Sceaux: Éditions Jacques Gabay). [VIEs with proportional limits of integration are treated on pp. 92–101.]

Volterra, V. (1916), Teoria delle potenze dei logaritmi e delle funzione di composizione, *Mem. Accad. Lincei Ser. 5* **XI**, 167–250; also in Volterra (1954), IV, pp. 118–199.

Volterra, V. (1954), *Opere Matematiche, Vol. I–V* (Rome: Accademia Nazionale dei Lincei, 1954, 1956, 1957, 1960, 1962).

Volterra, V. (1959), *Theory of Functionals and of Integral and Integro-Differential Equations* (New York: Dover Publications). [Based on lectures given at the University of Madrid in 1925 and first published in Spanish in 1927. An English translation with corrections appeared in 1930.]

Wagner, E. (1978), Über die Asymptotik der Lösungen linearer Volterrascher Integralgleichungen 2. Art vom Faltungstyp, *Beiträge Anal.* **11**, 165–183.

Walter, W. (1967), On nonlinear Volterra integral equations in several variables, *J. Math. Mech.* **16**, 967–985.

Walther, H.-O. (2013, On Poisson's state-dependent delay, *Discrete Contin. Dyn. Syst.* **33**, 365–379.

Waltman, P. (1974), *Deterministic Threshold Models in the Theory of Epidemics*, Lecture Notes in Biomath. **1** (Berlin-Heidelberg: Springer-Verlag).

Wang, H.Y. and S.H. Xiang (2011), Asymptotic expansion and Filon-type methods for a Volterra integral equation with a highly oscillatory kernel, *IMA J. Numer. Anal.* **31**, 469–490.

Webb, G.F. (1985), *Theory of Nonlinear Age-Dependent Population Dynamics* (New York: Marcel Dekker).

Whitley, R. (1987), The spectrum of a Volterra composition operator, *Integral Equations Operator Theory* **10**, 146–149. [Compare also Lyubich (1984) and Domanov (2007, 2008).]

Willé, D.R. and C.T.H. Baker (1992), The tracking of derivative discontinuities in systems of delay-differential equations, *Appl. Numer. Math.* **9**, 209–222.

Willé, D.R. and C.T.H. Baker (1994), Stepsize control and continuity consistency for state-dependent delay-differential equations, *J. Comput. Appl. Math.* **53**, 163–170.

Wiman, A. (1905), Über die Nullstellen der Funktionen $E_\alpha(x)$, *Acta Math.*, **29**, 217–234.

Witte, G. (1997), *Die analytische und die numerische Behandlung einer Klasse von Volterraschen Integralgleichungen im Hilbertraum*, Dissertation (Berlin: Freie Universität Berlin).

v. Wolfersdorf, L. (1965), Abelsche Integralgleichungen und Randwertprobleme für die verallgemeinerte Tricomi-Gleichung, *Math. Nachr.* **25**, 161–178.
   [$c(x) \int_0^x (x - t)^{-\alpha} \varphi(t) dt + d(x) \int_x^1 (t - x)^{-\alpha} \varphi(t) dt = f(x)$.]

v. Wolfersdorf, L. (1994), On identification of memory kernels in linear theory of heat conduction, *Math. Methods Appl. Sci.* **17**, 919–932.

v. Wolfersdorf, L. (1995), A class of multi-dimensional nonlinear Volterra equations of convolution type, *Demonstratio Math.* **28**, 807–820. [See in particular for first-kind VIEs associated with the Darboux problem.]

v. Wolfersdorf, L. (2000), Einige Klassen quadratischer Integralgleichungen, *Sitzungsber. Sächs. Akad. Wiss. Leipzig Math.-Nat.wiss. Kl.* **128**, No. 2, 34pp. [Contains a comprehensive list of (114) references on the development of the theory of quadratic integral equations.]

v. Wolfersdorf, L. (2007), On the theory of convolution equations of the third kind, *J. Math. Anal. Appl.* **331**, 1314–1336; Part II (with J. Janno): **342** (2008), 838–863. [Includes a result on the representation of solutions to $t^\alpha u(t) = f(t) + \lambda \int_0^t (t - s)^{\alpha-1} u(s) ds \ (\alpha > 0)$.]

v. Wolfersdorf, L. (2008), Autoconvolution equations and special functions, *Integral Transforms Spec. Funct.* **19**, 677–686; Part II: **21** (2010), 295–306.

v. Wolfersdorf, L. (2011), Autoconvolution equations of the third kind with Abel integral, *J. Integral Equations Appl.* **23**, 113–136. [See also Janno & von Wolfersdorf (2005) on a general class of third-kind auto-convolution equations.]

v. Wolfersdorf, L. and B. Hofmann (2008), A specific inverse problem for the Volterra convolution equation, *Appl. Anal.* **87**, 59–81.

v. Wolfersdorf, L. and J. Janno (1995), On a class of nonlinear convolution equations, *Z. Anal. Anwendungen* **14**, 497–508.

Wong, P.J.Y. and K.L. Boey (2004), Nontrivial periodic solutions in the modelling of infectious disease, *Appl. Anal.* **83**, 1–16.

Wouk, A. (1964), Direct iteration, existence and uniqueness, in Anselone (1964), pp. 3–31. [Comprehensive survey of early nonlinear VIEs.]

Wu, Z.J. (2011), Volterra operator, area integral and Carleson measures, *Science China Math.* **54**, 2487–2500.

Xiang, S.H. (2011), Efficient Filon-type methods for $\int_0^t f(x)e^{i\omega g(x)}dx$, *Numer. Math.* **105**, 633–658.

Xiang, S.H., Y.J. Cho, H.Y. Wang and H. Brunner (2011), Clenshaw-Curtis-Filon-type methods for highly oscillatory Bessel transforms and applications, *IMA J. Numer. Anal.* **31**, 1281–1314.

Xiang, S.H. and H. Brunner (2013), Efficient methods for Volterra integral equations with highly oscillatory Bessel kernels, *BIT Numer. Math.* **53**, 241–263.

Xie, H.H., R. Zhang and H. Brunner (2011), Collocation methods for general Volterra functional integral equations with vanishing delays, *SIAM J. Sci. Comput.* **33**, 3303–3332.

Yang, Z.W. (2015), Second-kind linear Volterra integral equations with noncompact operators, *Numer. Funct. Anal. Optim.* **36**, 104–131.

Yang, Z.W. and H. Brunner (2013a), Blow-up behavior of Hammerstein-type delay Volterra integral equations, *Front. Math. China* **8**, 261–280.

Yang, Z.W. and H. Brunner (2013b), Blow-up behavior of collocation solutions to Hammerstein-type Volterra integral equations, *SIAM J. Numer. Anal.*, **51**, 2260–2282.

Yang, Z.W. and H. Brunner (2014), Quenching behaviors of Volterra integral equations, *Dyn. Contin. Discrete Impulsive Syst. Ser. A Math. Anal.* **21**, 507–529. [This paper contains an extensive list of references on theory and applications of quenching.]

Yatsenko, Yu. (1995), Volterra integral equations with unknown delay time, *Methods Appl. Anal.* **2**, 408–419. [Compare also the monograph by Hritonenko & Yatsenko (1996) and its bibliography.]

Zhang, W.K. and H. Brunner (1999), Primary discontinuities for delay integro-differential equations, *Methods Appl. Anal.* **6**, 525–533.

Zhang, R., H. Liang and H. Brunner (2016), Analysis of collocation methods for generalized auto-convolution Volterra integral equations, *SIAM J. Numer. Anal.* **54**, 899–920.

Zhang, X.C. (2010), Stochastic Volterra equations in Banach spaces and stochastic partial differential equations, *J. Funct. Anal.* **258**, 1361–1425. [Contains extensive list of references on stochastic VIEs.]

Zhao, X.-Q. (2003), *Dynamical Systems in Population Biology*, CMS Books in Mathematics **16** (New York: Springer-Verlag).

Zima, M. (1992), A certain fixed point theorem and its applications to integral-functional equations, *Bull. Austral. Math. Soc.* **46**, 179–186.

Index

Lighthill's equation, 267, 311
Lighthill's Volterra integral operator, 251

maximal solution, 170
minimal solution, 170
Mittag-Leffler function, 19, 28, 54
    completely monotone, 20

Nemytzkiĭ (1934), 170
Nemytzkiĭ operator, 111
Neumann series, 4, 292
    convolution kernel, 12
    system of VIEs, 15
nilpotency, 299
non-linear second-kind VIE
    existence of solution, 105
    uniqueness of solution, 105
non-linear VFIE
    state-dependent delay, 163
non-linear VIE
    auto-convolution equation, 146
    continuation of solution, 170
    first kind, 157
    multiple solution, 116
    non-standard form, 145
    non-trivial solution, 117
    Picard iteration, 106
    standard form, 103
    trivial solution, 117
    weakly singular kernel, 107
non-vanishing delay, 74, 84
norm of operator, 330
null space, 330

ODE
    quenching of solution, 166
open mapping theorem, 331
operator
    adjoint, 338
    bounded, 330
    closed, 330
    compact, 332, 333
    graph, 330
    inverse, 331
    nilpotent, 284
    norm, 330
    projection, 332
    quasi-nilpotent, 284
    self-adjoint, 289, 338
    substitution, 111
orthogonal projection, 332
oscillator

Fox–Li, 183
    other types, 192
    separable, 179

Paley–Wiener theorem, 225
partial VIEs
    semi-discretisation, 14
PDE
    quenching of solution, 137
    quenching time, 135
Picard iteration, 3, 11, 53, 78
    auto-convolution VIE, 148
    non-linear VIE, 104
    VFIE, 82
    weakly singular VIE, 23
point spectrum, 337
population growth VFIE
    state-dependent delay, 309
primary discontinuity points, 75, 85
    separation property, 85
projection operator, 332
proportional delay, 159
    non-vanishing, 75
    VFIE, 76

quasi-nilpotent operator, 339
quenching
    differential equation, 134
    VIE, 139
quenching time, 134, 139

regularity of solution
    cordial VIE, 254
    second-kind VIE, 58
    weakly singular VFIE, 90
    weakly singular VIE, 58
renewal equation, 11, 13
    vector form, 16
resolvent equation, 5, 88
    abstract, 286
    complementary, 5, 286
    convolution kernel, 12
resolvent kernel, 4, 88
    abstract, 286
    first kind, 287
    for non-smooth kernel, 64
    for weakly singular kernel, 22
    integrability, 225
resolvent theory, 301

Schauder fixed point theorem, 342
Schauder–Leray fixed point theorem, 342